COMMUNICATING PROCESS ARCHITECTURES 2007

Concurrent Systems Engineering Series

Series Editors: M.R. Jane, J. Hulskamp, P.H. Welch, D. Stiles and T.L. Kunii

Volume 65

Previously published in this series:

ISSN 1383-7575

Communicating Process Architectures 2007

WoTUG-30

Edited by

Alistair A. McEwan
University of Surrey, Guildford, United Kingdom

Steve Schneider
University of Surrey, Guildford, United Kingdom

Wilson Ifill
Atomic Weapons Establishment Aldermaston, Reading, Berks, United Kingdom

and

Peter Welch
University of Kent, Canterbury, United Kingdom

Proceedings of the 30th WoTUG Technical Meeting,
8–11 July 2007, University of Surrey,
Guildford, United Kingdom

IOS *Press*

Amsterdam • Berlin • Oxford • Tokyo • Washington, DC

ISBN 978-1-58603-767-3
Library of Congress Control Number: 2007929667

Publisher
IOS Press
Nieuwe Hemweg 6B
1013 BG Amsterdam
Netherlands
fax: +31 20 687 0019
e-mail: order@iospress.nl

Distributor in the UK and Ireland
Gazelle Books Services Ltd.
White Cross Mills
Hightown
Lancaster LA1 4XS
United Kingdom
fax: +44 1524 63232
e-mail: sales@gazellebooks.co.uk

Distributor in the USA and Canada
IOS Press, Inc.
4502 Rachael Manor Drive
Fairfax, VA 22032
USA
fax: +1 703 323 3668
e-mail: iosbooks@iospress.com

Communicating Process Architectures 2007
Alistair McEwan, Steve Schneider, Wilson Ifill and Peter Welch (Eds.)
IOS Press, 2007

Preface

The University of Surrey is delighted to host the *Communicating Process Architectures 2007* conference. There are many reasons why this University, and Guildford in particular, is appropriate for a conference about Computer Science and models of Concurrency. Not least is the connection with one of the most famous Computer Scientists of them all: Alan Turing, who grew up only a short distance from the site of the University. A statue of him erected in the main plaza overlooks the conference venue, serving as an inspiration and reminder to the strong theoretical and mathematical basis of the topic of this conference. Delegates may also have paused for enlightenment at the village of Occam (now spelt *Ockham*), famously the birthplace of William of Occam, as they approached Guildford.

This is the 30th meeting of this conference series. The first was a single day workshop, organised by Inmos, and took place in Bristol in 1985. With the success of the Transputer, the meeting grew into an international conference series, with the proceedings formally published by IOS press since March 1988. The fact that the conference series is growing in strength as technology evolves shows that its founding ideas still push the boundaries of Computer Science and are as relevant as ever.

Since inception, the CPA conference series has always had strong interest from industry, and this year is no exception with the conference being co-sponsored by AWE UK. This year, there is a particular emphasis on hardware/software co-design, and the understanding of concurrency that results from these systems. A range of papers on this topic have been included, from the formal modeling of buses in co-design systems through to software simulation and development environments.

Industrial relevance is further reflected in the achievements of this year's invited speakers. Professor Sir Tony Hoare, FRS, is the founding father of the theoretical basis upon which much of the work in this series is based. The organisers are delighted that he has accepted the invitation to address the conference on his new thoughts on fine-grained concurrency. Professor David May, FRS, has been one of the leading lights of this, and other, communities for many years. He was chief architect for the Transputer and the occam programming language. The organisers are also delighted that he has accepted the invitation to address the conference on his latest work on communicating process architecture for massively multicore processors and how to program them.

We hope you will find the meeting exciting, invigorating, and motivating. We trust you will find the published proceedings informative and inspirational – and the informal Fringe presentations fun and thought provoking. This year, we have published abstracts of several fringe presentations that were offered in advance of the conference.

Finally, the editors would like to thank the Programme Committee, friends of the conference, and reviewers, for all their diligent hard work in reviewing papers, the staff of the University of Surrey – especially Sophie Gautier-O'Shea – for their assistance in organizing the event, and the *Systems Assurance Group* at the AWE for all their support.

Alistair McEwan *(University of Surrey)*
Steve Schneider *(University of Surrey)*
Peter Welch *(University of Kent)*
Wilson Ifill *(AWE UK)*

Programme Committee

Dr Alistair McEwan, *University of Surrey, UK (Chair)*
Prof Steve Schneider, *University of Surrey, UK*
Prof Peter Welch, *University of Kent, UK*
Mr Wilson Ifill, *AWE UK*
Dr Alastair Allen, *Aberdeen University, UK*
Dr Fred Barnes, *University of Kent, UK*
Dr John Bjorndalen, *University of Tromso, Norway*
Dr Jan Broenink, University of Twente, *The Netherlands*
Dr Barry Cook, *4Links Ltd., UK*
Dr Ian East, *Oxford Brookes University, UK*
Mr Marcel Groothuis, *University of Twente, The Netherlands*
Dr Gerald Hilderink, *Eindhoven, The Netherlands*
Prof Jon Kerridge, *Napier University, UK*
Dr Adrian Lawrence, *Loughborough University, UK*
Dr Jeremy Martin, *GSK Ltd., UK*
Dr Denis Nicole, *University of Southampton, UK*
Dr Jan Pedersen, *University of Nevada, Las Vegas*
Ir Herman Roebbers, *Philips TASS, The Netherlands*
Dr Marc Smith, *Vassar College, New York, USA*
Prof Dyke Stiles, *Utah State University, USA*
Dr Johan Sunter, *Philips Semiconductors, The Netherlands*
Mr Oyvind Teig, *Autronica Fire and Security, Norway*
Dr Brian Vinter, *University of Southern Denmark, Denmark*
Prof Alan Wagner, *University of British Columbia, Canada*
Mr David Wood, *University of Kent, UK*

Additional Reviewers

Dr Andrew Butterfield, *Trinity College Dublin, Ireland*
Dr Bill Gardner, *University of Guelph, Canada*
Dr Michael Goldsmith, *Formal Systems (Europe) Ltd., Oxford, UK*
Prof Jim Woodcock, *University of York, UK*
Dr Mike Poppleton, *University of Southampton, UK*
Prof David May, *Bristol University, UK*
Dr Neil Evans, *AWE UK*
Mr Bojan Orlic, *University of Twente, Netherlands*
Dr Richard Paige, *University of York, UK*
Prof Ian Marshall, *University of Lancaster, UK*
Dr Dominique Cansell, *Loria, France*
Dr Steve Dunne, *University of Teeside, UK*
Mr Adam Sampson, *University of Kent, UK*
Mr Kevin Chalmers, *Napier University, UK*
Dr Leonardo de Freitas, *University of York, UK*
Dr Soritis Moschoyiannis, *Univeristy of Surrey, UK*
Mr Damien Karkinsky, *University of Surrey, UK*
Dr Mike Shields, *Malta*
Mr Charles Crichton, *University of Oxford, UK*
Prof Susan Stepney, *University of York, UK*
Dr Jon Saul, *SystemCrafter, UK*

Contents

Preface   v
Alistair McEwan, Steve Schneider, Peter Welch and Wilson Ifill

Programme Committee   vi

Additional Reviewers   vii

Part A. Invited Speakers

Fine-Grain Concurrency   1
Tony Hoare

Communicating Process Architecture for Multicores   21
David May

Part B. Languages, Tools, Models, Platforms and Patterns

Lazy Exploration and Checking of CSP Models with CSPsim   33
Phillip J. Brooke and Richard F. Paige

The Core Language of Aldwych   51
Matthew Huntbach

JCSProB: Implementing Integrated Formal Specifications in Concurrent Java   67
Letu Yang and Michael R. Poppleton

Components with Symbolic Transition Systems: A Java Implementation of Rendezvous   89
Fabricio Fernandes, Robin Passama and Jean-Claude Royer

Concurrent/Reactive System Design with Honeysuckle   109
Ian East

CSP and Real-Time: Reality or Illusion?   119
Bojan Orlic and Jan F. Broenink

Testing and Sampling Parallel Systems   149
Jon Kerridge

Mobility in JCSP: New Mobile Channel and Mobile Process Models   163
Kevin Chalmers, Jon Kerridge and Imed Romdhani

C++CSP2: A Many-to-Many Threading Model for Multicore Architectures   183
Neil Brown

Design Principles of the SystemCSP Software Framework   207
Bojan Orlic and Jan F. Broenink

PyCSP – Communicating Sequential Processes for Python   229
John Markus Bjørndalen, Brian Vinter and Otto Anshus

A Process-Oriented Architecture for Complex System Modelling   249
    Carl G. Ritson and Peter H. Welch

Concurrency Control and Recovery Management for Open e-Business
Transactions   267
    Amir R. Razavi, Sotiris K. Moschoyiannis and Paul J. Krause

trancell – An Experimental ETC to Cell BE Translator   287
    Ulrik Schou Jørgensen and Espen Suenson

A Versatile Hardware-Software Platform for In-Situ Monitoring Systems   299
    Bernhard H.C. Sputh, Oliver Faust and Alastair R. Allen

High Cohesion and Low Coupling: The Office Mapping Factor   313
    Øyvind Teig

A Process Oriented Approach to USB Driver Development   323
    Carl G. Ritson and Frederick R.M. Barnes

A Native Transterpreter for the LEGO Mindstorms RCX   339
    Jonathan Simpson, Christian L. Jacobsen and Matthew C. Jadud

Integrating and Extending JCSP   349
    Peter Welch, Neil Brown, James Moores, Kevin Chalmers and
    Bernhard Sputh

Part C. Hardware/Software Co-Design

Hardware/Software Synthesis and Verification Using Esterel   371
    Satnam Singh

Modeling and Analysis of the AMBA Bus Using CSP and B   379
    Alistair A. McEwan and Steve Schneider

A Step Towards Refining and Translating B Control Annotations to Handel-C   399
    Wilson Ifill and Steve Schneider

Towards the Formal Verification of a Java Processor in Event-B   425
    Neil Grant and Neil Evans

Advanced System Simulation, Emulation and Test (ASSET)   443
    Gregory L. Wickstrom

Development of a Family of Multi-Core Devices Using Hierarchical
Abstraction   465
    Andrew Duller, Alan Gray, Daniel Towner, Jamie Iles,
    Gajinder Panesar and Will Robbins

Domain Specific Transformations for Hardware Ray Tracing   479
    Tim Todman and Wayne Luk

A Reconfigurable System-on-Chip Architecture for Pico-Satellite Missions   493
    Tanya Vladimirova and Xiaofeng Wu

Part D. Fringe Presentation Abstracts

Transactional CSP Processes   503
Gail Cassar and Patrick Abela

Algebras of Actions in Concurrent Processes   505
Mark Burgin and Marc L. Smith

Using **occam**-π Primitives with the Cell Broadband Engine   507
Damian J. Dimmich

Shared-Memory Multi-Processor Scheduling Algorithms for CCSP   509
Carl G. Ritson

Compiling **occam** to C with Tock   511
Adam T. Sampson

Author Index   513

Communicating Process Architectures 2007
Alistair A. McEwan, Steve Schneider, Wilson Ifill, and Peter Welch
IOS Press, 2007

Fine-grain Concurrency

Tony HOARE

Microsoft Research, Cambridge

Abstract. I have been interested in concurrent programming since about 1963, when its associated problems contributed to the failure of the largest software project that I have managed. When I moved to an academic career in 1968, I hoped that I could find a solution to the problems by my research. Quite quickly I decided to concentrate on coarse-grained concurrency, which does not allow concurrent processes to share main memory. The only interaction between processes is confined to explicit input and output commands. This simplification led eventually to the exploration of the theory of Communicating Sequential Processes.

Since joining Microsoft Research in 1999, I have plucked up courage at last to look at fine-grain concurrency, involving threads which interleave their access to main memory at the fine granularity of single instruction execution. By combining the merits of a number of different theories of concurrency, one can paint a relatively simple picture of a theory for the correct design of concurrent systems. Indeed, pictures are a great help in conveying the basic understanding. This paper presents some on-going directions of research that I have been pursuing with colleagues in Cambridge – both at Microsoft Research and in the University Computing Laboratory.

Introduction

Intel has announced that in future each standard computer chip will contain a hundred or more processors (cores), operating concurrently on the same shared memory. The speed of the individual processors will never be significantly faster than they are today. Continued increase in performance of hardware will therefore depend on the skill of programmers in exploiting the concurrency of this multi-core architecture. In addition, programmers will have to avoid increased risks of race conditions, non-determinism, deadlocks and livelocks. And they will have to avoid the usual overheads that concurrency libraries often impose on them today. History shows that these are challenges that programmers have found difficult to meet. Can good research, leading to good theory, and backed up by good programming tools, help us to discharge our new responsibility to maintain the validity of Moore's law?

To meet this challenge, there are a great many theories to choose from. They include automata theory, Petri nets, process algebra (many varieties), separation logic, critical regions and rely/guarantee conditions. The practicing programmer might well be disillusioned by the wide choice, and resolve to avoid theory completely, at least until the theorists have got their act together. So that is exactly what I propose to do. I have amalgamated ideas from all these well-known and well-researched and well-tested theories. I have applied them to the design of a structured calculus for low-overhead fine-grain concurrent programming. My theory of correctness is equally well-known: it is based on flowcharts and Floyd assertions. They provide a contractual basis for the compositional design and verification of concurrent algorithms and systems.

The ideas that I describe are intended to be an aid to effective thinking about concurrency, and to reliable planning of its exploitation. But it is possible to imagine a future in which the ideas can be more directly exploited. My intention is that a small collection of primitive operations will be simple enough for direct implementation in hardware, reducing the familiar overheads of concurrency to the irreducible minimum. Furthermore, the correctness of the designs may be certified by future programming tools capable of verifying the assertions that specify correctness. And finally, the pictures that I draw may help in education of programmers to exploit concurrency with confidence, and so enable all users of computers to benefit from future increases in hardware performance. But I leave to you the judgement whether this is a likely outcome.

1. Sequential Processes, Modeled by Flowcharts

I will start with a review of the concept of a flowchart. It is a graph consisting of boxes connected by arrows. Each box contains basic actions and tests from the program. On its perimeter, the box offers a number of entry and exit ports. Each arrow connects an exit port of the box at its tail to an entry port of the box at its head.

Execution of the program is modelled by a control token that passes along the arrows and through the boxes of the flowchart. As it passes through each box, it executes the actions and tests inside the box. In a sequential program there is only one token, so entry of a token into a box strictly alternates with exit from the box. Furthermore, there is no risk of two tokens passing down the same arrow at the same time. We will preserve an analogue of this property when we introduce concurrency.

The example in Figure 1 shows the familiar features of a flowchart. The first box on the left has two exits and one entry; it is the purpose of the test within the box to determine which exit is taken by the token on each occasion of entry. The two arrows on the right of the picture fan in to the same head. After the token has passed through a fan-in, it is no longer known which of the two incoming arrows it has traversed.

As shown in Figure 2, the execution control token starts at a designated arrow of the flowchart, usually drawn at the top left corner of the diagram. We regard the token as carrying the current state of the computer. This includes the names and current values of all the internal variables of the program, as well as the state of parts of the real world that are directly connected to the computer. In this simple example, we assume the initial state on entry of the token ascribes the value 9 to x.

Figure 1. A flowchart	**Figure 2.** A flowchart with token – 1

As shown in Figure 3, execution of the test in the first box causes the token to exit on the lower port, without changing the value of x. In Figure 4, execution of the code in the next box increases the value of x by 1.

Figure 3. A flowchart with token – 2	**Figure 4.** A flowchart with token – 3

In this sequence of diagrams, I have taken a snapshot of the passage of the token along each arrow. There is actually no storage of tokens on arrows, and conceptually, the emergence of a token from the port at the tail of an arrow occurs at the same time as entry of the token into the port at the head of the arrow.

The previous figures showed an example of a conditional command, selecting between the execution of a **then** clause and an **else** clause. Figure 5 shows the general structure of a conditional command. It is general in the sense that the boxes are empty, and can be filled in any way you like. Notice that all the boxes now have one entry and two exits. The exit at the bottom of each of each box stands for the throw of an exception, implemented perhaps by a forward jump.

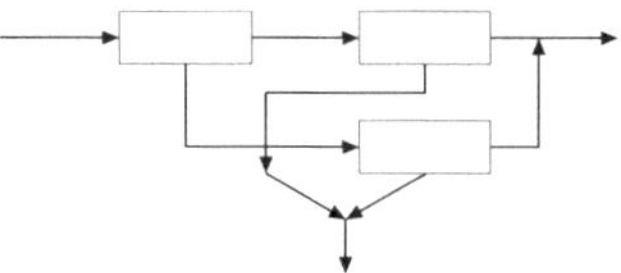

Figure 5. Conditional flowcharts

Figure 6 shows another useful generalisation of the concept of a flowchart, the structured flowchart: we allow any box to contain not only primitive commands of a program but also complete flowcharts. The pattern of containment must be properly nested, so the perimeters of different boxes do not intersect. Wherever an arrow crosses the perimeter between the interior and the outside of a containing box, it creates an entry or exit port, which is visible from the outside. Connections and internal boxes enclosed within the perimeter are regarded as externally invisible. Thus from the outside, the entire containing box can be regarded as a single command. The sole purpose of structuring is to permit flowcharts to be composed in a structured and modular fashion. The containing boxes are entirely ignored in execution.

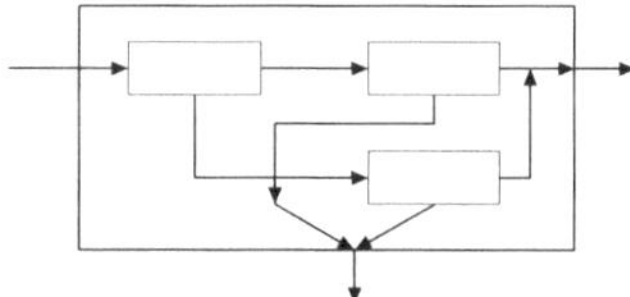

Figure 6. A structured flowchart

For convenience of verbal description, we will give conventional names to the entries and exits of each box as shown in Figure 7. The names are suggestive of the purpose of each port. In our simple calculus there will always be a start entry for initial entry of the token, a finish exit for normal termination, and a throw exit for exceptional termination. The names are regarded as local to the box. In pictures we will usually omit the names of the ports, and rely on the position of the arrow on the perimeter of the box to identify it.

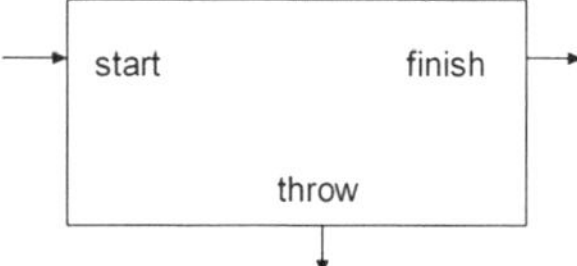

Figure 7. Port names

The ports of the enclosing boxes also have names. In fact, we generally use the same names for the enclosing box as well as the enclosed boxes. This is allowed, because port names inside boxes are treated as strictly local. The re-use of names emphasises the structural similarity of the enclosing box to the enclosed boxes. For example, in Figure 8, the enclosing box has the same structure and port names as each of the enclosed boxes. In fact, the whole purpose of the calculus that we develop is to preserve the same structure for all boxes, both large and small.

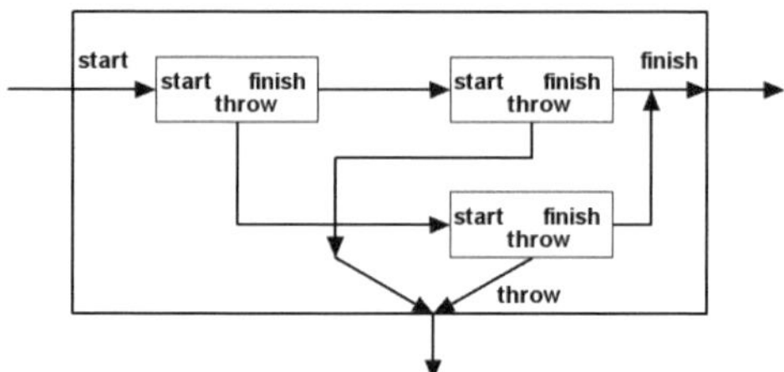

Figure 8. Structured naming

The notion of correctness of a flowchart is provided by Floyd assertions, placed on the entries and exits of the boxes. An assertion is a boolean condition that is expected to be true whenever a token passes through the port that it labels. An assertion on an entry port is a precondition of the box, and must be made true by the environment before the token arrives at that entry. The assertion on an exit port is a post-condition of the box, and the program in the box must make it true before sending the token out on that exit. That is the criterion of correctness of the box; and the proof of correctness is the responsibility of the designer of the program inside the box.

Figure 9 shows our familiar example of a flowchart, with assertions on some of the arrows. The starting precondition is that x is an odd number. After the first test has succeeded, its postcondition states that x is still odd and furthermore it is less than 10. After adding 1 to x, it is less than 11, and 1 more than an odd number. The postcondition of the other branch is obviously that x is 0. On both branches of the conditional, the postcondition on the extreme right of the flowchart states that x is even, and less than 11.

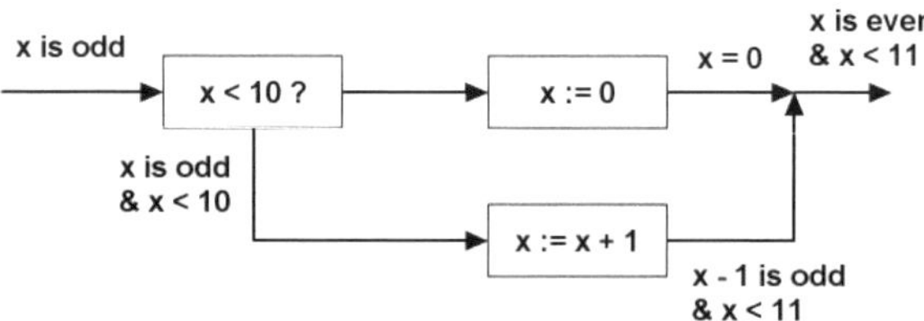

Figure 9. Flowchart with assertions

Let us examine the principles that have been used in this informal reasoning. The criterion of correctness for an arrow is very simple: the assertion at the tail of the arrow must logically imply the assertion at the head. And that is enough. As Floyd pointed out, a complete flowchart is correct if all its boxes and all its arrows are correct. This means that the total task of correctness proof of a complete system is modular, and can be discharged one arrow and one box at a time.

There is a great advantage in Floyd's method of formalising program correctness. The same flowchart is used both for an operational semantics, determining the path of the token when executed, and for an axiomatic semantics, determining the flow of implication in a correctness proof. There is no need to prove the consistency of the two presentations of semantics.

Figure 10. Arrow.   The arrow is correct if $P \Rightarrow R$

We allow any number of arrows to be composed into arbitrary meshes. But we are not interested in the details of the internal construction of the mesh. We are only interested whether any given arrow tail on the extreme left has a connection path of arrows leading to a given arrow head on the extreme right. We ignore the details of the path that makes the connection. Two meshes are regarded as equal if they make all the same connections. So the mesh consisting of a fan-in followed by a fan-out is the same as a fully connected mesh, as shown in Figure 11. Wherever the mesh shows a connection, the assertion at the tail on the left must imply the assertion at the head on the right. The proof obligation can be abbreviated to a single implication, using disjunction of the antecedents and conjunction of the consequents.

Figure 11. Equal meshes.   The mesh is correct if $P \vee Q \Rightarrow R \,\&\, S$

We will now proceed to give a definition of a little calculus of fine-grain concurrent programs. We start with some of the simplest possible boxes and flowcharts. The first example in Figure 12 is the simple skip action which does nothing. A token that enters at the start passes unchanged to the finish. The throw exit remains unconnected, with the result that it is never activated.

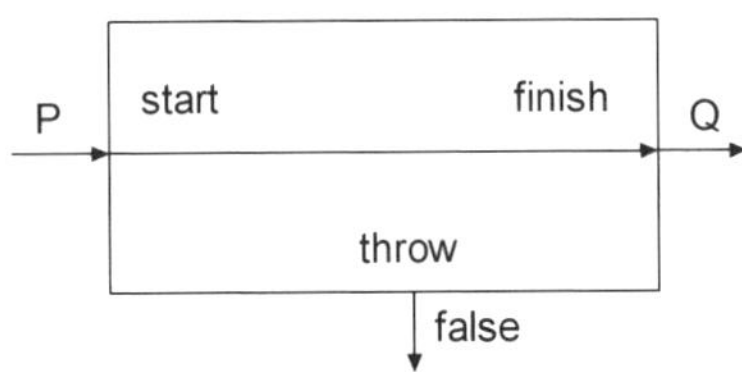

Figure 12. Skip action.   The box is correct if $P \Rightarrow Q$

The proof obligation for skip follows directly from the correctness condition of the single arrow that it contains. The false postcondition on the throw exit indicates that this exit will never be taken. Since false implies anything, an exit labelled by false may be correctly connected to any entry whatsoever.

The purpose of a throw is to deal with a situation in which successful completion is known to be impossible or inappropriate. The throw is usually invoked conditionally. Its definition is very similar to that of the skip, and so is its correctness condition. A flowchart for the throw action is shown in Figure 13.

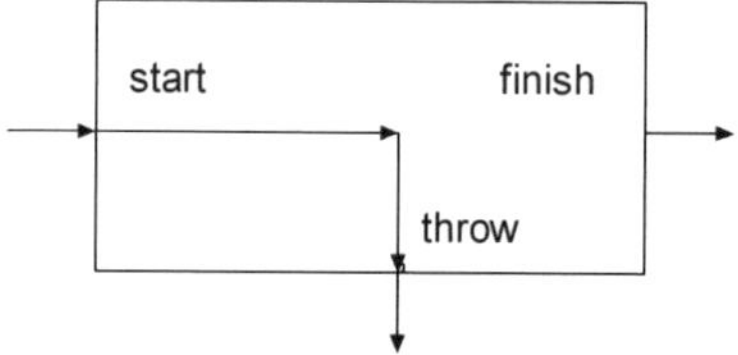

Figure 13. Throw action

The operators of our calculus show how smaller flowcharts can be connected to make larger flowcharts. Our first operator is sequential composition. We adopt the convention that the two operands of a composite flowchart are drawn as boxes inside an enclosing box that describes the whole of the composed transaction. The behaviour of the operator is determined solely by the internal connections between the ports of all three boxes. It is essential in a compositional calculus that the definition does not depend on the contents of its operand boxes. This rule is guaranteed if the internal boxes contain nothing, as shown in Figure 14.

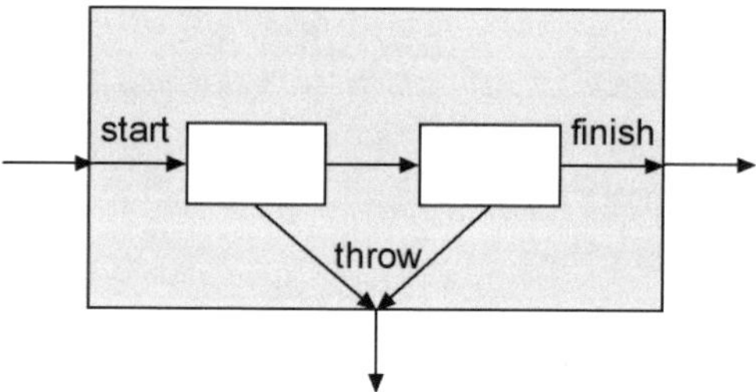

Figure 14. Sequential composition

To assist in proof of correctness, there should in principle be assertions on each of the arrows. However, the permitted patterns for these assertions are completely determined by the correctness principle for the arrows of a flowchart, so there is no need to mention them explicitly.

Sequential composition has many interesting and useful mathematical properties. For example, it is an associative operator. All the binary operators defined in the rest of this presentation will also be associative. Informal proofs of these and similar algebraic properties are quite simple. Just draw the flowcharts for each side of the equation, and then remove the boxes that indicate the bracketing. The two flowcharts will then be found to be identical. They therefore have identical executions and identical assertions, and identical correctness conditions.

Figure 15 shows the sequential composition of three transactions, with the gray box indicating that the brackets are placed to the left.

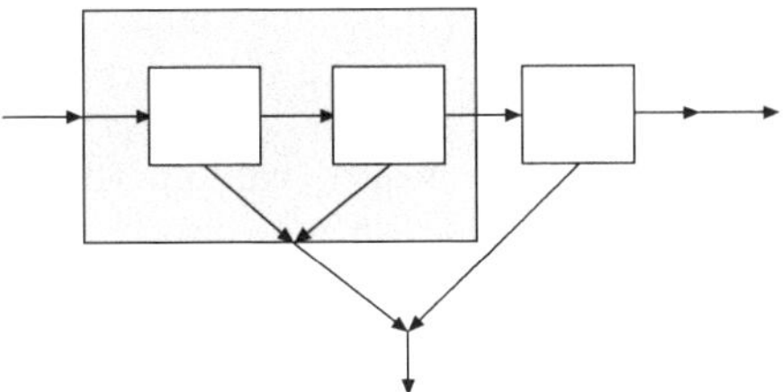

Figure 15. Asssociativity proof (left association)

And Figure 16 shows the same three processes with bracketing to the right. You can see that the flowcharts remain the same, even when the enclosing gray box moves. The apparent movement of the throw arrow is obviously not significant, according to our definition of equality of meshes of arrows.

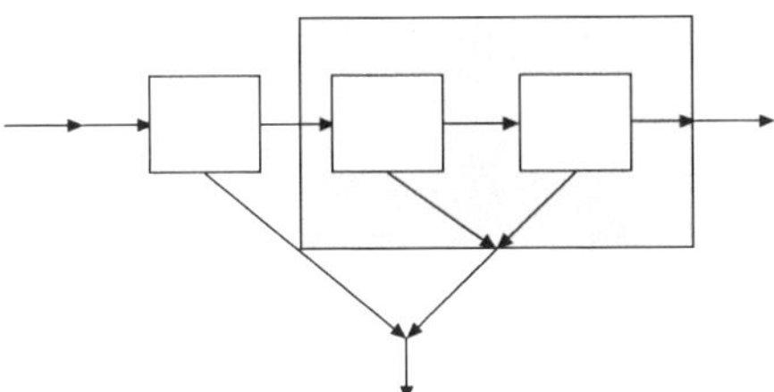

Figure 16. Asssociativity proof (right association)

In conventional flow-charts, it is prohibited for an arrow to fan out. Thus the thick arrow in Figure 17 would not be allowed. But we will allow fan-out, and use it to introduce non-determinism into our flowchart. When the token reaches a fan-out, it is not determined which choice it will make. This fact is exploited in the definition of a structured operator for non-deterministic choice between two operands. Whichever choice is made by the token on entry to the enclosing gray box, the subsequent behaviour of the program is wholly determined by the selected internal box. The other one will never even be started. The programmer must be prepared for both choices, and both must be correct. Non-determinism can only be used if the programmer genuinely does not care which choice is made. This is why non-determinism is not a useful operator for explicit use by programmers. We define it here merely as an aid to reasoning about the non-determinism that is inevitably introduced by fine-grain concurrency.

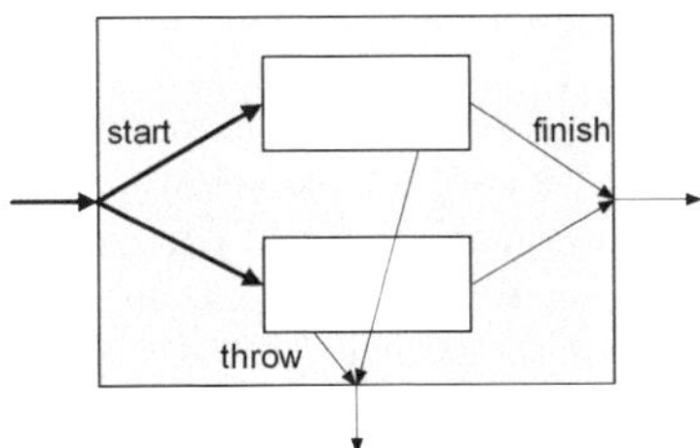

Figure 17. Non-determinism

Note that non-determinism is associative, but it has no unit. It is symmetric: the order in which the operands are written does not matter. It is idempotent: a choice between two identical boxes is the same as no choice at all. Finally, sequential composition, and most other forms of composition distribute, through nondeterminism. The proof of this uses Floyd's principle, that two flowcharts which have identical correctness conditions have the same meaning.

2. Concurrent Processes, Modeled by Petri Nets

We now extend our notation for flowcharts to introduce concurrency. This is done by one of the basic primitives of a Petri net, the transition. As shown in Figure 18, a transition is drawn usually as a thick vertical bar, and it acts as a barrier to tokens passing through. It has entry ports on one side (usually on the left) and exit ports on the other. The transition transmits tokens only when there are tokens ready to pass on every one of its entry ports. These tokens are then replaced by tokens emerging simultaneously from every one of the exit ports. Note that transitions in themselves do not store tokens: the firing of a transition is an atomic event. We will later introduce Petri net places as primitive devices to perform the storage function.

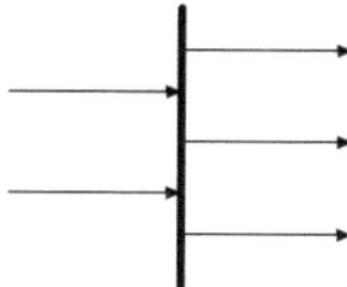

Figure 18. Petri net transition

As shown in Figure 19, if there is only one entry arrow, the transition is geometrically like a fan-out, since it contains two (or more) exit arrows. It is used to transmit a token simultaneously to a number of concurrent threads. It is therefore called a fork.

The other simple case of a transition is a join, as shown in Figure 20. It has only one exit port, and two or more entries. It requires tokens on all its inputs to pass through it simultaneously, and merges them into a single token. It thereby reduces the degree of concurrency in the system.

Figure 19. Petri net fork   **Figure 20.** Petri net join

The simple cases of forks and joins are sufficient to reconstruct all the more complicated forms of a Petri net transition. This is done by connecting a number of transitions into a mesh, possibly together with other arrows fanning in and fanning out. A mesh with transitions is capable of absorbing a complete set of tokens on some subset of its entry arrows, delivering tokens simultaneously to some subset of its exit arrows. These two subsets are said to be connected by the mesh. In the case of a mesh without transitions, the connection is made between singleton subsets. Two general meshes are regarded as equal if they make exactly the same connections between subsets. So the mesh shown in Figure 21 is equal to the mesh shown in Figure 22.

Figure 21. Petri net mesh – 1   **Figure 22.** Petri net mesh – 2

An inappropriate mixture of transitions with fan-in and fan-out of arrows can lead to unfortunate effects. Figure 23 shows an example corner case. A token at the top left of the mesh can never move through the transition. This is because the fan-out delivers a token at only one of its two heads, whereas the transition requires a token at both of them. As a result, the whole mesh has exactly the same effect as a mesh which actually makes only one connection. We will design our calculus of concurrency to ensure that such corner cases will never arise.

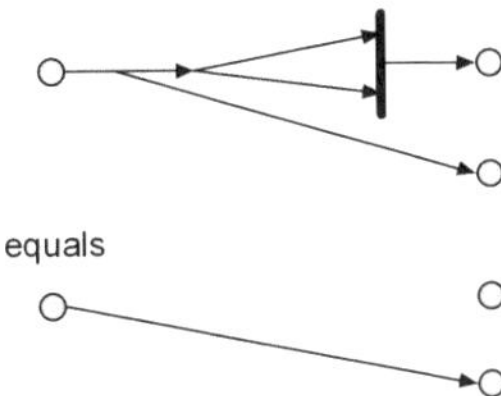

Figure 23. A corner case

In the design of fine-grain concurrent programs, it is essential to keep account of the ownership of resources by the threads which update them. We will therefore regard each to-ken as carrying with it a claim to the ownership (i.e., the write permissions and read permissions) for just a part of the state of the computer; though for simplicity, we will largely ignore read permissions. Obviously, we will allow a box to access and update only the resources carried by the token that has entered the box. The addition of ownership claims to the tokens helps us to use Petri nets for their initial purpose, the modelling of data flow as well as control flow through the system.

In Figure 24, the ownership of variables x and y is indicated by writing these names on the token which carries the variables. Figure 25 is the state after firing the transition. The resources claimed by the token are split into two or more disjoint parts (possibly sharing read-only variables); these parts are carried by the separate tokens emerging from the fork.

Figure 24. Token split: before   **Figure 25.** Token split: after

In Figure 24 and Figure 25, token at entry carries whole state: $\{x\ y\}$; at the exits, each sub-token carries a disjoint part of the state.

The Petri net join is entirely symmetric to the fork. Just as the fork splits the ownership claims of the incoming token, the join merges the claims into a single token. In Figure 26 and

Figure 27, each sub-token carries part of the state at entry; at exit, the token carries whole state again.

Figure 26. Token merge: before   **Figure 27.** Token merge: after

What happens if the incoming tokens make incompatible claims on the same resource? Fortunately, in our structured calculus this cannot happen. The only way of generating tokens with different ownership claims is by the fork, which can only generate tokens with disjoint ownership claims. As a result, the claims of each distinct token in the entire system are disjoint with the claims of all the others. The join transition shown above preserves this disjointness property. So no resource is ever shared between two distinct tokens.

We allow the assertion on an arrow of a Petri net to describe the ownership claims of the token that passes along the arrow. For simplicity, we will just assume that any variable mentioned in the assertion is part of this claim. In reasoning with these assertions, it is convenient to use a recently introduced extension of classical logic, known as separation logic; it deals with assertions that make ownership claims.

Separation logic introduces a new associative operator, the separated conjunction of two predicates, usually denoted by a star ($P \star Q$). This asserts that both the predicates are true, and furthermore, that their ownership claims are disjoint, in the sense that there is no variable in common between the assertions. The ownership claim of the separated conjunction is the union of the claims of its two operands.

In a program that uses only declared variables without aliasing, the disjointness of the claims can be checked by a compiler, and separation logic is not necessary. The great strength of separation logic is that it deals equally well with pointers to objects in the heap. It allows any form of aliasing, and deals with the consequences by formal proof. However, our example will not illustrate this power of separation logic.

The axiom of assignment in separation logic is designed to prevent race conditions in a fine-grain concurrent program. It enforces the rule that the precondition and the postcondition must have the same claim; furthermore, the claim must include a write permission for the variable assigned, and a read permission for every variable read in the expression that delivers the assigned value. In the displayed axiom of assignment (Figure 28) we have exploited the common convention that a proposition implicitly claims all variables that it mentions. So the precondition and postcondition claim x and y. Because of disjointness, R must not claim x or y. For simplicity, we have failed to distinguish read and write permissions.

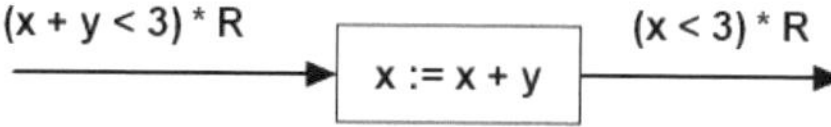

Figure 28. Axiom of assignment

Separated conjunction is used to express the correctness condition for Petri net transitions. The assertion at the entry of a must imply the separated conjunction of all the assertions at the exits. In Figure 29, the disjointness of P and Q represents the fact that the outgoing tokens will have disjoint claims.

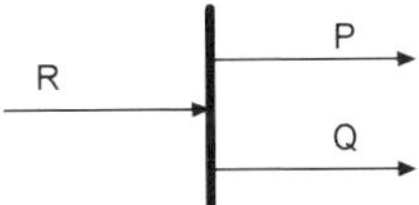

Figure 29. Correctness condition of fork: $R \Rightarrow P \star Q$

As mentioned before, the join is a mirror image of the fork. Accordingly, the correctness condition for a join is the mirror image of the correctness condition for a fork.

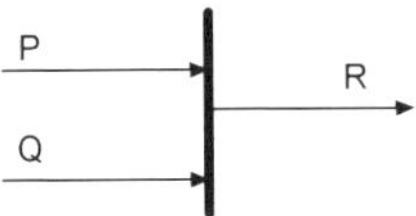

Figure 30. Correctness condition for join: $P \star Q \Rightarrow R$

There is a problem here. What happens if $P \star Q$ is false, even though both P and Q are both true? This would mean that the execution of the program has to make falsity true when it fires. But no implementation can do that – it is a logical impossibility. Fortunately, the rule of assignment ensures that P and Q must be consistent with each other. The details of the consistency proof of separation logic are beyond the scope of this paper.

The first example of the use of transitions in our calculus is the definition of the kind of structured (fork/join) concurrency introduced by Dijkstra. In Figure 31, the fork on the left ensures that both the threads labelled T and U will start together. The join on the right ensures that they will finish together. In between these transitions, each of the threads has its own token, and can therefore execute concurrently with the other. By definition of the fork and join, the tokens have disjoint claims. Since a thread can only mention variables owned by its token, the rule of assignment excludes the possibility of race conditions. It also excludes the possibility of any interaction whatsoever between the two threads.

In Figure 31, I have not allowed any possibility of a throw. The omission will be rectified shortly.

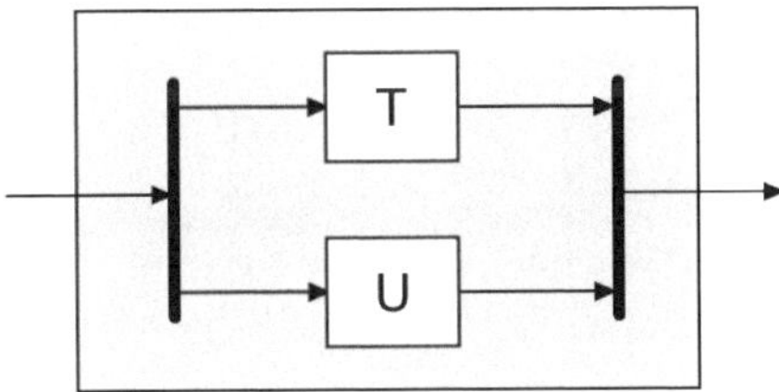

Figure 31. Concurrent composition.   There is no connection between T and U

Figure 32 is a simple example of a concurrent program. The precondition says that x and y have the same parity. One thread adds 2 to x, and the other multiplies y by 7. Both these operations preserve parity. So the same precondition still holds as a postcondition. Although this is obvious, the proof requires a construction, as shown in Figure 33. The construction introduces an abstract or ghost variable z to stand for the parity of x and y. A ghost variable may appear only in assertions, so it remains constant throughout its scope. For the same reason, a ghost variable can be validly shared among threads (though it may not be either read or written). When it has served its purpose, the ghost variable may be eliminated by existential quantification in both the precondition and the postcondition.

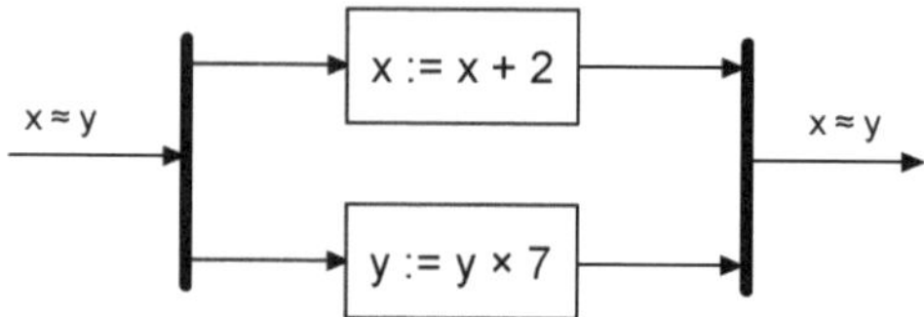

Figure 32. A concurrent composition example.　$x \approx y$ means $(x - y) \bmod 2 = 0$ (their difference is even)

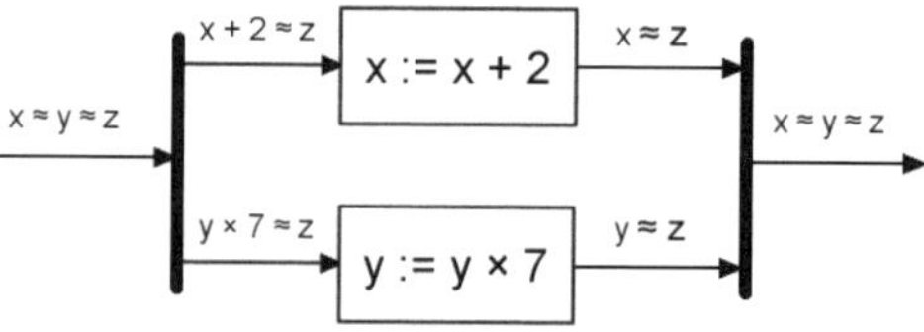

Figure 33. Ghost variable z

Proof:
$$x \approx y \;\Rightarrow\; x + 2 \approx y \times 7$$
$$x \approx y \approx z \;\Rightarrow\; (x + 2 \approx z) \star (y \times 7 \approx z)$$

We now return to the example of the structured concurrency operator and remove the restriction on throws. In Figure 34, the throw exits of T and U are connected through a new join transition to the throw exit of the composition. As a result, the concurrent combination throws just when both the operands throw. This still leaves an unfortunate situation when one of the operands attempts to throw, whereas the other one finishes normally. In an implementation, this would manifest itself as a deadlock.

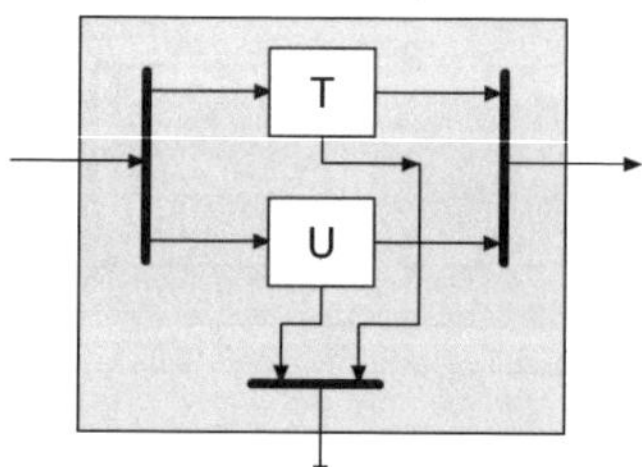

Figure 34. Concurrency with throw.　To avoid deadlock, T and U must agree on their exits

A solution is to adopt an even more complicated definition of concurrent composition. It ensures that a throw will occur when either of the operands throws, even if the other one finishes. As shown in Figure 35, this is achieved by additional joins to cover the two cases when the threads disagree on their choice of exit port.

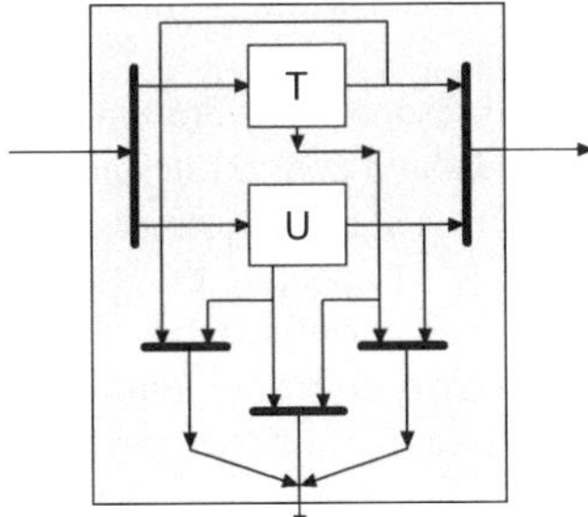

Figure 35. Deadlock avoided.　Disagreement on exit leads to throw

In Figure 36, note the four encircled fan-outs in the arrows at the exits of the operands T and U. Each of these introduces non-determinism. However, it is non-determinism of the external kind that is studied in process algebras like CCS and CSP. It is called external, because the choice between the alternatives is made at the head of the arrow rather than at the tail. On reaching the fan-out, the token will choose a branch leading to a transition that is ready to fire, and not to a transition that cannot fire. In Figure 36, we have ensured that at most one of the alternative transitions can be ready to fire. Thus the diagram is in fact still completely deterministic, in spite of the four fan-outs.

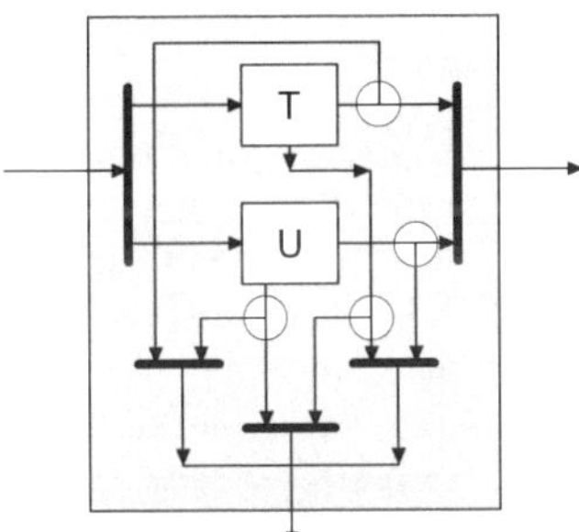

Figure 36. Fan-out gives external non-determinism

The calculus that we have described so far is not capable of exploiting fully the power of multi-core architecture. The reason is that the same rules that prohibit race conditions also prohibit any form of communication or co-operation among the threads. To relax this restriction, it is necessary to establish some method of internal communication from one thread to another. For the purpose of exploiting multi-core architecture, the highest bandwidth, the minimum overhead and the lowest latency are simultaneously achieved by use of the resources of the shared memory for communication. Communication takes place when one thread updates a variable that is later read by another.

Of course, race conditions must still be avoided. This is done by the mechanism of a critical region, which enables the programmer to define a suitable level of granularity for the interleaving of operations on the shared resource by all the sharing threads. A critical region starts by acquiring the shared resource and ends by releasing it, through new entry ports introduced into our calculus for this purpose. Inside a critical region, a thread may freely update the shared resource together with the variables that it owns permanently. Race conditions are still avoided, because the implementation ensures that at any time at most one thread can be in possession of the critical region. A simple implementation technique like an exclusion semaphore can ensure this.

In our Petri net model, a shared resource is represented by a token which carries ownership of the resource. In order to access and update the shared resource, a thread must acquire this token, which is done by means of a standard join between the control token and a token carrying ownership of the resource. After updating the shared state within the critical region, the thread must release the token, by means of a standard fork. The standard rules of ownership are exactly appropriate for checking critical regions defined in this way, since the token that travels through the region will carry with it the ownership of both the local variables of the thread and the variables of the shared resource. These can therefore be freely updated together within the critical region.

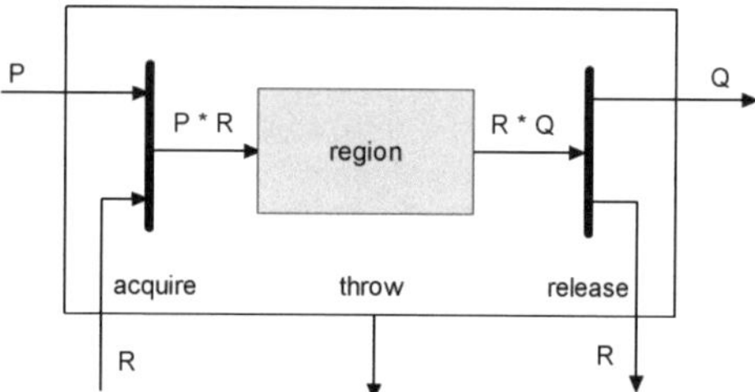

Figure 37. Critical region.   R is the resource invariant

Note that the body of the critical region has no acquire or release ports. This intentionally prohibits the nesting of critical regions. Furthermore, I have disallowed throws from within a critical region. To allow throws, the definition of a critical region requires an additional fork transition to ensure that the resource token is released before the throw exit. This means that the programmer must restore the resource invariant before the throw.

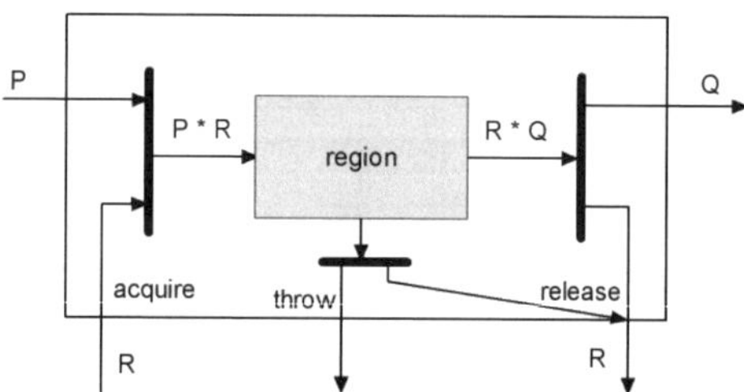

Figure 38. Critical region with throw

Addition of new ports into a calculus requires extension of the definition of all the previously defined operators. In the case of the new acquire and release ports, the resource is equally accessible to all the operands, and the standard extension rule is to just connect each new entry port of the enclosing block for the operator by a fan-out to the like-named new entry ports of both the operands; and connect every new exit port of each operand via a fan-in to the like-named port on the enclosing block. Figure 39 shows only the new ports and additional arrows that are to be added to every operator defined so far. It ensures that the new ports can be used at any time by either of the operands.

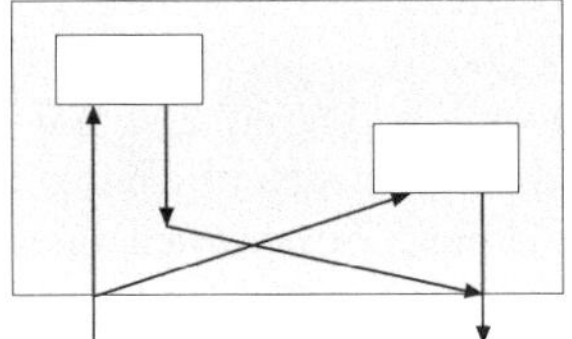

Figure 39. New ports

A shared resource is introduced by exactly the same operator which introduces multiple threads. The token that owns the resource is created by the fork on the left of Figure 40. It then resides at a place (denoted by a circle) specially designated for it within the Petri net. The resource token is acquired by its users one at a time through the acquire entry at the beginning of each critical region, and it is released after use through the release exit at the end of each critical region. It then returns to its designated place. If more than one user is simultaneously ready to acquire the resource token, the choice between them is arbitrary; it has to be made by the semaphore mechanism that implements exclusion. This is the way that shared memory introduces don't-care non-determinism into a concurrent program.

The assertion R in this diagram stands for the resource invariant. As shown in Figure 39, it may be assumed true at the beginning of every critical region, and must be proved true at the end. It thus serves the same role as a guarantee condition in the rely/guarantee method of proving concurrent programs.

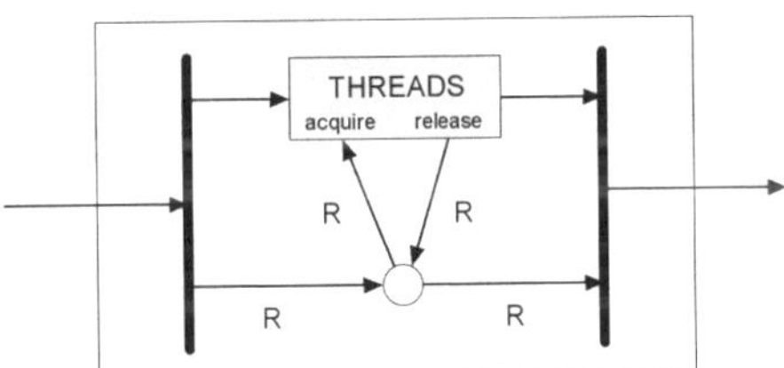

Figure 40. Resource declaration.   Petri net place: ◯ stores a token

Figure 41 caters for the possibility of a throw, in the usual way.

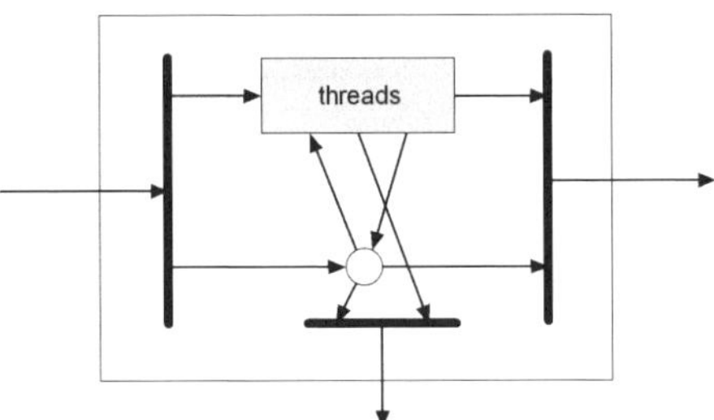

Figure 41. Resource declaration with throw

Figure 42 is an extremely simple example of concurrency with critical regions. Two threads share a variable x. One of them assigns to it the value 2, and the other one assigns the value 7. Because the variable is shared, this has to be done in a critical region Each thread is nothing but a single critical region. As a result, the two critical regions are executed in arbitrary order, and the final value of x will be either 2 or 7. The easiest proof is operational: just prove the postcondition separately for each of the two interleavings. But in general, the number of interleavings is astronomical. So we want to ask whether our assertional proof system capable of proving this directly in a more abstract way?

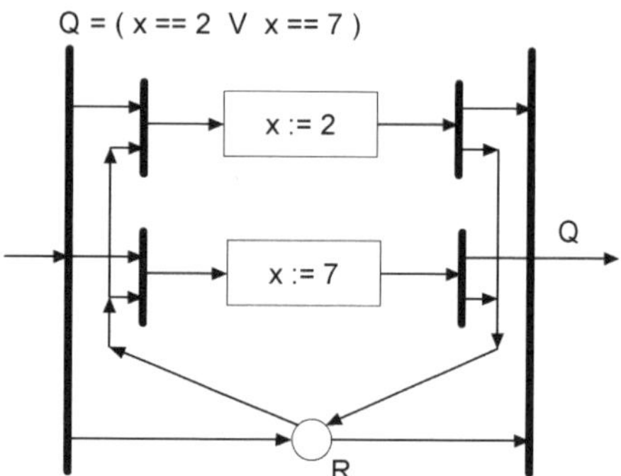

Figure 42. Example – 1

The answer seems to be yes, but only with the help of a ghost variable t, introduced to record the termination of one of the threads. The variable obviously starts false. By conditioning the resource invariant on t, its truth is assured at the beginning. Both critical regions leave the resource invariant R true. And one of them sets t true. Thus at the end, both t and R are true. Thus Q is also true at the end.

But the question arises, who owns t? It has to be joint ownership by the resource and the first thread. Such jointly owned variables can be updated only in a critical region, and only by the thread that half-owns it. The resource owns the other half. When the resource and the thread have come together in the critical region, full ownership enables the variable to be updated. This is adequate protection against race conditions. Fractional ownership is a mechanism also used for read-only variables in recent versions of separation logic.

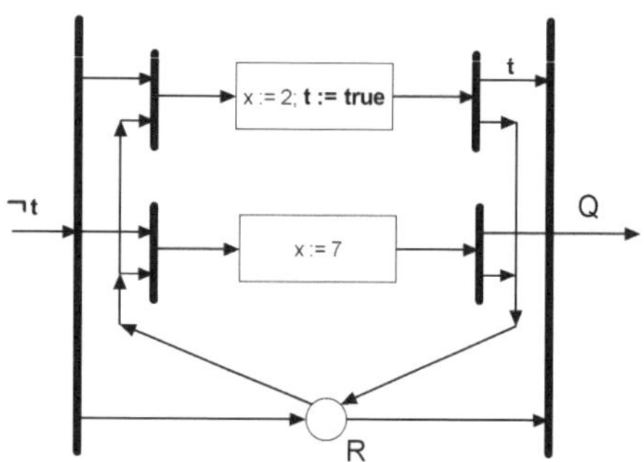

Figure 43. Example – 2.   Q = x ∈ {2, 7} and R = t ⇒ Q

3. Other Features of a Calculus

Recursion is the most important feature of any programming calculus, because it allows the execution of a program to be longer than the program itself. Iteration is of course an especially efficient special case of recursion. Fortunately, Dana Scott showed how to introduce recursion into flowcharts a long time ago. Just give a name X to a box, and use the same name as the content of one or more of the interior boxes. This effectively defines an infinite net, with a copy of the whole box inserted into the inner box. For this reason, the pattern of entry and exit ports of the recursive call must be the same as that of the outer named box. That is a constraint that is easily enforced by use of a calculus like one we have described.

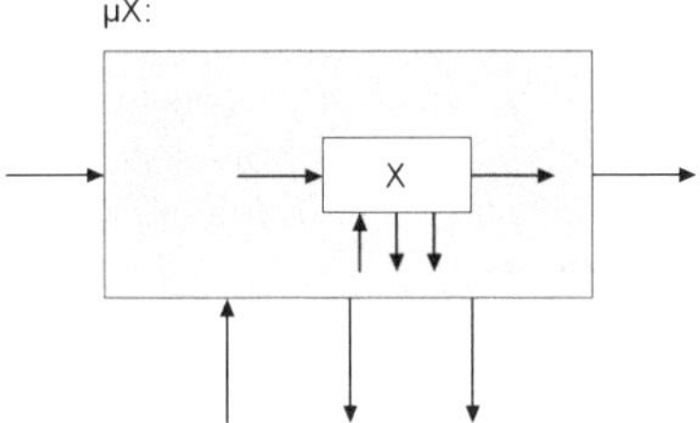

Figure 44. Scott recursion

A variable can be represented by a place pre-loaded with a token that owns the variable. This token joins the main control token on entry to the block, which can use the variable as required. It is forked off again on exit from the block, so that it is never seen from the outside. A place is needed at the finish to store the token after use. Let us use the same place as stored the token at the beginning.

The assertions on the arrow leading from and to the place should just be the proposition true, which is always true. This means that nothing is known of the value of the variable immediately after declaration. It also means that its value on termination is irrelevant. This permits an implementation to delay allocation of storage to the variable until the block is entered, and to recover the storage or exit.

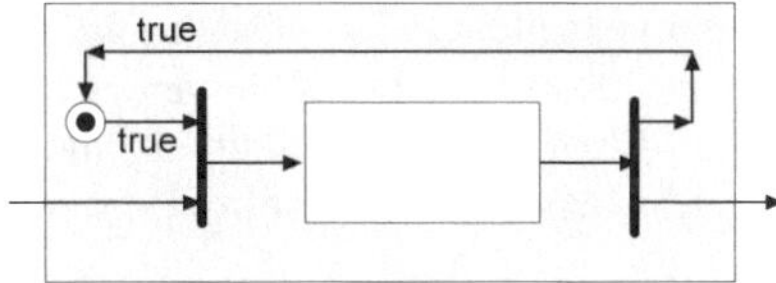

Figure 45. Variable declaration – 1

Figure 46 extends the diagram to show what happens on a throw. The variable still needs to be retained inside the box after an exception.

The Petri net fork is a direct implementation of an output from one thread of a system to another. It simply transfers ownership of the message (together with its value) to the inputting process. It does not copy the value. It does not allocate any buffer. Overhead is therefore held to a minimum. If buffers are desired, they can be modelled as a sequence of Petri net places.

Just as output was a fork, input is a join at the other end of an arrow between two threads. Note that the output is synchronised with the inputting process. In a sympathetic architecture (like that of the transputer), the operations of input and output can be built into the instruction set of the computer, thereby avoiding software overhead altogether.

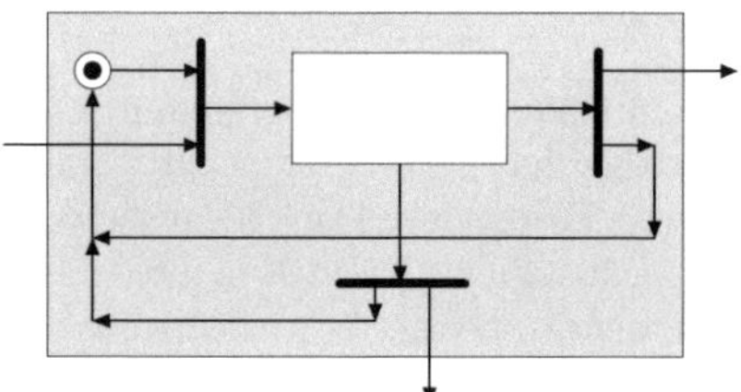

Figure 46. Variable declaration – 2

The introduction of arbitrary arrows communicating ownership among threads can easily lead to deadlock. Absence of deadlock can be proved by the methods of process algebra, and we will not treat it here. Fortunately, the use of non-nested critical regions is a disciplined form of communication which is not subject to deadlock. A simple hierarchy of regions can extend the guarantee to nested regions.

Figure 47. Output   **Figure 48.** Input

4. Conclusion

The main conclusions that may be drawn from this study are:

1. Flow-charts are an excellent pictorial way of defining the operational semantics of program components with multiple entry and exit points. Of course, they are not recommended for actual presentation of non-trivial programs.
2. Floyd assertions are an excellent way of defining and proving correctness of flowcharts. Consistency with an operational semantics for flowcharts is immediate.
3. Petri nets with transitions extend these benefits to fine-grain concurrent programs. The tokens are envisaged as carrying ownership of system resources, and permissions for their use.
4. Separation logic provides appropriate concepts for annotating the transitions of a Petri net. The axiom of assignment provides proof of absence of race conditions.
5. Critical regions (possibly conditional) provide a relatively safe way of using shared memory for communication and co-operation among threads.
6. Although they are not treated in this paper, rely/guarantee conditions provide a useful abstraction for the interleaving of critical regions.
7. Pictures are an excellent medium for defining the operators of a calculus. They are readily understood by programmers who are unfamiliar with programming language semantics (some of them even have an aversion to syntax).

Of course, there is abundant evidence, accumulated over many years, of the value of each of these ideas used separately. The only novel suggestion of this presentation is that their combined use may be of yet further value in meeting the new challenges of multi-core architecture.

Acknowledgment

Thanks to Robert Floyd, Carl Adam Petri, Cliff Jones, Simon Peyton Jones, Tim Harris, Viktor Vafeiadis, Matthew Parkinson and Wolfgang Reisig. And thanks to Steve Schneider and Zhe Xia for preparing this paper. Even though there are no references, it is a pleasure to express my thanks to those who have inspired this work, or helped its progress.

Communicating Process Architectures 2007
Alistair A. McEwan, Steve Schneider, Wilson Ifill, and Peter Welch
IOS Press, 2007

Communicating Process Architecture for Multicores

David MAY

*Department of Computer Science, University of Bristol,
BS8 1UB, United Kingdom.*

David.May@bristol.ac.uk

Abstract. Communicating process architecture can be used to build efficient multi-core chips scaling to hundreds of processors. Concurrent processing, communications and input-output are supported directly by the instruction set of the cores and by the protocol used in the on-chip interconnect. Concurrent programs are compiled directly to the chip exploiting novel compiler optimisations. The architecture supports a variety of programming techniques, ranging from statically configured process networks to dynamic reconfiguration and mobile processes.

Keywords. Multicores, Concurrency, On-chip interconnect, Message routing, Process scheduling, Programmed input-output

Introduction

It has become practical to construct multiprocessor chips scalable to hundreds of processors per chip. This paper outlines an architecture for these chips based on communicating processes, following the principles originated in CSP [1], occam [2] and developed in [3]. The processors used are general purpose in the sense that they can execute conventional sequential programs. Together they form a general purpose concurrent processor with hardware resources for scheduling a collection of concurrent processes and for performing synchronisation, communication and input-output.

The processors are connected by an *interconnect* which provides scalable throughput and low latency throughout the chip. Data can be streamed through virtual circuits at high speed or packets can be dynamically routed with low delay. Computations can rapidly bring more processors into use, and can move programs to data and interfaces to minimise latency and power.

Concurrent programs are compiled directly to the processor instruction set; no kernel is needed and no microcode is used in the processor implementation. The cost of communications, synchronisations, inputs and outputs is reduced to that of loads, stores, branches and arithmetic!

Multicores offer the possibility of computer systems scaling to millions of processors. But perhaps more important is that they offer a new way to construct programmable and configurable systems based on software design; in contrast to an FPGA which has an array of look-up tables which communicate via a statically configured interconnect, a multicore has an array of processors which communicate via a dynamic message routing network.

1. Outline

A multicore is constructed as a set of *tiles*. Each tile contains a processor, memory and communications system. There may be one or more chips in a system, and communication between all of the tiles in a system is provided by an interconnect which transports data, programs and control information.

It is practical for a single chip to contain from 1 to 1024 tiles, depending on relative processing and memory requirements, and on cost, power and area targets. The performance of multicore chips over the next five years is likely to range from 10^9 to 10^{12} instructions per second at a clock speed of around 500MHz.

The interconnect provides communication between all tiles on the chip (or system if there is more than one chip). It must be able to provide throughput scaling with the number of processors whilst maintaining bounded communication latency. Sufficient *latency hiding* can then be provided by the processors to conceal interconnect delays and allow pipelining of communications. In conjunction with simple programs, the interconnect can also be used to support access to the memory on any tile in the system from any other tile, and to enable any tile to initiate computations on any other tile in the system.

A key issue when combining a large number of processors is power efficiency. The synchronising communication of CSP provides a natural way to express event-driven systems, and event-driven architecture and software enables processors to be powered down when they have nothing to do. It addition, it is practical to use multiple time-domains within a chip to localise high speed clocks and eliminate the need for high speed clock distribution.

An important architectural requirement is compact program and data representation, as memory access consumes a significant proportion of the processor area and power. Interprocessor communication also consumes significant power and introduces latency; consequently it is often worth using application specific protocols for communication, compressing data or moving programs to data instead of moving data to programs. This can be facilitated by compact, position-independent code.

One of the potential uses of multicores is to perform many concurrent input and output operations, or to use processors to implement functions more commonly performed by hardware. System design using communicating processes can be used in many situations which would normally require hardware design. By closely coupling the input and output pins to the processor instruction execution, very low latency can be achieved allowing input and output operations to be programmed in high level software. From this perspective, the processor can be seen as a programmable state machine able to handle many concurrent input and output interfaces.

2. Interconnect

Each tile has a number of bidirectional links which connect it to a switch; it can communicate simultaneously in both directions via all of these links allowing data to be streamed through the tile under software control. A 500MHz processor can support several simultaneous streams of around 100Mbytes/second and modern manufacturing technologies with many layers of metal interconnect have made switches for links of this speed implementable even for hundreds of links. For small interconnects a single fully-connected switch is adequate; for larger networks Clos networks [4] are feasible. In larger systems or where there are significant physical constraints n-dimensional grids can be used. A discussion of various networks and their performance can be found in [5].

For example, for 128 processors a Clos network can be formed of 32×32 switches. The core of the network is formed from 16 of these, each connecting to one link on every one of 32 edge switches. The remaining 16 links on each of the edge switches are used to connect to the processors, providing 512 links in total. Each processor therefore has four links into the interconnect and the bisection throughput of the network is sufficient to maintain full speed on all of the links. Routing via the core switches is done by selecting the first available route from an edge switch to (any) core switch. A 32×32 switch is estimated at under $0.5mm^2$ on a 90nm manufacturing process. In the 128 processor example here, the network occupies $24mm^2$ and connects 128 processors. Assuming that a processor and memory occupy $2mm^2$, this interconnect represents less than 10% of the chip.

An alternative approach which involves more routing stages but less complex interconnections is to use n-dimensional grids as suggested in [6]; a simple scheme allows routing decisions to be made very rapidly. The incoming destination address at the start of each message is compared with the tile address, bit-by-bit. If all of the pairs of bits match, the tile is the message destination and the rest of the message is routed to the tile processor(s). If not, the number of the first pair of non-matching bits is used to select an entry in a lookup table; this table determines the direction to be used to forward the incoming message. Surprisingly, this very simple scheme - in which the lookup tables have one entry per bit of the message header and each table entry has only enough bits to identify an outgoing link - is sufficient to perform efficient deadlock-free routing in all n-dimensional arrays.

An example table configuration for a two dimensional array of 16 processors is shown in the table below.

Table 1. Table entries for routing in two dimensions

processor	entry	processor	entry	processor	entry	processor	entry
0	rrdd	4	rldd	8	lrdd	12	lldd
1	rrdu	5	rldu	9	lrdu	13	lldu
2	rrud	6	rldu	10	lrdu	14	lldu
3	rruu	7	rldu	11	lrdu	15	lldu

Each table entry selects either a right link (r), left link (l), up link (u) or down link (d). The routing takes all messages first right or left; when they have arrived at the correct column, they then move up or down to reach their destination. Although this example uses a two dimensional array, it is practical to use this scheme for higher dimensional on-chip interconnects especially in manufacturing technologies with several layers of metal interconnect; also in multi-chip systems it is common practice to use higher dimensional interconnects. Notice that the scheme uses shortest routes and is deadlock-free; it can be enhanced by using multiple links along each path. For scalable routing in very large networks, some form of load-distribution such as randomisation can also be added [5].

The protocol used for communication between processes provides *control* and *data* tokens which can be used by software to construct applications optimised protocols. It allows the interconnect to be used under program control to establish virtual circuits which stream data or transport a series of packets; alternatively it can be used for dynamic packet routing by establishing and disconnecting circuits on a packet-by-packet basis.

3. Processes

It is practical for each processor to include hardware support for a number of processes, including:

- a set of registers for each process
- a process scheduler which dynamically selects which process to execute
- a set of channels used for communication with other processes
- a set of ports used for input and output
- a set of timers to control real-time execution
- a set of clock generators to enable synchronisation of the input-output with external time domains

The set of processes on each tile can be used to allow communications or input-output to progress together with processing. There is no need for DMA controllers or specialised hardware interfaces as processes can be used to perform these functions. The process scheduling also provides latency hiding in the interconnect by allowing some processes to continue whilst others are waiting for communication to or from remote tiles.

The set of hardware processes in each tile can also be used to implement a kernel for a much larger set of virtual processes. In this case, some threads are dedicated to kernel functions such as managing communications to and from the virtual processes and system-wide resource allocation whilst others are used to actually execute the virtual processes.

Note that these uses of hardware process scheduling differ from the now common practice of using process - or *thread* - scheduling to hide latency in memory access originating from its use in shared memory multiprocessors [7].

4. The Processor Instruction Set

To equip each process with its own set of registers, the number of registers for each process must be small enough so that the processor's register file does not become big and slow. However, it must also be large enough to allow a process to execute efficiently. Another consideration is that it is not possible to use small instructions to address three operands in a large register file. Even with 16 registers, 12 bits are required to specify three register operands leaving only 4 opcode bits if the instruction length is 16 bits.

A good compromise is to provide dedicated *access* registers to access program, stack and data regions in memory together with a set of 12 *operand* registers for general purpose use. The three register operands can then be encoded using 11 bits (as $12 \times 12 \times 12 < 2048$) leaving 5 opcode bits. One or two opcodes can be used to extend the coding to include 32 bit instructions; one to extend the range of immediate values within instructions and the other to extend the number of opcodes. Careful choice of instructions within this framework results in most instructions being 16-bit and provides space to add opcodes for new instructions to extend the scope of the architecture.

Multicore architectures have potential uses in many areas, so it is important to provide encoding space for new instructions to be added in a systematic way. By decoding the second 16-bit part of a 32-bit instruction in the same way as the first 16-bit part, instructions with up to six register operands can be encoded. This is enough to efficiently support double length operations, long integer arithmetic for cryptography and multiply-accumulate for signal and image processing.

Each processor manages a number of different types of physical *resource*. These include processes, synchronisers, channels, timers, locks and clock generators. For each type of resource a set of available items is maintained; processes can *claim* and *free* resources using special instructions. Resources interact directly with the process scheduler and instructions such as inputs and outputs can potentially result in a process pausing until a resource is ready and then continuing. Information about the state of a resource is available to the scheduler within a single processor cycle.

5. Process Scheduler

Execution of instructions from each process is managed by the hardware process scheduler. This maintains a set of runnable processes, run, from which it takes instructions in turn. When a process is unable to continue, it is *paused* by removing it from the run set. The reason for this may be any of the following:

- Its registers are being initialised prior to it being able to run.

- It is waiting to synchronise with another process before continuing or terminating.
- It has attempted an input from a channel which has no data available, or a port which is not ready, or a timer which has not reached a specified time.
- It has attempted an output to a channel or a port which has no room for the data.
- It has executed an instruction causing it to wait for one of a number of events or interrupts which may be generated when channels, ports or timers become ready.

In many applications, it is important that the performance of an individual process can be guaranteed. Potential problems arise if, for example, all of the processes require memory accesses for data or instruction fetch at the same time or if several input-output events occur simultaneously. In these situations, one process may be delayed waiting for all of the other processes to complete their accesses.

An ideal scheduling system would allow any number of processes to share a single unified memory system and input-output system whilst guaranteeing that with n processes able to execute, each will get at least $1/n$ processor cycles. The set of n processes can then be thought of as a set of virtual processors each with clock rate at least $1/n$ of the clock rate of the processor itself. From a software design standpoint, this makes it possible to calculate the minimum performance of a process by counting the number of concurrent processes at a specific point in the program. In practice, performance will almost always be higher than this because individual processes will sometimes be delayed waiting for input or output and their unused processor cycles will be used to accelerate other processes.

Each process has a short instruction buffer sufficient to hold at least four instructions. Instructions are issued from the instruction buffers of the runnable processes in a round-robin manner, ignoring processes which are not in use or are paused waiting for a synchronisation or an input or output operation. The execution pipeline has a memory access stage which is available to *all* instructions. The rules for performing an instruction fetch are:

- Any instruction which requires memory access performs it during the memory access stage.
- Branch instructions fetch their branch target instructions during the memory access stage unless they also require a data access (in which case they will leave the instruction buffer empty).
- Any other instruction uses the memory access stage to perform an instruction fetch. This is used to load the process's own instruction buffer unless it is full, in which case it is used to load the buffer of another process.
- If a process's instruction buffer is empty when an instruction should be issued, a special *fetch no-op* is issued; this will use its memory access stage to load the process's instruction buffer.

There are very few situations in which a *fetch no-op* is needed, and these can often be avoided by simple instruction scheduling in compilers or assemblers. An obvious example is to break long sequences of loads or stores by interspersing arithmetic or logical operations.

The effect of this scheduling scheme is that, for example, a chip with 128 processors each able to execute 8 processes can be used as if it were a chip with 1024 processors operating at one eighth of the processor clock rate. Each of the 128 processors behaves in the same way as a symmetric multiprocessor with 8 processors sharing a memory with no access collisions and with no caches needed.

6. Concurrency and Process Synchronisation

A process may initiate execution of one or more concurrent processes, and can subsequently synchronise with them to exchange data or to ensure that all of them have completed before continuing. Process synchronisation is performed using hardware *synchronisers*, and processes using a synchroniser will move between running states and paused states. The advantage of using hardware to perform synchronisation is that it can be made very fast, allowing compilers to replace a join-fork pair by a much more efficient synchronisation. In the following example the fork and join operations can be moved outside the loop, with the result that only one fork and one join operation needs to be executed.

```
while true
{ par { inarray(inchan, a) || outarray(outchan, b) };
  par { inarray(inchan, b) || outarray(outchan, a) }
}
```

In terms of occam-pi [9]; the resulting optimised program can be expressed using SYNC operations on a synchroniser c, as shown below.

```
par
{  while true
     { inarray(inchan, a); SYNC c; outarray(outchan, b); SYNC c}
|| while true
     { inarray(inchan, b); SYNC c; outarray(outchan, a); SYNC c}
}
```

Another use of synchronisation is allow a sequential process to be split into a small set of concurrent ones, as this allows high performance to be achieved with a simple pipeline architecture [8]. An example of this is to break an operation on the elements of an array of length n into two concurrent operations on arrays of length $n/2$.

To start a slave process a master process must first use a *get synchroniser* instruction to acquire a synchroniser. The *get process* instruction is then used to get a synchronised process. It is passed the synchroniser identifier and returns a free process, having associated it with the synchroniser. When a process is first created, it is in a paused state and its access registers can be initialised using special instructions. The master process can repeat this operation to create a group of processes which all synchronise together. To start the slave processes the master executes a *master synchronise* instruction using the synchroniser identifier.

The group of processes can synchronise at any point by the slaves executing a *slave synchronise* and the master a *master synchronise*. Once all the processes have synchronised, each of them is allowed to continue executing from its next instruction. Instructions are provided to transfer data directly between the operand registers of two processes, avoiding the need to use memory when ownership of variables changes at a synchronisation point. To terminate all of the slaves and allow the master to continue the master executes a *master join* instruction instead of a *master synchronise*. When this happens, the slave processes are all freed and the master continues.

7. Communication

Communication between processes is performed using *channels*, which provide full-duplex data transfer between channel *ends*, whether the ends are both in the same processor, in different processors on the same chip or in processors on different chips. The channels therefore provide a uniform method of communication throughout a system with multiple tiles or multiple chips. Further, data can be passed via channels without the use of memory, supporting fine grained computations in which the number of communications is similar to the number of operations; this is characteristic of pipelined signal processing algorithms. For communication of large data objects between processes in a single tile, it is possible to implement communication by using the channels to pass addresses.

Channels carry messages constructed from data and control *tokens* between the two channel ends. Each token includes a single bit to identify it as a data or control token, and eight further bits. The control tokens are used to encode communication protocols and although most of them are available for software use, a small number are reserved for encoding the protocol used by the interconnect hardware.

In order to perform bidirectional communication between two processes, two channel ends need to be allocated, one for each process. This is done using the *get channel* instruction. The identifier of the channel end for process $p1$ must then be given to process $p2$, and vice versa. The channel end identifiers are system wide addresses and can be used by the two processes to transfer messages using output and input instructions. When they are no longer required, the channel ends can be freed using *free channel* instruction; otherwise the channel can be used for another message.

Each message starts with a *header* containing the identifier of the destination channel end. This is usually followed by a series of data or control tokens, ending with an end of message (EOM) control token. Tokens are sent and received using *out token* and *in token* instructions; to optimise the common case of communicating data words, the *out* and *in* instructions are used. It is possible to test whether an incoming token is a control token or not, allowing control tokens to be used to terminate variable length data sequences.

A channel end can be used as a destination by any number of processes which will be served on a round-robin basis. In this case the sender will normally send an identifier of a channel end which can be used to send a reply, or to establish bi-directional communication. The connection, once established, will persist until an EOM token has been sent, so that it is possible to perform a series of communications in both directions once a connection is established. This technique can also be used to establish a circuit between two processes to ensure that throughput or latency requirements can be met.

Channel ends have a buffer able to hold sufficient tokens to allow at least one word to be buffered. If an output instruction is executed when the channel is too full to take the data then the process which executed the instruction is paused. It is restarted when there is enough room in the channel for the instruction to successfully complete. Likewise, when an input instruction is executed and there is not enough data available then the process is paused and will be restarted when enough data becomes available.

Synchronised communication is implemented by the receiver explicitly sending an acknowledgement to the sender, usually as a message consisting of a header and an EOM. As most messages are built up from many individual data items, there is no need for all of these individual communications to be acknowledged. Indeed, it is impossible to scale interconnect throughput unless communication is pipelined and this requires that the use of end-to-end synchronisations is minimised.

A convenient way to express sequences of communications on the same channel is with a *compound communication*, which groups together the sequence of communications ensuring synchronisation only on completion. The examples below define procedures which can be used anywhere that a simple input or output could be used, including as input guards in alternatives (which is described in section 10).

```
proc inarray(chan c, []int a) is
  ?{ for i = 0 for 10 do c ? a[i] ?}

proc outarray(chan c, []int a) is
  !{ for i = 0 for 10 do c ! a[i] !}
```

The synchronisations at the end of each of these compound communications ensure that each compound output is matched by exactly one compound input. If the number of bytes transferred by the output is not the same as that expected by the input, the two processes will stop. This means that there is no need to associate data types or protocols [2] with the channels to ensure that the inputs and outputs of two processes correspond.

8. Timers and Clocks

Each tile has a 32-bit free-running clock and a set of timers which can be used by processes to read the current time or to wait until a specified time.

It is possible to synchronise input and output operations with either an internally generated clock or with an externally supplied clock. A set of clock generators is provided to do this and each of them can use either the free-running tile clock or an external 1-bit *port* as its clock source. The clock generator can be configured to divide this reference input to produce the required output frequency. Once a clock generator has been configured the ports which are to be clocked from that clock generator can be attached to it.

When an output port is driven from a clock, the data on the pin(s) changes state synchronously with the clock. If several output ports are driven from the same clock, they will appear to operate as a single output port, even though the processor may be supplying data to them at different times. Similarly, if an input port is driven from a clock, the data will be sampled synchronously with the clock; if several input ports are driven from the same clock they will appear to operate as a single input port even though the processor may be taking data from them at different times. It is also possible to set a 1-bit port into a special mode in which it outputs its source clock, allowing synchronisation of external devices to an internally generated clock.

The processes executed by a processor can therefore handle external devices at several different rates determined by clocks supplied externally or generated internally. The use of clocked ports decouples the internal timing of input and output program execution from the synchronous operation of the input and output interfaces. The processor can operate using its own clock, or could potentially be asynchronous.

9. Ports, Input and Output

Ports provide interfaces to physical pins. They can be used in *input* or *output* mode. They can be direct interfaces to the pins, or they can be *clocked* or *timed* to provide precise timing of values placed on or captured from the pins. In input mode a *condition* can be used to filter the data passed to the process. When the port is clocked or has a condition set then the captured data can be *time stamped* with the time at which it was captured.

The input and output instructions used for channels, *input* and *output* can also be used to transfer data to and from ports. In this case, the *input* instruction inputs and zero-extends the n least significant bits from an n-bit port and the *output* instruction outputs the n least significant bits.

A port in input mode can be configured to perform conditional input; this means that an input instruction will pause until the data meets the condition. Conditional input can also be used to wait for transitions on a pin. When the port becomes ready, the data value which satisfied the condition is captured so that the input instruction will return the data which met the condition even if the value on the port has subsequently changed. The following program fragment illustrates the use of conditional input in the input of data from a communications link coded in a dual-rail non return to zero form. Note that the conditional input $p\ ?\ =\ e\ :\ v$ waits until the value on port p is equal to e and then inputs it to v.

```
proc linkin(port in_0, port in_1, port ack, int token) is
var state_0, state_1, state_ack;
{ state_0 := 0; state_1 := 0; state_ack = 0;
  token := 0;
  for bitcount = 0 for 10 do
  { token := token >> 1;
    select
    { case in_0 ?= ~state_0: state_0 => skip
      case in_1 ?= ~state_1: state_1 => token := token | 512
    };
    ack ! state_ack; state_ack := ~ state_ack
  }
}
```

Two further instructions, *inshift* and *outshift*, optimise the transfer of data. The *inshift* instruction shifts the contents of the destination register right by n bits, filling the left-most n bits with the data input from the n-bit port. The *outshift* instruction outputs the n least significant bits of data from the source register to the n-bit port and shifts the contents of the register right by n bits. These instructions are useful for serialising and deserialising data at high data rates. They are especially useful when combining two or more ports clocked by the same clock to form a wider port.

Timed ports allow data to be captured from pins, or presented to pins at times specified by the program; they also allow timestamping of input data by recording the time when the port becomes ready. In input mode data is captured from the pins when the current time matches the specified time; this data can subsequently be input. In output mode data supplied to the port by an output instruction is placed on the pins when the current time matches the specified time.

Time stamping is used for ports in input mode which are clocked or have a condition set. The time at which data is captured from the input is recorded in a time stamp register associated with the port and can subsequently be accessed by the program. The following program fragment illustrates the use of timed ports in the software implementation of a UART. It first waits for the start bit signalled by a transition of the input to 0, then samples the input in the midst of each data bit; the duration of each bit is `bittime` ticks of the tile clock. Note that $p\ ? =\ e$ at v waits until the value on port p is e; then stores the current time in v; also p at $e\ ?\ v$ waits until the current time is e, then inputs to v.

```
proc uartin(port uin, byte b) is
{ var starttime;
  uin ?= 0 at starttime;
  sampletime := starttime + bittime/2;
  for i = 0 for 7
  { t := t + bittime; (uin at t) ? >> b };
  (uin at (t + bittime)) ? nil
}
```

10. Events and Alternative Input

In general, the implementation of a set of alternative input guards is performed by enabling a number of events selected by guard conditions, waiting for one of them, determining which event has occurred, disabling the enabled events and finally transferring control to a corresponding entry point of the guarded body. However there are many possible optimisations especially in situations where the guarded bodies within a set of alternatives do not themselves involve the use of alternatives.

Careful design of the instructions to implement alternatives together with compiler optimisations can minimise the time to enable and disable events, and reduce the time from an event becoming ready to completion of the corresponding instruction sequence to just a few cycles.

Event handling instructions allow resources to automatically transfer control to an associated entry point when they become ready. This entry point is specified by a *setvector* instruction prior to enabling the event. Event generation by a specific resource can then be enabled using an *event enable* instruction and disabled using an *event disable* instruction. The ability of a process to accept events is controlled by information held in a process status register and may be explicitly set and cleared by instructions. Having enabled events on one or more resources, a process can use a *wait* instruction to wait for at least one event; this instruction automatically enables the process to accept events.

When one of the events occurs, the ability of the process to accept events is automatically disabled and control is transferred to the entry point associated with the event. Finally, all of the events which have been enabled by a process can be disabled using a single *clear events* instruction. This disables event generation in all of the ports, channels or timers which have had events enabled by the process.

The event handling system must allow compound communications and calls to procedures defining compound communications to be used in guards in the same way as simple inputs; otherwise an additional communication would be needed to determine which guarded alternative is selected.

```
select
{ case inarray(c, a) => P(a)
  case inarray(d, a) => Q(a)
}
```

This is done by a *setcontext* instruction which is used in a manner similar to the *setvector* instruction but which initialises a *context* register in the port, usually to the stack pointer value at the time the event is enabled by the guarding procedure. In the above example, this is done by the `inarray` procedure when it is called by the `select` to enable its input channel (after which it returns). When the event occurs, the value of the context register is copied to one of the process registers and can then be used to re-establish the stack pointer and continue to execute the remainder of the body of the `inarray` procedure. The same mechanism can be used to allow procedures which define alternatives to themselves be used as components of alternatives. For example, a process can offer to input from either of two communications links using the procedure `linkin` defined earlier.

```
select
{ case linkin(x0, x1, a, t) => P(a)
  case linkin(y0, y1, a, t) => Q(a)
}
```

One important hardware optimisation is to provide dedicated paths connecting the ports, timers and channels to the processor, to enable scheduling decisions to be made within one cycle of the event becoming ready and to minimise the time to fetch the instructions at the entry point associated with the event.

The most important instruction set and compiler optimisations aim to optimise repeated alternatives in inner loops where the process is effectively operating as a programmable state machine. The guard bodies in these cases usually consist only of short instruction sequences possibly including inputs and outputs; they do not normally include nested use of alternatives. It is important that the guard bodies can perform input and output operations (even involving the resource which gave rise to an event) whilst leaving some or all of the event information unchanged. This allows the process to complete handling an event using one of a set of alternative guard bodies and immediately wait for another similar event.

The setting of event vectors and other invariant conditions associated with the resources can be moved outside the inner loops using normal compiler optimisation techniques. Conditional versions of the event enable instructions shorten the instruction sequences to implement the guard conditions. The *event enable true* instruction enables the event if its condition operand is true and disables it otherwise; conversely the *event enable false* instruction enables the event if its condition operand is false and disables it otherwise.

Finally, conditional versions of the wait instruction allow the loop terminating condition to be implemented as a conditional wait, eliminating the loop-closing branch. The *wait true* instruction waits only if its condition operand is true, and the *wait false* waits only if its condition operand is false.

In order to optimise the responsiveness of a process to high priority resources the *set enable* instruction can be used to enable events before starting to enable the ports, channels and timers. This may cause an event to be handled immediately, or as soon as it is enabled. An enabling sequence of this kind can be followed either by a *wait* instruction to wait for one of the events, or it can simply be followed by a *clear enable* to continue execution when no event takes place. The *wait true* and *wait false* instructions can also be used in conjunction with a *clear enable* to conditionally wait or continue depending on a guarding condition. These instructions provide an efficient implementation of prioritised alternatives such as those of occam [2].

The provision of dedicated registers for each process means that a process can be dedicated to handling an individual event or to an alternative handling multiple events. For each process most if not all of the data needed to handle each event will be instantly available when the event occurs, having been initialised prior to waiting for the event. This is in sharp contrast to an interrupt-based system in which context must be saved and the interrupt handler context restored prior to entering the handler - and the converse when exiting.

11. Summary

Communicating process architecture can be used to design and program efficient multicore chips with performance scaling to thousands of processes, or virtual processors. Each process can be used to run conventional sequential programs, as a hardware emulation engine, to implement input and output operations or as a DMA controller with an application-optimised protocol.

Communicating processes provide a natural way of expressing event-driven programs. Event-driven processes, triggered by communications, synchronisations and ports enable power to be minimised within a multicore. Clocked and timed ports allow the interface timing to be matched to external needs, decoupling on-chip operation from the interface. Programmed input-output rates can be sustained up to the process instruction rate.

The interconnect architecture scales to hundreds of cores per chip. Inter-processor channels support over 100Mbytes in both directions simultaneously, and multiple concurrent channels between the processors support streaming applications. The interconnect supports both virtual circuits and packet switching under software control, and system-wide channel addressing simplifies system design and programming.

The architecture is tuned to compiler and software needs supporting direct execution of concurrent software; no software kernel is needed. It supports conventional programs, message passing programs, synchronised and timer driven programs, or any combination. The processor instruction set enables optimisation of concurrent and event-driven programs. The compact instruction representation, position independent code and high speed interconnect enables software mobility using the techniques similar to those described in [9] and [10]. This reduces latency and power, and can also be used to support efficient remote process initiation and dynamic re-use of processors at runtime.

References

[1] C. A. R.Hoare: *Communicating Sequential Processes*, Communications of the ACM, **21(8)** (August 1978), 666–677.

[2] Inmos: *Occam-2 Reference Manual*, Prentice Hall, 1998.

[3] D. May: *The transputer revisited* in Millenial perspectives in Computer Science, Palgrave, 2000, 215–228.

[4] C. Clos: *A study of non-blocking switching networks*, Bell System Technical Journal, **32** (1953), 406–424.

[5] D. May, P. H. Welch, P Thompson: *Networks, Routers and Transputers*, IOS Press, 1993.

[6] W. J. Dally, C. L. Seitz: *Deadlock free routing in multiprocessor interconnection networks*, in IEEE Transactions on Computers **36(5)** (1987) 547–553.

[7] J. S. Kowalik, editor: *Parallel MIMD Computation*, MIT Press, 1985.

[8] D. Towner, D. May: *The Uniform Heterogeneous Multi-threaded processor architecture*, in Communicating Process Architectures, IOS Press, 2001, 103–116.

[9] F. R. M. Barnes, P. H. Welch, A. T. Sampson: *Communicating Mobile Processes: introducing occam-pi* in 25 Years of CSP, LNCS 3525, April 2005.

[10] D. May, H. Muller: *A simple protocol to communicate channels over channels* in EURO-PAR 1998, LNCS 1470, Springer-Verlag, 1998, 591–600.

Communicating Process Architectures 2007
Alistair A. McEwan, Steve Schneider, Wilson Ifill, and Peter Welch
IOS Press, 2007

Lazy Exploration and Checking of CSP Models with CSPsim

Phillip J. BROOKE [a,1] and Richard F. PAIGE [b]

[a] *School of Computing, University of Teesside, U.K.*
[b] *Department of Computer Science, University of York, U.K.*

Abstract. We have recently constructed a model, and carried out an analysis, of a concurrent extension to an object-oriented language at a level of abstraction above threads. The model was constructed in CSP. We subsequently found that existing CSP tools were unsuitable for reasoning about and analysing this model, so it became necessary to create a new tool to handle CSP models: CSPsim. We describe this tool, its capabilities and algorithms, and compare it with the related tools, FDR2 and ProBE. We illustrate CSPsim's usage with examples from the model. The tool's on-the-fly construction of successor states is important for exhaustive and non-exhaustive state exploration. Thus we found CSPsim to be particularly useful for parallel compositions of components with infinite states that reduce to finite-state systems.

Keywords. CSP, Simulation, Lazy

Introduction

This paper describes the early stage of a tool, CSPsim, initially created to analyse a CSP model of a concurrent extension (SCOOP) to an object-oriented language (Eiffel).

The Simple Concurrent Object-Oriented Programming (SCOOP) mechanism has been proposed as a way to introduce inter-object concurrency into the Eiffel programming language [1,2]. SCOOP extended the Eiffel language by adding one keyword, **separate**, which can be applied to classes, entities, and formal routine arguments. SCOOP allows introduction of both conceptual threads of control and synchronisation through a uniform syntax. As a result, the semantics of SCOOP is complicated, and understanding it would be easier with a formal model.

We modelled SCOOP systems using CSP [3]; this model is presented in detail in [4]. The process of constructing the models, and identifying points of potential semantic variation, was informative: we found ambiguities and questions relating to lock passing and when locks should be released. Further, we desired to mechanically analyse the systems to compare and contrast different policies.

The construction of the CSP model resulted in systems that were hard for the leading tools, FDR2 [5] and ProBE [6], to handle. The model comprised the parallel composition of ten components, some of which are very large. Asking FDR2 to process the full system resulted in a futile attempt to construct its internal representation: we ran out of memory. As well, the obvious compressions and optimisations applicable to the CSP model, in order to accommodate FDR2 and ProBE's internal optimisations, led to a revised CSP model that was difficult to understand and, particularly to use for

---

[1]Corresponding Author: Phil Brooke, University of Teesside, Middlesbrough, TS1 3BA, U.K.
`pjb@scm.tees.ac.uk`

analysis. Thus a new tool, CSPsim, was created. CSPsim's main feature is the on-the-fly, or *lazy*, evaluation of state enabling it to explore processes that are relatively small when composed, even if the individual component subprocesses are large, thus making it possible to explore and analyse processes that are beyond what is easily possible using FDR2 and ProBE.

Although initially constructed to solve a single problem – that of how to provide automated capabilities to explore and simulate process algebra specifications of complex, layered models of concurrent object-oriented systems – CSPsim's scope is more general. Thus, the focus of the work in this paper is not directly on SCOOP; instead, we describe CSPsim, the design decisions, and its abilities when compared to other tools, as well as future plans.

To this end, we commence with a description of our motivating problem in Section 1, including an outline of the CSP model in Section 1.5. We describe CSPsim in Section 2 and compare it with related work in Section 3. The paper ends with our conclusions in Section 4. The appendix explains where the tool and examples can be found online.

1. Motivation: Eiffel and SCOOP Modelled in CSP

We briefly describe our motivating problem to illustrate the underlying complexity. Smaller test examples have been created during the development of CSPsim. The later examples are drawn from the work carried out on this motivating problem.

1.1. Eiffel

Eiffel is a pure object-oriented (OO) programming language [1,2] that provides constructs typical of the OO paradigm, including classes, objects, inheritance, associations, composite ("expanded") types, polymorphism and dynamic binding, and automatic memory management. Novelties with Eiffel include its support for full multiple inheritance, generic types (including constrained generics), agents (closures and iterators over structures), and strong support for assertions, via preconditions and postconditions of routines, and invariants of classes.

Routines may have pre- (**require** clauses) and postconditions (**ensure** clauses). The former must be true when a routine is called (i.e., it is established by the caller) while the latter must be true when the routine's execution terminates. Classes may have invariants specifying properties that must be true of all objects of the class at stable points in time, i.e., after any valid client call on the object. An exception is raised if an assertion (precondition, postcondition or invariant) evaluates to false.

For more details on the language, see [1] or [2].

1.2. SCOOP

SCOOP introduces concurrency to Eiffel by the addition of the keyword **separate**. The **separate** keyword may be applied to the definition of a class or the declaration of an entity (a variable) or formal routine argument.

Access to a separate object, whether via an entity or formal argument indicates different semantics to the usual sequential Eiffel model. In the sequential model, a call to a routine causes execution to switch to the called object whereupon the routine executes; on completion, execution continues at the next instruction of the original object. In SCOOP, procedure calls are asynchronous. The called object can queue multiple calls, allowing callers to continue concurrent execution. Function calls and reference access to attributes are synchronous — but may be subject to *lazy evaluation* (also known as *wait-by-necessity*).

Races are prevented by the convention that a separate formal argument causes the object to be exclusively locked ('reserved') during that routine call. However, there are complications with locking, in that deadlocks may arise, or concurrency may not be maximised, unless some form of *lock passing* [7] is used.

1.3. SCOOP Processors

SCOOP introduces the notion of a *processor* (not to be confused with real CPUs). When a separate object is created, a new processor is also created to *handle* its processing. This processor is called the object's *handler*. (Objects created as non-separate are handled by the creator's handler.) Thus, a processor is an autonomous thread of control capable of supporting *sequential* instruction execution [1]. A system in general may have many processors associated with it.

Compton [8] introduces the notion of a *subsystem*: a model of a processor and the set of objects it operates on. In his terminology, a separate object is any object that is in a different subsystem. In this paper, we will refer to subsystems rather than processors (to avoid possible confusion with real CPUs).

1.4. SCOOP Assertions

Eiffel uses **require** and **ensure** clauses for specifying the pre- and postconditions of routines. In sequential programming, a **require** clause specifies conditions that must be established and checked by the client of the routine; the **ensure** clause specifies conditions on the implementer of the routine. If a precondition or postcondition evaluates to false, an exception is raised.

In SCOOP, a **require** clause on a routine belonging to a separate object specifies a *wait* condition: if the routine's **require** clause evaluates to false, the processor associated with that object waits until the precondition is true before proceeding with routine execution.

1.5. Outline of CSP Model

We constructed a model of SCOOP in CSP [4]. We initially chose CSP as our working language because of the existence of good tools (FDR2 and ProBE), and because the most interesting (and least understood) aspects of the problem at hand related to concurrency and synchronisation.

SCOOP causes a number of distinct components to interact: objects and subsystems coordinate locking, freeing and execution of features. Additionally, new objects and subsystems are created as needed. Our CSP model comprises the alphabetised parallel of ten different components to simulate these behaviours. Some of these components are trivial book-keeping processes (e.g., $CALLCOUNT$ records the number of calls made) whereas others encode the intended behaviours; they are relatively complicated with a very large state space. Indeed, some of these processes would be infinite if we did not specifically restrict the length of particular sequences within their definitions.

The model is parametrised by:

- *CLASSES*, a list of all possible classes, and *FEATURES*, the names of all possible features in the system.
- *MaxCallCount*, the maximum number of calls the system will execute.
- *MaxInstances*, the number of distinct objects within each class.
- *MaxSubsystems*, the maximum number of subsystems (other than the initial subsystem).
- *MaxParams*, the number of parameters each call may record.
- *MaxLocals*, the number of local variables for each object.

Clearly, keeping these values as small as possible for each example is important to reduce the overall state space. However, even a small example —four classes, with nine distinct features, $MaxCallCount = 9$, $MaxInstances = 1$, $MaxSubsystems = 1$, $MaxParams = 3$, $MaxLocals = 3$— proves difficult to analyse and explore in model checkers that need to fully construct component processes before combining them. This motivates the need for an on-the-fly (lazy) model checker and simulator for CSP – something that, to the best of our knowledge, does not exist.

The essential problem we face is that our model is constructed to make it (relatively) obvious, with the result that the individual components have a very large (potentially infinite) state space. But in composition, the state space is relatively small. Certainly at the beginning of the simulations we have explored so far, there are often only a small number of events available. Thus we wish to avoid exploring the behaviour of any individual component that is strictly necessary to answer the immediate question of 'what events are available now?'

However, we suspect that the size of the overall state space of these models is sufficiently large that it is unreasonable to expect any model checking technology to easily cope. Being able to manually or semi-automatically examine systems can provide useful results to us, although fully automatic, exhaustive search is clearly beneficial.

2. CSPsim

CSPsim was originally constructed to allow us to explore, in a systematic and sometimes interactive manner, the state space of SCOOP programs. However, CSPsim has developed into a more general-purpose CSP explorer and simulation tool, applicable to many CSP modelling and analysis problems (specific restrictions are discussed in the sequel).

A feature of CSPsim is its lazy, or on-the-fly, evaluation of state. This is the major factor that enables us to explore some models that have potentially infinite components, and a key distinguishing characteristic of CSPsim over FDR2 and ProBE.

2.1. Representation

CSPsim is implemented in Ada (due to author familiarity), and uses tagged (object oriented) types to describe processes. A CSP system is described by a set of acyclic graphs, where the nodes in each graph are objects of concrete types (of an abstract `Process`) representing any of

- calling a named process;
- guards (i.e., if b then P else *Stop*);
- *Stop*, *Skip* and sequential composition;
- prefix (see below);
- external choice (but not internal choice at this time);

- interleaving, generalised (interface) parallel and alphabetised parallel;
- renaming and hiding; and
- 'dynamic' (dynamically-created) processes (see below).

The arcs in the graph connect nodes to their immediate successors. Nodes may also contain other information; for example, the prefix operator needs information about the event(s) offered. The following process[1]

$$T = \quad (a \rightarrow Stop \,\square\, b \rightarrow Stop \,\square\, a \rightarrow Stop \,\square\, b \rightarrow Stop \,\square\, b \rightarrow Stop)$$
$$\square\, (a \rightarrow Stop \,\square\, b \rightarrow Stop \,\square\, a \rightarrow Stop \,\square\, b \rightarrow Stop \,\square\, b \rightarrow Stop)$$
$$\square\, c \rightarrow Stop \,\square\, d \rightarrow S$$
$$S = d \rightarrow S$$

has two graphs, one for T and one for S

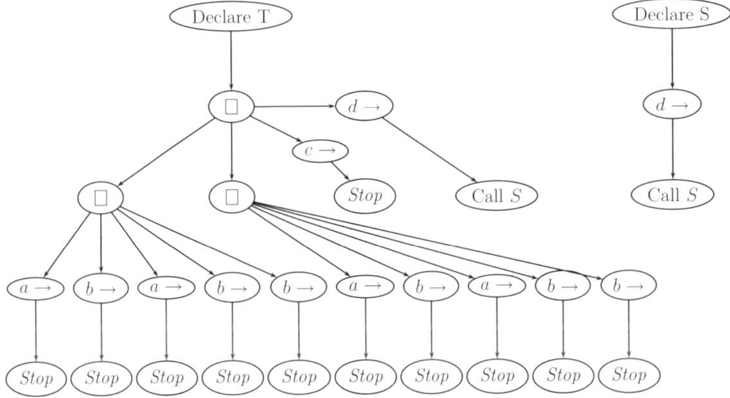

We illustrate the Ada encoding of this example in Section 2.9.

A catalogue or register is kept of all named processes, known as *declared processes*. Thus although the graphs defining the processes are themselves acyclic, a 'call' can access any named process.

The appendix gives a list of the functions and CSP operators supported by CSPsim.

2.2. Parameters

'Blackboard' presentations of CSP are often parametrised, e.g.,

$$A(i) = foo \rightarrow A(i+1) \,\square\, bah \rightarrow A(2i)$$

Similarly, CSP_M (as used in FDR2 and ProBE) has a rich functional-style for expressions.

CSPsim, too, handles parameters. Each declared process contains a list of expected parameters. Each parameter is described by its name (a string) and a 'type', e.g., Decl("A", + Integer_Formal("i"), ... The types currently available are integers, strings, arrays of strings and two-dimensional arrays of strings. The two-dimensional arrays do not need to be rectangular.

---

[1] This example corresponds to TEST2b in the CSPsim distribution.

An illustration of the process $A(i)$ above is

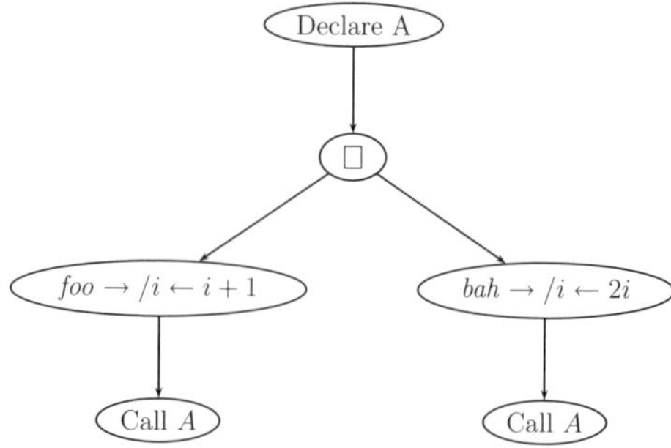

The value of a parameter during an execution is only modified when an event is taken due to a prefix process or when a named process is called. For example, the Prefix functions (available via the Factory package) include a PU argument that, if non-null, is called when an event is taken. PU has type

```
type Params_Update is
    access procedure (PS           : in out Parameters.Params_Access;
                      Last_Event : in      Event;
                      CB           : in      Parameters.Params_Access);
```

and can alter the list of parameters (PS), often depending on the specific event, Last_Event.

2.3. Prefix

Prefix is the only way to introduce events. The simplest form is

$$a \rightarrow P$$

i.e., a becomes P.

There are four types of prefix process in CSPsim.

1. A single, fixed event is offered by $a \rightarrow P$.
2. Any event from a fixed, given set is offered by $a : A \rightarrow P$.
3. Any event is offered from a set that is calculated lazily: $a : F \rightarrow P$. In this case, F is a function with signature

```
type FS_Available is
    access function (PS  : Parameters.Params_Access;
                     CBF : Parameters.Params_Access)
           return Event_Array_Access;
```

where PS is the current parameter set (e.g., i in our examples above). Functions of this type return a set listing all the offered events. The usual place to use this form of prefix is where the events offered cannot be known prior to computation, e.g.,

$$A(i) = read.i \rightarrow \ldots$$

where the event $read.i$ cannot be known until i is known.

4. Any event is offered that is acceptable to the a function of the form

```
type FE_Available is
   access function (E   : Event;
                    PS  : Parameters.Params_Access;
                    CBF : Parameters.Params_Access) return Boolean;
```

This returns True if th event E is acceptable to the function given the current parameters PS. The purpose here is that whereas in the three forms of prefix above, we already know or can easily calculate the events offered, it may be very expensive to calculate a full set of offered events (for example, there could be an infinite number of acceptable events, say, any integer).

Note that any of the forms may cause parameters to be updated when the event is taken as described above.

When the last form of prefix is asked to list all possible events instead of merely being asked if a particular event is acceptable, it returns a 'hint' — a textual prompt intended for the user explaining that there are too many events. This hint is propagated upwards through the user interface.

We plan that a future update to CSPsim will enable intersection of 'large' event sets in parallel composition, to result in a small set of events that can be listed. For example, if i and j are both to be drawn from the set of integers, then $a.i.1 \rightarrow P$ offers many events as does $a.2.j \rightarrow Q$, but the parallel composition offers only $a.2.1$. We can already express events of this form, but the infrastructure for calculating the intersection as part of the simulation is not complete. If such symbolic descriptions are generally available, then we can avoid manipulating large sets of events generally (which is currently one of the issues for our motivating example).

2.4. Dynamically-created Processes

Two structures in the CSP model of SCOOP motivated the inclusion of dynamically created processes in CSPsim. Firstly, CSP_M has an if-then-else construct. Secondly, processes that naturally take many component processes —external choice, interleave, generalised parallel and alphabetised parallel— are difficult to describe at compile-time if the component processes depend on a parameter. The latter is illustrated by

$$|||i : S \bullet P(i)$$

i.e., the interleaving of process $P(i)$ where i is drawn from the set S — which may be a parameter rather than a fixed constant.

When created, an entry is made in a register to record the parameters used. A later request to create from the same prototype with the same parameters results in the previous dynamic process being re-used. This both reduces memory consumption and reduces the state space (as individual processes are distinguished by their instance). Thus these processes are denoted 'dynamic' because they result in the creation of new processes during the computation rather than prior to the computation.

2.5. Loops and Recursive Processes

The recursive operator, $\mu X \bullet f(X)$, is not directly supported. Instead, loops and recursion are introduced by 'calling' a named process. So the typical clock process is represented as

$$TICKING = tick \rightarrow TICKING$$

rather than $TICKING = \mu X \bullet tick \rightarrow X$.

2.6. States

Whereas the CSP process is defined by descendants of `Process`, descendants of `State` form an analogous structure representing the state of a particular execution of the process. For example, a `State_Prefix` object will be created for the `Process_Prefix`.

This state contains a link to the `Process` concerned, the parameters for that state, as well as information specific to that type of process, e.g., a flag indicating if an event has been taken in $a \rightarrow P$. Additionally, the state contains information for caching (for speed).

With the exception of dynamic processes (above), `Process` objects are neither created nor modified during computation. However, there may be multiple state objects for each process object due to recursion and loops.

Finally, compression and deletion of state occurs each time an event is taken, including the following cases:

- a resolved external choice causes the unchosen arms to be deleted;
- the left side of a sequential composition successfully completing causes the entire sequential composition to be replaced by the right side;
- a prefix process where the event is taken is replaced by its successor process;
- the rule $P|||Stop = P$ is applied to interleaving; and
- guard 'processes' are replaced by the guarded process if the guard is true or *Stop* if the guard is false (because once a state has been created, the parameters will not change).

To a large extent, this mimics what would be expected in a language with direct support for lazy evaluation and automatic garbage collection.

The 'execution' of a call results in the creation of a new state. The parameters of the new state are only those listed in the relevant declared process, and an exception is raised if any parameters are missing or of the wrong type. In our motivating example, this trapped a number of mistakes during the construction of the CSPsim version of the model, clearly indicating the process and the events leading up to the error. By contrast, we have found FDR2 and ProBE's error messages to be unhelpful at times, particularly when mismatched types are involved.

2.7. Nondeterminism and Hiding

There are complications with this operational approach due to both nondeterminism and hiding.

Nondeterminism can arise from processes such as $a \rightarrow P \,\square\, a \rightarrow Q$: the environment has no control over 'which' a is chosen. However, a simulator needs to resolve this decision, so dummy events are inserted into the trace recording which arm was selected. The selection may be manual (the tool can ask) or random.

Hidden events are renamed by prefixing the string _tau_. This means that a trace actually indicates the event that was selected, even though for other rules (such as parallel combination) the event is treated simply as τ.

The tool provides an option to take hidden events immediately (`-eat`). The intention is that a process with internal interleavings may offer visible events at the same time as offering hidden events: taking these hidden events may cause further visible events to become available. By immediately taking the hidden events, we reduce the possible state space. This is naively implemented at this time: it should check that it does not cause events to become refused (which would invalidate the search by wrongly removing possible traces) — but this is not an issue for our particular examples.

2.8. Differential Updates

The interleaving, generalised parallel and alphabetised parallel operators can optionally update the events offered from a cache rather than recalculating completely each time. Suppose we have the process

$$P = \|_A i : S \bullet Q(i)$$

The first time that this process is accessed, each component process $Q(i)$ supplies the events it will offer as a set $D(i)$. The events offered are calculated, i.e., all $Q(i)$ must agree on events in A, all other events can be offered if at least one $Q(i)$ will offer it; successful completion must be agreed by all.

All of these sets $D(i)$ can be cached, as can the result. If an event is taken, then the cache is marked as invalid, but not deleted. The new sets $D'(i)$ are calculated, some of which may be cached themselves, so this is fast when there are few changes. We then determine which events have been added and removed and make changes to the previous result. Brief experimentation suggests that this is effective for some problems, but there needs to be some threshold whereby it falls back to complete recalculation when there are too many changes to make.

2.9. CSP Input and Ada Encoding

The first user action is to encode CSP into an Ada program that creates the structure of **Processes**. The example in Section 2.1 is encoded thus:

```
with Factory;     use Factory;
with Processes;   use Processes, Processes.PXA;


procedure Test2b is
begin
   Decl("TEST2b",
        ExtChoice(+ ExtChoice(+ Prefix("a", Stop)
                              + Prefix("b", Stop)
                              + Prefix("a", Stop)
                              + Prefix("b", Stop)
                              + Prefix("b", Stop))
                  + ExtChoice(+ Prefix("a", Stop)
                              + Prefix("b", Stop)
                              + Prefix("a", Stop)
                              + Prefix("b", Stop)
                              + Prefix("b", Stop))
                  + Prefix("c", Stop)
                  + Prefix("d", Call("SLOOP")))));
   Decl("SLOOP",Prefix("d", Call("SLOOP")));
   Explore;
end Test2b;
```

Another partial example is given in Section 2.13. The program should 'with' the package **Factory**. This provides a façade to the rest of CSPsim. The appendix lists some of the functions available for introducing CSP operators.

This program is then compiled and run. A call to **Factory.Explore** accesses the interface and search features of CSPsim.

This is very user unfriendly at this time: we envisage a proper parser front-end reading something similar to CSP_M. Alternatively, we could consider using Graphical Timed CSP [9].

2.10. User Interface

Simple explorations can be carried out within the text interface. At its simplest, the user can be given a list of available events unless the set is 'infinite' (in which case, the user is given a hint and can type in an event). 'Tab completion' is available for events (particularly useful with the long and complex event names we require for our motivating problem).

The interface can be told to take events immediately if they are the only event available; or even to randomly walk through the system (which can be useful when attempting an intuition of the system behaviour).

There are a number of other commands, including load and save stored states (although these are only valid within a particular instance of the tool).

The model in our motivating problem attempts to capture a wide range of a system's behaviours, but sometimes we do not care about the behaviours following particular events. A command line option, `-avoid`, can be given (the prefix of) events to avoid unless they are the only event available.

2.11. Refinement, Traces, Refusal and Nondeterminism

Refinement is not directly implemented, although we can easily simulate trace-refinement by extracting all possible traces of P and checking that they are all valid traces of the specification, S. (Replaying traces in CSPsim is faster than the normal exploration modes.) This could be automated.

Failure-refinement cannot be easily achieved at this time, as refusal information is not directly recorded. However, the tool constructs acceptance sets (in most cases) and could therefore determine the relevant refusals and thus failures for later checking.

Since we do not record refusals, it makes no sense to offer a distinct internal choice operator: in the traces model, this is equivalent to external choice.

2.12. Search and Post-processing

Exhaustive searches can be attempted via command line options. The most useful, `-exhaustive2` attempts a simple state-based exploration. This has proved effective in our motivating problem. (Note also, that we did not have nondeterminism due to visible events: thus the search algorithm currently chooses only the first arm.)

The output of `-exhaustive2` is a directory of state files. A post-processing tool, `states2dot`, can turn these files into input suitable for `dot` from the Graphviz distribution [10]. `states2dot` can additionally remove edges that duplicate other edges.

2.13. Example

CSPsim and examples are available online (see the appendix). We illustrate this with part of `e3.adb` (itself part of the much larger motivating problem)

```
Decl("class_B_feature_m",
    + String_Formal("c")
    + String_Formal("i"),
   Prefix(+C("getHandler") +P("i") +B_Str("hn", +C_Alias(SUBSYSTEMS)),
    Prefix(+C("newSubsystem")+B_Str("ha",+C_Alias(SUBSYSTEMS))+P("c"),
     Prefix(+C("createObject")
            +B_Str("a",+C("Object.cl_A")+C(1,MaxInstances))
            +P("ha")+P("c"),
       Prefix(+C("setLocal")+P("i")+C("1")+P("a"),
        Prefix(+C("callCount")+P("c")+B_Str("c1",+C_Alias(CALLS)),
```

```
          Prefix(+C("setParam")+P("c1")+C(1)+P("a"),
           Prefix(+C("setSepParam")+P("c1")+S(+C("{")+P("a")+C("}")),
           SeqComp(SCall("ADDCALL",
                   +String_Param("f", "f_B_o"),
                   +"c"+"c2"+"i"+"i2",
                   +"c"+"c1"+"i"+"i"),
               Call("ENDFEATURECALLS",
                   PI => PI1'Access)))))))))));
```

which represents the CSP

$$\text{class\_B\_feature\_m}(c, i) = \text{getHandler}.i.(h_n : \text{SUBSYSTEMS})$$
$$\rightarrow \text{newSubsystem}.(h_a : \text{SUBSYSTEMS}).c$$
$$\rightarrow \text{createObject}.(a : \text{Object.cl\_A}.\{1, \ldots, \text{MaxInstances}\}).h_a.c$$
$$\rightarrow \text{setLocal}.i.1.a$$
$$\rightarrow \text{callCount}.c.(c_1 : \text{CALLS})$$
$$\rightarrow \text{setParam}.c_1.1.a$$
$$\rightarrow \text{setSepParam}.c_1.\{a\}$$
$$\rightarrow (\text{ADDCALL}(c \leftarrow c, c_2 \leftarrow c_1, i \leftarrow i, i_2 \leftarrow i, f \leftarrow \text{f\_B\_o});$$
$$\text{ENDFEATURECALLS}(C \leftarrow \{c_1\}))$$

where both c and i are strings. We write $(x : S)$ to represent binding x to something from set S, even if part of a compound event, e.g., $a.(x : S).b$, although S may itself be compound (as in createObject above). An example run of the process class_B_feature_m follows:

```
$ ./example3.exe −nowait −indirect
 [...]

Explore> .switchProcess
Available processes:  [...]

Process name> class_B_feature_m
Resetting ...
class_B_feature_m has formal parameters!  You must supply them.
   1. c : PARAMETERS.STRINGS.STRING_PARAMETER
   2. i : PARAMETERS.STRINGS.STRING_PARAMETER
S−expression> (2 (c Str Call.1) (i Str Object.cl_B.1))

Exploring process a:FS −> (a:FS −> (a:FS −> (a:FS −> (a:FS −> (a:FS −> (a:FS −>
(( ADDCALL()) ; (ENDFEATURECALLS() )))))))))

Events available: getHandler.Object.cl_B.1.Subsystem.0 getHandler.Object.cl_B.1.Subsystem.1
    (2 event(s))
Type the event to take or a '.' command (try '.help').
Explore> getHandler.Object.cl_B.1.Subsystem.0
Taking event 'getHandler.Object.cl_B.1.Subsystem.0'
 1 event(s) in trace so far

Events available: newSubsystem.Subsystem.0.Call.1 newSubsystem.Subsystem.1.Call.1 (2 event(s))
Type the event to take or a '.' command (try '.help').
Explore> newSubsystem.Subsystem.1.Call.1
Taking event 'newSubsystem.Subsystem.1.Call.1'
 2 event(s) in trace so far

Events available: createObject.Object.cl_A.1.Subsystem.1.Call.1    (1 event(s))
```

```
Type the event to take or a '.' command (try '.help').
Explore> createObject.Object.cl_A.1.Subsystem.1.Call.1
Taking event 'createObject.Object.cl_A.1.Subsystem.1.Call.1'
 3 event(s) in trace so far

Events available: setLocal.Object.cl_B.1.1.Object.cl_A.1    (1 event(s))
 [...]
```

An exhaustive search of the full example (not just class_B_feature_m) is triggered by

```
./example3.exe −nowait −indirect −eat −stq /tmp/S1/ \
−notick −steps 300 −exhaustive2 2500
```

or if we wish to avoid exploring traces involving preconditionsFail.,

```
./example3.exe −nowait −indirect −eat −avoid preconditionsFail. −stq /tmp/S2/ \
−notick −steps 300 −exhaustive2 2500
```

The first search takes one minute (76 states) and the second 50 seconds (63 states)[2]. We produce state diagrams using `states2dot` and `dot`.

From left to right in Figure 1, the first and third graphs are derived from the first run, while the second and fourth are from the second run. The first and second graphs have had no edges removed, whereas the third and fourth had the argument

```
−sr addCall big blocked create free new preconditions reserve schedule unreserved
```

applied to `states2dot` to remove edges that we consider irrelevant for the purposes of this simulation. We can then see that these are linear in terms of gross progression, as they represent a series of local calls in SCOOP.

A more interesting example (`example1` in the examples available online) was expected to offer more parallelism, depending on the choice of options to the model. The pair of graphs in Figure 2 was constructed without using `-avoid`, and then postprocessed removing edges as above.

The only difference is due to the semantic model chosen, and this clearly shows that the options for the left side result in a linear progression of calls, whereas the right side allows greater parallelism. This provides useful evidence regarding a key semantic variation point in SCOOP, and which choice of semantics helps to promote maximal parallelism.

We note that the left example took around 9 minutes to create 123 states while the right took 54 minutes for 699 states[23]. For a CSP_M version of the right example, FDR2, by comparison, was aborted after four hours and examining 7.21 million transitions and consuming 2198 MB of virtual memory, at which point performance rapidly degraded due to swapping.

2.14. Validation, Correctness and Robustness

The individual behaviours of CSP processes are relatively simple. However, bugs are common, and to identify them, the output of particular explorations and test cases was compared to the behaviour obtained under FDR2 and ProBE. Although this is not

---

[2]The computer concerned is a Linux server running Fedora Core 5 with 4 2.66 GHz Intel Xeon processors and 2 GB RAM. CSPsim currently only makes use of one of the processors.

[3]The performance for these examples is now substantially improved with recent updates to CSPsim. The 54 minutes is down to 8 minutes.

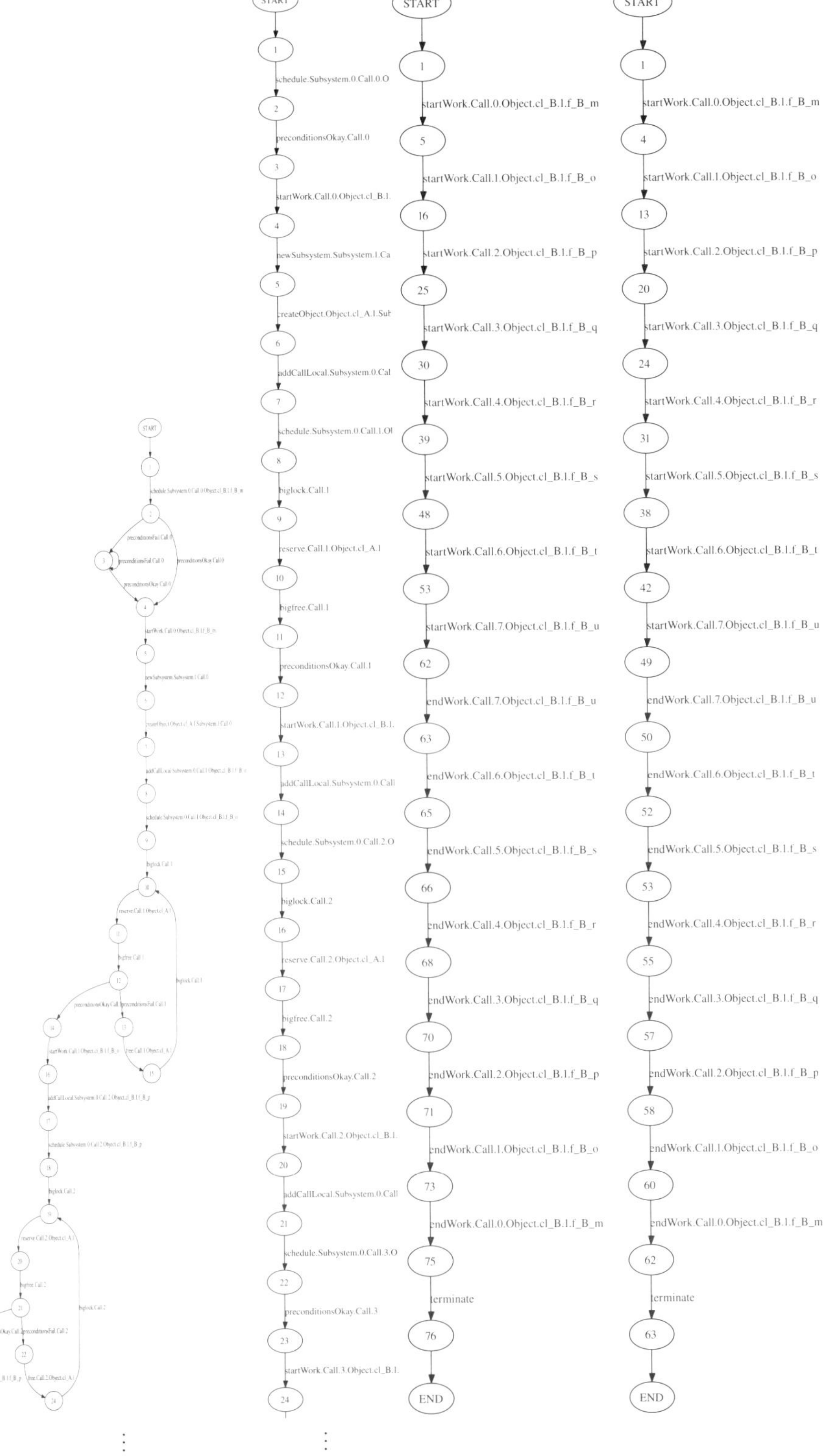

Figure 1. Output from `example3`.

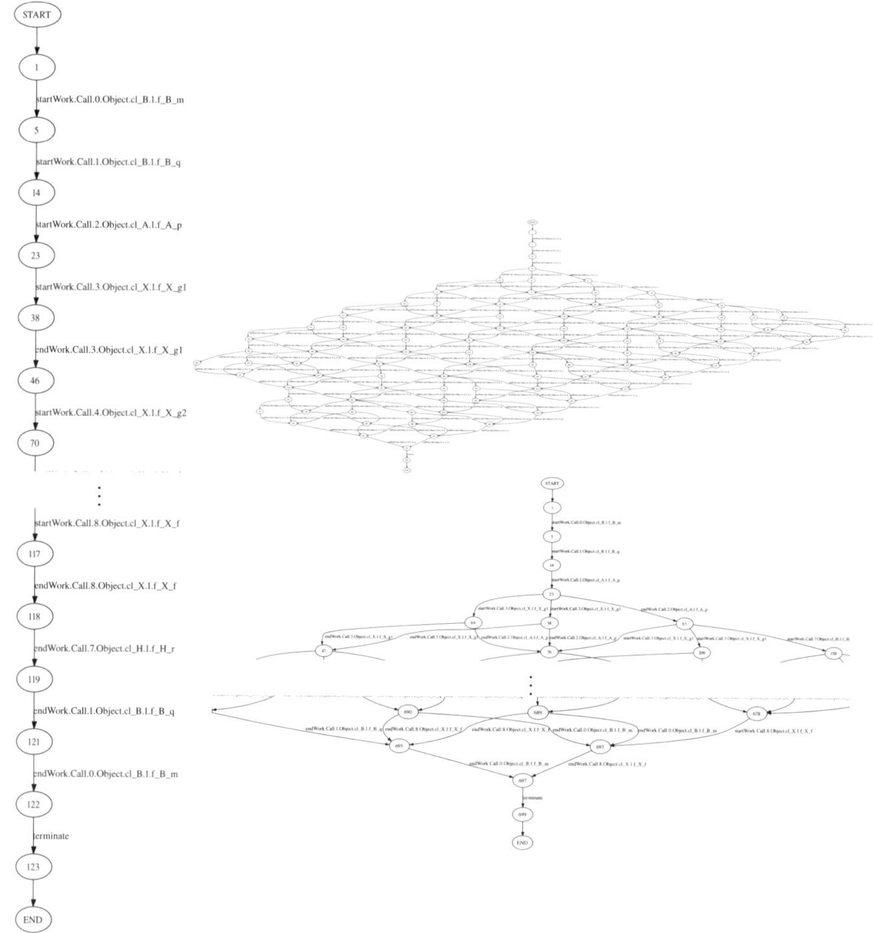

Figure 2. Output from `example1`.

systematic or broad coverage testing, the correspondence of results to those obtained in these better-known tools gives confidence that the implementation is correct.

The additional features added for improving performance (e.g., caching, updating) have added much complexity. The current CSPsim is best viewed as a prototype that now needs re-implementation to provide a better experimental base as well as correctness and robustness.

3. Related Work and Comparison

Other CSP tools already exist. The most well known is FDR2 (a model checker) and ProBE, both from Formal Systems [5,6]. Both read the notation CSP_M.

FDR2's state exploration facilities are very efficient, provided that the problem size is sufficiently small for the internal representation to be constructed initially. Expert users can construct their CSP_M code to assist FDR2. In this case, we suspect that our motivating problem is simply too big regardless (as opposed to lack of sufficient skill with FDR2).

Similarly, ProBE offers a GUI interface for examining CSP_M described processes. However, deep explorations have proved difficult for us: we cannot load and restore states, execute particular (saved) sequences of events, and simply finding an event in a long (unordered) listing is difficult.

For CSPsim, FDR2 and ProBE, there are similarities of approach directly due to the common CSP heritage. For example, CSPsim's `-eat` switch is similar to FDR2's tau-loop compression.

By contrast, each tool has a specific niche: FDR2's model checking is substantially faster than CSPsim's exhaustive state exploration. ProBE has a significant advantage in terms of reading CSP_M, but the user interface issues were part of the motivation for a new tool. The error messages from CSPsim are, we claim, more useful: typing of parameters makes it difficult to wrongly call declared processes. Thus CSPsim occupies a middle-ground between FDR2 and ProBE, but without the common notation. However, integrating CSP_M syntax with CSPsim's facilities is a straightforward, mechanical problem (typing and parameters excluded).

Other tools exist: there is the CSP-specific CSP++ [11], which is aimed at converting CSP_M specifications into C++ rather than providing further intuition about the system at hand. Similarly, Communicating Sequential Processes for Java (JCSP) [12] provides CSP-style concurrency and distribution for Java programmers, but is not intended directly for state exploration and refinement, though it can certainly be used to support such tasks.

There are other model checkers which can be applied generally, such as Bogor [13], the Jack and FMC toolsets [14], and SPIN [15]. SPIN in particular has a number of similarities to CSPsim: it constructs models on-the-fly, and does not require a global state graph or Kripke structure to be generated. Its specification language, PROMELA, is not specifically targeted to concurrent systems modelling; our view is that CSP is better suited to our initial motivating problem. Particularly, a custom CSP tool has the advantage of being able to apply CSP's algebraic rules for compression and other comparisons. We note that Bogor in particular is intended to be extensible to other domains: it might be interesting to attempt a CSP extension.

4. Conclusion

CSPsim has significantly aided our work in concurrent Eiffel using CSP models. In particular, it has helped us analyse SCOOP models, which was not possible with other tools. The way in which we are able to analyse SCOOP models has proven particularly helpful, as we are able to produce compressed views of the state space of a SCOOP program. This in turn allows us to experiment with different semantics (e.g., lock passing, lock releasing) and generate evidence that one or the other semantics should be preferred on the grounds that they increase the overall level of concurrency in the system.

4.1. Summary

We can summarise CSPsim's strengths as follows:

- lazy (on-the-fly) evaluation of CSP processes,
- typed parameters, and
- direct access to Ada for imperative calculations (though these must be deterministic);

and its limitations:

- it is very slow for exhaustive search, and
- requires Ada compilation.

4.2. Future Work

Future work involves a number of areas. In terms of semantic correctness, we can improve the treatment of hidden processes: essentially, mimicking FDR2's tau-loop compression.

A better front-end to avoid the need to write and compile Ada is useful. This would hopefully generate input suitable for FDR2 and ProBE to enable different aspects of the the same problem to be dealt with by the most appropriate tool. Similarly, we could generate output for theorem provers based on previous work on CSP in PVS [16].

Modifying the internal interfaces to pass symbolic descriptions of sets of events, rather than simply listing the events, would allow intersection of 'large' event sets in parallel composition (as described in Section 2.3).

We can also consider

- direct support for refinement checking;
- recording refusal information (and adding the internal choice operator); and
- improving process and state comparison by applying more algebraic rules to reduce state spaces further.

Overall, we plan a re-implementation of the current prototype to enable further development; this should include stronger statements of correctness of the CSP semantics. As remarked in Section 2.14, a number of features have been added for performance. Thus a convincing demonstration is needed that the CSP semantics are correctly honoured, even in the face of loading and saving of states.

Notably, a language with direct support for lazy evaluation and automatic garbage collection may be more suitable.

Acknowledgements

CSPsim uses code for SHA-1 written by John Halleck, which is being used by his permission. CSPsim also uses the libedit library, which is software developed by the NetBSD Foundation, Inc. and its contributors.

We thank the CPA referees for their helpful and encouraging comments.

Availability of Code and Examples

CSPsim is available from

    http://www.scm.tees.ac.uk/p.j.brooke/cspsim/

for particular versions of the GNAT compiler. Example source code, state output, and dot files are available at

    http://www.scm.tees.ac.uk/p.j.brooke/ce1/.

CSP Operators

We list the CSP operators we support, and give the function(s) that introduce them. These functions are found in the façade package, `Factory`. Many of these functions are overloaded to allow different patterns of usage.

'calling' a process, e.g., C	`Call` and `SCall`
guard	`Guard`
Stop	`Stop`
Skip	`Skip`
sequential composition	`SeqComp`
prefix	`Prefix`, `APrefix`, `FSPrefix` and `FEPrefix`
external choice	`ExtChoice`
interleaving	`Interleave`
generalised parallel	`GParallel`
alphabetised parallel	`AParallel`
renaming	`Rename`
hiding	`Hide`
'dynamic' creation	`Dynamic`

This façade also provides a procedure, `Explore`, that accesses the simulation interface and handles searches.

References

[1] B. Meyer. *Object-Oriented Software Construction*. Prentice Hall, 2nd edition, 1997.

[2] ECMA-367: Eiffel analysis, design and programming language. ECMA International, June 2005.

[3] C.A.R. Hoare. *Communicating Sequential Processes*. Prentice-Hall International UK, 1985.

[4] Phillip J. Brooke, Richard F. Paige, and Jeremy L. Jacob. A CSP model of Eiffel's SCOOP. To appear in Formal Aspects of Computing, accepted 2007.

[5] FDR2 model checker. http://www.fsel.com/software.html, last visited 10th October 2006.

[6] ProBE — CSP animator. http://www.fsel.com/software.html, last visited 10th October 2006.

[7] Phillip J. Brooke and Richard F. Paige. A critique of SCOOP. In Richard F. Paige and Phillip J. Brooke, editors, *Proc. First International Symposium on Concurrency, Real-Time, and Distribution in Eiffel-like Languages (CORDIE)*, number YCS-TR-405. University of York, July 2006.

[8] M. Compton. SCOOP: an investigation of concurrency in Eiffel. Master's thesis, Australian National University, 2000.

[9] Phillip J. Brooke and Richard F. Paige. The design of a tool-supported graphical notation for Timed CSP. In Michael Butler, Luigia Petre, and Kaisa Sere, editors, *Integrated Formal Methods*, number 2335 in LNCS, 2002.

[10] Graphviz — Graph Visualization Software http://www.graphviz.org/, last visited 10th October 2006.

[11] CSP++. http://www.uoguelph.ca/ gardnerw/csp++/index.html, last visited 10th October 2006.

[12] Communicating Sequential Processes for Java. http://www.cs.kent.ac.uk/projects/ofa/jcsp/, last visited 10th October 2006.

[13] Bogor — Software Model Checking Framework. http://bogor.projects.cis.ksu.edu/, last visited 10th October 2006.

[14] S. Gnesi and F. Mazzanti. On the Fly Model Checking of Communicating UML State Machines. *Proc. Second Int. Conference on Software Engineering Research, Management, and Applications (SERA) 2004*, May 2004.

[15] SPIN — model checker. http://spinroot.com/spin/whatispin.html, last visited 10th October 2006.

[16] Phillip J. Brooke. *A Timed Semantics for a Hierarchical Design Notation*. DPhil thesis, University of York, 1999.

Communicating Process Architectures 2007
Alistair McEwan, Steve Schneider, Wilson Ifill and Peter Welch (Eds.)
IOS Press, 2007

The Core Language of Aldwych

Matthew HUNTBACH

Department of Computer Science, Queen Mary University of London,
Mile End Road, London E1 4NS, UK

mmh@dcs.qmul.ac.uk

Abstract. Aldwych is a general purpose programming language which we have developed in order to provide a mechanism for practical programming which can be thought of in an inherently concurrent way. We have described Aldwych elsewhere in terms of a translation to a concurrent logic language. However, it would be more accurate to describe it as translating to a simple operational language which, while able to be represented in a logic-programming like syntax, has lost much of the baggage associated with "logic programming". This language is only a little more complex than foundational calculi such as the pi-calculus. Its key feature is that all variables are *moded* with a single producer, and some are *linear* allowing a reversal of polarity and hence interactive communication.

Keywords. Concurrency, logic programming, linear variables, single-assignment variables.

Introduction

It has been noted since the observations of Landin [1], that a complex programming language can be understood by showing a translation into a tiny core language which captures the essential mechanisms of its programming style. This idea has been most influential in the field of functional programming languages which can be considered as just "sugared lambda-calculus". Modern computing, however, tends to be about interaction as much as calculation. An early attempt to build a programming language based on an abstract model of interaction was occam with its basis in CSP [2]. More recently, the pi-calculus [3] has received much attention as the suggested basis for a model of interactive computing. Unlike CSP, the pi-calculus is a name-passing calculus, meaning that communication channels can themselves be passed along communication channels, leading to the communication topology changing dynamically as code is executed. There have been some attempts to build languages which are "sugared pi-calculus", for example PICT [4], but even when sugared this model seems to be difficult for programmers to use practically.

We have been working on building a programming language with an abstract concurrent model which uses the concept of shared single-assignment variables rather than pi-calculus's channels. It is another name-passing calculus, since a variable may be assigned a value which is a tuple containing variables. Our work in this area sprang from earlier work in concurrent logic languages [5]. Although these languages have been proposed as practical programming languages in their own right, with an application area in parallel programming [6], our experience with them suggested they had serious defects. Firstly, their lack of structure meant it was difficult to scale them up from toy examples to large scale use. Secondly, in their attempt to emulate the logic programming style of Prolog, they led to programs where the data flow could not easily be detected. This was despite the fact that in reality programmers in them almost always had an intended *mode* for every variable, with a single producer [7].

We considered building a programming language which compiles to an underlying concurrent logic form, but which has a rich set of "derived forms" to enable more practical programming. Although this is not a new idea (see [8] for a survey), unlike previous attempts to build logic-programming based object-oriented languages, our intention was not to "combine" object-orientation with logic programming. Rather, we felt the very simple underlying operational model of the concurrent logic languages would be a good core language for developing a richer language which made no claims itself to be logic-oriented but which enabled practical programs to be written in a style where concurrency is a natural feature rather than an awkward add-on extra.

This language is being developed under the name "Aldwych" [9] and we describe some of its features elsewhere [10]. Early in the development of Aldwych it became clear that a key feature would be for all variables to be *moded*, that is with a single producer identified clearly by the syntax and one or more consumers. Another key feature was the division of variables into *linear* and *non-linear*, where linear variables have a single consumer as well as a single producer. This enables the consumer-producer relationship to be reversed with ease.

Many of the complexities of implementing concurrent logic languages disappear when moding can be guaranteed, and when there is also a clear indication of which variables are linear, implementation can be even more efficient [11]. Since the modes and linearity of variables in the language to which Aldwych compiles can be guaranteed, there is no need for any mechanisms to analyse it. In fact the underlying language represents such a dramatic simplification of committed choice logic languages, which in turn are a dramatic simplification of the logic programming model (see [12] for a discussion of the paring down of logic programming features or "de-evolution" of logic programming in the search for efficient concurrent implementation) that it no longer makes sense to give references to it which emphasise and elaborate on its more complex logic programming ancestry.

The purpose of this paper, therefore, is to describe the underlying operational model of the language into which Aldwych compiles in a way that does not attempt to link it to more general concepts of logic programming or describe stages in its de-evolution which are no longer relevant to its current state. The model can be described in terms of a few reduction rules. Full Aldwych is described in terms of "derived forms" which translate to the simpler model given here, thus this paper complements our previous papers which describe those derived forms.

Section 1 of this paper introduces the model in terms of a first-order functional language with explicit output variables and the potential for parallel execution. Section 2 notes that the model, unlike conventional functional programming, handles non-determinacy naturally enabling it to make decisions based on the order of interaction between concurrent processes. Section 3 introduces the key principle of "back communication", which enables two-way interaction between concurrent processes and also can be used to simulate higher-order functions. Section 4 develops a set of reduction rules which fully describe the operational behaviour of the model. Section 5 indicates the syntactic requirements to ensure every variable has a single producer. Section 6 gives an extended example which shows how the model provides dynamic communication topology between interacting processes. Section 7 concludes, and notes links with other more theoretical work.

1. A Relational Language

In a conventional imperative language, the computation:

```
f(g(x),y)
```

is taken as a command that the code for **g** with argument **x** is fully evaluated and gives a value which becomes the first argument to **f**. The same applies in a strict functional language. We can regard the construct as a shorthand for evaluating **g(x)**, putting the result in a variable and using that variable as the first argument for **f**:

```
z<-g(x); f(z,y)
```

where the semi-colon is a sequencing operator, all code before the semi-colon is completed before code after it is executed.

Suppose we replace the sequencing operator by one which doesn't have the property of ordering computations, let us use a comma:

```
z<-g(x), f(z,y)
```

We could now, given a suitable architecture, evaluate **g(x)** and **f(z,y)** in parallel. The variable **z** could be regarded as a "future" [13]. The computation **f(z,y)** may use it as a placeholder, passing it to other computations or incorporating it into data structures while **g(x)** is still computing its value. The computation **f(z,y)** will suspend, however, if it needs to know the actual value of **z** in order to progress.

We could flatten code out, replacing all embedded calls $g(x_1, \ldots, x_n)$ by a variable **z** with the call $z\text{<-}g(x_1, \ldots, x_n)$ occurring beforehand. If the call $z\text{<-}g(x_1, \ldots, x_n)$ is always placed before the call which has **z** as an argument, and the calls are executed in order we will obtain strict order evaluation in functional programming terms. We could however execute the calls in some other order, possibly involving some parallelism. If we use the comma operator as above, then the only necessary sequencing is that imposed by the necessity for a computation to suspend if it needs to know the value of a variable until that value has been computed.

Note that the possibility of parallelism does not imply its necessity. The extreme alternative to the conventional sequential execution of **z<-g(x),f(z,y)** is that **f(z,y)** is evaluated while **g(x)** is suspended, and **g(x)** is only evaluated if **f(z,y)** suspends due to the need to know the value of **z**. Suppose we employ the convention that the rightmost computation is picked for progression, but if it needs the value of a variable and that variable has not yet been given a value, the computation that gives that variable a value is picked, and if that one is also suspended due to the need to know the value of another variable, the computation which gives that variable a value is picked and so on. We will then obtain what is known in functional programming as "call-by-need" or "lazy evaluation". The advantage to this is that if a computation gives a value to a variable but there is no computation left which has that variable as an argument, that computation can be abandoned without evaluation.

Clark and Gregory [14] suggested a "relational language" which worked like the above description of a flattened functional language, with no suggested ordering on computations apart from that imposed by computations suspended while waiting for others to bind variables. The intention, however, was that the potential parallelism would be exploited. This language was influenced by logic programming in that the function call assignment to a variable $y\text{<-}f(x_1, \ldots, x_n)$ was written as a relation $r(x_1, \ldots, x_n, y)$. Logic programming in its Prolog form and most other varieties does not have a concept of a direction on variables, whereas in functional programming all variables have a direction with just one computation that can write to them. Clark and Gregory's relational language imposed a direction on variables by giving a mode declaration to its relations, so the above relation r would have mode $r(m_1,...,m_n,m_{n+1})$ with m_1 to m_n being ? meaning "input" and m_{n+1} being ^ meaning "output". Furthermore, it insisted that each variable have just one producer, though it may have many consumers. So a computation would consist of a collection of relations which shared variables, but each variable had to occur exactly once in the position in a relation which had mode declaration ^.

The result of this directionality was that the arguments were "strong" [15]. That is, for each argument of a relation call, the argument was either completely constructed by that call (if an output argument) or by another call (if an input argument). The right to set a variable to a value remained solely with one relation call, that relation call might set the variable to a structure containing further variables, but it had either to take those variables from its own input variables or set up new relation calls to construct their values.

The result appeared to be a rather restricted syntax for first order functional programming. Lacking embedded function calls there was a proliferation of variables introduced merely to take the result of one call and make it an argument to another. The lack of facilities for higher order functions might be considered a serious weakness given the importance which many advocates of functional programming give them [16]. However, the making of all variables used for communication explicit, and the flat structure of computations with a single environment of variables, led to an attractively simple programming model. As we shall show below, it also had the advantage of being able to handle non-determinism with ease whereas this is a problem in functional programming.

2. Non-Determinism

Since the programming model does not rely on the lambda-calculus of functional programming it can cope with situations where there is more than one way of progressing a computation and the outcome will differ depending on the choice made. As an example, consider the following declaration:

```
#p(x,y)->z
{
 x=a  ||  z=b;
 y=c  ||  z=d;
 :
      ||  z=e
}
```

The syntax used here is not the Prolog-like one of Clark and Gregory's relational language, but one we have developed and will describe further in this paper. It is used so that programs in this core language will be a subset of the full Aldwych language. The # is used to denote the introduction of a new procedure name (we will use this term rather than "relation"). Input and output modes are denoted by separating them in the procedure heading, so that if the heading is $\#p(u_1,...,u_m)->(v_1,...v_n)$ then $u_1,...,u_m$ have input mode, and $v_1,...,v_n$ have output mode; we omit the brackets around $v_1,...,v_n$ when n is 1, and we omit -> and the brackets when n is 0.

The description of a procedure consists of a list of sets of rules. A set of rules is enclosed by braces, with a semicolon as the separator between rules, and for convenience } : { denoting the end of one set and the start of another may be written ; :. Each rule consists of two parts, a left-hand side (*lhs*) and a right-hand side (*rhs*). The lhs contains tests of variable values ("asks" in terminology introduced by Saraswat [17] for concurrent logic programming) and the rhs contains variable assignments (in Saraswat's terminology "tells"). So **x=a**, where **x** is a variable and **a** is a constant means "wait until **x** is given a value and test that it is **a**" when it is on the lhs, and "set **x** to **a**" when it is on the rhs.

The first set of rules in the procedure **p** above means that a call **p(u,v)->w** will set **w** to **b** if **u** gets set to **a**, and will set **w** to **d** if **v** gets set to **c**. If both **u** is set to **a** and **v** is set to **c**, **w** could be set to either **b** or **d**. A functional programming computation, whether strict or lazy, would insist that **u** be evaluated before proceeding either to assign **b** to **w** or to go on to test the value of **v**. In a parallel setting, however, we may have **u** and **v** being computed in parallel and be content to react accordingly depending on which computation finishes first without being forced to wait for the other. Having received the news that **u** is set to **a**, we could kill off the computation of **v** if there is no other computation that has **v** as an input argument, and similarly we could kill the computation of **u** if we receive the news that **v** is set to **b** [18].

The multiple sets of rules in our notation mean that the conditions for rules to apply can be tested sequentially if that is required. If and only if the conditions for none of the rules in the first set applies, the second set is used, and so on. In the above example there are only two sets, and the last set has a single rule with an empty lhs meaning no conditions are required for its application. So in our call **p(u,v)->w** if both **u** becomes set to something other than **a**, and **v** becomes set to something other than **c**, the final set of rules is used and causes **w** to be set to **e**.

If there is always a final rule set consisting of a single unconditional rule, a relation call in our notation can never fail. This contrasts with Prolog where failure due to no applicable rules is a natural part of the system and causes computation to backtrack to the point where a previous non-deterministic choice was made, and to change the choice made there. Such a backtracking may be practical in the single-processor sequential computation like Prolog, but is impractical in a concurrent or parallel system where one non-determinate choice may have sparked off several distributed computations, and impossible if the variable is linked to an effect on some physical system in the real world: the real world does not backtrack. We discuss in more detail the arguments against non-determinism combined with backtracking (termed "don't know non-determinism" [19]) in an earlier work [5], although doubt over the usefulness of automated backtracking in programming languages can be found much earlier than that [20].

3. Back Communication

Handling non-determinism is one aspect where a relational as opposed to functional approach to programming languages gives increased power, particularly in a concurrent setting. Another is the "logic variable" [21] used to provide "back communication" [15]. Building on the relational language described above, the idea here is that the input-output modes are weakened. In particular, a computation may bind a variable to a structure containing further variables, but leave a computation which inputs that structure to provide a value for some of those variables. The relational language of Clark and Gregory was developed into Parlog which provided such back communication, at the same time a number of similar languages were developed which were given the general name "committed choice logic languages" [22].

Given back communication, the mode system of the relational language broke down. Parlog's mode system applied only to the arguments of a relation at top level, and not to the individual components of structured arguments. It existed only to enable Parlog programs to have a more superficial appearance to Prolog programs where assignment to variables is done through pattern matching with clause heads. The other committed choice logic languages used explicit assignment operators to give values to variables, as did Parlog when the variables were arguments to tuples used for back communication. The languages ended up as modeless – there was no syntactic way of discovering which computation could actually assign a value to a variable, in fact the possibility of several different computations being able to bind a single variable was opened up, and handled in a variety of different ways. This then necessitated elaborate mechanisms to rediscover intended modes in code, since practice revealed that programmers almost always intended every variable to have just one computation that could write to it, and knowledge of the intended moding could greatly improve the efficiency of implementation [7].

In our notation, we extend moding to the terms of compound arguments. On the lhs we have $x=t(i_1,...,i_m)$ meaning a test that x is bound to a tuple with tag t and m arguments $i_1,...,i_m$, all of which are taken to have input mode, that is, they will be assigned by the computation which is giving a value to x. On the rhs $x=t(i_1,...,i_m)$ means that x is assigned a tuple with tag t and m arguments $i_1,...,i_m$, all with input mode, that is the computation which does the assignment must have i_k as an argument or must provide another computation which gives i_k a value for $1{\leq}k{\leq}m$. We also allow $x=t(i_1,...,i_m)->(o_1,...,o_n)$ on the lhs, where $o_1,...,o_n$ are output variables, meaning that the computation which has this test must provide values for $o_1,...,o_n$ in the rhs of the rule. We allow $x=t(i_1,...,i_m)->(o_1,...,o_n)$ on the rhs, meaning that $o_1,...,o_n$ will be used in the rhs, but that a computation which takes in the value of x will give values to $o_1,...,o_n$.

As an example, consider the following:

```
#map(xs)->(ys,f)
{
  xs=cons(x,xs1)  ||  f=ask(x,cont)->y,
                      map(xs1)->(ys1,cont),
                      ys=cons(y,ys1);
  xs=empty || ys=empty, f=done
}

#square(queries)
{
  queries=ask(u,cont)->v || v<-u*u, square(cont);
  queries=done ||
}
```

with the following initial computations:

```
map(list1)->(list2,stream), square(stream)
```

The result of executing this will be that a list in variable **list1** composed of tuples with tag **cons**, first argument an integer and second argument a further list (with **empty** indicating the empty list), is taken as input, and a square function is mapped onto it to produce the list in **list2**. This shows how back communication can be used to obtain a higher-order function effect. The input of a function is represented by the output of a stream of queries taking the form **ask(i,cont)->o**, where **i** is the argument to the function, **o** the result, and **cont** the rest of the stream giving further queries to the same function, or set to **done** if the function is not to be used any more. The code is not elegant, but the point is the higher order effect can be achieved within this model, and could be incorporated into a language which is based on this model but uses derived forms to cover commonly used patterns at a more abstract level for use in practical programming.

However, this back communication leads to the problem that since a variable may occur in several input positions, if it is set to a tuple which includes output arguments, those output arguments will be become duplicated. Each of the computations which takes the tuple as an input could become a writer to its output arguments. One way of avoiding this, adopted for example in the logic programming language Janus [23], was to insist that every variable must be linear, that is occur in exactly one input position and one output position. This however acts as a considerable constraint on the power of the language, meaning that we cannot use variables as "futures" in the Multilisp way [13].

Our solution to the problem is to adopt a system which involves both modes and linearity. So arguments to a procedure or to a tuple may be one of four types: input-linear, output-linear, input-non-linear and output-non-linear. Only a linear variable may be assigned a tuple value which contains output arguments or linear arguments either input or output. A non-linear variable may only be assigned constants or tuples all of whose arguments are input-non-linear. In the above example, the arguments **f** to **map** and **queries** to **square** should be denoted as linear, as should the variable **cont** in the first rule for **map** and the first rule for **square**.

4. Computation

We can now describe our operational model in more detail. A computation in our notation consists of a set of procedure calls which take the form $p(i_1,\ldots,i_m) \rightarrow (o_1,\ldots,o_n)$ with $m,n \geq 0$, where each i_h and o_k, $1 \leq h \leq m$, $1 \leq k \leq n$, are variable names, and a set of variable assignments which take the form either $v=t$ or $v<-u$. In the variable assignments, v and u are variables, and t is a term which takes the form $s(i_1,\ldots,i_m) \rightarrow (o_1,\ldots,o_n)$, $m,n \geq 0$, where each i_h and o_k, $1 \leq h \leq m$, $1 \leq k \leq n$ are variable names, and s is a "term tag", that is an atomic value. For notational convenience in a term, if n is 1 the second set of brackets are omitted, if n is 0 the $\rightarrow$ is also omitted, and if m is 0 the first set of brackets is omitted.

The moding is used to ensure that every variable occurs exactly once in an output position, where an output position is v in $v=t$ or $v<-u$, or o_k, $1 \leq k \leq n$, in $p(i_1,\ldots,i_m) \rightarrow (o_1,\ldots,o_n)$, or o_k, $1 \leq k \leq n$, in $v=s(i_1,\ldots,i_m) \rightarrow (o_1,\ldots,o_n)$. A non-linear variable may occur in any number of input positions, but every linear variable must occur in exactly one input position, where an input position is i_k, $1 \leq k \leq m$, in $p(i_1,\ldots,i_m) \rightarrow (o_1,\ldots,o_n)$, or i_k, $1 \leq k \leq m$, in $v=s(i_1,\ldots,i_m) \rightarrow (o_1,\ldots,o_n)$, or u in $v<-u$. We can regard a more general procedure call $p(t_1,\ldots,t_m) \rightarrow (v_1,\ldots,v_n)$ where $t_1,\ldots,t_m$ are terms and $v_1,\ldots,v_n$ are variables, as a shorthand for $p(i_1,\ldots,i_m) \rightarrow (o_1,\ldots,o_n)$, $i_1<=t_1,\ldots,i_m<=t_m$, $v_1<-o_1,\ldots,v_n<-o_n$. Here, $i_h<=t_h$ is $i_h<-t_h$ if t_h is a variable, otherwise it is $i_h=t_h$, and each i_h and o_k is a variable which does not otherwise occur in the computation. The point of this is to give each procedure call a fresh set of variables as its arguments.

Similar to the more general procedure call, assignment to a variable of a tuple which contains non-variable arguments can be regarded as shorthand for an assignment which contains only variable arguments with separate assignments of terms to the arguments where necessary. So $u=s(t_1,\ldots,t_m) \rightarrow (v_1,\ldots,v_n)$ is considered shorthand for $u=s(i_1,\ldots,i_m) \rightarrow (v_1,\ldots,v_n)$, $i_1<=t_1,\ldots,i_m<=t_m$. An output argument in a tuple or procedure call can only ever be a variable.

The first computation rule is that v<-u,u=t transforms to v=t,u=t, written:

$$v\text{<-}u,u\text{=}t \longrightarrow v\text{=}t,u\text{=}t$$

There is no concept of ordering on the computations, so u=t,v<-u also transforms to v=t,u=t. Note that if u is a linear variable we can say v<-u,u=t transforms to v=t, since v<-u is the one input occurrence of u and u=t the one output occurrence, and the variable u thus occurs nowhere else. If u is a linear variable, then v<-u is only allowed if v is also a linear variable, although v<-u is allowed if v is a linear variable but u is not. Since u=t where t contains linear variables and/or output positions is only allowed if u is a linear variable, this ensures that output positions of any variable and input positions of linear variables do not get duplicated.

We also have v<-u,u<-w transforms to v<-w,u<-w or:

$$v\text{<-}u,u\text{<-}w \longrightarrow v\text{<-}w,u\text{<-}w$$

Similar to above, if u is linear, we can transform v<-u,u<-w to just v<-w.

We also allow v<-e as an expression in the language, where e is an arithmetic expression involving variables. The computation rule for this is that the expression transforms to v=n when there are assignments u_i=m_i for all variables in e, and replacing each variable u_i by m_i in e and evaluating e gives n.

A procedure call $p(i_1,\ldots,i_m)\text{->}(o_1,\ldots,o_n)$ is linked to a set of rules initially given by the procedure declaration for p, and we assume there is a universal fixed set of named procedure declarations. Each procedure call produces a new copy of these rules, where if the procedure heading is #$p(u_1,\ldots,u_m)\text{->}(v_1,\ldots,v_n)$, any occurrence of u_h, $1{\le}h{\le}m$, in the rules is replaced by i_h, any occurrence of v_k, $1{\le}k{\le}n$, in the rules is replaced by o_k, and any other variable in the rules but not in the header is replaced by a fresh variable. The replacement of a procedure call by a set of rules initialised with entirely fresh variables can be regarded as a step in the computation.

The basis of the rule for procedure rewrite, which we develop in more detail later, is that given x=a and a set of rules including the rule x=a||$body$, where a is a constant, we rewrite the set of rules to $body$. Note that, unlike the pi-calculus, the assignment is not consumed once used, and the variable may never be re-used in an assignment. We can show this by the computation rule:

$$x\text{=}a,\{\ldots;x\text{=}a||body;\ldots\}:\ldots \longrightarrow x\text{=}a,body$$

We allow more than one test on the lhs, so we can generalise this to:

$$x_1\text{=}a_1,\ldots,x_n\text{=}a_n,\{\ldots;x_1\text{=}a_1,\ldots,x_n\text{=}a_n||body;\ldots\}:\ldots \longrightarrow x_1\text{=}a_1,\ldots,x_n\text{=}a_n,body$$

The ordering of the assignments and the ordering of the tests is irrelevant, as is the ordering of the rules in $\{\ldots;x_1\text{=}a_1,\ldots,x_n\text{=}a_n||body;\ldots\}$. However, the rules following : in the rule set cannot be employed at this stage.

A rule is discarded if there is an assignment to a constant other than the one being tested for in the rule:

$$x\text{=}a,\ \{\ldots;\ \ldots,x\text{=}b,\ldots||body;\ \ldots\}:\ldots \longrightarrow x\text{=}a,\{\ldots;\ldots\}:\ldots \quad \textit{if } a\neq b$$

If all rules in the first set have been discarded, we can go on to consider the rules in the second set:

$$\{\}:\{rule_1;\ldots;rule_n\}:\ldots \longrightarrow \{rule_1;\ldots;rule_n\}:\ldots$$

We allow rules with an empty lhs which rewrite unconditionally, so:

```
{…;  ||body; …}:… ⟶ body
```

Another way of thinking of this is as individual assignments picking off tests from the lhs of rules until a lhs becomes empty and the above rule applies, in which case we have:

```
x=a,{…; …,x=a,…||body; …}:… ⟶ x=a,{…;…,…||body;…}:…
```

Or, since ordering of rules and tests does not matter:

```
x=a,{x=a,T||body;R}:S ⟶ x=a,{T||body;R}:S
```

where **T** is a set of tests, **R** a set of rules, and **S** a list of sets of rules, and the existence of computation rules to reorder **T** and **R** (but not **S**) is assumed. We have also (indicating discarding a rule when one test fails, moving to a second set of rules when all rules are discarded, and using a rule when all its tests succeed):

```
x=a,{x=b,T||body;R}:S ⟶ x=a,{R}:S   if a ≠ b
```

```
{}:S ⟶ S
```

```
{T||body;R}:S ⟶ body   if T is the empty set.
```

Here **body** consists of further procedure calls and assignments which application of the last computation rule above causes to be added to the top-level set of procedure calls and assignments. The assignments in **body** will then cause other sets of rules to become rewriteable.

We allow ordering tests on the lhs of rules, which fail if their arguments are orderable:

```
x=a,y=b,{x>y,T||body;R}:S ⟶ x=a,y=b,{T||body;R}:S   if a > b
```

```
x=a,y=b,{x>y,T||body;R}:S ⟶ x=a,y=b,{R}:S   if a ≤ b
```

```
x=a,y=b,{x>y,T||body;R}:S ⟶ x=a,y=b,{R}:S
```
$$\text{if } \mathbf{a} \text{ and } \mathbf{b} \text{ are not orderable by } >.$$

The precise definition of "orderable" (whether numerical, or applying more widely, for example, alphabetic ordering of tags) is not relevant for this paper.

Also a wait test allows suspension until a variable is bound to any value:

```
x=a,{wait(x),T||body;R}:S ⟶ x=a,{T||body;R}:S
```

and type tests give dynamic typing:

```
x=a,{integer(x),T||body;R}:S ⟶ x=a,{T||body;R}:S   if a is an integer.
x=a,{integer(x),T||body;R}:S ⟶ x=a,{R}:S   if a is not an integer.
```

Our computation rules, as given so far, have not taken account of variables being assigned or tested for tuples containing further variables. In the full rules we allow tests on the lhs of a rule of the form `x=s(i₁,…,iₘ)->(o₁,…,oₙ)`, where $m \geq 0$ and $n \geq 0$. The notational convenience for omitting brackets described previously may again be used. The variable names `i₁,…,iₘ` and `o₁,…,oₙ` must all be new variable names, with rule scope so they can be re-used in other rules. For the purposes of the describing the operational behaviour, all arguments to tuples in tests must be variables. However, for notational convenience we can write a input tuple argument in a test as a non-variable, and take this as being shorthand for introducing a separate variable and testing it, so `x=s(…,t,…)->(…)` is shorthand for `x=s(…,y,…)->(…),y=t` where **y** is a new variable name, and **t** a term.

Given this, the computation rule for matching an assignment against a test is:

$$x=s\,(u_1,\dots,u_m)\,\text{->}\,(v_1,\dots,v_n)\,,\,\{x=s\,(i_1,\dots,i_m)\,\text{->}\,(o_1,\dots,o_n)\,,T\,|\,|\,body;R\}:S\;\longrightarrow$$

$$x=s\,(u_1,\dots,u_m)\,\text{->}\,(v_1,\dots,v_n)\,,\,\{i_1\text{<-}u_1,\dots,i_m\text{<-}u_m,v_1\text{<-}o_1,\dots,v_n\text{<-}o_n,T\,|\,|\,body;R\}:S$$

If x is a linear variable, we could at this point add to **body** on the rhs an indication that if the lhs becomes empty and the rule is chosen, the assignment can be removed as this is the one permitted reading of the variable.

Now we need to deal with $x\text{<-}y$ occurring on the lhs of rules (which can only occur temporarily after the application of the above computation rule). If we are testing that x has a particular tuple value, and x is matched against variable y, then we are testing that y has that pattern, replacing an internal variable in the test with an external one. So:

$$\{x\text{<-}y,x=t,T\,|\,|\,body;R\}:S\;\longrightarrow\;\{x\text{<-}y,y=t,T\,|\,|\,body;R\}:S$$

We must also take account of the other tests that may occur on the lhs, for example:

$$\{x\text{<-}y,\texttt{wait}(x),T\,|\,|\,body;R\}:S\;\longrightarrow\;\{x\text{<-}y,\texttt{wait}(y),T\,|\,|\,body;R\}:S$$

If the lhs of a rule consists only of variable assignments, there are no further tests, so the rule can be applied but the variable assignments must be retained for use with **body**:

$$\{x_1\text{<-}y_1,\dots,x_n\text{<-}y_n\,|\,|\,body;R):S\;\longrightarrow\;x_1\text{<-}y_1,\dots,x_n\text{<-}y_n,body$$

This ensures the internal variables of **body** are linked with external variables. Note that since a linear variable cannot be assigned to a non-linear variable, this rule is conditional on there being no $x_i\text{<-}y_i$ where y_i is denoted as linear but x_i is not. If there is such a match, the rule becomes inapplicable:

$$\{x\text{<-}y,T\,|\,|\,body;R\}:S\;\longrightarrow\;\{R\}:S\quad \textit{if}\,y\text{ is linear and }x\text{ is non-linear.}$$

A rule also becomes non-applicable if it involves matching tuples of differing arities, or tuples of differing tags:

$$x=s\,(u_1,\dots,u_m)\,\text{->}\,(v_1,\dots,v_n)\,,\,\{x=s\,(i_1,\dots,i_p)\,\text{->}\,(o_1,\dots,o_n)\,,T\,|\,|\,body;R\}:S\;\longrightarrow$$
$$\{R\}:S\quad\textit{if}\,p\neq m$$

$$x=s\,(u_1,\dots,u_m)\,\text{->}\,(v_1,\dots,v_n)\,,\,\{x=s\,(i_1,\dots,i_m)\,\text{->}\,(o_1,\dots,o_p)\,,T\,|\,|\,body;R\}:S\;\longrightarrow$$
$$\{R\}:S\quad\textit{if}\,p\neq n$$

$$x=s_1\,(u_1,\dots,u_m)\,\text{->}\,(v_1,\dots,v_n)\,,\,\{x=s_2\,(i_1,\dots,i_m)\,\text{->}\,(o_1,\dots,o_n)\,,T\,|\,|\,body;R\}:S\;\longrightarrow$$
$$\{R\}:S\quad\textit{if}\,s_1\neq s_2$$

5. Procedure Structure

As already noted, a procedure consists of a header giving a name and two lists of arguments, one for input, and one for output, followed by a list of sets of rules. Each rule consists of a set of tests forming the lhs and a set of computations forming the rhs. A test on the lhs takes the form $x=s\,(u_1,\dots,u_m)\,\text{->}\,(v_1,\dots,v_n)$, with a small number of other tests permitted, such as the comparison tests $x\text{>}y$, the wait test $\texttt{wait}(x)$ and dynamic type tests. The computations on the rhs consist of assignments $x\text{<-}y$, $x=s\,(u_1,\dots,u_m)\,\text{->}\,(v_1,\dots,v_n)$, and procedure calls $p\,(u_1,\dots,u_m)\,\text{->}\,(v_1,\dots,v_n)$. Here x, y, each u_i and v_j are variable names. No other syntax is required, though a certain amount of syntactic sugar may be used to make the notation more readable, such as using a term instead of a variable, so that $p\,(\dots,t,\dots)\,\text{->}\,(v_1,\dots,v_n)$ is shorthand for $p\,(\dots,y,\dots)\,\text{->}\,(v_1,\dots,v_n)\,,y=t$ on either the lhs or the rhs, with y an otherwise unused variable.

In order to ensure correct moding, with variables having exactly one producer, and in the case of linear variables exactly one consumer, the following conditions apply in use of variables:

1) In any test $x=s(u_1,...,u_m)->(v_1,...,v_n)$ on the lhs, if $n>0$ or any u_i is indicated as linear, x must be linear. There will be a notational indication to show which variables are to be treated as linear.

2) In any test $x=s_1(u_1,...,u_m)->(v_1,...,v_n)$ on the lhs, x must be either an input argument to the procedure, or occur as one of the w_is in another test $y=s_2(w_1,...,w_p)->(z_1,...,z_q)$ on the same lhs.

3) No variable occurring as u_i or v_j in $x=s_1(u_1,...,u_m)->(v_1,...,v_n)$ on the lhs may occur in the procedure header, or as w_h or z_k in another test $y=s_2(w_1,...,w_p)->(z_1,...,z_q)$ on the same lhs, or occur more than once in the same test.

4) Every output variable to the procedure, and every extra output variable in a rule, that is one of the v_is in any $x=s(u_1,...,u_m)->(v_1,...,v_n)$ on the lhs, must be used in exactly one output position on the rhs. An output position is x in $x=s(u_1,...,u_m)->(v_1,...,v_n)$ or in $x<-y$, or any v_i in $x=s(u_1,...,u_m)->(v_1,...,v_n)$ or any v_i in $p(u_1,...,u_m)->(v_1,...,v_n)$.

5) If a linear variable occurs as x in a test $x=s(u_1,...,u_m)->(v_1,...,v_n)$ on the lhs, it must not occur at all on the rhs.

6) Any input linear variable either from the procedure heading or occurring as one of the u_is in a test $x=s_1(u_1,...,u_m)->(v_1,...,v_n)$ on the lhs which does not occur in a test as y in $y=s_2(w_1,...,w_k)->(z_1,...,z_h)$ on the lhs must be used exactly once in an input position on the rhs. An input position is y in $x<-y$ or any u_i in $x=s(u_1,...,u_m)->(v_1,...,v_n)$ or any u_i in $p(u_1,...,u_m)->(v_1,...,v_n)$.

7) Any variable that occurs only on the rhs of a rule must occur in exactly one output position. If it is a linear variable, it must also occur in exactly one input position, otherwise it can occur in any number of input positions.

A new variable is introduced under condition 7 when a procedure call rewrites using one of its rules. We refer to this as a "local variable". If the variable is introduced with its output occurrence as one of the v_is in $p(u_1,...,u_m)->(v_1,...,v_n)$, the procedure has itself set up a computation to give the variable a value. If, however, it is introduced as one of the v_is in $x=s(u_1,...,u_m)->(v_1,...,v_n)$ in the rhs where x is not itself a local variable, the variable will be given its value by the procedure which has x as an input. This is a form of what is called "scope extrusion" in pi-calculus. Scope extrusion of read access to a variable is given when it is used as one of the u_is in $x=s(u_1,...,u_m)->(v_1,...,v_n)$. If the procedure heading is $\#p(i_1,...,i_m)->(o_1,...,o_n)$, write access to a variable x can also be passed out of the procedure by $x<-i_k$ and read access passed out by $o_h<-x$. Also if we have $y=s(u_1,...,u_m)->(v_1,...,v_n)$ as a test on the lhs, write access to a variable x can also be passed out of the procedure by $x<-u_i$ on the rhs and read access passed out by $v_j<-x$. Otherwise, access to a variable remains private within the procedure where it was created and it cannot be interfered with by another procedure.

Although values given to variables are not rescinded, condition 5 can be seen as dictating consumption of a value sent on a linear variable considering it as a channel. If a rule with a linear variable test is used to progress computation, that linear variable cannot be used again, so in practice the assignment to it could be deleted. If a reference count is kept to record the number of readers of a non-linear variable, the assignment to the non-linear variable could in practice be deleted if that reference count drops to zero.

6. Dynamic Communication Topology

If a procedure call has output access to two variables **X** and **Y** (from here we will adopt the convention that linear variables are indicated by an initial capital letter), with input access to **X** and **Y** being two separate procedures, a direct communication channel can be made between those two procedures. **X=t1->c,Y=t2(c)** will establish a one-way communication channel from the call which inputs **X** to the call which inputs **Y**. If this linking variable is itself linear, as in **X=t1->C,Y=t2(C)**, a channel which may be reversed in polarity is established.

Let us consider an extended example. We have a dating agency which for simplicity has access to just one girl client and one boy client, shown by computation:

```
agency(Girl,Boy), girl()->Girl, boy()->Boy
```

Here an **agency** call has two input linear variables, and a **girl** and **boy** call each produce one linear output variable. The **agency** call must wait until both the girl and the boy request an introduction. The boy's request contains a channel on which he can send his first message to the girl he is put in contact with, while the girl will send a request which sends back a channel on which a message from a boy will be received. This is programmed by:

```
#agency(Boy,Girl)
{
 Boy=ask(Channel1), Girl=ask->Channel2 || Channel2<-Channel1
}
```

The output linear variable of the **girl** call is set to the input linear variable of the **boy** call. Now we can set up code to let them communicate:

```
#girl()->Dating
{
 || Dating=ask->Channel, goodgirl(Channel)
}

#boy()->Dating
{
 || Dating=ask(Channel), Channel=hello->Reply, goodboy(Reply);
 || Dating=ask(Channel), Channel=hello->Reply, badboy(Reply)
}
```

Sending a message on a channel and waiting for a reply is implemented by binding the channel variable to a tuple containing just one variable of output mode, and then making a call with that variable as input which suspends until the variable is bound. It can be seen that the message a **girl** call sends on the **Dating** channel reverses the polarity of that channel with the reversed channel renamed **Channel**, while the message a **boy** call sends on **Dating** keeps the polarity with **Channel** being a continuation of the same channel in the same direction.

For the sake of interest, we will let the **boy** call become non-deterministically either a **goodboy** call or a **badboy** call.

A **goodboy** call sends the message **hello**, waits for the reply **hi** back, then sends a **kiss** message and waits for a **kiss** message back. When that happens it sends another **kiss** message in reply and so long as a **kiss** message is replied with a **kiss** message this continues forever.

A **badboy** call sends a **bed** message when it receives a **kiss** message. We show here a **girl** call which can only become a **goodgirl** call, where a **kiss** message is replied with a **kiss** message, but a **bed** message is replied with a **no** message that has no reply variable, thus ending communication.

Either type of **boy** call, on receiving a **no** message can do no more, the call is terminated. Otherwise, the recursive calls represent a continuation of the call.

Here is how this is all programmed:

```
#goodboy(Channel)
{
 Channel=hi->Me || Me=kiss->Her, goodboy(Her);
 Channel=kiss->Me || Me=kiss->Her, goodboy(Her);
 Channel=no ||
}

#badboy(Channel)
{
 Channel=hi->Me || Me=kiss->Her, badboy(Her);
 Channel=kiss->Me || Me=bed->Her, badboy(Her);
 Channel=no ||
}

#goodgirl(Channel)
{
 Channel=hello->Me || Me=hi->Him, goodgirl(Him);
 Channel=kiss->Me || Me=kiss->Him, goodgirl(Him);
 Channel=bed->Me || Me=no
}
```

In the first two rules of each procedure here, **Channel** is an input channel on which is received a message which causes a reversal of polarity, so a message can be sent out on it which again reverses its polarity to receive a further message in reply. Effective two-way communication is established. A recursive call turns a transient computation into a long-lived process, the technique introduced by Shapiro and Takeuchi [24] to provide object-based programming in a concurrent logic language.

An alternative way of setting up this scenario would be for the **agency** call to take the initial initiative and send the **boy** and **girl** call a channel on which they communicate rather then them having to request it. In this case, the **agency**, **boy** and **girl** procedures will be different although the **goodboy**, **badboy** and **goodgirl** procedures will remain the same. The initial set-up is:

```
agency->(Girl,Boy), girl(Girl), boy(Boy)
```

with procedures:

```
#agency->(Girl,Boy)
{
 || Girl=tell(Channel),Boy=tell->Channel
}
```

```
#girl(Dating)
{
 Dating=tell(Boy) || goodgirl(Boy)
}

#boy(Dating)
{
 Dating=tell->Girl || Girl=hello->Her,goodboy(Her);
 Dating=tell->Girl || Girl=hello->Her,badboy(Her)
}
```

A third way of setting it up would be for the **boy** call to take the initiative while the **girl** call waits for the agency to communicate:

```
agency(Boy)->Girl, boy()->Boy, girl(Girl)
```

with the code for the **agency** procedure:

```
#agency(Boy)->Girl
{
 Boy=ask(Channel) || Girl=tell(Channel)
}
```

Here the **boy** procedure used will be the same as the first version given above, and the **girl** procedure the same as the second.

These examples show how the communication topology can be dynamic. We initially have a **boy** and **girl** call which both have a communication link with an **agency** call, but have no direct communication with each other. We show three different ways in which a direct communication link can be obtained, one in which the **boy** and **girl** call take the initiative jointly, another in which the **agency** call takes the initiative, and the third in which only the **boy** call takes the initiative.

Note that the examples shown here have no final default rule, thus it could be argued the whole program could fail if a call bound a variable to a value which its reader had no rule to handle. However, moding means we can always add an implicit default rule to prevent failure. In this rule, all output variables of the procedure are set to a special value indicating an exception. All input linear variables become the input variable to a special exception-handling procedure, which for any tuple the variable becomes bound to sets all output variables of the tuple to the special value indicating exception and makes all input linear variables the argument to another call to this procedure.

7. Conclusions and Related Work

The work described here can be considered a presentation of the work done by Reddy [25] oriented towards a language that can be used for practical programming. Reddy's work is inspired by Abramsky's computational interpretation [26] of linear logic [27]. We extend Reddy's typed foundation by allowing non-linear as well as linear variables, but our typing extends only as far as is necessary for modes to establish the single-writer multiple-reader property. Other attempts to build practical programming languages which add linearity to concurrent logic programming, such as Janus [23], have insisted that all variables be linear.

Our language could also be considered as a re-presentation of a committed choice logic language [22] which avoids logic programming terminology or the attempt to maintain some backward compatibility with Prolog that we argue elsewhere [28] was a contributing factor to these languages gaining little acceptance. The formalisation of modes and insistence that every variable is moded is new, however. Our strong moding expressed in the syntax of the language makes programs much easier to understand since it is always clear from where a variable receives its binding. It also means that the problem of dealing with the rare possibility of more than one computation wishing to bind a variable, which led to many of the variations discussed in [22], does not occur.

Another computation model related to ours is Niehren's delta-calculus [29]. Like our notation, the delta-calculus represents functions as relations with an explicit output variable and an assignment operator. The delta-calculus also uses linear types to enforce single assignment to variables. Unlike our language, the delta-calculus is higher order, that is variables may be assigned procedure values and used as operands. Although our language is first-order, we have shown elsewhere [10] how the effect of higher order functions can be obtained using the standard techniques for representing objects in committed choice logic languages [24], a function can be considered as just an immutable object which has only one method (application).

As with functional programming, our language works with a small and fully-defined set of reduction rules, which can be implemented to give eager evaluation, lazy evaluation, or some mixture including parallel evaluation. Like functional programming with its lambda calculus basis, our terse underlying notation can be made more easy to use by syntactic sugar. Unlike functional programming, it handles non-determinacy and interaction with ease. Our language enforces a single-assignment property on variables which removes all the complex issues of concurrent handling of mutable variables in conventional imperative languages. Single-assignment variables can be viewed as channels when we bind them to a pair consisting of a message and a variable for the continuation of the channel. In some cases variables are indicated as linear, allowing messages to be replied to (or synchronisation to be achieved) and channels to be reversed without interfering with the single-assignment property. This is done by "back communication", where a linear variable is bound to a tuple containing a variable which the tuple's consumer binds. Our use of linear variables arose from practical necessity, but its closeness to Reddy's work establishes a stronger theoretical justification for it.

Our work originates from attempts to build an object-oriented language on top of concurrent logic programming under the name "Aldwych" [9]. Previous attempts to do so [8] had been criticised for losing some of the flexibility of concurrent logic programming [30]. However these languages did have the benefit of being much less verbose than the equivalent code expressed directly as concurrent logic programming. Our intention was to have a syntax in which common patterns of using the underlying concurrent logic language were captured, as little as possible of the operational capability was lost, and the direct translation into concurrent logic programming kept in order to maintain a clear operational semantics. During the process of this work it became clear that the particularly simple form of concurrent logic programming to which Aldwych translates deserved attention and proper operational explanation as a language in its own right: "the core language of Aldwych".

Full Aldwych has subsets which appear as functional programming, object-oriented programming, and communicating process programming. It can be fully described in terms of the simple language covered in this paper, with sections 4 and 5 giving a description of its syntax and operational semantics. The language can be considered as doing for concurrent programming what Landin's ISWIM [1] did for sequential programming in the early days of high-level languages: providing a simple framework for a whole family of languages. Perhaps we can look forward to it providing the foundation for the next 700 concurrent programming languages.

References

[1] P.J.Landin The next 700 programming languages. *Comm. ACM* 9 (3):157-166 (1966).

[2] D.Q.M.Fay Experiences using Inmos proto-OCCAM™ *SIGPLAN Notices* 19 (9) (1984).

[3] R.Milner, J.Parrow and D.Walker. A calculus of mobile processes. *J of Information and Computation*, 100:1-77 (1992).

[4] B.C.Pierce and D.N.Turner. Pict: a programming language based on the pi-calculus. *Proof, Language and Interaction: Essays in Honour of Robin Milner*, MIT Press (2000).

[5] M.M.Huntbach and G.A.Ringwood. *Agent-Oriented Programming.* Springer LNCS 1630 (1999).

[6] I.Foster and S.Taylor *Strand: New Concepts in Parallel Programming*, Prentice-Hall (1989).

[7] K.Ueda. Experiences with strong moding in concurrent logic/constraint programming. *Proc. Int. Workshop on Parallel Symbolic Languages and Systems (PSLS'95)*, Springer LNCS 1068:134-153. (1996).

[8] A.Davison. A survey of logic programming based object oriented languages. In *Research Directions in Concurrent Object Oriented Programming.* G.Agha, P.Wegner, A.Yonezawa (eds) MIT Press (1993).

[9] M.Huntbach. The concurrent language Aldwych. *Proc. 1st Int. Workshop on Rule-Based Programming (RULE 2000)* (2000).

[10] M.Huntbach. Features of the concurrent language Aldwych. *ACM Symp. on Applied Computing (SAC'03)* 1048-1054 (2003).

[11] K.Ueda. Linearity analysis of concurrent logic programs. *Proc. Int Workshop on Parallel and Distributed Computing for Symbolic and Irregular Applications.* T.Ito and T.Yuasa (eds) World Scientific Press (2000).

[12] E.Tick. The de-evolution of concurrent logic programming languages. *J. Logic Programming* 23(2): 89-123 (1995).

[13] R.H.Halstead. Multilisp: a language for concurrent symbolic computation. *ACM Trans. Prog. Lang and Sys*, 7(4):501-538. (1985).

[14] K.L.Clark and S.Gregory. A relational language for parallel programming. In *Proc. ACM Conf. on Functional Programming Languages and Computer Architecture.* 171-178. (1981).

[15] S.Gregory. *Parallel Logic Programming in PARLOG.* Addison-Wesley. (1987).

[16] J.Hughes. Why functional programming matters. *Computer Journal*, 32(2):98-107. (1989).

[17] V.A.Saraswat, M.Rinard and P.Panangaden. Semantic foundations of concurrent constraint programming. *Principles of Prog. Lang. Conf. (POPL'91)*, 333-352 (1991).

[18] D.H.Grit and R.L.Page. Deleting irrelevant tasks in an expression-oriented multiprocessor system. *ACM Trans. Prog. Lang and Sys*, 3(1):49-59. (1981).

[19] R.A.Kowalski. *Logic for Problem Solving.* Elsevier/North Holland (1979).

[20] G.J.Sussman and D.V.McDermott. From Planner to Conniver – a genetic approach. *Proc AFIPS Fall Conference* 1171-79 (1972).

[21] S.Haridi, P Van Roy, P.Brand, M.Mehl, R.Scheidhauser and G.Smolka. Efficient logic variables for distributed computing. *ACM Trans. Prog. Lang and Sys*, 21(3):569-626. (1999).

[22] E.Y.Shapiro. The family of concurrent logic programming languages. *ACM Computing Surveys* 21(3):413-510 (1989).

[23] V.A.Saraswat, K.Kahn and J.Levy. Janus: a step towards distributed constraint programming. *Proc. 1990 North American Conf. on Logic Programming.* MIT Press. 431-446 (1990).

[24] E.Y.Shapiro and A.Takeuchi. Object oriented programming in Concurrent Prolog. *New Generation Computing* 1:25-48 (1983).

[25] U.S.Reddy. A typed foundation for directional logic programming. *Proc. 3rd Int. Workshop on Extensions of Logic Programming.* Springer LNCS 660:282-318 (1993).

[26] S.Abramsky. A computational interpretation of linear logic. *Theoretical Computer Science* 111:3-57 (1993).

[27] J.-Y.Girard. Linear logic. *Theoretical Computer Science* 50:1-102 (1987).

[28] M.Huntbach, The concurrent language Aldwych. *World Multiconference on Systemics, Cybernetics and Informatics (SCI 2001)* XIV:319-325.

[29] J.Niehren. Functional computation as concurrent computation. *Proc. 23rd Symp. on Principles of Programming Languages (PoPL '96)* 333-343 (1996).

[30] K.M.Kahn. Objects – a fresh look. *Proc. 3rd European Conf. on Object-Oriented Programming (ECOOP 89).* S.Cook (ed), Cambridge University Press.

Communicating Process Architectures 2007 67
Alistair A. McEwan, Steve Schneider, Wilson Ifill, and Peter Welch
IOS Press, 2007

JCSProB: Implementing Integrated Formal Specifications in Concurrent Java

Letu YANG and Michael R. POPPLETON

Dependable Systems and Software Engineering,
Electronics and Computer Science, University of Southampton,
Southampton, SO17 1BJ, UK.

{ly03r, mrp} @ecs.soton.ac.uk

Abstract. The ProB model checker provides tool support for an integrated formal specification approach, combining the classical state-based B language with the event-based process algebra CSP. In this paper, we present a developing strategy for implementing such a combined ProB specification as a concurrent Java program. A Java implementation of the combined B and CSP model has been developed using a similar approach to JCSP. A set of translation rules relates the formal model to its Java implementation, and we also provide a translation tool JCSProB to automatically generate a Java program from a ProB specification. To demonstrate and exercise the tool, several B/CSP models, varying both in syntactic structure and behavioural/concurrency properties, are translated by the tool. The models manifest the presence and absence of various safety, deadlock, and bounded fairness properties; the generated Java code is shown to faithfully reproduce them. Run-time safety and bounded fairness checking is also demonstrated. The Java programs are discussed to demonstrate our implementation of the abstract B/CSP concurrency model in Java. In conclusion we consider the effectiveness and generality of the implementation strategy.

Keywords. ProB, JCSP, Integrated formal methods, Code generator

Introduction

Formal approaches to modelling and developing concurrent computer systems, such as CSP [1] and CCS [2], have been in existence for more than thirty years. Many research projects and a number of real world systems [3] have been developed from them. However, most programming languages in industry, which support concurrency, still lack formally defined concurrency models to make the development of such systems more reliable and tractable. The Java language has a painful history inasmuch as it lacks explicit and formal definitions of its concurrency model. Before Java 5.0, the JMM (Java Memory Model) didn't explicitly define the read/write order that needs to be preserved in the memory model. This confused the developers of JVMs (Java Virtual Machines). The different JVMs developed under the old JMM can represent different behaviours, and lead to different results from running the same piece of Java code. To clarify this issue, Java 5.0 and the third version of the Java language specification had to redefine a new JMM.

Although the new defined JMM addressed the safety issue previously in Java concurrency, liveness and fairness issues, such as deadlock and starvation, still remain intractable, and depend totally on developers' skills and experience in concurrent systems development. Therefore, many approaches have been attempted to formalize the development of concurrent Java systems. Formal analysis techniques have been applied to concurrent Java programs. JML [4] and Jassda [5] provide strategies to add assertions to Java programs, and employ runtime verification techniques to check the assertions. Such approaches are concerned with the satisfaction of assertions, not explicit verification against a formal concurrency model. An explicit formal concurrency model, which can be verifiably transformed into a concurrent Java program, would represent a useful contribution.

Magee and Kramer [6] introduce a process algebra language, FSP (Finite State Processes), and provides a formal concurrency model for developing concurrent Java programs. Then the LTSA (Labelled Transition System Analyser) tool is employed to translate the formal model into a graphical equivalence. The tool can also check desirable and undesirable properties of the FSP model. However, there is still an obvious gap in this approach between the graphical equivalence and the Java implementation. To construct the Java application, the formal model is only provided as a guidance, while the developers still need to implement the model in Java through their own experience and skill in concurrency. That means there is no guarantee that the Java code would be a correct implementation of the formal model.

JCSP [7] is a Java implementation of the CSP/*occam* language. It implements the main CSP/*occam* structures, such as process and channel, as well as key CSP/*occam* concurrency features, such as parallel, external choice and sequential composition, in various Java interfaces and classes. It bridges the gap between specification and implementation. With all the Java facility components in the JCSP package, developers can easily construct a concurrent Java program from its CSP/*occam* specification. The correctness of the JCSP translation of the *occam* channel to a JCSP channel class has been formally proved [8]: the CSP model of the JCSP channel communication was shown to refine the CSP/*occam* concurrency model. Early versions of JCSP (before 1.0-rc6) targetted classical *occam* which only supported point-to-point communication, while recently, new versions of JCSP have moved on to support the *occam-pi* language, which extends classical *occam* with π-calculus. More CSP mechanisms, e.g. external choice over multiway synchronization, have been implemented in new JCSP (1.0-rc7). Our work is mainly based on JCSP 1.0-rc5, while we plan to move to 1.0-rc7. We will discuss this in Section 5.

Raju et al. [9] developed a tool to translate the *occam* subset of CSP/occam directly into Java with the JCSP package. Although in our experience the tool is not robust enough to handle complex examples, it provides a useful attempt at building automatic tool support for the JCSP package.

Recent research on integrating state- and event- based formal approaches has been widely recognized as a promising trend in modeling large-scale systems. State-based specification is appropriate when data structure and its atomic transition is relatively complex; event-based specification is preferred when design complexity lies in behaviour, i.e. event and action sequencing between system elements. In general of course, significant systems will present design complexity, and consequently require rich modeling capabilities, in both aspects. CSP-OZ [10], csp2B [11], CSP‖B [12] and Circus [13] are all existing integrated formal approaches. However, the lack of direct tool support is a one of the most serious issues for these approaches. Proving the correctness of their combined specifications requires complex techniques, such as composing the verification results from different verification tools [14], or translating the combined specification back into a single specification language [11,15].

The implementation issue is another significant question mark over integrated formal methods. The more complex structures and semantics they commonly share usually create difficulty in developing a stepwise implementation strategy for the integrated specification. For the above integrated formal approaches, only CSP-OZ has considered the association with programming languages. The applied technique, Jassda [16], is a light-weight runtime verification approach based on the *Design-by-Contract* concept [17], and is really a verification technique, rather than an implementation strategy.

ProB [18] supports an integrated formal approach [19] which combines B [20] and CSP[1]. A composite specification in ProB uses B for data definition and operations. A CSP specification is employed as a filter on the invocations of atomic B operations, thus guiding their execution sequence. An operational semantics in [19] provides the formal basis for combining the two specifications. The ProB tool, which was designed for the classical B method, provides invariant checking, trace and single-failure refinement checking, and is able to detect deadlock in the state space of the combined model.

The main issue in developing an implementation strategy for ProB is how to implement the concurrency model of the B+CSP specification in a correct and straightforward way. Furthermore, we need an explicit formal definition, or even automatic tool support, to close the gap between the abstract specification and concrete programming languages. The structure of the JCSP package gives significant inspiration. We implement the B+CSP concurrency model as a Java package with similar process-channel structure to JCSP. Based on this implementation package, we formally define a set of translation rules to convert a useful and deterministic subset of B+CSP specification to Java code. To make the translation more effective and stable, an automatic translation tool is constructed as a functional component of the ProB tool. Run-time invariant checking and bounded fairness assertions checking are also implemented and embedded inside the Java implementation.

There are two main contributions of this paper. The first one is the Java implementation strategy for the B+CSP concurrency model. It implements basic features of the combined abstract specification, and provides the fundamental components for constructing concurrent Java programs. In Section 2 we introduce the combined B+CSP specification, and our restrictions on its semantics. We then discuss the Java implementation of the concurrency model. Several key Java classes are explained, and compared with the JCSP package. The translation rules and the tool are also presented. Section 3 discusses the translation rules that are implemented in the translation tool.

The second contribution is the experimental evaluation of this implementation strategy, discussed in section 4. We carry out a number of experiments, implementing some concurrent formal models. In order to exercise the coverage of the translation rules, these models differ syntactically, using both B and CSP elements differently. Beyond exercising the translation, there are three dimensions to the experiments:

- The models illustrate the presence and absence of various behavioural properties, including safety, deadlock freeness, and bounded fairness. ProB can be used to verify the presence or absence of a safety or deadlock freeness property. In this case, we run the translated Java to *check* the translation, and expect to see the property either manifested or not, depending on whether it is present/ absent in the model.
- In the case of properties that we think might hold in the model, or that we might not even have an opinion about - such as bounded fairness - we use the Java to *simulate* the model, using a number of diverse runs to estimate the presence or absence of the property.
- We also demonstrate a simple mechanism for generating a variety of timing and interleaving Java patterns for a given input model, and consider its utility.

This experimental evaluation of the implementation strategy gives confidence in the work, and provides a basis for addressing problems and for further development.

Finally section 5 discusses the ongoing work of this approach, including GUI development and scalability issues. A formal verification of the translation is briefly discussed as necessary future work.

---

[1]We will call this notation B+CSP for shorthand

1. The Combined B+CSP Specification

As our work is inspired by the development of JCSP, when we discuss the Java implementation in this section, we compare it with JCSP in various aspects. We first give a brief introduction to the B+CSP specification. Then we discuss the operational semantics of B+CSP, and the restricted semantics used in the work. Finally, we demonstrate how the semantics works.

Table 1 gives the B and CSP syntax supported in our approach. We use quote marks as well as boldface to denote BNF terminal strings.

Table 1. The main B and CSP specification supported in JCSProB

B Machine	*Machine*	**MACHINE** *Header* *Clause_machine** **END**
	Clause_machine	... | *Clause_variables* | *Clause_invariant* | *Clause_assertions* | *Clause_initialization* | *Clause_operations* | ...
B Operation	*Clause_operations*	**OPERATIONS** *Operation*$^{+}$";"
	Operation	*Header_operation* "=" *Level1_Substitution*
	Header_operation	[*ID*$^{+}$";" ←] *ID* ["(" *ID*$^{+}$";" ")"]
	Precondition	**PRE** *Condition* **THEN** *Substitution* **END**
B	*Block*	**BEGIN** *Substitution* **END**
Substitution	*If-Then-Else*	**IF** *Condition* **THEN** *Substitution* [**ELSIF** *Condition* **THEN** *Substitution*]* [**ELSE** *Substitution*] **END**
	Var	**VAR** *ID*$^{+}$";" **IN** *Substitution* **END**
	Sequence	*Substitution* ";" *Substitution*
	Parallel	*Substitution* || *Substitution*
	Assignment	*ID*[(*Expression*)] ":=" *Expression*
	Prefix	*ChannelExp* → *Process*
CSP	*Sequential Composition*	*Process* ";" *Process*
Process	*External Choice*	*Process* "[]" *Process*
and Channel	*Alphabetical Parallel*	*Process* "[|" *Ch_List* "|]" *Process*
	Interleaving	*Process* "|||" *Process*
	Process call	*Proc_Header*
	If-Then-Else	**if** *CSP_Condition* **then** *Process* [**else** *Process*]
	Skip	**SKIP**
	Stop	**STOP**
	ChannelExp	*ID* [*Output_Parameter**] [*Input_Parameter**]
	Output_Parameter	"!"*CSPExp* | "."*CSPExp*
	Input_Parameter	"?"*CSPExp*

The B part of the combined specification language supported in our approach is mainly from the B0 subset. B0 is the concrete, deterministic subset of the B language describing operations and data of implementations. It is designed to be mechanically translatable to programming languages such as C and Ada. A B machine defines data variables in the **VARIABLES** clause, and data substitutions in the **OPERATIONS** clause. Possibly subject to a **PRE**condition - all of whose clauses must be satisfied to enable the operation - an operation updates system state using various forms of data substitution. Although the B specification used in our approach is the B0 subset, we do support some abstract B features which are not in B0, e.g. precondition. These features are implemented to provide extra functions for rapidly implementing and testing some abstract specification in Java programs. In the implementation, preconditions are interpreted as guards, which will block the process if the precondition is not satisfied.

A B operation may have input and/or output arguments. For an operation *op* with a header *rr* ← *op(ii)*, *ii* is a list of input arguments to the operation, while *rr* is a list of return arguments from it. The **INITIALIZATION** clause establishes the initial state of the system.

The **INVARIANT** clause specifies the safety properties on the data variables. These properties must be preserved for all the system states. Figure 1 shows a simple lift example in a B machine. It has a variable *level* which indicates the level of the lift, and two operations, *inc* and *dec* to move the lift up and down.

> ***MACHINE*** *lift*
> ***VARIABLES*** *level*
> ***INVARIANT*** *level : NAT & level $\geq$ 0 & level $\leq$ 10*
> ***INITIALIZATION*** *level := 1*
> ***OPERATIONS***
> 　　*inc =* ***PRE*** *level $<$ 10* ***THEN*** *level := level + 1* ***END***;
> 　　*dec =* ***PRE*** *level $>$ 0* ***THEN*** *level := level - 1* ***END***
> ***END***

Figure 1. An example of B machines: lift

Table 1 also defines the supported CSP process and channel structures. A detailed definition of supported CSP syntax can be found in the ProB tool.

Currently ProB only supports one paired B and CSP combination. Although ProB supports trace refinement checking for the combined specification, it still hasn't provided a refinement strategy for composing or decomposing an abstract B+CSP model into a concrete distributed system. The CSP∥B approach does provide a refinement strategy [14] for composing combined B and CSP specifications. However, it is unlikely that this approach can be directly used in ProB. Therefore our work here focusses on one concrete B and CSP specification pair. All the processes in the CSP specification are on a local machine.

1.1. The ProB Combination of B and CSP Specification

We have seen that B is essentially an action system. The system state is shared by a number of guarded atomic actions, i.e. B operations, in the system model. The actions can change the state of the system by updating the values of system variables. Whether an action is enabled is determined by its guard, a predicate on the system state. State-based formal approaches give an explicit model of data definitions and transitions. However, as behaviour is only defined by the pattern of enablement over time of the guards, any such behaviour is only observable in a runtime trace, and not explicitly in the model syntax.

An event-based approach, on the other hand, explicitly defines the behaviours of the system. The actions in the system are regarded as stateless events, i.e. the firing of CSP channels. A process, a key concept, is defined in terms of possible behaviour sequences of those events. In CSP, traces, failures and divergences semantics are used to interpret system behaviours. Thus although event-based approaches are good at explicitly defining system behaviour, they lack strength in modelling data structure and dynamics. In event-based approaches like CSP, state is nothing more than local process data, communicated through channels or by parameter passing through processes. There is no explicit way to model system states on globally defined data. a An early integration [21] of state- and event-based system models provided the theoretical correspondence between action systems and process algebras. Many approaches [10,11,12,13,19] have been made at combining existing state- and event-based formal methods. It is clearly essential, however, to provide a semantics for any proposed combined model.

The operational semantics of the B+CSP specification is introduced in [19] and provides a formal basis for combining the B and CSP specification. The B machine can be viewed as a special process which runs in parallel with CSP processes. The system state is maintained by the B machine in that process, while CSP processes only maintains their local states and cannot directly change the system state. The execution of a B operation need to synchronize with a CSP event which has the same identical name. In this way, CSP can control the firing of B operations.

The combination of a CSP event and a corresponding B operation is based on the operational semantics. The operational semantics of the combined B+CSP channel are: (σ,P) $\rightarrow_A (\sigma',P')$. σ and σ' are the before and after B states for executing B operation O, while P and P' are the before and after processes for processing CSP channel ch. The combined channel A is a unification of the CSP channel $ch.a_1, ..., a_j$ and the B operation $O = o_1,...,o_m$ $\leftarrow op(i_1,...,i_n)$.

The operational semantics of B+CSP in ProB [19] provides a very flexible way to combine B operations and CSP channels. This flexibility is in handling the arguments on the combined channel. As a model checking tool, ProB is relatively unrestricted in combining the B operation arguments and CSP channel arguments. There is no constraint on input/output directions of the arguments. CSP processes can be used to drive the execution of B machines by providing the value of the arguments, or vice-versa. It is even possible that neither B and CSP provide values for channel arguments, or that the numbers of arguments on the combined B operations and CSP channel differ. ProB can provide the values by enumerating values from the data type of the arguments. This gives more power to ProB to explore the state space of system models. However, as our target is generating concrete programs, it is not possible to allow such flexibility in the implementation semantics.

1.2. The Restricted B+CSP Semantics for JCSProB

As a model checking tool, ProB aims to exhaustively explore all the states of an abstract finite state system, on the way enumerating all possible value combinations of operation arguments. The flexibility in combining the two formal models provides more power to the ProB tool to model check the state space of a model. However, for concrete computer programs, it is not realistic to support the same flexible and abstract semantics as model checkers. We need a more restricted and deterministic semantic definition.

We thus define a restricted B+CSP operational semantics as follows. For a B operation $o = o_1,...,o_m \leftarrow op(i_1,...,i_n)$, its corresponding CSP channel must be in the form of $ch!i_1...!i_n?o_1...?o_m$. At CSP state P, a CSP process sends channel arguments $i_1,...,i_n$ through the channel to a B operation. After the data transitions of the channel complete - taking B state from σ to σ' - the CSP state changes to P'. The arguments $o_1,...,o_m$ represent the data returned from B to CSP. The new restricted semantics can be expressed as $(\sigma,P,in) \rightarrow_A (\sigma',P',out)$, where $in = i_1,...,i_n$, and $out = o_1,...,o_m$.

Furthermore, the flexible ProB semantics also supports CSP-only channels without B counterparts. These channels preserve the semantics of CSP/*occam*. We handle them separately from the combined B+CSP channel, and implement them in the Java application using the JCSP package. However, the CSP semantics supported by ProB is still larger than JCSP/*occam*. The allowed argument combinations in this work are showed in Table 2, although some of them have not been fully implemented yet.

Table 2. The allowed arguments combination for B operations and CSP channels

JCSProB	B: input arguments ($c(x)$)	B: return arguments ($y \leftarrow c$)	B: no argument (c)
CSP output ($c!x$, $c.x$)	$\checkmark$ (multi-way sync)	$\times$	$\times$
CSP input ($c?y$)	$\times$	$\checkmark$ (multi-way sync)	$\times$
CSP none (c)	$\times$	$\times$	$\checkmark$ (multi-way sync)
JCSP	CSP input ($c?y$)	CSP output ($c!x$)	CSP none (c)
CSP output ($c!x$)	$\checkmark$ (p2p sync)	$\times$	$\times$
CSP input ($c?y$)	$\times$	$\checkmark$ (p2p sync)	$\times$
CSP none (c)	$\times$	$\times$	$\times$

The top half of the table shows the argument combinations for the restricted B+CSP semantics. If a CSP channel *c!x* outputs an argument *x*, the argument is combined with an input argument *x* in the corresponding B operation *c(x)*. A return argument *y* from a B operation *y ← c* is combined with an input argument *y* in the corresponding CSP channel *c?y*. These two kinds of arguments provide two-way data flow between the B and CSP models:

- In B state σ, CSP passes data in CSP→B arguments to invoke the execution of a B operation with these arguments. This will change system state in the B model from σ to σ'. We can see this as the CSP model reusing a stateful computation, rather like an abstract subroutine call.
- In B state σ', the return data in CSP←B arguments returns the B state to the CSP process. This can be seen as a subroutine call to read internal state, used to influence behaviour in the CSP model.

If an invocation of a B operation requires that the arguments be fixed, synchronization on the combined B+CSP channel is not only defined by the name of the channel, but also by the value of the arguments. Two processes calling a combined B+CSP channel with different argument values cannot synchronize, because the two calls represent two different data transitions in the B model. This is multi-way CSP-out-B-in synchronization.

In a similar way, multi-way B-out-CSP-in synchronization is defined, this time on the channel name only. In this case the synchronization represents one call to the B operation, returning one result, which is read by mutiple CSP input channels.

The bottom half of the table demonstrates ProB's support of the JCSP/*occam* channel in a pure JCSP semantics. As the communication in JCSP is between two processes, a call of channel output (*c!y*) corresponds to one or more channel input call (*c?x*) from other processes. The standard channel model of JCSP/occam provides point-to-point communication between a writer and read process: synchronisation happens as a by-product, since these channels provide no buffering capacity to hold any messages.

1.3. How the Restricted Semantics Works

The CSP part of the combined specification defines the behaviours of the system model. It is used to drive the execution of the B machine. It controls system behaviour by defining the execution sequence of combined channels in CSP processes, and using the channels to fire data operations in the B model. Therefore, the execution of a combined channel is guarded by a call from CSP, as well as the B precondition on the channel.

In Figure 2, process *Q* defines system behaviour by giving the execution order of channel *m* and *n*. When process *Q* calls the execution of channel *n*, whether the call will enable the data transition in the corresponding B operation *n*, is still guarded by:

- the synchronization strategy in the CSP part. In this case, as process *Q* needs to synchronize with process *R* on channel *n*, the channel is only enabled when process *R* also calls the channel.
- the precondition on the corresponding B operation *n*.

As defined in the restricted semantics, the synchronization on a combined B+CSP channel is determined by both channel name and CSP→B arguments. Multiple processes synchronize on the execution of data transitions inside a combined channel. The combined channel performs a barrier synchronization with state changes inside the barrier. Processes *Q* and *R* synchronize on channel *n*, with arguments *X+1* and *Y* on the channel respectively. The two processes will wait, and only be invoked if *X+1* = *Y*, i.e. the channel arguments match. If they do not match then the calls on channel *n* will block.

On the other hand, as discussed in Section 1.2, the B state model can use CSP←B arguments to modify CSP system behaviour. In Figure 2, the B operation *m* returns an argument *rr* through the combined channel. In the CSP part, as process *P* and *Q* interleave on channel *m*, both of them can receive data from the channel but on interleaving calls. Return data from different calls on the channel represent different system states in the B model. In particular, process *Q* uses the returned argument *X* afterwards to invoke the other channel *n*. This shows B state data affecting the behaviour of the combined system model.

Figure 3 shows a very simple example of one-to-one channel communication model of JCSP/*occam*, which is also supported in our semantics. Process *P* sends data *X* through channel *c* to process *Q*. When one of the processes is ready, it needs to confirm that the process on the other side of the channel is also ready for the communication. The communication only involves one reader and one writer. JCSP/occam also support multiple writers and/or readers interleaving with each other to use a shared *any-to-any* channel. Note that the writers (respectively readers) do not synchronise with each other – only one reader with one writer.

Thus two distinct concurrency models are supported, but because of their differences in synchronization, must be treated separately in translation.

2. The Java Implementation of B+CSP

2.1. JCSP and JCSProB

The JCSP package enables the implementation of a formal specifications in CSP/*occam* in Java. Our combined B+CSP specifications are expressed in a much larger language than the classical *occam* subset of CSP. Although the *occam-pi* language extended the *occam* language and supports multi-way synchronization, the semantics of it are still different from that of CSP+B. However, it is possible to use *occam-pi* to express the semantics of B+CSP.

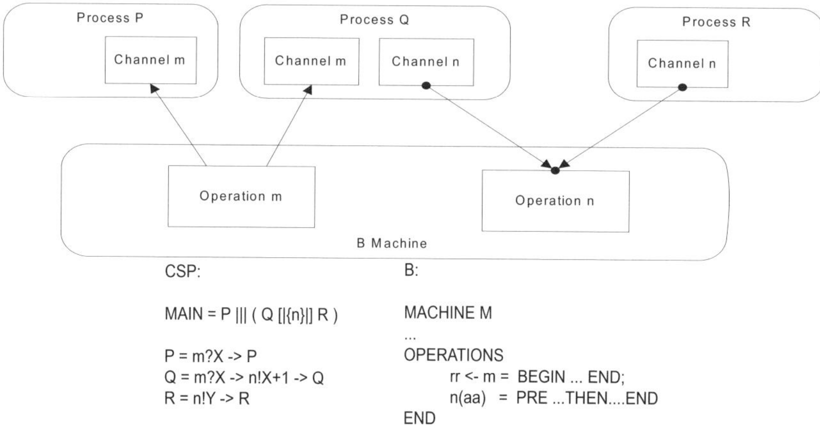

Figure 2. Concurrency Model of B+CSP in ProB

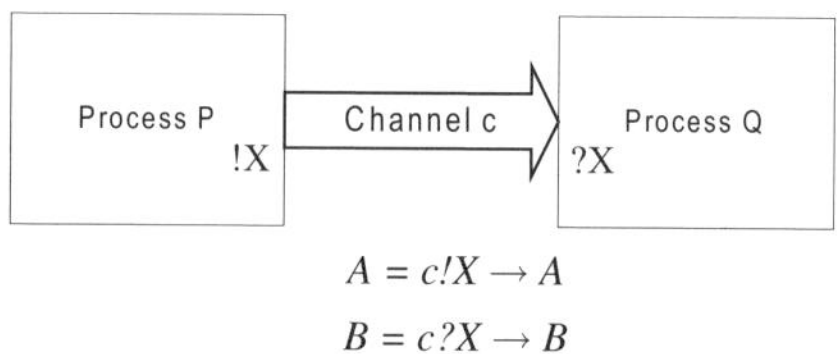

$$A = c!X \rightarrow A$$
$$B = c?X \rightarrow B$$

Figure 3. One-to-one Channel Model of JCSP

That means it is also possible to use the new JCSP package to construct the implementation of B+CSP.

When we started this work, the new JCSP package (1.0-rc7) had not been published. There were no facilities for multi-way synchronization on external choice (*AltingBarrier*), or atomic state change during an extended (rendezvous). This why we augment the point-to-point communication of previous JCSP/*occam* with a new concurrency model, called *JCSProB*. Like *occam*-π, the old JCSP package (before 1.0-rc6) implements a *barrier* class, which supports the synchronization of more than two processes. However, there is still no state change mechanism inside the *barrier* class. State change is the other issue concerned. JCSP channels are mainly used for communication and synchronization. The state change can only happen in JCSP process objects, while in B+CSP, only the B part of combined channels can access the system variables and change the system state. Therefore, we need to implement the data transitions on system states inside the implementation of combined channels.

To deal with these limitations, we construct a new Java package, JCSProB, to implement the B+CSP semantics and concurrency. This package provides infrastructure for constructing concurrent Java programs from B+CSP specifications. In this section, we will discuss several fundamental classes from the JCSProB package. We inherit the process-channel structure from JCSP, as well using a part of its interfaces and classes. As a Java implementation of *occam* language, JCSP provides severak kinds of Java interfaces and classes:

- *CSProcess* interface, which implements the *occam* process. All the process classes in the JCSP package and in the Java application need to implement this interface.
- Some process combining classes, e.g. *Parallel*, *Sequence* and *Alternative*. They provide direct implementation of the key process structures, e.g. PAR, SEQ, and ALT in *occam*.
- Channel interfaces and classes. JCSP provides a set of channel interfaces and classes for implementing the point-to-point communication in *occam*.
- Barriers, alting barriers and call channels. These are not used in the work reported here, but may become useful in future developments 5.
- Timers, buckets, etc. These are not relevant here.

The JCSProB package is developed for implementing the restricted B+CSP semantics and concurrency model. In JCSProB it is mainly the channel interfaces and classes that are rewritten, as well as the process facilities which interact with the execution of channel classes, e.g. *external choice*. Figure 4 illustrates the basic structure of the JCSProB package, and its relation with JCSP, and how to build the target Java application upon these two packages.

The figure shows that there are three kinds of classes that need to be developed to construct a Java application:

- At least one process class, ⟨*process*⟩_*procclass* ("_procclass" is the suffix of a process class name), which implements the JCSP *CSProcess* interface. Each process in the CSP part of the combined specification is implemented in a process class.
- JCSProB channel classes, ⟨*channel*⟩_*chclass* ("_chclass" is the suffix of a channel class name), which extends the new *PCChannel* class from JCSProB. The *PCChannel* class implements the semantics of the combined B+CSP channel. It is an abstract class which has synchronization and precondition check mechanism implemented inside. Every channel class needs to extend this class, and override the abstract *run* method of it (If a B operation has precondition, the channel class also needs to override the *preCondition* method).
- A *MaVar* class (Machine Variable), which extends the *JcspVar* class of JCSProB. It implements the B variables, as well as the invariant and assertions on them.

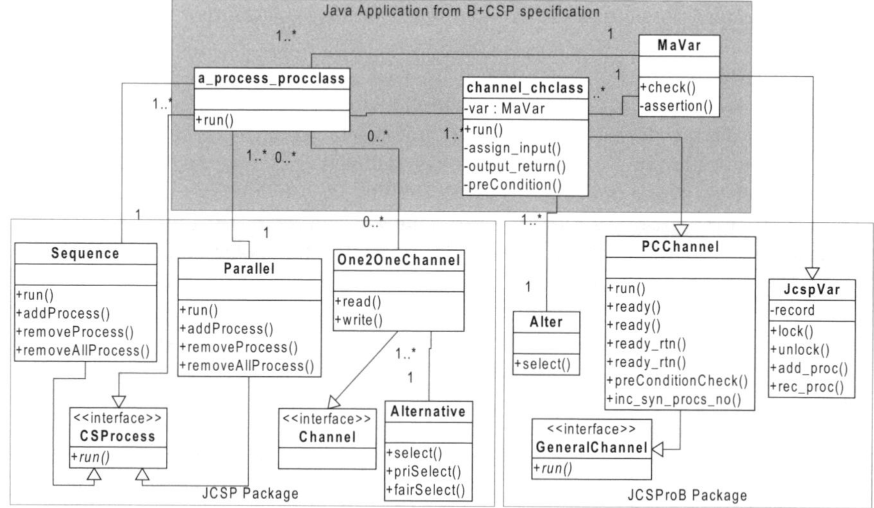

Figure 4. The structure of Java Application developed from JCSP and JCSProB packages

The JCSP/*occam* semantics are implemented in the JCSP package. As the semantics is also supported in ProB, the JCSP package is also used in the implementation. As our approach supports both the combined semantics of B+CSP and JCSP/*occam*, the differences between the two semantics and concurrency model result in two modes of translation for two kinds of channels. B+CSP channels are translated to subclasses of *PCChannel* class from JCSProB, while JCSP/*occam* channels are translated to JCSP channel classes.

Process classes in this work implement the process interface *CSProcess* from JCSP. Some JCSP process classes, e.g. *Parallel* and *Sequence*, are also directly used to construct concurrent Java applications. Because these classes are concerned with execution orders over a set of processes, they are not concerned with internal process behaviour. For example, the *Parallel* class takes an array of process objects, and runs all the them in parallel. *Parallel* class is not involved in implementing synchronization. The synchronization strategies are implemented in the channel classes. Changing to channel classes with different synchronization strategies does not affect the functions of these process classes. Therefore, both JCSP and JCSProB channels can be used in a process class.

There is a restriction on the use of *external choice* for the two kinds of channels: B+CSP and JCSP channels cannot be used in *external choice* at the same time. The *Alternative* class from JCSP is used to implement *external choice* for JCSP channels, while *Alter* class in JCSProB implements it for JCSProB channels.

Some key JCSProB classes are discussed in the following few sections.

2.2. Channel Classes

The channel class in JCSProB is *PCChannel*; all the channel classes in the Java application need to extend this class to obtain the implemented B+CSP semantics and concurrency. The data transitions of a channel should be implemented in the *run()* method of the channel class.

The allowed argument combinations for the restricted semantics are shown in Table 2. The *PCChannel* class provides four methods to implement this semantics policy. All the input and output arguments are grouped into objects of Java *Vector* class (*java.util.Vector*):

- **void ready()**: there is no input/output on the combined channel
- **void ready(Vector InputVec)**: CSP process passes arguments to B operation

- ***Vector ready_rtn()***: CSP process receives arguments from B operation
- ***Vector ready_rtn(Vector InputVec)***: CSP process passes arguments to B operation, and receives arguments from B operation

Implementing the synchronization in the restricted B+CSP concurrency is another important issue for the *PCChannel* class. When there is more than one process synchronizing on a channel, the *run()* method will not be invoked until the condition from the concurrency model is satisfied. In the *PCChannel* class, we implement the synchronization illustrated in Section 1.2. The *inc_syn_procs_no(int)* method from *PCChannel* class is used to indicate the number of processes which synchronize on it. For example, in Figure 2 the *inc_syn_procs_no(int)* method of channel *n* is called to indicate that process *Q* and *R* synchronize on this channel, before the two processes are initialized in the *MAIN* process. The following Java code shows how this mechanism is implemented:

```
n_ch.inc_syn_procs_no(2);
new Parallel(
    new CSProcess[]{
        new P_procclass(var,m_ch),
        new Q_procclass(var,m_ch,n_ch),
        new R_procclass(var,n_ch),
    }
).run();
```

Process classes *P_procclass*, *Q_procclass* and *R_procclass* are running in parallel. An instance of *Parallel* class from JCSP package groups all three of them together, and uses the *run()* method to run the three processes in parallel. The *inc_syn_procs_no(int)* method of channel object *n_ch* is called to inform the channel that there are two processes, *Q_procclass* and *R_procclass*, synchronizing on it. Although channel object *m_ch* is also shared by two processes *P_procclass*, *Q_procclass*, the two processes interleave with each other, and do not synchronize on it.

There are two other issues concerning the *PCChannel* class. One is the precondition check, which can guard conditions on the data transitions inside a B operation. The *PCChannel* class provides a method *preConditionCheck()* for checking the precondition on the data transition, and blocking the caller process when the condition is not satisfied. The actual precondition should be implemented in the *preCondition()* method of the channel class. The default *preCondition()* method in *PCChannel* guards on no condition, and always indicates the precondition is satisfied. The concrete channel subclass needs to override the *preCondition()* method to implement the precondition. The other issue is the implementation of atomic access by the B operations; this is discussed with the *JcspVar* class in Section 2.3.

2.3. Global B Variables Class

In B-method, the data transitions of a B operation must be kept atomic in order to preserve the consistency of the state model. The combined B+CSP model also has this requirement. The JCSProB packages provide a *JcspVar* class for implementing this feature in the Java implementation. It explicitly implements an exclusive lock to control the access to the B variables. Only one channel object can have the lock at a time. When a subclass of *PCChannel* tries to override the *run()* method, it is forced to use *lock()* method from *JcspVar* class to obtain the access authorization first, and release it by calling *unlock()* method after data transitions. When constructing a Java implementation from its formal specification, the *JcspVar* class need to be extended, and all the global B variables should be implemented in the new constructed class.

2.4. External Choice Class

In current JCSP, external choice is implemented in the *Alternative* class. As the decision of choices is based on the first events of all possible paths, *Alternative* needs to cooperate with JCSP channel objects to make the choice. Because the JCSP channel implements the *occam* point-to-point communication, the guard on the channel is based on the state of two communicating processes. In JCSP, only the guard on channel input is implemented, because guarding on channel output can cause major system overhead.

The *Alter* class from JCSProB package implements external choice for the extended channel classes *PCChannel*. It queries the *preCondition()* method of channel objects for their the preconditions, and then makes choice on the paths with ready channels. However, for the B+CSP semantics, the guard on a combined channel includes not only the precondition from the B part, but also, for the shared channel, the availabilities of all the synchronizing processes, all of whom may back off any time after offering to synchronize, having chosen something else.

Even for point-to-point communication in JCSP (1.0-rc5), the previous two-phase commit protocol for implementing guarding on channel output has been considered complex and costly. Therefore, only the guard on channel input is allowed. Implementing the guard on multi-way synchronized channel is expected to be even harder. In [22,23,24], a fast algorithm for implementing external choice with multi-way synchronization is discussed. Furthermore, for the combined B+CSP channel, the synchronization guards need to coordinate with the precondition on B operations for guarding the execution of the channel. Therefore, we are still working on multi-way synchronization guard; that work is currently under testing. We also consider using the *AltingBarrier* class in JCSP 1.0-rc7 to implement the multi-way synchronization for B+CSP channel, and it would be very interesting to compare the two implementations.

3. Translation: From B+CSP to JCSP

3.1. Translation Rules

The JCSProB package provides basic facilities for constructing concurrent Java applications from B+CSP models. However, there is still a big gap between the specification and the Java implementation. Manually constructing the Java implementation with the package is still very complex, and cannot guarantee whether the Java application is correctly constructed. To close the gap, a set of translation rules are developed to provide a formal connection between the combined specification and the target Java application. The translation rules can be recursively used to generate a concurrent Java application from a B+CSP model.

To define the translation rules *Tr*, we first use the BNF (Backus Naur Form) notation to define the subset of B and CSP specification that can be implemented in the Java/JCSProB programs. Then the allowed target Java/JCSProB language structures are also defined in a recursive notations. With the help from a set of interpretative assumptions A, the translation rule *Tr* relates the definitions of B+CSP and Java/JCSProB:

$$B + CSP \stackrel{A}{\Longrightarrow}_{Tr} Java/JCSProB$$

The assumptions are introduced to express the B+CSP semantic features which are not obvious from the BNF-style definition. For example, in *external choice*, the channel of all possible paths can be obtained by continuously deducing the B+CSP language rules, but it is not very convenient to express them explicitly in the translation rule. Therefore, we introduce an assumption, which clearly say that $a_0,...,a_n$ are the first channels on all the paths.

The translation rules can be classified into three parts:

- Rules for generating process classes
- Rules for generating channel classes
- Rules for generating B variable classes, invariants and assertions

In Table 3, the translation rules concerned the translation of *external choice* are listed, as well as the B+CSP syntactic structures involved. The items with angle brackets in B and CSP language specification, e.g. $\langle$BOp_PreCondition$\rangle$, are expandable B+CSP language syntax, and the items with fat font in translation rules, e.g. Ch_PreC_Check, are names of expandable translation rules.

Table 3. Translation rules concerning external choice: *ProcE:Ext_Choice*

CSP spec involved	$\langle$ProcE$\rangle \Longrightarrow$ $\qquad\langle$ProcE:Ext_Choice$\rangle \mid \langle$ProcE:Parallel$\rangle \mid ....$ $\langle$ProcE:Ext_Choice$\rangle \Longrightarrow$ $\qquad\langle$ProcE$\rangle$ [] $\langle$ProcE$\rangle$
B spec involved	$\langle$BOp_Substitution$\rangle \Longrightarrow$ $\qquad\langle$BOp_PreCondition$\rangle \mid \langle$BOp_Begin$\rangle \mid ....$ $\langle$BOp_PreCondition$\rangle \Longrightarrow$ $\qquad$**PRE** $\langle$B_Condition$\rangle Con$ **THEN** $\langle$BOp_Substitution1$\rangle$ **END**
Additional assumptons	A1: $P_0,....,P_N$ are all the paths for external choice A2: $a_0,....,a_n$ are the first channels on all the external choice paths $P_0,....,P_N$.

Rule name	ProcE : Ext_Choice
Rule function	Implements: $\langle$ProcE:Ext_Choice$\rangle$ Uses: $\langle$ProcE$\rangle$, $\langle$BOp_Substitution$\rangle$, A1, A2

Rule content	`PCChannel[] in = {`Ch_Name_List$\lceil a_0, ..., a_n \rfloor$`};`
	`Vector⟨Vector⟩ inputVec = new Vector⟨Vector⟩=();`
	Ec_Add_Args$\lceil a_0, ..., a_n \rfloor$
	`Alter alt = new Alter(in,choiceVec);`
	`switch (alt.select()){`
	$\qquad$Ec_Choice_Paths$\lceil P_0...P_n \rfloor$
	`}`

Rule name	Ch_PreC_Check
Rule function	Implements: $\langle$BOp_PreCondition$\rangle$

Rule content	`public synchronized boolean preCondition(){`
	$\qquad$`return` B_Conditions$\lceil Con \rfloor$`;`
	`}`

The translation rule ProcE : Ext_Choice is in the rule set ProcE, which handles all the CSP process structures. The *rule function* indicates the B+CSP syntactical structures or assumptions that the rule implements, and uses to obtain information. The *rule content* shows the Java code that the rule generates. A very abstract lift specification in Figure 5 has an *external choice* with two paths. Note that there is a deliberate bug in the definition of the B machine in Figure 1.

The Java code in Figure 6 demonstrates how the *external choice* in the *MAIN* process is implemented in Java. Inside the rule ProcE : Ext_Choice a channel array which includes the first channel objects on all choice paths, is initialized first. The channel name list *inc_ch,dec_ch* is generated by rule Ch_Name_List. The Java *Vector choiceVec* stores all the argument values of the first channels of all the choices. The translation rule Ec_Add_Args generates the Java code to add arguments on channel $a_0, ..., a_n$ to *choiceVec*. As the two chan-

```
MACHINE lift
VARIABLES level
INVARIANT level : NAT & level ≥ 0 & level ≤ 10
INITIALIZATION level := 1
OPERATIONS
        inc = BEGIN level := level + 1 END;
        dec = BEGIN level := level - 1 END
END
```

$$MAIN = inc \longrightarrow inv\_check \longrightarrow MAIN \; [] \; dec \longrightarrow inv\_check \longrightarrow MAIN \;;;$$

Figure 5. Combined Specification of lift

```
PCChannel[] in = {inc_ch,dec_ch};
Vector<Vector> choiceVec = new Vector<Vector>();
{Vector<Object> inputVec = new Vector<Object>();
choiceVec.addElement(inputVec); }
{Vector<Object> inputVec = new Vector<Object>();
choiceVec.addElement(inputVec); }
Alter alt = new Alter(in,choiceVec);
switch(alt.select()){
    case 0 : { {inc_ch.ready(); }
            var.check(); }
    case 1 : { { dec_ch.ready(); }
            var.check(); }
}
```

Figure 6. Java code implementing *external choice* in the lift process class

nels *inc* and *dec* in the example have no argument, they just pass two null *Vector* objects to *choiceVec*.

The channel array *in*, as well as an arguments array from *choiceVec* are used to construct the *Alter* class. The *select()* method of *Alter* class chooses between the ready channel objects. Whether a channel is ready may further depend on the precondition of the B operation and the synchronization ready state, all of which depends on the argument values on that channel. Rule Ec_Choice_Paths generates all the possible paths inside the choice structure. In the generated Java program, the two possible paths are represented by the two cases in the Java *switch* structure.

The implementation of external choice may also depend on the semantics implemented in other Java classes, for example, the pre-condition check. Although the precondition check mechanism is provided by the *preConditionCheck()* method of *PCChannel* class, the actual conditions are defined in the subclass of *PCChannel* through translation rules. In Table 3, translation rule Ch_PreC_Check is used to generate the *preCondition()* method which implement the precondition.

3.2. Translation Tool

The automatic translation tool is constructed as part of the ProB tool. Our translation tool is also developed in *SICStus* Prolog, which is the implementation language for ProB. In ProB, the B+CSP specification is parsed and interpreted into Prolog terms, which express the operational semantics of the combined specification. The translation tool works in the same environment as ProB, acquires information on the combined specification from the Prolog terms, and translates the information into the Java program.

4. Examples and Experiments

In this section, experimental evaluation of the implementation strategy is discussed. We first test the usability and syntax coverage of the translation tool by using different syntactic structures to construct various formal models. Then the models are put into the translation tool. The target Java programs from different models are tested.

How the behavioural properties of the formal models are implemented in the Java programs is the other experimental target. Generally, there are two kinds of properties:

- **Known properties**. These properties, e.g. safety and deadlock, can be checked in the ProB tool for the system model. The test is whether verified properties are also preserved in the Java implementation. This provides partial evidence for the correctness of the Java implementation strategy.
- **Unknown properties**. For other properties, e.g. fairness, which cannot be verified in ProB, we provide alternative experimental means to evaluate them in the Java programs at runtime. In these circumstances, the generated Java program runs simulator for the B+CSP specification. It generates traces, and experimentally demonstrates the properties on the traces.

4.1. Invariant Check: Simple Lift Example

Figure 5 specifies an abstract lift model. We use this simple example to demonstrate the implementation of invariant check, which demonstrates the safety properties.

Invariants in a B machine demonstrate safety properties of the system model. In the ProB model checking, the B invariants are checked on all the states of the state model. The violation of the invariants indicate an unsafe state of the system model. Implementing the invariant check in the target Java programs can provide a practical correctness demonstration for the translation strategy on safety properties.

The Java implementation of our approach supports invariants checking at runtime. The invariants supported by the translation are mainly from the B0 language conditions. The subclass of *JcspVar* class needs to implement an abstract *check* to support the checking. There are two ways to process the invariant check in the translation and implementation. The first one uses the same semantics of invariant checking as B+CSP. It forces the *check()* method to be called in every channel object after it finish its data transition. That means the invariant checking is processed at all the states of the system. However, this may seriously degrade performance in some Java applications. An alternative lightweight solution requires users to indicate the invariant check explicitly at some specific positions; the *lift* specification falls into this class. The CSP-only channel *inv_check* is used to indicate a runtime invariant check. As it has no B counterpart, it has no effect on system state. When handling this channel, the translator generates the Java code which calls the *check()* method from the subclass of *JcspVar* class. However, the alternative solution cannot guarantee all the violations of invariants being found, or discovered promptly. With a weak check, the system can run all the way through without noticing the existing violated state.

The unguarded B operations *inc* and *dec* can freely increase or decrease the B variable *floor*. That would easily break the invariant on *floor* (*level* $\geq$ *0 & level* $\leq$ *10*). In the ProB model checking, the violated state can be quickly identified from the state model.

Runtime results of the target Java application demonstrate that the check mechanism can find violation of invariant conditions, and terminate the system accordingly. Therefore, we correct the model to that of Figure 1, by adding preconditions.

The Java programs generated from the modified specification find no violation of invariants.

4.2. Bounded Fairness Assertions

ProB also provides a mechanism to detect deadlock in the state space. When the system reaches a state where no further operation can progress, it is deadlocked. Stronger liveness properties, such as livelock-freeness and reachability, are difficult to detect in model checking, and are not supported by ProB. Fairness, which involves temporal logic, is an even more complex property for model checking. Many approaches [25,26,27] have been attempted for extending model checking of B or CSP specifications to temporal logics. However, none of these approaches can be directly supported in the B+CSP specification.

The bounded fairness assertions check is used informally to address some limited fairness properties on bounded scales. In the specification, a sequence *record* is specified. A special combined channel *rec_proc* is built to add runtime history to the *record* sequence. A CSP process can call the *rec_proc* channel with a specific ID number to record its execution. The fairness assertions are specified on a limited size of the *record* sequence.

The assertions check here is only used in Java, not ProB. Such properties cannot be model-checked in ProB because of state explosion; even an assertion with a very short window on the *record* sequence could easily explode the state space.

The translation tool and the target Java application support three kinds of bounded fairness assertions. For example:

> ***Frequency Assertion:***
> $!(i).(i \in ProcID\ \&\ card(record) > 24 \Rightarrow$
> $\qquad card(card(record)\text{-}24..card(record) \lhd record \rhd \{i\}) > 2)$
> ***Duration Assertion***
> $!(j).(j \in ProcID\ \&\ card(record) > 12 \Rightarrow$
> $\qquad j \in ran(card(record)\text{-}12..card(record) \lhd record))$
> ***Alteration Assertion***
> $card(record) > 3 \Rightarrow$
> $\qquad record(card(record)) \notin ran(card(record)\text{-}3..card(record)\text{-}1 \lhd record)$

Figure 7. Bounded fairness assertions in JCSProB

The symbol !(i) here means "for all *i*", *card()* is a cardinality operator, and *ran()* returns the range of a function. The symbol $\lhd$ represents domain restriction, while the symbol $\rhd$ represents range restriction. In the example assertions, six processes are monitored. The frequency assertions try to make sure that for *n* (= 6) processes, in the last *4n* record steps, each concerned process should progress more than twice. The duration assertions check the last *2n* steps to make sure each concerned process should progress at least once. The alternation assertion check that the last progressed process does not occurs in the last three steps before that.

As our translation targets the concrete and deterministic subset of the combined specification, generally, we only support the B0 subset of B language. Many predicates and expressions in B-method are too abstract to be implemented in Java. Our bounded fairness assertions, which are defined with syntax beyond B0, are restricted to very limited formats.

In the Java application, the B sequence *record* is implemented in an array of *jcspRecord* objects. When the Java application terminates, the runtime trace is automatically saved in a log file for further investigation.

4.3. Fairness: Wot-no-chickens

The *Wot,no chickens?* example [28] was originally constructed for emphasizing possible fairness issues in the wait-notify mechanism of Java concurrent programming. There are five philosophers and one chef in this story. The chef repeatedly cooks four chickens each time, puts the chicken in a canteen, and notifies the waiting philosophers. On the other hand, philosophers, but not the greedy one, recursively continue the following behaviours: think, go to canteen for chickens, get a chicken, and go back to think again. The greedy philosopher doesn't think, and goes to the canteen directly and finds it devoid of chickens. The Java implementation in [28] employs the Java *wait-notify* mechanism to block the philosopher object when there are no chickens left in the canteen. The chef claims the canteen monitor lock (on which the greedy philosopher is waiting), takes some time to set out the freshly cooked chickens and, then, notifies all (any) who are waiting. During this claim period, the diligent philosophers finish their thoughts, try to claim the monitor lock and get in line. If that happens before the greedy philospher is notified, he finds himself behind all his colleagues again. By the time he claims the monitor (i.e. reaches the canteen), the shelves are bare and back he goes to waiting! The greedy philosopher never gets any chicken.

4.3.1. Two Formal Models

To test the syntax coverage of the JCSProB package and the translation, several formal models of this example are specified. We use various synchronization strategies and recursion patterns to explore the syntax coverage of the B+CSP specification in the JCSProB package, as well as in the translation tool. Furthermore, we also want to compare fairness properties of different formal models, in order to evaluate the behaviour of the generated Java programs in practice.

```
MACHINE chicken
VARIABLES
        canteen,record, ...
INVARIANT
        canteen: NAT & record: seq(NAT) ......
INITIALISATION
        canteen := 0 || record := <> ...
OPERATIONS
        ......
        getchicken(pp) =
                PRE pp:0..4 & canteen > 0 THEN
                canteen := canteen - 1 || ...
                END;
        ......
        put =
                BEGIN canteen := canteen + 4 || ... END
END
        ————————
MAIN = Chef ||| XPhil ||| PHILS ;;
PHILS = |||X:0,1,2,3@Phil(X);;
Phil(X) = thinking.X → waits.200 → getchicken.X → rec_proc.X →
        backtoseat.X → eat.X → Phil(X);;
XPhil = getchicken.4 → rec_proc.4 → backtoseat.4 → eat.4 → XPhil;;
Chef = cook → waits.200 → put → rec_proc.5 → Chef ;;
```

Figure 8. Formal specification of Wot-no-chicken example, Model 1

The first combined B+CSP model of this example is presented in Figure 8. The CSP part of the specification in Figure 8 only features some interleaving processes. However, the atomic access control on the B global variables, and the precondition on the *get_chicken* channel actually require synchronization mechanisms to preserve the consistency of the concurrent Java programs. As all the features concerning the concurrency model are implemented in the JCSProB package, users can work with the high-level concurrency model without noticing the low-level implementation of synchronization.

An alternative model is specified in Figure 9. As the B machine is very similar to the first one in Figure 8, only the CSP specification is given here. This model explicitly uses a multi-way synchronization on the *put* channel to force all the philosophers and the chef to synchronize.

$$
\begin{aligned}
&MAIN = Chef \, [|\{put\}|] \, XPhil \, [|\{put\}|] \, PHILS \,;; \\
&PHILS = [|\{put\}|]X:\{0,1,2,3\} @ Phil(X) \,;; \\
&Phil(X) = thinking.X \rightarrow waits.200 \rightarrow PhilA(X) \,;; \\
&XPhil = PhilA(4) \,;; \\
&PhilA(X) = put \rightarrow PhilB(X) \,;; \\
&PhilB(X) = waits.100 \rightarrow PhilA(X) \, [] \, getchicken.X \rightarrow rec\_proc.X \rightarrow \\
&\qquad\qquad \mathbf{if}(X == 4) \\
&\qquad\qquad \mathbf{then} \, XPhil \\
&\qquad\qquad \mathbf{else} \, Phil(X) \\
&\qquad\qquad ;; \\
&Chef = waits.300 \rightarrow cook \rightarrow waits.200 \rightarrow put \rightarrow Chef \,;;
\end{aligned}
$$

Figure 9. Formal specification of Wot-no-chicken example, Model 2

4.3.2. Experiments and Results

The experimental evaluation test is based on the two models specified above. In the first part of the evaluation, we test the safety and deadlock-freeness properties on the two channels. In Table 4, the test results on these properties are demonstrated. The Timing column indicates how many different timing configurations are tested with the model, and the Steps column shows the lengths of the runtime records we concerned. As the concurrent Java applications constructed with the JCSProB package preserve the same safety and deadlock-freeness properties as their formal models, it partially demonstrates the correctness of the JCSProB package, as well as the translation tool.

Table 4. The experimental result: Safety and Deadlock-freeness

Model Name	Property	Processes	Timing	Steps	Result
Model 1	Safety/Invariant	-	15	1000	√
Model 1	Deadlock-freeness	-	15	1000	√
Model 2	Safety/Invariant	-	15	1000	√
Model 2	Deadlock-freeness	-	15	1000	√

To test the bounded fairness properties on the target Java programs at runtime, we first need to generate various traces from the concurrent Java programs. Currently, we use the *waits* channel in the CSP part specification to define various timing configurations for generating traces for the target Java programs. The *waits* channel forces the calling process to sleep for a fixed time period. In this way, we can explicitly animate formal models with specific timing settings for experimental purposes. Then we employ the bounded fairness assertions check on Java programs embedded with timing settings. The target of this experiment is to practically animate the Java/JCSProB applications, and evaluate their runtime performances with the bounded fairness properties.

In Table 5, we show the experimental results of the two models with bounded fairness properties. For each property, we use five different timing settings; and for each timing setting, the Java program is tested in five runs. In the result column of the table, *18P7F* means in 25 runs, the check passes 18 times and fails 7 times.

Table 5. The experimental result: Bounded Fairness Properties

Model Name	Property	Processes	Timing	Steps	Result
Model 1	Frequency 1	All	5	150	4P21F
Model 1'	Frequency 2	Phils+XPhil	5	150	1P24F
Model 1"	Frequency 3	Phils	5	150	23P2F
Model 1	Duration 1	All	5	300	20P5F
Model 1'	Duration 2	Phils+XPhil	5	300	18P7F
Model 1"	Duration 3	Phils	5	300	25P0F
Model 2	Frequency 1	All	5	150	5P20F
Model 2'	Frequency 2	Phils+XPhil	5	150	0P25F
Model 2"	Frequency 3	Phils	5	150	24P1F
Model 2	Duration 1	All	5	300	5P20F
Model 2'	Duration 2	Phils+XPhil	5	300	5P20F
Model 2"	Duration 3	Phils	5	300	25P0F

In Section 4.2, three kinds of bounded fairness assertions were introduced. In the testing, frequency and duration assertions on the formal models are checked at runtime. The assertions check also concerns different process groups. In the tests on *Model 1* and *Model 2*, both the philosophers and the chef processes are recorded for assertions check. In model *Model 1'* and *Model 2'*, only the philosopher processes are run. In *Model 1"* and *Model 2"*, the greedy philosopher is removed and only normal philosopher processes are tested.

A number of points are summarized from the testing results:

- The unnecessary group synchronization in *Model 2* brings particular fairness problems to the system. The fairness properties in this model heavily depend on the timing setting. For example, all five passes for the frequency check on *Model 2* are from the same timing configuration, while the other 20 check runs on the other four different timing configurations all failed. It is mainly caused by the *wait* channel in the *PhilB(X)* process. As the greedy philosopher does not wait as other philosophers in *Phil(X)*, it enters *PhilB(X)* first and may find there is chickens there. A specific timing setting may make the greedy one waiting in *PhilB(X)*, while other philosophers take all the chickens in this time gap. In this way, we can even starve the greedy philosopher for a period of time.
- In *Model 1*, as long as the chef does not run too much faster than normal philosophers, different timings won't make the results very irregular.
- The duration assertion check also demonstrates that *Model 2* has a more serious fairness problem than *Model 1*, even with a very short trace.
- As we expect, *Model 1"* and *Model 2"*, which have no greedy philosopher, demonstrate better fairness properties than the other models.
- Further analysis of the experimental results shows that the number of channels in a process is the main factor which affects the progress of processes. For example, if we remove all the timing configurations in *Model 1*, the chef process, which has fewer combined channels than the philosopher processes, runs much faster than the five philosopher processes in the first model. The *backtoseat* and *eat* channel classes, which actually have just very simple data transitions inside the channel, result in differences in the performance. The chef keeps on producing far more chickens than the five philosophers can actually consume.

A generated Java program provides a useful simulation for its formal model. It is used to explore and discover the behaviour properties which cannot be verified in ProB model checking.

5. Conclusion and Future Work

Our implementation strategy is strongly related to a similar approach in the *Circus* development. In [29], a set of translation rules is developed to formally define the translation from a subset of the *Circus* language to Java programs that use JCSP. As the JCSP package only supports point-to-point communication, and does not allow state change inside the channel, the supported *Circus* language subset in the translation is very limited. In [30], an ongoing effort develops an extended channel class to support multi-way synchronization. Moreover, an automatic translation tool and a brief GUI program are constructed using these translation rules. CSP/*occam* is used to model multi-way synchronization, and then JCSP to implement that model.

The JCSP package (1.0-rc5) does not provide support of external choice with multi-way synchronization or output guard. As an alternative approach, we have implemented multi-way synchronization for external choice for the JCSProB channel class. As the implementation is still under test, we will report it in the future. Our plan was always to re-implement the JCSProB package with the facilities from the new JCSP package (1.0-rc7). New JCSP features, such as *AltingBarrier* and rendezvous, can be used directly to construct the implementation classes of the combined B+CSP channel. The current JCSProB implementation of combined channels has a *run()* method inside the channel class. The data transitions on system states are inside the method. That is actually very similar to a JCSP process class. The JCSProB based on JCSP 1.0-rc7 or later would see that the combined channels are implemented as special JCSP processes. They communicate with JCSP process objects, which implement CSP processes, through JCSP channels. The synchronization on the combined channel would be resolved using the *AltingBarrier* class from the new JCSP library. The data transitions would be put in the *run()* method of the process. However, although many JCSP channel classes have been formally proved in [8], the correctness proof of the *AltingBarrier* class has still to be completed. Therefore, we regard the re-implementation of JCSProB with new JCSP package as a future work.

Since the current JCSProB package implements and hides the B+CSP semantics and concurrency model inside the package, the Java application generated by the translation is clear and well structured. The disadvantage is that the implementation of the B+CSP semantics and concurrency inside JCSProB still requires a formal proof of correctness of the translation. The current JCSProB is hard to prove because it is hard to build a formal model for it. The new JCSProB channel implementation will be based on JCSP, and many JCSP channels have already been formally proved. Thus we expect that it will be modelled in CSP/*occam* and proved by FDR as before.

The other issue in the JCSProB implementation is recursion. Classical *occam* does not support recursion[2] and a `WHILE`-loop must be used for simple *tail* recursion. However, in CSP, it is very common to see a process calling other processes or itself to perform linear or non-linear recursions. In JCSP, we can employ a Java `while`-loop for any *tail* recursion in the CSP. Continually constructing and running a new process object from within the existing one to implement a never unwinding recursion must eventually cause a Java `StackOverflow` exception. To support the CSP-style recursion used in B+CSP, we implemented the existing *CSProcess* interface with a new process class. As for the multi-way synchronization classes, this recursion facility is not ready to be reported in this paper.

Considering the results of the experiments, we find that atomic access to the objects of *JcspVar* class is the most significant problem affecting the performance of the Java implementation. The exclusive lock in the subclass of *JcspVar* provides safety and consistency. We explicitly defined it because it is not only used for accessing the data, but also for our

---

[2] *occam-π* does support recursion.

implementation of multi-way synchronization on external choice. However, it heavily effects the performance. Applying advanced read-write techniques to replace the exclusive lock on a variable's access control may improve the concurrency performance of the Java implementation. The pragmatic solution to this problem is to provide guidance to the specifier as to how B variables may be interpreted as local CSP process variables, thus not requiring locking. For example, a member of an array of CSP processes $ProcX(i)$ might call B operation $Op(i)$, allowing B to index an array of variables, one per process. This reduces the number of global variables and thus locking load. Furthermore, future work is to implement the B+CSP channel with the new JCSP package. That means the access of the data variables and the implementation of multi-way synchronization would be separated. In this case, we could simplify the lock implementation to reduce the performance overload.

There are further outstanding issues to be resolved. We are aware that the special channels (*rec_proc, wait, inv_check*) in the invariants and assertion checking are not the best way to animate the generated Java programs and generate test cases from them. Although the channels do not affect the state of the system, this solution mixes implementation detail with the formal specification. Three solutions are under consideration:

- Configuration File. A configuration file, along with the B+CSP specification, would be used to generate the Java programs. The setting in the configuration file would guide the target Java programs to produce specific or random timing delays on the selected channels, and output system state at runtime. This can be seen as a form of specialization of the model mapping that the translation represents.
- User Interaction. A GUI interface for the target Java programs would allow users to manually manipulate the programs at runtime, producing different traces on each run.
- Traces from ProB. As an animator and model checker for the B+CSP specification, ProB can provide traces satisfying certain properties in a specific format. Using these ProB traces to guide the execution of target Java programs would be very useful.

Scalability is another significant issue. The JCSProB package, as well as the translation, should be applied to bigger case studies to evaluate and improve its flexibility and scalability. Currently, only one B+CSP specification pair is allowed in ProB. A proven refinement strategy for producing a concrete B0+CSP implementation from an abstract specification, as well as a technique for composing B+CSP specification pairs, are still unavailable. Therefore, the JCSProB application is now restricted on a single machine. An abstract B+CSP specification cannot currently be refined and decomposed into a distributed system. In [27], an approach for composing combined B and CSP specification CSP‖B is presented. Whether a similar technique is applicable for B+CSP in ProB remains to be seen.

References

[1] C.A.R Hoare, *Communicating Sequential Processes*, Prentice Hall International, 1985.

[2] R. Milner, *A Calculus of Communicating Systems*, Springer Verlag, 1980.

[3] G. Guiho and C. Hennebert, "SACEM software validation", In *Twelfth International Conference on Software Engineering*, 1990.

[4] G. T. Leavens and E. Poll and C. Clifton and Y. Cheon and C. Ruby and D. Cok and P. Müller and J. Kiniry, *JML Reference Manual*, 2005.

[5] M. Brörken and M. Möller, "Jassda Trance Assertions: Runtime Checking the Dynamic of Java Programs", In *International Conference on Testing of Communicating Systems*, 2002.

[6] J.Magee and J. Kramer, *Concurrency: State Models & Java Programs*, John Wiley and Sons, 1999.

[7] P. H. Welch and J. M. Martin, "A CSP Model for Java Multithreading", In *ICSE 2000*, pages 114-122, 2000.

[8] P.H. Welch and J.M. Martin, "Formal Analysis of Concurrent Java System" In *Communicating Process Architectures 2000*, 2000.

[9] V. Raju and L. Rong, and G. S. Stiles, "Automatic Conversion of CSP to CTJ, JCSP, and CCSP", In *Communicating Process Architectures 2003*, pages 63-81, 2003.

[10] C. Fischer, "CSP-OZ: A combination of Object-Z and CSP", Technical report, Fachbereich Informatik, University of Oldenburg, 1997.

[11] M.J. Butler, "csp2B: A practical approach to combining CSP and B", In *World Congress on Formal Methods*, pages 490-508, Springer, 1999.

[12] H. Treharne and S. Schneider, "Using a Process Algebra to Control B Operations", In *IFM 1999*, pages 437-456, 1999.

[13] J. C. P. Woodcock and A. L. C. Cavalcanti, "A concurrent language for refinement", In *IWFM01: 5th Irish Workshop in Formal Methods, BCS Electronic Workshops in Computing*, 2001.

[14] S.A. Schneider and H.E. Treharne and N. Evans, "Chunks: Component Verification in CSP∥B", In *IFM 2005*, Springer, 2005.

[15] C. Fischer and H. Wehrheim, "Model-checking CSP-OZ specifications with FDR", In *IFM 1999*, pages 315-34, Springer-Verlag, 1999.

[16] M. Brörken and M. Möller, "Jassda Trance Assertions: Runtime Checking the Dynamic of Java Programs", In *International Conference on Testing of Communicating Systems*, 2002.

[17] B. Meyer, "Applying 'design by contract'", In *Computer*, volume 25, pages 40-51, 1992.

[18] M. Leuschel and M.R. Butler, "ProB: A model checker for B", In *FME 2003*, LNCS 2805, pages 855-874, Springer-Verlag, 2003.

[19] M. J. Butler and M. Leuschel, "Combining CSP and B for Specification and Property Verification", *FM 2005*: 221-236, Springer, 2005

[20] J.-R. Abrial, *The B-Book: Assigning Programs toMeanings*, Cambridge University Press, 1996.

[21] C.C. Morgan, "Of wp and CSP", In *Beauty is our business: a birthday salute to Edsger W. Dijkstra*, SpringerCVerlag, 1990.

[22] P.H. Welch, F.R.M. Barnes, and F.A.C. Polack, "Communicating Complex Systems", In *ICECCS 2006*, IEEE, 2006

[23] P.H. Welch, "A Fast Resolution of Choice Between Multiway Synchronisations", CPA-2006, IOS Press, IBSN 1-58603-671-8, 2006.

[24] P.H. Welch and Neil Brown and James Moores and Kevin Chalmers and Bernhard Sputh. "Integrating and Extending JCSP", CPA-2007, IOS Press, 2007.

[25] M. R. Hansen and E.R. Olderog and M. Schenke and M. Fränzle and B. von Karger and M. Müller-Olm and H. Rischel, "A Duration Calculus semantics for real-time reactive systems", Technical Report, Germany, 1993.

[26] M. Leuschel and T. Massart and A. Currie, "How to make FDR spin LTL model checking of CSP by refinement", In *FME'01: Proceedings of the International Symposium of Formal Methods Europe on Formal Methods for Increasing Software Productivity*, pages 99-118, Springer-Verlag, 2001.

[27] H. Treharne and S. Schneider, "Capturing timing requirements formally in AMN", Technical report, Royal Holloway, Department of Computer Science, University of London, Egham, Surrey, 1999.

[28] P.H. Welch, "Java Threads in the Light of occam/CSP", In Architectures, Laugages and Patterns for Parallel and Distributed Applications 1998, pages 259-284, IOS Press, 1998.

[29] M. Oliveira and A. Cavalcanti, "From *Circus* to JCSP", In *ICFEM 2004*, pages 320-340, 2004.

[30] A. Freitas and A. Cavalcanti, "Automatic Translation from *Circus* to Java", In *FM 2006*, pages 115-130, Springer, 2006.

Communicating Process Architectures 2007
Alistair A. McEwan, Steve Schneider, Wilson Ifill, and Peter Welch
IOS Press, 2007

Components with Symbolic Transition Systems: a Java Implementation of Rendezvous

Fabricio FERNANDES, Robin PASSAMA and Jean-Claude ROYER

OBASCO Group, École des Mines de Nantes – INRIA, LINA
4 rue Alfred Kastler, 44307 Nantes cedex 3, France.

{Fabricio.Fernandes, Robin.Passama, Jean-Claude.Royer} @emn.fr

Abstract. Component-based software engineering is becoming an important approach for system development. A crucial issue is to fill the gap between high-level models, needed for design and verification, and implementation. This paper introduces first a component model with explicit protocols based on symbolic transition systems. It then presents a Java implementation for it that relies on a rendezvous mechanism to synchronize events between component protocols. This paper shows how to get a correct implementation of a complex rendezvous in presence of full data types, guarded transitions and, possibly, guarded receipts.

Keywords. Component-Based Software Engineering, Behavioural Interfaces, Explicit Protocols, Symbolic Transition Systems, Rendezvous, Synchronization Barriers

Introduction

Component-Based Software Engineering (CBSE) is becoming an important approach for system development. As large distributed systems become always more critical, the use of formal analysis methods to analyze component interactions arises as a crucial need. To this end, explicit protocols have been integrated to component interfaces to describe their behaviour in a formal way. Behavioural interface description languages are needed in component models to address architectural analysis and verification issues (such as checking component behavioural compatibility, finding architectural deadlocks or building adapters to compensate incompatible component interfaces) and also to relate efficiently design and implementation models. Nevertheless, explicit protocols are often dissociated from component codes: they are "pure" abstractions of the way components behave. This is really problematic, since nothing ensures component execution will respect protocols rules. So, a critical issue is to fill the gap between high-level formal models and implementation of protocols to ensure consistency between analysis and execution phases.

In this field, our long term-goal is to define a component programming language with explicit executable protocols, coupled with a formal ADL (Architectural Description Language) and associated analysis tools. To make a strong link between specification or design models and programming languages for implementation, there are two possible ways: (i) automated translation of models into programming code, and (ii) extraction of abstract model and protocol information from programming code. We focus on the first approach. The features of the target language are object-orientation, multi-threading and facilities for synchronization.

---

[1]**Acknowledgment.** This work was partly supported by the AMPLE project www.ample-project.net, and the CAPES grant from Brazil.

As an instance, we consider Java 1.5. The second way, from code to model, is a bit different both on concepts and tools; see for example [1,2]. Our development process is decomposed into two steps: the first one is the description of components and architectures with our ADL formalism, and the second one is to represent, in Java, the state machine, the synchronization and to implement the data type part with Java classes. In the realization of this process, our current objective is to provide support for implementing component protocols in such a way that their execution respects the semantics of the protocol description language.

The chosen protocol description language is the Symbolic Transition System (STS) formalism [3]. STSs are finite state and transition machines with unrestricted data types and guards. The STS formalism is a general model of computation which may be seen as a strict subset of UML statecharts or as a graphical model of a process algebra with value passing and guards. It is adequate for formal design and verification, although this latter is still a difficult challenge. Various ways to verify these systems already exist: using a general prover, calculating abstractions or interfacing with classical model-checkers (the interested reader may look at [4,5,6,7]). Our approach for the verification of these systems relies on the interaction with efficient model-checkers and on the use of specific techniques for symbolic systems. For several examples, the use of boundedness and decomposition techniques we developed are described in [8]. The STS formalism has many advantages for design: it improves readability and abstraction of behavioural descriptions compared to formalisms with restricted data types. It helps to control state explosion with the use of guards and typed parameters associated to the transitions. Lastly, it allows the description of message exchange with asynchronous or synchronous communication mode.

Implementing STS requires to manage different development steps: (i) implementing the data part (ii) representing the protocol (iii) gluing the data part and the protocol into a primitive component (intra-component composition), and iv) implementing components synchronization and communication mechanism (inter-component composition). The three first steps may be viewed from either a code generation or a code reutilization perspective. On the one hand, code generation from formal specification [9,10] is a problem related to compilation, but with a greater abstract gap between the source and the target language than for general purpose programming languages. On the other hand, code reutilization may be done with a more or less intrusive approach, a related reference is [11]. Whatever the way STSs are created, the central need is a generic mechanism to execute and synchronize the activities of STSs and to make them communicate.

The main proposal of this paper is to present this mechanism; we avoid here the discussion related to data type representation. Such a mechanism is important to get components that can be composed in a safe manner – with a direct link to the formal semantics level. More precisely, we focus on synchronous communication mode (see [12,11] for a related work on asynchronous communications). For the time being, we only consider one-to-many, one-way synchronous communications. As quoted in [13], during a synchronous communication the sender, to resume its own execution, waits for (i) completion by the receiver of the invoked method execution and then (ii) the return statement of the replier. It is opposed to asynchronous communication in the sense that the sender does not resume its execution as soon as the message has been sent. STS synchronous communication is a bit more sophisticated: a message transmission is bound to the service execution both on sender and receiver side. Semantics models of STS, as for process algebras or finite state machines use an advanced *rendezvous* that strongly glues several participants of a communication executing their guarded actions simultaneously. This can be seen as a generalization of synchronous communication modes of object-oriented programming languages with concurrent features. Coupled to guards on receipts, this allows complex interactions between components to be described. However, previous protocol implementations, for instance [14,15,16], only propose variants such as Remote Procedure Calls (RPCs) or synchronous message sending.

The STS formalism supports this composition semantics based on concurrency and event synchronizations [8]. The rendezvous is a synchronization point between several actions which may involve communications. The proposed mechanism implements a n-ary rendezvous, with receipt on guarded events and allows independent synchronizations to processes in the same time. We restrict communication to one sender of a value and several receivers, which provides a basic and powerful mechanism. Such a rendezvous requires two synchronization barriers, one for entering and one for leaving the rendezvous. The synchronization barrier principle for multiple threads is the following: all threads must arrive at the barrier before any of them are permitted to proceed past the barrier.

We show here how to get a correct synchronization mechanism build on top of a synchronization barrier, with respect to STS synchronization specificities. Our approach is achieved with four progressive steps: (i) we start with a simple rendezvous for Labelled Transition Systems (LTS) and a central arbiter, (ii) we then split the arbiter into several lock objects associated to the synchronizations, (iii) we improve the solution allowing independent synchronizations to enter simultaneously the barriers, and (iv) we add full data types, communications and guards.

The paper is organized as follows. Section 1 reviews related work. Section 2 presents the main features of our component model and an example of a system design. Section 3 introduces our hypotheses for the model implementation in Java. Section 4 describes the synchronization barrier principles and discusses how to implement communications and guards in the rendezvous. Finally, Section 5 draws concluding remarks and discusses future work.

1. Related Work

In the last decade, formal component models with behavioural descriptions have been proposed either on their own [17,18] or in the context of software architectures [19]. Different behavioural models have been used, such as process algebras [20,19] or automata-based formalisms [1,21]. However, if they propose different analysis mechanisms for component architectures, they do not address the issue of taking protocols into account within the implementation, which is a mandatory issue for seamless CBSE development. Discussions in this section focus on approaches with a strong coupling between message sending and service execution. Thus we do not discuss purely asynchronous approaches or synchronization by need (readers can refer to [13] for more details). Discussions are also directed towards approaches that propose a direct link between formal models and code.

The STS formalism [22,3] has initially been developed as a way to control the state and transition explosion problem in value-passing process algebras using substitutions associated to states and symbolic values in transition labels. The STS formalism we use, [8], is a generalization of this latter, associating a symbolic state and transition system with a data type description. The data type description is given using algebraic specifications [23,24]. The STS semantics provides concurrent composition of STSs with event synchronizations, namely the rendezvous notion introduced by CSP [25]. In a previous work, we extended the synchronous product of LTSs to STSs. The principles and the precise definition may be found in Section 2 and [23,8]. This formal basis for STS composition is helpful to implement a correct synchronization mechanism for STSs. We have previously done some experiments on translating formal behavioural specifications, namely LOTOS, in Java [9]. The code generation was based on a Java dialect providing condition activations and state notifications. It proposes a central monitoring mechanism with controllers for each node in the tree structure of the specification. In [11], we presented how to compound components with STS protocols, thanks to asynchronous communications links. The asynchronous communications are implemented with channels. Our current work extends these first proposals with a more precise proposition to glue the STS protocol and the data type part and introduces the possibility for complex synchronization mechanisms between components with STSs.

In the concurrent object-oriented community, the use of explicit behavioural protocols at the language level is not new. PROCOL [14], SOFA [15] and Cooperative Objects [16] are three representative proposals. To describe protocols, PROCOL and SOFA employ regular expressions denoting traces – *i.e.* sequences of events (required, provided, and internal calls). Cooperative Objects employs Object Petri-Net-like notations for the same reason. Both formalisms are less readable and user-friendly than STSs. PROCOL and Cooperative Objects protocols consider data types and guards. SOFA and Cooperative Objects synchronous communications can be reduced to simple 1-1 RPC calls. PROCOL allows basically 1-1 communication, it separates message sending and service execution, and only message sending implies synchronization. The receiver waits for the message and, once received, the sender resumes and then the receiver executes the associated service. STSs composition semantics, as in LOTOS [22], allows one to express the synchronization of actions executed by one sender and several receivers. As far as we know, current object or component-oriented languages do not provide such a native synchronization feature.

A related work is [26] that provides methods to link Finite State Processes (FSP) and Java constructions. FSP is a recent process algebra originally proposed to design software architectures and is based on CSP. FSP can define constants, ranges, sets and simple data types like integers and strings. It also provides the classic construction to define processes and to compose them. The synchronization is based on the rendezvous mechanism and the common actions of processes. Important facilities of FSP are renaming of actions and a powerful notation for labels. FSP is a different model from STS for several reasons. The most important one is that FSP considers only finite state systems. The semantics of STS is based on configuration graphs which are not necessarily finite labelled transition systems as in FSP. Knowing that a system is finite is useful to generate the state space exhaustively; this is not generally possible with STSs which provide more general semantics. STSs also support unrestricted data types and the synchronization uses an external vector of synchronization and no explicit renaming. As in LOTOS, we provide the notion of guard with receipt (post guard in LOTOS) as a primitive mechanism. There is no direct support in FSP for this kind of guard, there are only classical guards. FSP does not provide an interpreter of process algebras but the LTSA book details the Java implementation of rendezvous: it is a synchronous message, thus it is more basic than our rendezvous notion.

JCSP is a pure Java class library designed by Welch and Austin and provides a base range of CSP primitives and a rich set of extensions, see [27] for more details. One main interest is that it conforms to the CSP model of communications and there is a long experience of tools and many practical case studies. The Java monitor thread model is rather easy to understand; however, it is more difficult to use safely as soon as examples are not small. Thus JCSP is indeed a safer alternative than the built-in monitor model of Java threads. The use of explicit shared channels is a simple way to synchronize processes. We have no explicit channel. Processes synchronize on any service execution – not only on read and write operations. Our prototype is not strictly based on CSP but may be viewed as an operational framework for a LOTOS like model of concurrency. Other differences are, as with FSP, support for full data types and guards with receipt. Our approach is oriented to the development of a true language supporting components, rather than a library for Java. One other important reason to quote this work is that it provides a CSP model for the Java thread model [28,29]. This formal model has been used to prove the correctness of a non trivial example. Thus we expect to reuse this model as one of the tools to prove that our rendezvous mechanism is correct. With the same purpose, CTJ [30] is another Java library based on CSP concepts with the additional property of providing support for real-time software via a built-in kernel. Both libraries provide access to the CSP model and have some similarities (see [31] for a comparison).

The aim of our work being to implement the STS synchronization mechanism, we need to define a complex synchronization structure, based on a more classic synchronization barrier. There are many algorithms to implement synchronization barriers. While we are interested in software implementation in Java, the two relevant references are [32,33]. In [33], the principles of these algorithms are explained and an overview of their cost is given. Several proposals are limited to two threads thus they are not sufficient enough for us. In [32], a precise analysis of the performances of several barrier algorithms are compared. The authors note that **synchronized** is needed to get a safe barrier, but that this feature and the **wait-notify** mechanism reduce performance. The **wait-notify** is a safe contention-free wakeup method, but it is slow compared to the Butterfly or the Static f-way barriers. Our basic barrier mechanism (Section 4.1) is fundamentally the same used, for example in [34], to synchronize an aspect with its base code. However, our approach differs from this, not only on the formalism used and the context, but also on the additional synchronization mechanisms presented here.

2. STS-oriented Component Model

Our component model is a subset of the Korrigan model described in [35,36]. This model builds on the ADL ontology [37]: architectures or configurations made of components with ports, and connections between component ports. The specifics we are discussing here are the use of Symbolic Transition Systems and the rendezvous semantics.

There are two categories of component: primitive and composite. We will present the description and the implementation principles of primitive components in the next section. Composite components are reusable compositions of components (*i.e.* architectures). In this paper they are reduced to a simple assembly of primitive components without entering in the detail of hierarchical decomposition of architectures. The runtime support for compositions of components is the main focus of Section 4.

2.1. Formal Definition of Symbolic Transition Systems

An STS is a dynamic behaviour coupled with a data type description. In our previous formal definition, we use abstract data type (see [8]). In this section, the data type part is described with an informal algorithmic language which is refined in Java code in the next sections. A *signature* (or static interface) Σ is a pair (S, F) where S is a set of *sorts* (type names) and F a set of function names equipped with *profiles* over these sorts. If R is a sort, then Σ_R denotes the subset of functions from Σ with result sort being R. X is used to denote the set of all variables. From a signature Σ and from X, one may obtain *terms*, denoted by $T_{\Sigma,X}$. The set of *closed terms* (also called ground terms) is the subset of $T_{\Sigma,X}$ without variables, denoted by T_Σ. An *algebraic specification* is a pair (Σ, Ax) where Ax is a set of axioms between terms of $T_{\Sigma,X}$.

Definition 1 (STS) *An STS is a tuple $(D, (\Sigma, Ax), S, L, s^0, T)$ where: (Σ, Ax) is an algebraic specification, D is a sort called* sort of interest *defined in (Σ, Ax), $S = \{s_i\}$ is a finite set of states, $L = \{l_i\}$ is a finite set of event labels, $s^0 \in S$ is the initial state, and $T \subseteq S \times T_{\Sigma_{Boolean},X} \times Event \times T_{\Sigma_D,X} \times S$ is a set of transitions.*

Events denote atomic activities that occur in the components. Events are either: *i)* hidden (or internal) events: τ, *ii)* silent events: l, with $l \in L$, *iii)* emissions: $l!e$, with $e \in T_\Sigma$, or *iv)* receipts: $l?x : R$ with $x \in X$. Internal events denote internal actions of the components which may have an effect on its behaviour yet without being observable from its context. Silent events are pure synchronizing events, while emissions and receptions naturally correspond, respectively, to requested and provided services of the components. To simplify, we only consider binary communications here; but emissions and receptions may be extended to n-ary emissions and receptions. STS transitions are tuples $(s, \mu, \epsilon, \delta, t)$ for which s is called the source state, t the target state, μ the guard, ϵ the event and δ the action. Each action is denoted by a program with variables. A do-nothing action is simply denoted by -. In forthcoming figures, transitions will be labelled as follows: $[\mu]\ \epsilon\ /\ \delta$.

2.2. Connections and Synchronizations

A primitive component, for example the **server** component in Figure 1, is made of ports and a protocol described in the STS formalism. The STS has states and transitions between states. The general syntax of an STS transition is **[guard] event / action**, where **guard** is a condition to trigger the transition, **event** is a dynamic event (possibly with emission ! or receipt ?) and **action** is the action performed. An action corresponds to the call of a sequential operation. An event corresponds to the (external) notification of action execution. Ports are component connection points, each port externalizes the triggering of a given event in the STS protocol.

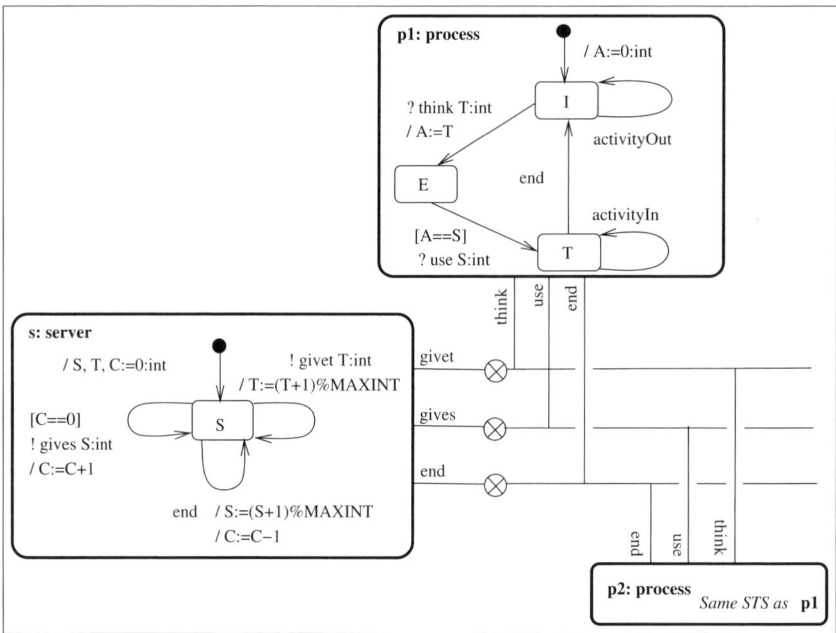

Figure 1. The Ticket Protocol Architecture with Two Processes

Connections are primitive bindings between ports rather than complex connectors. They denote synchronous communications between components. When ports are connected, their corresponding events are synchronized. Synchronizing several events means triggering them in any real order, but in the same logical time: this is the rendezvous principle. In case of communication (! and ? events), the rendezvous takes place but the sender necessarily initiates a value computation which is communicated to receivers during the rendezvous. An STS of a primitive component already involved in a synchronization cannot trigger any other event during this synchronization. This rendezvous provides execution of actions of all the participants as well as a 1 to n communication.

This composition model proposes three ways for components to interact: (i) asynchronous activity: one component executes an action independently (*i.e.* without interaction), (ii) rendezvous without communication: n components execute a given action in the same logical time, and (iii) rendezvous: in addition to the latter case, a component emits a value and the others receive it during the rendezvous. In this case, we consider that every receiver guard may check the emitted value, that is we have a *guard with receipt* (see Section 2.5).

2.3. Global Semantics

One way to define the global semantics of such a system is to compute the synchronous product of STSs [38] or the concurrent composition of processes [26]. These computations rely both on primitive component protocols and connections, so they can be automated from an architecture specification. They take as input STSs defining protocols and *synchronization vectors* defined by connections to produce semantic models. Fig. 2 shows the synchronous product of the three STSs in Fig. 1.

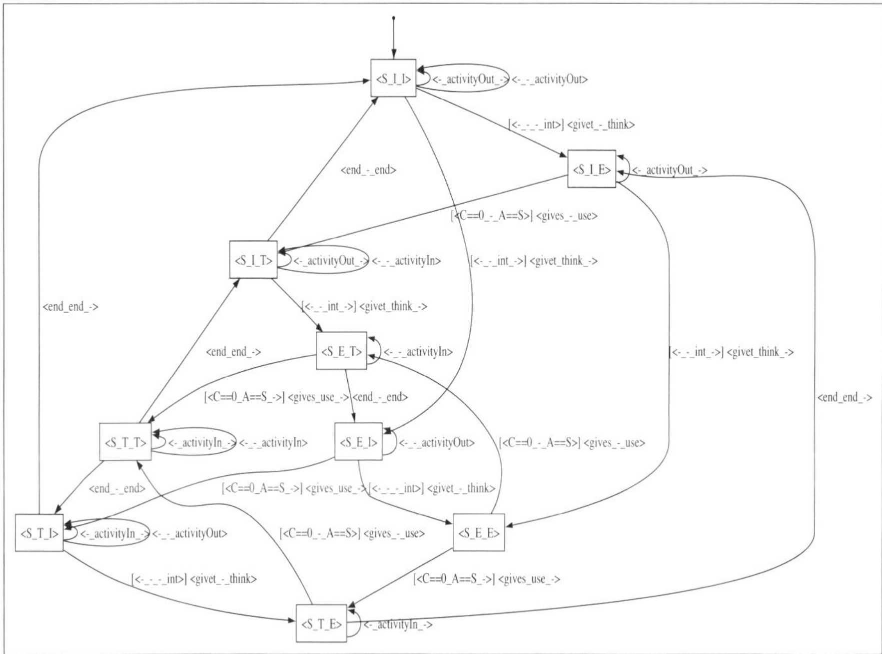

Figure 2. The STS Global Product of the Ticket Protocol (s × p1 × p2)

A synchronization vector is a *vector of events* that denotes a possible synchronization, at runtime, between a set of events. Synchronization vectors are computed according to the connections between component ports and defined according to an arbitrary ordering of primitive components. Each connection defines a given computation of synchronization vectors depending on connected ports. The three connections in the architecture of Figure 1 use the same communication operator. For example, one of these connections connects **think** ports of **processes** components with **givet** port of **server** component. It defines synchronizations that are binary between server and processes and exclusive between processes (denoted by the ⊗ symbol in Fig. 1). If we admit that possible synchronizations are denoted using synchronization vectors with ordering (**s, p1, p2**), then this connection produces two synchronization vectors: (**givet, think, -**), (**givet, -, think**). The - symbol is the stuttering notation to denote asynchronous (*i.e.* independent) activities of components. So, the resulting vectors express that the **givet** event of **server s** synchronizes with **think** event of **process p1** or **process p2**, but not with these two **think** events in the same time. We have to notice that many different connections can be described to produce various computations of synchronization vectors.

Once all synchronization vectors are computed for a given architecture, they are used to compute the semantic model of the system, by combining STSs. Then, verification methods can be used to check the semantic model, but this is out of the scope of this paper. What we show in this paper, is that synchronization vectors are also useful for configuring runtime support of components.

Concurrent communicating components can be described with protocols modelled by STS and synchronous products, adapted from the LTS definition [38], can be used to obtain the resulting global system. Given two STSs with sets of event labels L_1 and L_2 and a set V of synchronization vectors, there is a set of pairs (l_1, l_2), called synchronous events, such that $l_1 \in L_1$ and $l_2 \in L_2$. Hidden events cannot participate in a synchronization. Two components synchronize at some transition if their respective events are synchronous (*i.e.* belong to the vector) and if the *event offers* are compatible. Offer compatibility follows simple rules: type equality and emission/receipt matching. An event label l such that there is no pair in V which contains l, is said to be asynchronous. Corresponding transitions are triggered independently.

Definition 2 (Synchronous Product) *The synchronous product (or product for short) of two STS $d_i = (D_i, (\Sigma_i, Ax_i), S_i, L_i, s_i^0, T_i)$, $i = 1, 2$, relatively to a synchronization vector V, denoted by $d_1 \otimes_V d_2$, is the STS $(D_1 \times D_2, (\Sigma_1, Ax_1) \times (\Sigma_2, Ax_2), S, L_1 \times L_2, s^0, T)$, where the sets $S \subseteq S_1 \times S_2$ and $T \subseteq S \times T_{\Sigma_{Boolean},X} \times (Event_1 \times Event_2) \times T_{\Sigma_D,X} \times S$ are inductively defined by the rules:*

- $s^0 = (s_1^0, s_2^0) \in S,$
- *if* $(s_1, s_2) \in S$, $(s_1, \mu_1, \epsilon_1, \delta_1, t_1) \in T_1$, *and* $(s_2, \mu_2, \epsilon_2, \delta_2, t_2) \in T_2$, *then*

  * *if* $(l_1, l_2) \in V$ *then* $((s_1, s_2), \mu_1 \wedge \mu_2, (\epsilon_1, \epsilon_2), (\delta_1, \delta_2), (t_1, t_2)) \in T$ *and* $(t_1, t_2) \in S.$
  * *if* l_1 *is asynchronous then* $((s_1, s_2), \mu_1, (\epsilon_1, \tau), (\delta_1, Self_{D_2}), (t_1, s_2)) \in T$ *and* $(t_1, s_2) \in S.$
  * *if* l_2 *is asynchronous then* $((s_1, s_2), \mu_2, (\tau, \epsilon_2), (Self_{D_1}, \delta_2), (s_1, t_2)) \in T$ *and* $(s_1, t_2) \in S.$

The synchronous product operator can be extended to an n-ary product and to any depth.

2.4. The Ticket Protocol Example

The example depicted in Figure 1 illustrates an architecture of primitive components with a mutual exclusion protocol inspired by the ticket protocol [5]. Processes and Server components are organized following a client-server architectural style. However, our version differs from the one in [5] since we deal with distributed components communicating by rendezvous, and not processes operating on a shared memory. We also distinguish entering (use event) and leaving (end event) the critical section.

In the example, there are six synchronization vectors computed according to connections between component ports: (givet, think, -), (givet, -, think), (gives, use, -), (gives, -, use), (end, end, -), and (end, -, end). Note that, whenever an event of a component does not occur in any synchronization vector, it is an asynchronous event, which can be triggered independently of others. Here, we note that processes p1 and p2 have asynchronous activities, either outside the critical section (activityOut) or inside it (activityIn). The server gives a ticket number to the process which memorizes it in its variable A. This synchronization step is represented by synchronization vectors (givet, think, -) or (givet, -, think), depending on respectively p1 or p2 enters the critical section. Then, to enter in critical section, the process p1 or p2 checks if its variable A is equal to the ticket S of the server. This synchronization step is represented by synchronization vectors (use, gives, -) or (use, -, gives), depending on respectively p1 or p2 enters the critical section. If all guards succeed, then the one process enters in critical section (state T). Then the process leaves critical section on the end event. This synchronization step is represented by synchronization vectors (end, end, -) or (end, -, end), depending on respectively p1 or p2 enters the critical section.

Figure 2 was calculated with our STS tool to illustrate the global behaviour of our example. The picture is simplified since actions are not depicted, but they may be easily inferred from the component STS. Note that something like [<C==0_A==S_->] is a compound guard expressing that s and p1 evaluate their guards while p2 has a default true guard. The same thing applies for the compound events which glues three events each one coming from a component. The reader may see that processes have asynchronous activities which are expressed by transitions like <-_activityOut_-> or <-_-_activityIn>. The semantics provides concurrent composition of components with event synchronizations, namely the rendezvous notion introduced by CSP [25]. This synchronization mode is not generally what we find in programming languages, for instance in the PROCOL, SOFA or Cooperative Objects approaches. Thus to relate the formal level with the operational one we want to implement the concurrent composition of STSs. This construction takes several STSs and the synchronization vectors which link events of the input STSs.

2.5. *Guard with Receipt*

One reason to introduce the ticket example is that it shows a complex communication with guarded receipt during the (gives, use, -) or (gives, -, use) synchronization. Guards with possible receipt is an important construction with a specific semantics: components can conditionally receive and synchronize on a value in the same logical time. They correspond to post guards in the LOTOS language. One benefit is to increase abstraction and reduce the size of the finite state machine. Note that, in such a communication, the emitter must have a guard without receipt.

Some translations of guarded transitions are possible. The [A=S] ? use S:int transition of the STS process has a guard with receipt and no action. This complex transition may for example be split into three steps: a receipt, a guard checking and a null action. However this decomposition should be used with care since in place of a single event we get a sequence of three events. In other words, hiding for instance the guard checking is not preserving the observational semantics (it is only a strict behavioral abstraction). From a practical point of view, the consequence on the synchronization mechanism is that when a rendezvous occurs the sequences of these three steps have to be synchronous, not only one of them. This last point raises a major implementation issue to keep the model semantics and components execution consistent.

3. Model Implementation Overview

In this section, we detail our hypotheses related to the description of primitive components in Java. A global picture of intra-component implementation is depicted in Figure 3. It represents the different elements defining a primitive component.

In the component model of the Korrigan formal ADL, the finite state machine notations are mixed with the data type part description. This is convenient when we want an integrated model suited for verification purposes. However for the operational side, we think that it is better to separate the finite state machine and the data part. This simplifies a little the implementation and, moreover, separates the two aspects which makes the implementation more reusable. For example, we can reuse a given state machine with another data type implementation provided that some compatibility rules are ensured. The Java representation of the finite state machine is thus reduced to the states, the transitions and some names. These names represent the guards, the events, the receipt variables, the senders and the actions. The

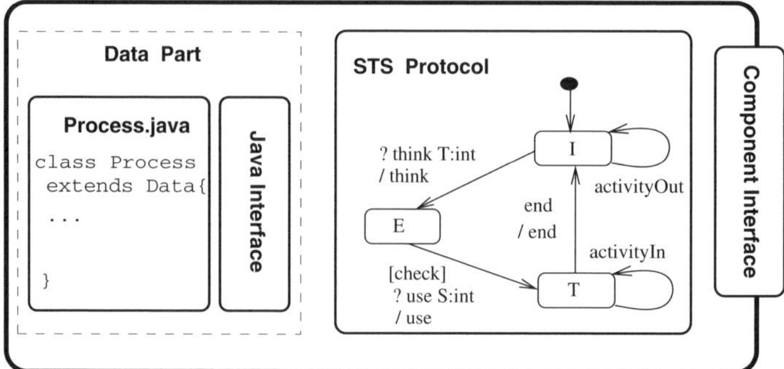

Figure 3. Implementation of the Process Primitive Component

data part is a Java class implementing the formal data type part. The exact role of the class is to give a real implementation, with methods, of the names occurring in the state machine part. Thus, both parts are glued thanks to a normalized Java interface which is automatically computed from the STS. An *emitter* is a pure function computing the emitted value in a given state of the component. Similarly, a *guard* is a boolean function implementing a condition.

So, in our current scenario, a primitive component results from the combination of a protocol and existing Java code (henceforth referred to as the *data part*), more precisely, a passive Java class implementing a specific Java interface. Each primitive component is implemented with an active object (thread in Java) in charge of both the STS protocol execution and the call to the passive object implementing the component data part. We choose to rely on an active object since it may simulate a passive one (a usual class), the reverse being false. Thus from now on, an STS defines the events, guards, emitters, and actions names related to the Java interface of the data part class.

The code may be either automatically generated from an explicit and formal description ([9,10]) or provided by the user writing some programs or reusing some classes. One important issue is the compatibility or coherence between the data part intrinsic protocol (*i.e.* the execution protocol) and the externally defined STS protocol. One way to address this issue is to provide a method that extracts a compatible data type from the STS description [9,24]. Another way is to develop the data part and the protocol separately and then to check compatibility between both parts. Behavioural compatibility has been addressed in process algebra [39] and in state machine [38,40] approaches. There exists related work on component compatibility (for instance [2,41]). We assume to rely on the technique presented in [41] which is compatible with the STS behavioural semantics. As an example, a Java interface and a Java class compatible with the STS process presented are described in Figures 4 and 5.

```
public interface IProcess {

  public void think (int T);

  public boolean check (int S);  // check for guard (A == S)

  public void use (int S);

  public void end ();

}
```

Figure 4. Java Interface for the Process STS

```
public class Process extends Data implements IProcess {

  protected int A;

  public Process () {
    this.A = 0;
  }

  public void think (int T) {
    this.A = T;
  }

  // guard with receipt
  public boolean check (int S) {
    return this.A == S;
  }

  // use action with receipt
  public void use (int S) {
    System.out.println ("Enter critical section");
  }

  public void end () {
    System.out.println ("Leaving critical section");
  }

}
```

Figure 5. Java Class for the Process STS

[guard] event !emitter:Type / action	⎧ public boolean guard(); ⎨ public Type emitter(); ⎩ public void action(Type var);
[guard] event ?var:Type / action	⎧ public boolean guard(Type var) ⎩ public void action(Type var);

Figure 6. Rules to Generate Interfaces

Figure 6 presents the translation rules for emission and receipt labels. Note that, in case of receipt, the guard and the action signatures of the receiver transition have to accept the received argument. However the methods have the possibility to forget this parameter if useless. Formally, the syntactic compatibility between the STS label information and the Java interface can be checked on the basis of the rules presented in Figure 6. The syntactic compatibility between the Java interface and the data part class follows the Java 1.5 type checking rules. In Figure 6, guard, action and emitter are Java method identifiers, var is a Java variable identifier and Type is a Java type identifier.

Architecture or component assembly relies on primitive and composite components and a glue mechanism to synchronize them. A direct way to compose properly components is to build their synchronous product according to synchronization vectors. This product represents the global execution of the component interactions as a global STS (*e.g.* Figure 2) and a compound data part may be built from the subcomponent data parts. However, one important drawback of this solution is the computation cost of the synchronous product, which is exponential in the size of the STS components. Another problem is that the resulting application will be centralized if we consider the global STS as a primitive component's STS, since it will be executed on a single active object. Lastly, although this provides an equivalent simulation (as with [9]) of the compound system, the original components are not really reused.

That is why we choose to implement the concurrent composition of STSs. This construction takes as input several STSs and synchronization vectors that bind their events. It configures STS runtime support in such a way that STS execution conforms to the semantic model. The direct consequences are that each STS has its own execution thread and that all STSs have to be synchronized depending on synchronization vectors. In this implementation, a primitive component corresponds, at runtime, to a unique thread and a composite component corresponds to a collection of interacting threads. Synchronization of threads is supported by a specific rendezvous mechanism, presented in the next section.

4. A Java Implementation of Rendezvous

In this section, we present the principles to implement our rendezvous mechanism for Java components with STSs. While our solution is a general one we suggest to use it only in local networks since it may be a bottleneck in wide area networks due to communication delays. In wide area networks, asynchronous communications have to be used in place of synchronous communications [13]. Nevertheless this latter communication mode can be implemented by synchronous communication and channel or intermediate component, but this is out of the scope of this paper. We choose to implement our proper rendezvous mechanism in Java 1.5 using monitors. The two other alternatives were the join method or the CyclicBarrier class. Technically, when using the join method, threads are exiting and we need to start new ones; handling persistent state for data is more complex. A second remark is that the implementation of the rendezvous would require constructions similar to those we introduce later to cope with guards and communications. The CyclicBarrier seems to be a perfect candidate to synchronize the threads associated to our STSs. However, the problem is still the implementation of guards which are conditions to enter into the synchronization barrier. One may have one thread which is waiting on the barrier and another one which cannot since its guard is false. Thus we have to check all the involved guards before reaching the barrier. Except for the exiting barrier, the use of the CyclicBarrier does not really simplify our implementation. Lastly, we need to know precisely the synchronization mechanism since this first approach will be optimized later. In the following subsections, we present the implementation of component runtime in four progressive steps, from a simple barrier to the rendezvous with receipt on guards.

4.1. The Basic Barrier Principles

The basic mechanism described in this subsection is nearly the same as in [34]. In this first setting, a mechanism was implemented to synchronize LTS. As in FSP [26], a synchronization is possible between two actions if they have the same name. A central object, the arbiter, controls that synchronizations are correctly handled. The principle is to use a Java monitor to implement two synchronization barriers. Note that one synchronization barrier is generally not sufficient enough to ensure the correct rendezvous between actions. With only one barrier, an asynchronous action of an STS may be triggered in the same logical time as a synchronous action of another component. This would be inconsistent with the STS composition semantics. The right solution requires one barrier for entering in the synchronization area and another one for all participants to leave it.

Figure 7 gives the static class diagram of the solution. Actions and states are encoded by integers. An LTS[1] is encoded with a list of actions and a matrix. In this matrix, for each state we have a vector (indexed by actions) of the target states. The LTS has also a reference

---

[1]These details of implementations are provided to give to the reader an understanding of the synchronization mechanism. However, in the real implementation, things are much more complex and based on hash mappings.

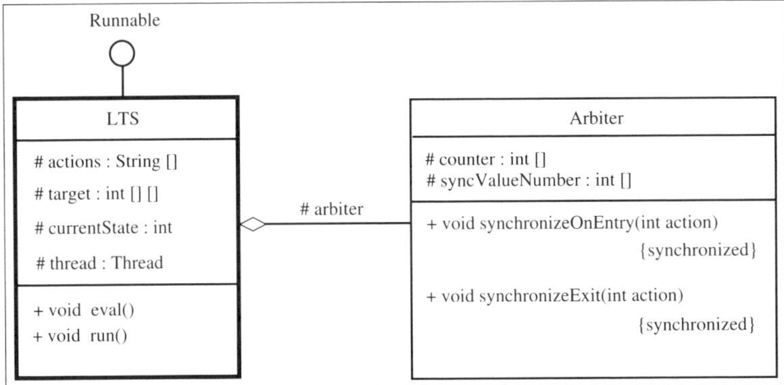

Figure 7. UML Class Diagram for the Basic Barrier

to an **Arbiter** instance. The **LTS** class is active by implementing the **Runnable** interface and owning an instance of class **Thread**. The **run** method evaluates (using **eval**) an action until the thread is interrupted or the LTS reaches a state without outgoing transitions. The **eval** method of the **LTS** class checks whether the transition is synchronous. If the action is asynchronous, the LTS evolves independently of others. If synchronous, the arbiter is called with a **synchronizeOnEntry** for this action. Then, **currentState** is updated with the **target** matrix and the arbiter finishes the rendezvous with a **synchronizeOnExit** call.

The arbiter is a shared passive object which is called to synchronize actions. Its **syncValueNumber** variable defines, for each synchronization, the number of actions (and consequently the number of LTSs) to synchronize. The **counter** variable defines, for each synchronization, the number of LTS that have passed the entry barrier and wait for others LTS involved in the synchronization. The entry and exit barriers are implemented with two **synchronized** methods. The code of the entry barrier is shown in Figure 8, the **synchronized** qualifier ensures that only one thread is executing this call. The exit barrier has a similar implementation.

```
synchronized public void synchronizeOnEntry (int action) {
  if (counter[action] < syncValueNumber[action] - 1) {
    counter[action]++;                        // we are not the last thread
    try {                                     // so block
      wait ();
    } catch (InterruptedException e) {}
  } else {
    counter[action]=0;                        // we are the last thread
    notifyAll ();                             // so wake up all
  }
}
```

Figure 8. The Synchronization Barrier

All synchronized LTSs query the entry barrier in any order and their supporting threads then wait. When the last LTS queries the entry barrier, all threads are woken (**notifyAll**) and the synchronization counter is reset to 0. Then, all LTSs concurrently execute their respective actions. When an LTS ends its action, it queries the exit barrier and then waits. When the last LTS queries the exit barrier, all threads are woken and all LTSs can continue their execution independently: the synchronization of LTSs' actions is ended. Actions require the barrier in any ordering and have to wait before starting to execute their actions until the last action is also ready to synchronize.

In fact, the wait should be enclosed in a while loop – see [26] for details. Here, this should not (logically) be needed since, once awakened, a sleeping thread simply exits from the barrier. However, because of the *spurious wakeup* problem (whereby a waiting thread can be awoken for no reason whatsoever), this actually is necessary! For simplicity of presentation, this is not programmed in Fig. 8.

Our implementation also takes into account protocol non-determinism by simulating a random choice of actions. It is not too difficult to write such a solution. However, we have to minimize the number and the size of the synchronized parts to increase concurrency between threads, while getting a correct solution.

4.2. Synchronization Vectors Representation

A first improvement is to relax the restriction on names for synchronization. Often design and component languages do not decouple the behavioural description from the communications, for instance PROCOL, FSP or UML. To reuse components, they have to be synchronized into various environments and there is no reason for port naming to be a global knowledge. To fight against name mismatch, the two classic solutions are renaming (as in FSP) or component adapter. We think that a solution based on synchronization vectors is most general since it does not need code modification or any additional programmable entity (*i.e.* adapter).

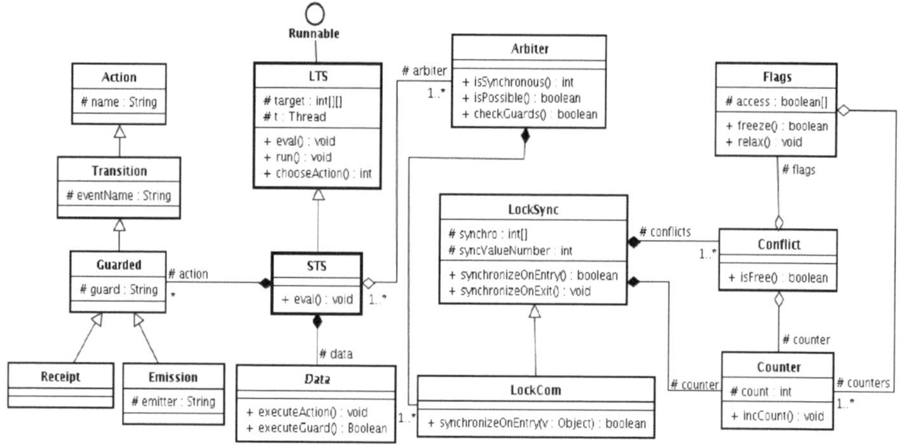

Figure 9. Partial UML Class Diagram

In this new setting, a set of synchronization vectors (cf. Section 2) is declared, each one representing a possible synchronization between some component events. An event name and an action name are associated inside a transition (class Transition in Fig. 9). A synchronization vector, denoting a set of synchronous events, indirectly defines a set of synchronous actions. The LockSync class, that represents a synchronization vector, is then introduced to the diagram. The synchronization barrier methods are moved from the Arbiter to this new class, and there are now two barriers for each synchronization vector.

The eval method is also changed. It first asks the arbiter to get the LockSync instance which concerns the current action. It uses the isSynchronous method to choose one LockSync object. Then a synchronizeOnEntry call is made and returns a boolean indicating if entering in the barrier succeeds or fails (see Fig. 10).

The first thread entering in the barrier must process two specific tests, isPossible and isFree, which are implemented in the Arbiter class. The isPossible method checks if a syn-

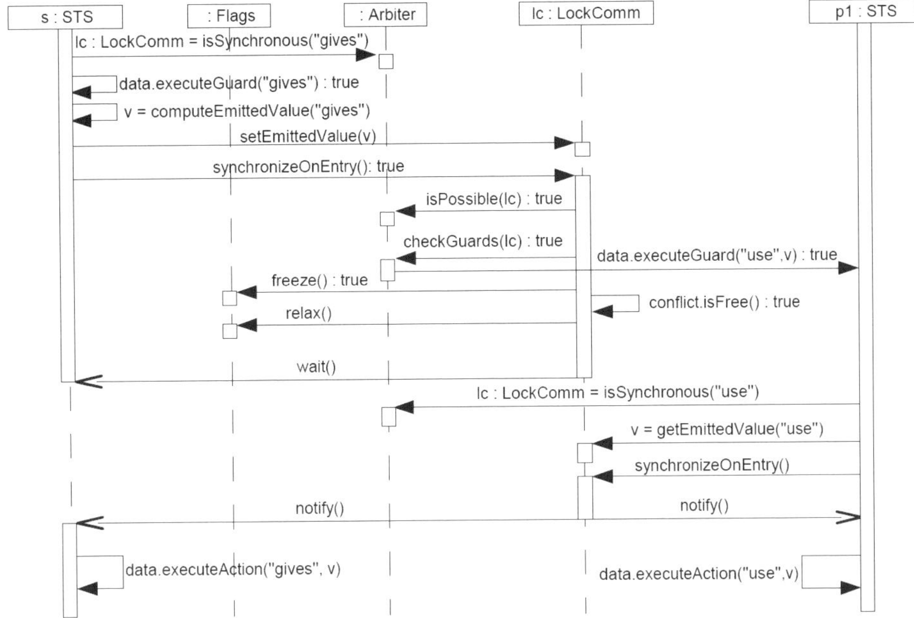

Figure 10. Message Sequence Chart: Entering the Barrier for s.gives and p1.use

chronization can occur from the current global state. Method **isFree** tests if a synchronization is possible – *i.e.* if a component is not already involved in another synchronization. If true, the synchronization counter of the **LockSync** object is incremented. This latter method is critical and depends on the current state of the threads, so has the **synchronized** qualifier. This method uses a global boolean table which contains a truth value if two synchronizations are synchronizing the same thread. It is important since it avoids initiating two conflicting synchronizations that would lead to a deadlock.

4.3. Independent Synchronizations

One may observe that, in the basic barrier, two distinct synchronization entries or exits are always serialized since there is a single arbiter and the methods to enter and leave the barriers are **synchronized**. The definition of the **LockSync** class is a first attempt to break this centralized control.

The *conflict* of a synchronization is defined as the set of synchronizations which synchronize on, at least, a common component. In our example, all the synchronizations are mutually conflicting because of the central server component. A synchronization is independent from another one *iff* it does not belong to its conflict set. The improvement, here, consists in implementing the conflicts set (class **Conflict** in Fig. 9) of each synchronization and to allow two (or more) independent synchronizations to enter simultaneously in the barrier (or to leave it).

We define a **Flag** class which contains, for each synchronization, a reference on the corresponding synchronization counter and a boolean (**access**) representing the possibility of accessing this counter (cf. Fig. 7). Two methods (**freeze** and **relax**) are defined with the **synchronized** qualifier; they have the responsibility to implement exclusive access to the vector of shared counters, by testing and setting the value of the **access** attribute. Now the **isFree** method is no longer synchronized. It tests if conflicting synchronizations are not already entering a barrier and, if OK, allows the current synchronization to proceed. The first thread entering the barrier – and only this thread – has to freeze the counters conflicting with the current synchronization; then isFree is called and, finally, the counters are released (cf. Fig. 10). This current solution minimizes the bottleneck access to the vector of counters with two small methods, defined as tests and assignments on a vector of boolean objects.

4.4. Guards and Communications Management

Since STS transitions are more complex than those represented until now, we need a richer class diagram to manage STS properties not already taken into account. Classes Guarded, Emission and Receipt are defined to represent the corresponding transitions (see Fig. 9). An abstract class Data contains methods, based on the reflexive Java API, to execute guards, emitters and actions on an instance. This class is specialized by the specific data part class of each component (cf. Fig. 3). The method run tries to trigger a possible transition if there is one. There is no busy waiting loop to check the guards, they are evaluated only when needed at the entry in the synchronization barrier.

The management of communication has to be implemented to conform to the STS model (cf. Section 2). Since there are guards with receipt, communications have to be evaluated before any synchronization and even before checking guards. Furthermore, all guards related to all synchronized actions have to be checked before the execution of these actions. So, the eval method is modified to manage synchronous actions with communication, in addition to the two previous cases (asynchronous actions and synchronous actions). A synchronous action with communication is initiated by the first thread entering the barrier, which is necessarily the sender. The local guard of the emitter transition (if any) is checked and the emitted value is then computed (see Fig. 10). The call to synchronizeOnEntry is then performed with the value communicated to the LockSync object (setEmittedValue(v) in Fig. 10). This object is an instance of the LockCom class (that specializes the LockSync class to introduce a specific version of the entry barrier for the communication case). In addition to LockSynch operations, it realizes a checkGuards method call to check if the guards associated with a synchronization vector are true, coping with value communicated to other STSs. The eval method of the STS class also retrieves the communicated value (getEmittedValue("use") in Fig. 10), to perform the execution of synchronized actions that use this value as argument.

4.5. Final Comments

The previous implementation provides an interpreter supporting rendezvous and allowing dynamic changes of STSs, data parts or even components (obviously with some care in stopping and restarting components). The current discussion is mainly directed to get a correct barrier with complex synchronization conditions allowing receipt on guards. Efficiency has been taken into account in two ways: distributing the central arbiter in several sets of objects (locks, conflicts and flags) and minimizing the synchronized parts. The guard checking, the emission computation (if needed) and freezing the flags are only done by the first thread that enters the synchronization barrier.

In this interpreter version, reflexivity is used to glue protocols and data parts. In the compiler version, protocols will do direct call to the data parts methods. Note also that exception handling, barrier delays and RMI have to be integrated to get a true usable system. The current version relies on a "wait and notify barrier". An optimization is to use result from [32], for instance, to replace it with a Static f-way barrier. However, a major problem will be the distribution of the shared objects and the limitation of remote communications. We have also to fight against the global synchronization problem (see [42]). Here, we partially addressed this problem with the introduction of conflicts and locks and we will feature the balance between synchronous and asynchronous communications. A more comprehensive analysis has to provide a solution that scales up to wide distributed systems.

5. Conclusion and Future Work

In this paper, we provide a mechanism to synchronize components with protocols. We consider complex protocols, namely symbolic transition systems, with full data types, guards and communications. We allow non-determinism in the protocols and we provide a flexible naming notation to define event synchronizations. One original and powerful feature is the possibility to define conditional rendezvous taking into account the communicated values. These protocols are adequate for the formal specification of systems and our approach gives a means to execute them – thus relating verification and execution of component systems. We describe an implementation of a complex rendezvous based on two synchronization barriers, each of them implemented with the monitor and wait/notifyAll facilities of Java. One delicate thing is synchronization in presence of communications and guards. We show how to proceed in four steps to get a correct solution. This solution is general in the sense that we do not constrain the ordering of processes to enter, execute their action and leave the critical section. We also propose a first optimization to allow several independent synchronizations to process the barrier. This is a first way to distribute the central arbiter mechanism used to synchronize the components. Currently, this work provides an operational interpreter to program primitive components in Java with STSs and a powerful way to compose them.

Until now we have done tests, implemented various small and middle size examples and checked with our verification tool some specific parts of the mechanism. We have also implemented a dynamic check which verifies that events generated by the runtime are according to the synchronization rules and compatible with each running state machine. This defines a dynamic checking which is able to alert the user if some synchronizations are not correct and if state changes or transition triggering are not occurring at the right moment. While this checking is useful it is not sufficient to prove that our mechanism respect its specifications.

One thing we would prove in our future work is the correctness of the solution. First, we already reused the work of [28,29] which gives a CSP view of the Java monitoring mechanism. Rather than a CSP view, we get an STS description of the mechanism. We model the simple barrier with our STS tool and we try to do verifications on some simple examples. We are able to verify that the mechanism allows a correct entry and exit in the rendezvous area, but with only LTS behaviour. One result of this was a simplification of the two barriers which are the base of our actual mechanism. This was a first step, the second, yet future, is to design the full mechanism with STSs integrating the guard and communication mechanisms. We have also to model the locks and flags features, but these are passive objects. Then we will prove that, from a temporal logic point of view, our two barriers define an area of synchronization. That is a logical time area where synchronous actions occur inside (in any ordering) and synchronous components have no other activities. Last, our locks and conflicts own the following properties: (i) two different threads with a same synchronization vector cannot compete for entering the barrier since the synchronizeOnEntry is a synchronized method, and (ii) two different threads with different synchronization vectors can simultaneously start an area of synchronization iff the synchronizations are not conflicting. We think that it is sensible to get a full manual proof, however our STS tool will be used to check some examples. One final improvement will be to translate our specifications into PVS (see [24] for a related work) and to run the manual proof.

Future work will also consider the definition of a Java based language with STS, asynchronous and synchronous communications. We have to make precise the compilation mechanism as well as some optimization aspects. Amongst these, we expect to propose a solution to choose automatically between passive and active object implementations. Another feature is to elaborate a splitting mechanism for the central flags based on the analysis of synchronizations and communications in the deployed architecture.

References

[1] T. Barros, L. Henrio, and E. Madelaine. Behavioural Models for Hierarchical Components. In *Proc. of SPIN'05*, volume 3639 of *LNCS*, pages 154–168. Springer-Verlag, 2005.

[2] P. Jezek, J. Kofron, and F. Plasil. Model Checking of Component Behavior Specification: A Real Life Experience. *Electronic Notes in Theoretical Computer Science*, 160:197–210, 2005.

[3] Anna Ingolfsdottir and Huimin Lin. *A Symbolic Approach to Value-passing Processes*, chapter Handbook of Process Algebra. Elsevier, 2001.

[4] Ph. Schnoebelen, B. Bérard, M. Bidoit, F. Laroussinie, and A. Petit. *Vérification de logiciels : Techniques et outils du model-checking*. Vuibert, 1999.

[5] G. Delzanno. An Overview of MSR(C): A CLP-based Framework for the Symbolic Verification of Parameterized Concurrent Systems. In *Proc. of WFLP'02*, volume 76 of *ENTCS*. Elsevier, 2002.

[6] S. Bardin, A. Finkel, and J. Leroux. FASTer Acceleration of Counter Automata in Practice. In *Proc. of TACAS'04*, volume 2988 of *LNCS*, pages 576–590. Springer, 2004.

[7] A. Bouajjani, P. Habermehl, and T. Vojnar. Abstract Regular Model Checking. In *Proceedings of CAV'04*, volume 3114 of *LNCS*, pages 372–386. Springer Verlag, 2004.

[8] Pascal Poizat, Jean-Claude Royer, and Gwen Salaün. Bounded Analysis and Decomposition for Behavioural Description of Components. In Springer Verlag, editor, *FMOODS*, number 4037 in Lecture Notes in Computer Science, pages 33–47, 2006.

[9] Christine Choppy, Pascal Poizat, and Jean-Claude Royer. From Informal Requirements to COOP: a Concurrent Automata Approach. In J.M. Wing and J. Woodcock and J. Davies, editor, *FM'99 - Formal Methods, World Congress on Formal Methods in the Development of Computing Systems*, volume 1709 of *Lecture Notes in Computer Science*, pages 939–962. Springer-Verlag, 1999.

[10] R. Guimarães and W. da Cunha Borelli. Generating java code for TINA systems. In *Symposium on Computer and High Performance Computing (SBAC-PAD)*, pages 68–74. IEEE Computer Society, 2002.

[11] Sebastian Pavel, Jacques Noyé, Pascal Poizat, and Jean-Claude Royer. A java implementation of a component model with explicit symbolic protocols. In *Proceedings of the 4th International Workshop on Software Composition (SC'05)*, volume 3628 of *Lecture Notes in Computer Science*, pages 115–125. Springer-Verlag, 2005.

[12] Jean-Claude Royer and Michael Xu. Analysing Mailboxes of Asynchronous Communicating Components. In D. C. Schmidt R. Meersman, Z. Tari and al., editors, *On the Move to Meaningful Internet Systems 2003: CoopIS, DOA, and ODBASE*, volume 2888 of *Lecture Notes in Computer Science*, pages 1421–1438. Springer Verlag, 2003.

[13] J. P. Briot, R. Guerraoui, and K. P. Lohr. Concurrency and Distribution in Object Oriented Programming. *ACM Computing Surveys*, 30(3):330–373, 1998.

[14] Jan van den Bos and Chris Laffra. PROCOL: A parallel object language with protocols. In Norman Meyrowitz, editor, *OOPSLA'89 Conference Proceedings: Object-Oriented Programming: Systems, Languages, and Applications*, pages 95–102. ACM Press, 1989.

[15] Frantisek Plasil and Stanislav Visnovsky. Behavior protocols for software components. *Transations on Software Engineering*, 28(11):1056–1076, November 2002.

[16] C. Sibertin-Blanc. Cooperative objects : Principles, use and implementation. In *Concurrent Object Oriented Programming and Petri Nets*, volume 1973 of *LNCS*, pages 216–246. Springer-Verlag, 2001.

[17] L. de Alfaro and T. A. Henzinger. Interface Automata. In *Proc. of ESEC/FSE'01*, pages 109–120. ACM Press, 2001.

[18] Mario Südholt. A model of components with non-regular protocols. In Thomas Gschwind, Uwe Assman, and Oscar Nierstrasz, editors, *International Workshop on Software Composition (SC)*, volume 3628 of *Lecture Notes in Computer Science*, pages 99–114. Springer-Verlag, April 2005.

[19] J. Kramer, J. Magee, and S. Uchitel. Software Architecture Modeling and Analysis: A Rigorous Approach. In *Proc. of SFM'03*, volume 2804 of *LNCS*, pages 44–51. Springer-Verlag, 2003.

[20] A. Bracciali, A. Brogi, and C. Canal. A formal approach to component adaptation. *Journal of Systems and Software*, 74(1), 2005.

[21] S. Moschoyiannis, M. W. Shields, and P. J. Krause. Modelling Component Behaviour with Concurrent Automata. *Electronic Notes in Theoretical Computer Science*, 141(3), 2005.

[22] Muffy Calder, Savi Maharaj, and Carron Shankland. A Modal Logic for Full LOTOS Based on Symbolic Transition Systems. *The Computer Journal*, 45(1):55–61, 2002.

[23] Christine Choppy, Pascal Poizat, and Jean-Claude Royer. A Global Semantics for Views. In T. Rus, editor, *International Conference on Algebraic Methodology And Software Technology, AMAST'2000*, volume 1816 of *Lecture Notes in Computer Science*, pages 165–180. Springer Verlag, 2000.

[24] Jean-Claude Royer. The GAT Approach to Specify Mixed Systems. *Informatica*, 27(1):89–103, 2003.

[25] C.A.R. Hoare. *Communicating Sequential Processes*. C.A.R Hoare Series. Prentice-Hall International, 1985.

[26] Jeff Magee and Jeff Kramer. *Concurrency: State Models and Java Programs*. Wiley, 2 nd edition, 2006.

[27] Peter Welch. Communicating Sequential Processes for Java (JCSP). `http://www.cs.kent.ac.uk/projects/ofa/jcsp`.

[28] P. H. Welch and J. M. R. Martin. A CSP Model for Java Multithreading. In P. Nixon and I. Ritchie, editors, *Software Engineering for Parallel and Distributed Systems*, pages 114–122. IEEE Computer Society Press, 2000.

[29] P.H. Welch and J.M.R. Martin. Formal Analysis of Concurrent Java Systems. In Peter H. Welch and Andrè W. P. Bakkers, editors, *Communicating Process Architectures 2000*, pages 275–301, 2000.

[30] G. Hilderink, A. Bakkers, and J. Broenink. A Distributed Real-time Java System Based on CSP. In *The Third IEEE International Symposium on Object-Oriented Real-Time Distributed Computing*, pages 400–407. IOS Press, 2000.

[31] N. C. Schaller, G. H. Hilderink, and P. H. Welch. Using Java for Parallel Computing: JCSP versus CTJ, a Comparison. In P.H. Welch and A.W.P. Bakkers, editors, *Communicating Process Architectures*, pages 205–226. IOS Press, 2000.

[32] Carwyn Ball and Mark Bull. Barrier Synchronization in Java. Technical report, High-End Computing programme (UKHEC), 2003.

[33] Torsten Hoefler, Torsten Mehlan, Frank Mietke, and Wolfgang Rehm. A Survey of Barrier Algorithms for Coarse Grained Supercomputers. Technical Report 3, University of Chemnitz, 2003.

[34] Rémi Douence, Didier Lebotlan, Jacques Noyé, and Mario Südholt. Concurrent aspects. In *"Generative Programming and Component Engineering (GPCE)"*, pages 79–88. ACM press, October 2006.

[35] Christine Choppy, Pascal Poizat, and Jean-Claude Royer. Specification of Mixed Systems in KORRIGAN with the Support of a UML-Inspired Graphical Notation. In Heinrich Hussmann, editor, *Fundamental Approaches to Software Engineering. 4th International Conference, FASE 2001*, volume 2029 of *LNCS*, pages 124–139. Springer, 2001.

[36] Pascal Poizat and Jean-Claude Royer. A Formal Architectural Description Language based on Symbolic Transition Systems and Modal Logic. *Journal of Universal Computer Science*, 12(12):1741–1782, 2006.

[37] Nenad Medvidovic and Richard N. Taylor. A classification and comparison framework for software architecture description languages. *IEEE Transactions on Software Engineering*, 26(1):70–93, 2000.

[38] André Arnold. *Finite Transition Systems*. International Series in Computer Science. Prentice-Hall, 1994.

[39] J. A. Bergstra, A. Ponse, and S. A. Smolka, editors. *Handbook of Process Algebra*. Elsevier, 2001.

[40] Daniel M. Yellin and Robert E. Strom. Protocol specifications and component adaptors. *ACM Transactions on Programming Languages and Systems*, 19(2):292–333, March 1997.

[41] Christian Attiogbé, Pascal André, and Gilles Ardourel. Checking component composability. In *Proceedings of the 5th International Workshop on Software Composition (SC'06)*, volume 4089 of *Lecture Notes in Computer Science*, pages 18–33. Springer-Verlag, 2006.

[42] Rachid Guerraoui and Luis Rodrigues. *Introduction to Reliable Distributed Programming*. Springer-Verlag, 2006.

Communicating Process Architectures 2007
Alistair A. McEwan, Steve Schneider, Wilson Ifill, and Peter Welch
IOS Press, 2007

Concurrent/Reactive System Design
with Honeysuckle

Ian EAST

Dept. for Computing, Oxford Brookes University, Oxford OX33 1HX, England

`ireast@brookes.ac.uk`

Abstract. Honeysuckle is a language in which to describe systems with *prioritized service architecture* (PSA), whereby processes communicate values and (mobile) objects deadlock-free under client-server protocol. A novel syntax for the description of service (rather than process) composition is presented and the relation to implementation discussed. In particular, the proper separation of design and implementation becomes possible, allowing independent abstraction and verification.

Keywords. Client-server protocol, compositionality, component-based software development, deadlock-freedom, programming language, correctness-by-design.

Introduction

Honeysuckle [1] is intended as a tool for the development of systems that are both concurrent and reactive (event-driven). Formal design rules govern the interconnection of components and remove the possibility of deadlock [2,3].

A model for abstraction is provided that is derived from *communicating process architecture* (CPA) [4]. Processes encapsulate information and communicate with each other synchronously. In place of the occam channel, processes send values or transfer objects to each other according to a *service* ("client/server" or "master-servant") protocol. Whereas a channel merely prescribes data type and orientation of data flow for a single communication, a service governs a series of communications and the order in which they can occur. It therefore provides for a much richer component interface [5].

In addition to describing service architecture, Honeysuckle also provides for the expression of reactive systems. A *prioritized alternation* construct [6] affords pre-emption of one process by another, allowing multiple services to interleave, while retaining *a priori* deadlock-freedom [3]. This allows the expression of systems with *prioritised service architecture* (PSA). One additional benefit of including alternation is that it overcomes the limitation of straightforward service architecture to hierarchical structure.

Honeysuckle also addresses certain short-comings of occam. It is possible to securely transfer objects between processes, rather than just copy values[1]. Provision is included for the expression of abstract data types (ADTs), and project-, as well as system-, modularity. Definitions of processes, services, and object classes, related by application, can be gathered together in a *collection*.

Previous papers have been concerned with the programming language and its formal foundation. This one is about Honeysuckle's support for proper engineering practice; in particular, how a PSA *design* may be expressed (and verified), independent of, but binding upon, any implementation. It is simple, yet powerful.

---

[1]Mobility has also been added in occam-π [7].

1. The Problem of Engineering Software

1.1. Engineering in General

In general, the term 'engineering' has come to mean a logical progression from specification through design to implementation, with each phase rendered both concrete and binding on the next. All successful branches of the discipline have found it necessary to proceed from a formal foundation in order to express the outcome of each phase with sufficient precision. Rarely, however, do engineers refer to that foundation. More common, and much more productive, is reliance upon design rules that embody necessary principles.

A common criticism of software engineering is that theory and practice are divorced. All too often, verification (of a design against specification) is applied *a posteriori*. This amounts to "trial and error" rather than engineering, and is inefficient, to say the least. Furthermore, verification typically requires formal analysis that is specific to each individual system. It requires personnel skilled in both programming and mathematics. In systems of significant scale, analysis is usually difficult and thus both expensive and error-prone.

The primary motivation behind Honeysuckle is to encapsulate analysis within the model for abstraction offered by a programming language. Adherence to formal design rules, proven *a priori* to guarantee security against serious errors, can be verified automatically at design-time ("static verification"). Both the cost and risk of error incurred by system-specific analysis can be thus avoided. "Trial and error" gives way to true engineering.

In order to serve as an engineering tool, Honeysuckle must fulfill a number of criteria.

1.2. Compositionality and the Component Interface

Design is a matter of finding an appropriate component composition (when proceeding "bottom-up") or decomposition (when proceeding "top-down"). In order to compose or decompose a system, we require:

- some components that are indivisible
- that compositions of components are themselves valid components
- that behaviour of any component is manifest in its interface, without reference to any internal structure

A corollary is that *any system forms a valid component*, since it is (by definition) a composition. Another corollary, vital to all forms of engineering, is that it is then possible to *substitute any component with another*, that possesses the same interface, without affecting either the design or its compliance with a specification.

Software engineering now aspires to these principles [8].

Components whose definition complies with all the above conditions may be termed *compositional* with regard to some operator or set of operators. Service network components (SNCs) may be defined in such a way as to satisfy the first two requirements when subject to parallel composition [3].

With regard to the third criterion, clearly, listing a series of procedures, with given parameters, or a series of channels, with their associated data types, does little to describe object or process as a component. To substitute one object (process) with another that simply sports the same procedures (channels) would obviously be asking for trouble. One way of improving the situation is to introduce a finite-state automaton (FSA) between objects (processes) to govern the order of procedure invocation (channel communication) and thus constrain the interface [9]. Such a constraint is often termed a *contract*. The notion of a service provides an intuitive abstraction of such a contract, and is implemented using a FSA [5].

Honeysuckle is thus able at least to reduce the amount of ancillary logic necessary to adequately define a component, if not eliminate it altogether.

1.3. Balanced Abstraction

It has long been understood that system abstraction requires an appropriate balance between data and control (object and process). This was reflected in the title of an important early text on programming — *Algorithms + Data Structures = Programs* [10]. Some systems were more demanding in the design of their control structure, and others in their data structure. An equal ability to abstract either was expected in a programming language.

Imperative programming languages emerging over the three decades since publication of Wirth's book have typically emphasized "object-oriented", while occam promoted "process-oriented", programming. While either objects or processes alone can deliver both encapsulation and a "message-passing" architecture, Honeysuckle offers designers the liberty to determine an appropriate balance in their system abstraction. This is intended to ease design, aid its transparency, and increase the potential for component reuse.

A programming language can obscure and betray abstraction. Locke showed how encapsulation, and any apparent hierarchical decomposition, can dissolve with the uncontrolled aliasing accepted in conventional "object-oriented" programming languages [11]. He also illustrated how the 'has' relation between two objects can become subject to inversion, allowing each to 'own' the other. State-update can be rendered obscure in a manner very similar to *interference* between two parallel or alternating processes.

Clearly, if modularity and transparency can break down even in simple sequential designs then it hardly bodes well for any extension of the model to include concurrency and alternation. The possibility then of multiple threads of control passing through any single object poses a serious threat to transparency and exponentially increases opportunity for error.

Honeysuckle applies strict rules upon objects: each object has but a single owner at any time, class structure is statically determined, and no reference is allowed between objects. All interaction is made manifest in their class definition, rendering interdependence explicit.

1.4. Separation of Design from Implementation

Electronic engineering typically proceeds with the graphical capture of a design as a parallel composition of components interconnected by communication channels, collectively governed by precisely-defined protocol. This provides for both intuition and a precise concrete outcome. Modularity and compositionality impart a high degree of scalability.

One important principle at work is the clear separation of design and implementation. This has allowed electronic design to remain reasonably stable while implementation has moved from discrete devices, wires, and soldering irons, to VLSI and the FPGA.

All this remains an aspiration for software engineering.

This paper reports how Honeysuckle facilitates the separation of design from implementation. A sub-language expresses the behaviour of component or system purely in terms of communication. Design may be thus delivered: concrete, verified, and binding.

2. Process (De)Composition

2.1. Direct (One-to-One) Connection

The simplest protocol between two processes may be expressed as a *simple service* [5]. A simple service is one comprising a single communication. It is equivalent to channel abstraction, where only data type and orientation of data flow is stipulated. As a result, anything that can be expressed using general communicating process architecture (CPA) and occam can be expressed using service architecture and Honeysuckle, but for a single constraint. There must be *no circuit* in the digraph that describes the system.

Since circuits can give rise to the possibility of deadlock, this is not a severe limitation. It does, however, remove the option to employ a very useful alternative design pattern for the proven denial of deadlock — *cyclic ordered processes* (COPs) [12,13,4]. The theoretical foundation for design rules that deny deadlock [14,15] allows for the composition of components, each guaranteed deadlock-free by adherence to a different rule. An appealing extension to Honeysuckle would be to allow the inclusion of ('systolic') COP arrays.

Service architecture, and especially *prioritised* service architecture, affords a much richer interface than channels allow. Much more information can be captured in a design. The behaviour of system or component can be expressed in terms of communication protocol alone, without reference to procedure.

It may seem odd to define a system with reference only to communication and not to 'physical' entities like objects or processes. But a system can be very well described according to the way it communicates. It can, in fact, be defined this way. An emphasis on communication in a specification often leads to concurrency and alternation in implementation. It is only natural to retain such emphasis within a design. Honeysuckle offers a simple way to abstract such behaviour to a degree intermediate between specification and implementation, and in a manner open to intuitive graphical visualization.

For example, suppose a system is built around a component that offers a single service, which is *dependent* upon the consumption of just one other (Figure 1).

Figure 1. A single service dependent on just one other.

We can express this simply:

```
network
  s1 > s2
```

Note that the symbol used to denote a dependency is suitably asymmetric, and one that also correctly suggests the formation of a partial order.

As it stands, the above forms a complete system, implemented as a parallel composition. The centre component, isolated, requires an INTERFACE declaration:

```
interface
  provider of s1
  client of s2
```

but no network definition. A complete system requires no interface declaration.

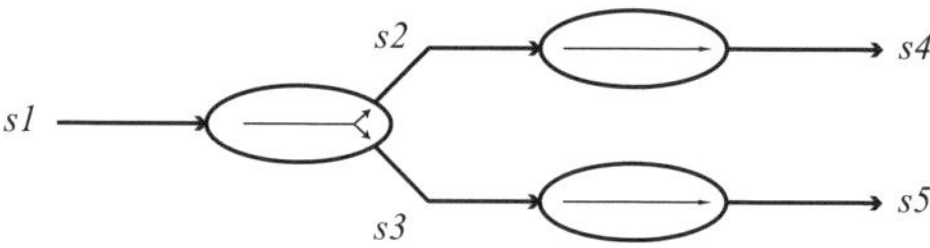

Figure 2. A tree structure for service dependency.

A tree-structured component (Figure 2) is easily described:

```
network
  s1 > s2, s3
  s2 > s4
  s3 > s5
```

Chains of identical services can be indicated via replication of a dependency:

```
network
  repeat for 2
    s1 > s1
```

Note that all reference to services has been solely according to their type. No instance of any service has yet needed distinction by name. Honeysuckle can connect processes correctly simply by their *port*[2] declarations and the network definition that governs their composition.

Naming might have become necessary should one component provide multiple identical services, except that such structure may be described more simply.

2.2. Sharing and Distribution

A common CPA design pattern is the consumption of a common service by multiple clients, which is why **occam 3** introduced *shared channels* [16]. Honeysuckle similarly permits the sharing of any service. For example, Figure 3 depicts a simple closed system where two components share a service.

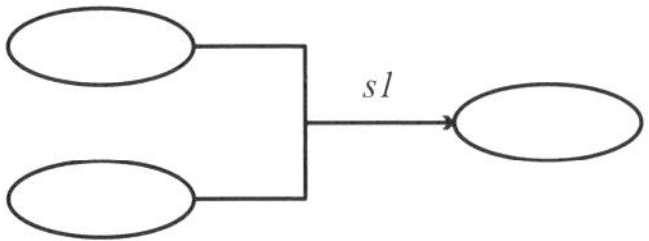

Figure 3. Sharing of a service between two clients.

Such a design is declared as follows:

```
network
  shared s1
```

As outlined in a previous paper [5], one-to-any, and any-to-any connection patterns are also supported via the DISTRIBUTED and SHARED DISTRIBUTED attributes, respectively. None of these options are the concern of implementation. They neither appear in nor have any effect upon the interface of any single component.

Within a design, there is still no need for naming instances of service. We have thus far presumed that every service is provided in precisely the same way, according to its definition only, and subject to the same dependencies.

2.3. Service Bundles, Mutual Exclusion, and Dependency

A design may require a certain *bunch* of services to be subject to mutual exclusion. If any member of the bunch is initiated then all of the others become unavailable until it completes.

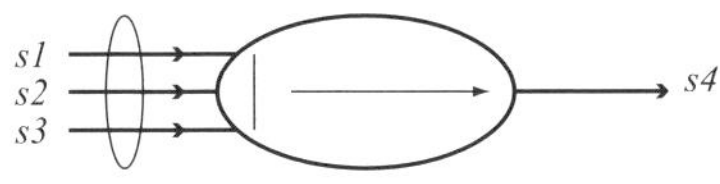

Figure 4. A service *bunch*, subject to mutual exclusion and a dependency.

Connections to the component depicted in Figure 4 can be expressed:

---

[2]A *port* is one end of a service, *i.e.* either a *client* or *server* connection.

```
network
  exclusive
    s1
    s2 > s4
    s3
```

A bunch of mutually exclusive services can be provided by a single, purely sequential, process. All that is required is selection between the initial communications of each. In oc-cam, an ALT construct would be employed. The body of each clause merely continues the provision of the chosen service until completion. An outer loop would then re-establish availability of the entire bundle. (In PSA, and in CPA in general, it is often assumed that processes run forever, without terminating.)

An object class, when considered as a system component, typically documents only procedures offered, within its interface. It does not usually declare other objects on which it depends. A service network component (SNC) documents both services provided *and* services consumed, together with the dependency between. Any interface thus has two 'sides', corresponding to provision and consumption, respectively. Honeysuckle requires documentation of dependency beginning with service provision and progressing towards consumption.

Suppose a system including the component shown in Figure 4 were to be extended with $s4$ being provided under mutual exclusion with another service, $s5$, and a dependency upon the consumption of yet another, $s6$. We would then write:

```
network
  exclusive
    s1
    s2 > s4
    s3
  exclusive
    s4 > s6
    s5
```

If $s4$ failed to reappear under the second EXCLUSIVE heading, $s5$ (and any other services in that bundle) would be listed together with $s1 - 3$. Mutual exclusion is fully associative.

2.4. Service Interleaving

An alternative to bunching services under mutual exclusion is to allow them to *interleave*. This allows more than one service in a group to progress together. Should two be ready to proceed at the same moment, the ensuing communication is decided according to service[3] *prioritization*. A service of higher priority will pre-empt one attributed lower priority.

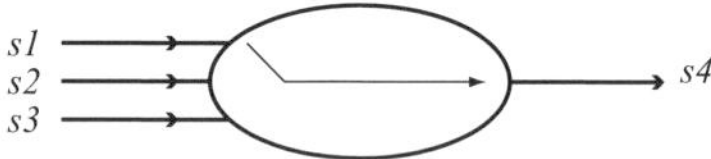

Figure 5. Interleaving services.

This too can be expressed as a feature of design, quite separate from implementation:

```
network
  interleave
    s1 > s4
    s2
    s3
```

---

[3]Each member of any bunch is attributed a common priority.

Prioritisation is indicated simply by the order in which services are listed (highest uppermost in both picture and text).

A process might interleave bunches. Each bunch would remain subject to mutual exclusion between its members:

```
network
  interleave
    exclusive
      s1 > s4
      s2
      s3
    exclusive
      s5
      s6
      s7 > s8
```

Again, implementation reduces to a programming construct; in this case, *prioritized alternation* (WHEN) [6]. Each clause in a Honeysuckle alternation may be a guarded process or a selection, according to whether a single service or a bunch is offered.

Interleaving several instances of a common service offers an alternative to sharing a single instance, where each client is effectively allocated the same priority. Replication may be used to indicate vertical repetitive structure, as it can horizontal:

```
network
  interleave for 2
    exclusive
      s1
      s2 > s4
      s3
```

Note that replication under mutual exclusion would add nothing to the notion of sharing.

3. Asymmetry in Service Provision

For many systems with PSA, it is enough to define their design without distinction between two instances of the same service type. Implementation could proceed with components whose interface can be defined with reference only to that type. If two different processes each declare the capability of providing that type of service, it would not matter which provides each instance of it.

Any departure from that scenario is termed an *asymmetry*. There are two kinds.

A *design asymmetry* is one where dependency in the provision of two services of the same type differs. An example might be formed were $s3$ in Fig. 2 replaced by a second use of $s2$. This would make it impossible to document dependency without ambiguity. Note that no such ambiguity would result upon implementation since component interface could be matched with dependency. A (reasonably intelligent) compiler will still be able to compose components correctly.

Note that any service is necessarily shared (or distributed) symmetrically, since no provider (or client) can distinguish one client (provider) from another.

An *implementation asymmetry* is where the provision of two instances of the same service are not interchangeable, even though there may be no design asymmetry. Some relationship between information exchanged is material to the system required. If so then a single instance may neither be shared nor distributed.

It is worth reflecting that, in traditional, typically purely sequential, programming, we commonly distinguish between "data-oriented" and "control-oriented" application design. Often, the orientation is inherent in the problem. Sometimes, it is a choice reflecting that of the designer. One might similarly identify "service-orientation" also. Business nature and organization has re-oriented itself towards service provision, to great effect. The same development in the design of software would arguably result in a greater reliance upon service architecture, with less asymmetry appearing.

At the cost of complicating the declaration of design a little, a mechanism is provided by Honeysuckle by which asymmetry may be introduced. For each asymmetric use of a service, a *service alias* ('renaming') is declared within the network declaration. It then becomes possible for the interface declaration of each process to distinguish one instance of service from another of the same type.

If we again refer back to Fig. 2 for an example, let us suppose that $s2$ and $s3$ are of the same type (share the same definition), and $s4$ and $s5$ are similarly alike. Suppose that we *care* that each instance of $s2/s3$ is provided separately, because there is some difference we cannot yet make explicit. All we need do is declare two service aliases within the network definition:

```
network
  named
    s2 : s3
    s4 : s5
  ...
```

Each component interface can now distinguish the desired connection.

4. Parametric and Dynamic Configuration

Modular software engineering calls for the ability to compose components whose utility is *not* restricted to a single application. Having renamed services in order to apply an implementation asymmetry in service provision, it should be possible to employ a component designed for wider use. While it must possess an interface appropriate to any design asymmetry, it will know nothing of any service alias. Its interface will refer only to the original service (type) names given in each corresponding definition.

An in-line component definition can match service and alias directly:

```
{
  ...
  network
    named
      s2 : s3
      ...

  parallel
    {
      interface
        provider of s2 alias s3

      ...
    }
    ...
}
```

while the option remains to state simply "`provider of s3`".

The interface of any 'off-line' process definition can indicate that it expects to be told which service it is to consume/provide via `alias ?`, in which case its reference (invocation) should provide a *configuration parameter*.

There is one other kind of configuration parameter, used by the network declaration of the recipient. A *configuration value* may be passed and used to limit replication. Since this may be computed upon passing, it allows the network of a parallel component to be configured dynamically.

A Honeysuckle process reference may thus include up to three distinct actual parameter lists, arranged vertically ("loo roll" style), and delimited by semi-colon. When each list has no more than one item, parameters can be arranged on the same line as the command (process invocation). For example, suppose a process *mediate* is defined separately (like a procedure in Pascal), and it expects one service alias and one configuration value. Definition would be as follows:

```
process mediate is
  {
    ...
    interface
      client of s1 alias ?
      ...

    network
      received Length
      interleave for Length
        ...
  }
```

An invocation might be simply:

```
mediate ; s2 ; 4
```

5. Conclusion

Honeysuckle began as a single-step method for the composition of concurrent/reactive software guaranteed free from the threat of deadlock. As such, it was either going to remain a simple academic exemplar, or grow into a tool suited to professional use. It was decided to take the latter path, which has inevitably proved long and arduous.

Here, elements of the language have been introduced that afford PSA *design*, separate from, and independent of, implementation. Design of system or component is expressed purely in terms of communication, as a composition of services rendered. Any such design may be compiled and verified independently, and automatically, using the same tool used for implementation. It will then remain binding as the implementation is introduced and refined. Every verified design, and thus implementation, is *a priori* guaranteed deadlock-free.

It has been shown how a design may be composed under service dependency, mutual exclusion, and interleaving, and how repetitive structure can be efficiently expressed. While prioritized service architecture alone may suffice to abstract some systems, especially when design is oriented that way, others may call for significant emphasis on process rather than communication. A mechanism has therefore been included whereby *asymmetry* in service implementation can be introduced.

Given that the parallel interface of each component is defined purely according to services provided and consumed, *configuration parameters* have proved necessary in order to allow the reuse of common components, and preserve modularity. They also afford limited dynamic configuration of components, allowing the structure of each invocation to vary.

With regard to the progress of the Honeysuckle project, another decision taken has been to complete a draft language manual before attempting to construct a compiler. A *publication language* would then be ready earlier, to permit experiment and debate. This is now complete, though the language (and thus manual) is expected to remain fluid for some time yet [17].

Work is now underway towards a compiler. A degree of platform independence will be facilitated by the use of *extended transputer code* (ETC) [18] as an intermediary[4].

While Honeysuckle has evolved into a rather ambitious project, it is nonetheless timely. The beginning of the twenty-first century has marked the rise of large embedded applications, that are both concurrent and reactive. Consumers demand very high integrity from both home and portable devices that command prices, and thus (ultimately) development costs, orders of magnitude below those of traditionally challenging applications, such as aerospace. Existing methods are inappropriate. While a sound formal foundation is an essential prerequisite for something new, proper support for sound engineering practice is also required.

Honeysuckle now offers both.

By clearly separating design from implementation, while rendering it inescapably formal and binding, Honeysuckle brings the engineering of software into closer harmony with that of electronic and mechanical systems, with which it must now co-exist.

References

[1] Ian R. East. The Honeysuckle programming language: An overview. *IEE Software*, 150(2):95–107, 2003.

[2] Jeremy M. R. Martin. *The Design and Construction of Deadlock-Free Concurrent Systems*. PhD thesis, University of Buckingham, Hunter Street, Buckingham, MK18 1EG, UK, 1996.

[3] Ian R. East. Prioritised Service Architecture. In I. R. East and J. M. R. Martin et al., editors, *Communicating Process Architectures 2004*, Series in Concurrent Systems Engineering, pages 55–69. IOS Press, 2004.

[4] Ian R. East. *Parallel Processing with Communicating Process Architecture*. UCL Press, 1995.

[5] Ian R. East. Interfacing with Honeysuckle by formal contract. In J. F. Broenink, H. W. Roebbers, J. P. E. Sunter, P. H. Welch, and D. C. Wood, editors, *Proceedings of Communicating Process Architecture 2005*, pages 1–12, University of Eindhoven, The Netherlands, 2005. IOS Press.

[6] Ian R. East. Programming prioritized alternation. In H. R. Arabnia, editor, *Parallel and Distributed Processing: Techniques and Applications 2002*, pages 531–537, Las Vegas, Nevada, USA, 2002. CSREA Press.

[7] Fred R. M. Barnes and Peter H. Welch. Communicating mobile processes. In I. R. East and J. M. R. Martin et al., editors, *Communicating Process Architectures 2004*, pages 201–218. IOS Press, 2004.

[8] Clemens Szyperski. *Component Software: Beyond Object-Oriented Programming*. Component Software Series. Addison-Wesley, second edition, 2002.

[9] Marcel Boosten. Formal contracts: Enabling component composition. In J. F. Broenink and G. H. Hilderink, editors, *Proceedings of Communicating Process Architecture 2003*, pages 185–197, University of Twente, Netherlands, 2003. IOS Press.

[10] Niklaus Wirth. *Algorithms + Data Structures = Programs*. Series in Automatic Computation. Prentice-Hall, 1976.

[11] Tom Locke. Towards a viable alternative to OO — extending the occam/CSP programming model. In A. Chalmers, M. Mirmehdi, and H. Muller, editors, *Proceedings of Communicating Process Architectures 2001*, pages 329–349, University of Bristol, UK, 2001. IOS Press.

[12] E. W. Dijkstra and C. S. Scholten. A class of simple communication patterns. In *Selected Writings in Computing*, Texts and Monographs in Computer Science, pages 334–337. Springer-Verlag, 1982. EWD643.

[13] Jeremy Martin, Ian East, and Sabah Jassim. Design rules for deadlock freedom. *Transputer Communications*, 2(3):121–133, 1994.

[14] A. W. Roscoe and N. Dathi. The pursuit of deadlock freedom. Technical Report PRG-57, Oxford University Computing Laboratory, 8-11, Keble Road, Oxford OX1 3QD, England, 1986.

[15] S. D. Brookes and A. W. Roscoe. Deadlock analysis in networks of communicating processes. *Distributed Computing*, 4:209–230, 1991.

[16] Geoff Barrett. *occam 3 Reference Manual*. Inmos Ltd., 1992.

[17] Ian R. East. *The Honeysuckle Programming Language: A Draft Manual*. 2007.

[18] Michael D. Poole. Extended transputer code — a target-independent representation of parallel programs. In P. H. Welch and A. W. P. Bakkers, editors, *Architectures, Languages and Patterns for Parallel and Distributed Applications*, pages 187–198. IOS Press, 1998.

---

[4]Subject to the kind permission of Prof. Peter Welch and his colleagues at the University of Kent

Communicating Process Architectures 2007
Alistair McEwan, Steve Schneider, Wilson Ifill and Peter Welch (Eds.)
IOS Press, 2007

CSP and Real-Time: Reality or Illusion?

Bojan ORLIC and Jan F. BROENINK
Control Engineering,
Faculty of EE-Math-CS, University of Twente
P.O.Box 217, 7500AE Enschede, the Netherlands
{B.Orlic, J.F.Broenink}@utwente.nl

Abstract. This paper deals with the applicability of CSP in general and SystemCSP, as a notation and design methodology based on CSP, in particular in the application area of real-time systems. The paper extends SystemCSP by introducing time-related operators as a way to specify time properties. Since SystemCSP aims to be used in practice of real-time systems development, achieving real-time in practice is also addressed. The mismatch between the classical scheduling theories and CSP paradigm is explored. Some practical ways to deal with this mismatch are presented.

Keywords. SystemCSP, CSP, real-time.

Introduction

Concurrency is one of the most essential properties of reality as we know it. We can perceive that in every complex system, many activities are taking place simultaneously. The main source of complexity in designed systems stems actually from the simultaneous (concurrent) existence of many objects, events and scenarios. Better control over the concurrency structure should therefore automatically reduce the problem of complexity handling. Thus, a structured way to deal with concurrency is needed. CSP theory [1, 2] is a convenient tool to introduce a sound and formally verifiable concurrency structure in designed systems. Our SystemCSP [3] graphical notation and design methodology, as well as its predecessor GML [4], is built on top of the CSP theory. SystemCSP is an attempt to put CSP into practical use for the design and implementation of component-based systems.

Various approaches attempt to introduce ways to specify time properties in CSP theory [1, 2]. SystemCSP as a design methodology based on CSP and intended to be suitable for the real-time systems application area, offers a practical application of those theories. The way in which time properties are introduced in SystemCSP also makes a connection between the two referenced approaches of theoretical CSP.

Specifying time properties is one part of the problem. It allows capturing time requirements and execution times. In practical implementations, resulting time behavior of processes is also the consequence of time-sharing of a processor or network bandwidth between several processes. This time-sharing implies switching the context of execution from one involved process to another, where the order of execution is based on some kind of priority assignment.

Classical scheduling theory offers recipes to give real-time guarantees for systems where several tasks share the same processing or network resource using some priority based scheme. However, as it will be illustrated in Section 2.1.3, there is an essential mismatch between the programming paradigm assumed by classical scheduling techniques and the one offered by the CSP way of design. This mismatch raises the fundamental question: are CSP-based systems suitable for usage in real-time systems or for this application area should one rely on some other method? This paper will attempt to show possible directions in solving the problem of achieving real-time in CSP-based systems. The first direction in addressing this problem is constructing CSP-based design patterns that can match the form required by the classic scheduling techniques. The second direction is oriented towards the creation of scheduling or real-time analysis theories specific for CSP-based systems.

1. Time Properties in Specification of CSP-based Systems

1.1 Discrete Time Event 'tock'

In [1], time properties are specified by introducing an explicit time event named 'tock'. This implicitly introduces the existence of a discrete clock that advances the time of the system for one step with each occurrence of the tock event. Time instants can thus be represented by a stream of natural numbers, where every occurrence of the tock event can be considered to increase the current time for one basic time unit. All processes with time constraints synchronize with the progress of time by participating directly in the tock event, or via interaction with processes that do. Advantages of this approach are that it is simple, easy to understand and flexible. It does *not* introduce any theoretical extensions to CSP theory and thus formal checking is possible using the same tools (FDR) as in untimed CSP.

1.2 Timed CSP

Timed CSP [2] extends CSP theory by introducing ways to specify time properties in CSP descriptions. There is, however, (yet) no tool that can verify designs based upon Timed CSP.

Times associated with events are non-negative real numbers, thus creating a dense continuous model of time. This assumption makes the verification process complicated and *not* practical. The difference between this approach and introducing the explicit time event ("tock") is comparable to the difference between continuous systems and their simulation on a computer using discretized time.

The method is also not related to real-time scheduling. It defines the operational semantics for introducing time properties in CSP-based systems. Several essential extensions to CSP are basis for making a system of proofs according to the ones that exist in basic CSP theory. Newly introduced operators include: observing time, evolution transition, timeout operator, timed interrupt operator and time delay.

Time can be observed at any event occurrence. The observed time can then be used in a following part of process description as a free variable.

The expression

$$P \ = \ ev1@t1 \ \rightarrow \ ev2@t2 \ \rightarrow \ display(t2\text{-}t1) \ \rightarrow \ SKIP$$

specifies that the time of occurrence of event ev1 is stored in variable t1 and the time of occurrence of ev2 is stored in variable t2. Afterwards a function is called that displays the time interval between the occurrences of event ev1 and event ev2.

The timeout operator is a binary operator representing the time-sensitive version of the *external choice* operator of CSP. It is offering the choice between the process specified as its first operand and the process specified as second operand. In case when a timeout event takes place before the process guarded via the timeout operator engages in some external event, the control is given to the process specified as second operand.

The expression:

$$(\text{ev1} \rightarrow \text{P1}) \overset{d}{\rhd} \text{Q}$$

specifies that if the event `ev1` takes place in d time units from the moment it is offered, then the process will subsequently behave as process P1. Otherwise, it will behave as process Q.

The *timed interrupt* is a binary operator representing the time-sensitive version of the *interrupt operator* of CSP. The main difference is that the event that triggers the interrupt is actually a timeout event. The process specified as the second operand will be executed after the timeout event signifies that the guarded process did not succeed to successfully finish its execution in the given time interval. As opposed to the timeout operator that uses timeouts to guard only a single event, the timed interrupt operator is guarding the completion of a process. If that process does not finish its execution in the predefined time interval, its further execution is abandoned.

The expression:

$$(\text{ev1} \rightarrow \text{P1}) \; \triangle_d \; \text{Q}$$

specifies that the process ev1->P1 will be granted a time interval of d time units to be performed. In the moment when the given time interval expires, further execution of the process ev1->P1 is aborted (interrupted) and the process Q is executed instead.

Introducing *time delay* (*delay event prefix* in Timed CSP) is a step from the world of ideal computing devices capable of infinitely fast parallel execution (as assumed by CSP) to the world of real target implementations. Time delay is used to extend process descriptions with the specification of execution times. In software implementations, the execution times get certain values depending on the processing node which executes a process. During this delay time, a process cannot engage in any event, that is, it acts as a STOP process. In fact, specifying the *delay event prefix* is equivalent to applying the timeout operator on a STOP process as the first operand and the rest of original process as the second operand. A *delay event prefix* is specified by augmenting an event prefix arrow with a time delay value. Instead of single number denoting a fixed execution time, it is possible to specify an interval for the expected time delay. In that case, a pair of values is grouped via square brackets.

The expression:

$$P \; = \; \text{ev1} \; \overset{10}{\rightarrow} \; \text{ev2} \; \overset{[10,20]}{\rightarrow} \; \text{SKIP}$$

specifies that after occurrence of event ev1, process P is for 10 time units unable to participate in any event. After the interval of 10 time units expires, process P will offer event ev2 to the environment. Then, after the event ev2 is accepted by the environment, it will take between 10 and 20 time units before process P can successfully finish its execution.

Evolution transition is a way to display an observed delay between events in some particular execution of the process description.

The expression:

$$
\begin{aligned}
P &= ev1 \rightarrow ev2 \rightarrow SKIP \\
&\quad \wr\ 10 \\
&\quad ev1 \rightarrow ev2 \rightarrow SKIP \\
&\quad \downarrow\ ev1 \\
&\quad ev2 \rightarrow SKIP \\
&\quad \wr\ 20 \\
&\quad ev2 \rightarrow SKIP \\
&\quad \downarrow\ ev2 \\
&\quad SKIP
\end{aligned}
$$

represents an execution in which the event ev1 has taken place 10 time units after it was initially offered to the environment, and where the event ev2 has taken place 20 time units after it was initially offered to the environment.

1.3 Specification of Time Properties in SystemCSP

SystemCSP recognizes that operators introduced in TimedCSP are practical for describing time properties of systems. However, we are aware that there is no real need to introduce a dense continuous model of time for modelling software and hardware implementation of processes. Therefore, in SystemCSP we start with the discrete notion of time as in [1] and introduce the basic event `tock` produced by the timing subsystem in regular intervals. Upon the `tock` event, we construct a process that implements a timing subsystem. This subsystem provides services used in the implementation of the higher-level design primitives that provide functionality analogue to the one defined by the timeout and the timed interrupt operators defined in TimedCSP [2]. In this way, it is possible to create designs using Timed CSP-alike operators, to describe them in basic CSP theory, making these designs amenable to formal verification equally as untimed CSP designs.

Section 1.3.1 introduces a notation for specifying time constraints and delays in SystemCSP. Section 1.3.2 provides design patterns for implementation the timing subsystem based on the `tock` event. Sections 1.3.4 and 1.3.5 provide graphical symbols for specification and design patterns for implementation of behaviours defined as in timeout and timed interrupt operators of TimedCSP.

1.3.1 Execution Times and Time Constraints

In the control flow oriented part of SystemCSP, a process description starts with its name label (e.g. name label P in Figure 1). Control flow operators match the CSP set of operators, and are used to relate event-ends and nested process blocks, specifying in that way the concurrency structure of the process. The prefix operator of CSP is represented with an arrow, parallel composition and external choice are represented with a pair of FORK and JOIN elements marked with the type of operator, and so on.

SystemCSP specifies time properties inside square brackets positioned in a separate node element or next to the element they are associated with (see Figure 1). In Figure 1, the first block specifies that the process P is to be triggered in precisely periodic moments of time, with the period equal to Ts. The occurrence of event `ev1` is a point when time is stored to the variable `t1`. The time when the event `ev2` occurs is stored in the variable `t2`. The keyword `time` is used to denote the current time in the system.

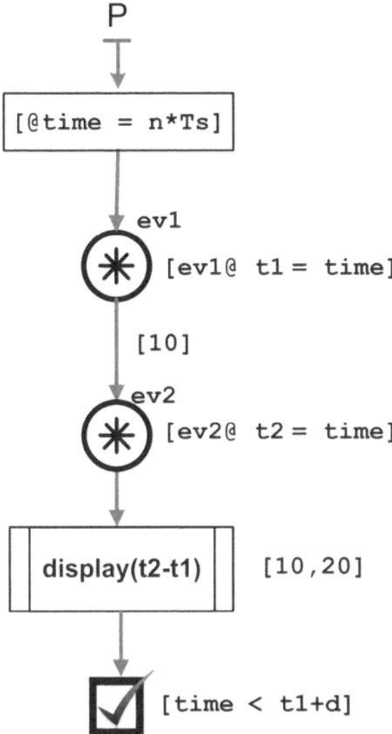

Figure 1 Specifying time requirements

Execution times can also be visualized on SystemCSP diagrams. In SystemCSP, as in Timed CSP, the time delay is specified inside square brackets and instead of a single value representing a fixed execution time, it is possible to display a pair of values that defines a range. The range of possible execution times is bounded by the *minimum execution time* (minET) and the *worst case execution time* (WCET). In addition, often it is useful to keep track of the *average execution time* (avET). In that case, a triple is specified.

The position of the time delay specification is related to the associated diagram element (e.g. next to the associated computation process block or the prefix arrow replacing it or the event that allows progress of the following computation block). The specified delay can be just a number, in which case the default time unit is implied. Otherwise, the specification of time delay should also include a time unit. Time delay can also be specified as a variable that evaluates to some time value.

The *evolution transitions* of Timed CSP are not represented in SystemCSP so far, since they are used for visualizing time delays in observed actual executions of processes. In future, when a prospective tool that can simulate or display execution is built, it can use the same symbol as in Timed CSP.

In addition to operators defined in Timed CSP, SystemCSP also introduces a notation for visual specification of time constraints. Those constraints are not directly translated to the CSP model of the system. These time constraints specify that certain events take place before some deadline or precisely at some time. A deadline can be set relative to some absolute time or as a maximally allowed distance in time between the occurrences of two events. The deadline constraints are independent of the platform on which they are executed. In Figure 1, process P is scheduled to be triggered periodically at precise moments in time. The time constraint associated with the termination of process P specifies that it should take place strictly less then d time units after `t1` moment in time, or, in other words, process P must finish successfully at most `d` time units after an occurrence of the event `ev1`.

1.3.2 Timing Subsystem

Figure 2 introduces one possible design of a timing subsystem. The purpose of this example is *not* to provide a ready-to-use design, but rather to illustrate the point of constructing a timing subsystem starting with the `tock` event. Note that in SystemCSP the symbol * is used to mark the event-end that can initiate an occurrence of the event, while the symbol # is used on the side that can only accept events. From a CSP point of view, this is irrelevant because event occurrence is symmetrical. However, in a design it is often very useful to provide additional insight by specifying the difference between the side that can initiate events and the side that can only accept events.

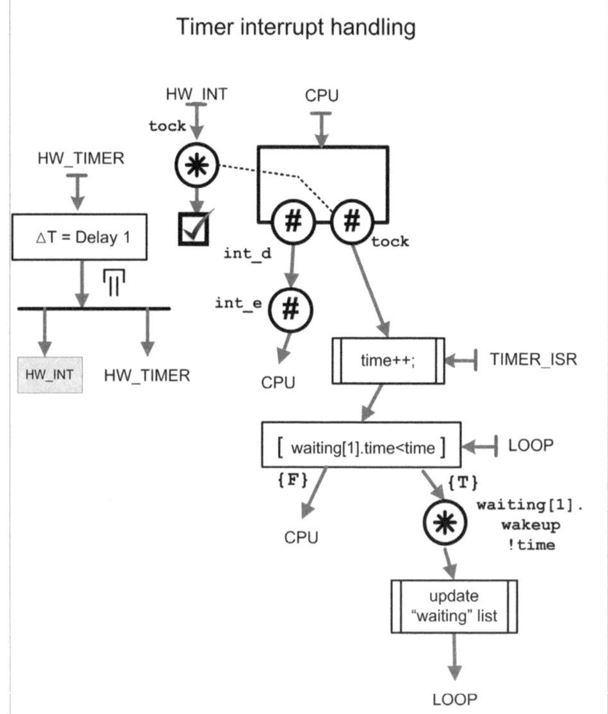

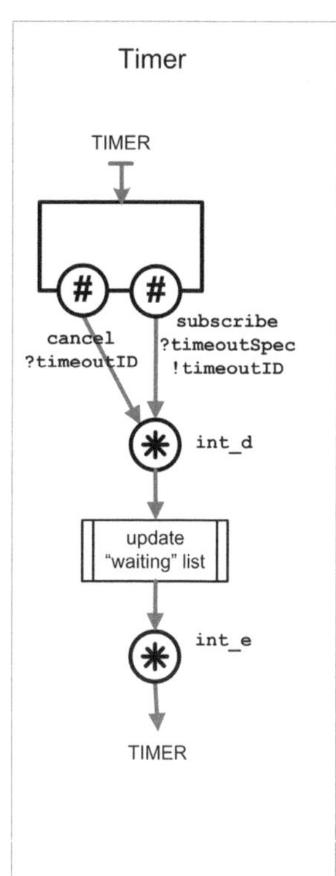

Figure 2 Timing susbsystem

The timing subsystem in Figure 2 contains several processes executed concurrently. HW_TIMER is implemented in hardware and forks instances of the hardware interrupt process, `HW_INT`, at regular intervals. The `HW_INT` process synchronizes with the CPU on the event `tock`, invoking the timer interrupt service routine (TIMER_ISR process). TIMER_ISR increments the value of the variable `time`. TIMER_ISR also maintains a sorted list of processes waiting on timeout events. Processes in this list, for which the time they wait for is less then or equal to the current time, will be awakened using the `wakeup` event. The awoken processes will be removed from the top of the list. In the case the awoken process is periodic, it is added again to a proper place in the waiting list.

The process CPU acts as a gate that can disable (event `int_d`) or enable (event `int_e`) the timer and other interrupts. When interrupts are enabled, event `tock` can take place and as a consequence the interrupt service routine TIMER_ISR will be invoked. The case of occurrence of the `int_d` event, as a next event only the `int_e` event is allowed and until then, the event `tock` cannot be accepted, and consequently interrupts are *not* allowed to happen.

Processes using services of the timing subsystem can, via the TIMER process, either subscribe (via event `subscribe`) to the timeout service or generate a `cancel` event to cancel a previously requested timeout service. Since these activities are actually updating the waiting list, this list must be protected from being updated in the same time by TIMER and TIMER_ISR processes. That is achieved in this case via disabling/enabling interrupts (`int_d` / `int_e` events).

1.3.3 Watchdog Design Pattern

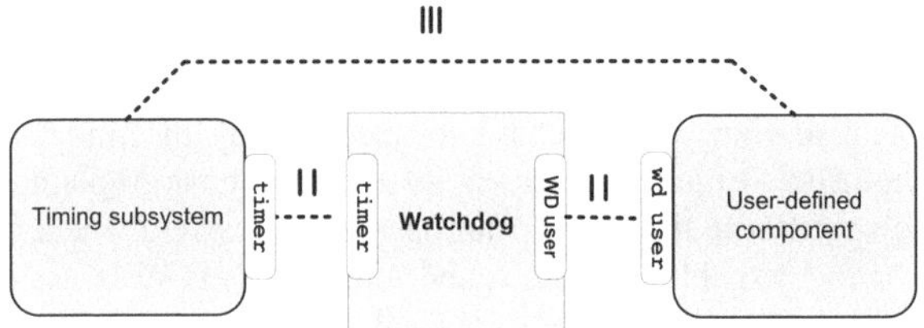

Figure 3 Interaction diagram: using a watchdog interaction contract

The interaction view specified in Figure 3 illustrates the interaction between a user-defined component and the timing subsystem component via the watchdog interaction contract. The watchdog pattern is used to detect timing faults and to initiate recovery mechanisms.

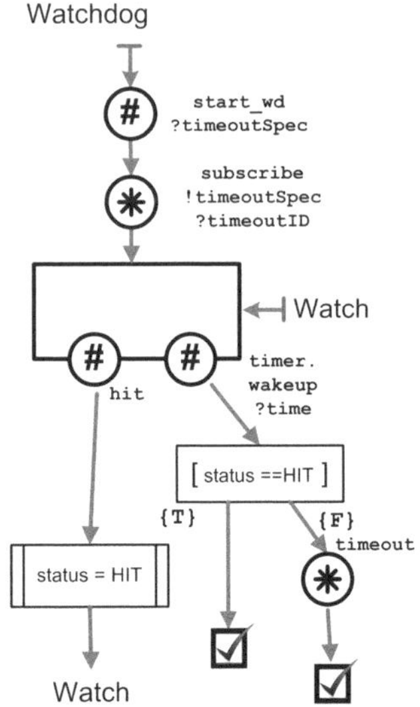

Figure 4 Watchdog design pattern

The design pattern for the watchdog process (see Figure 4) relies on services provided by the timing subsystem. A user initializes the watchdog using the `start_wd` event, which results in a watchdog request to be notified by the timing subsystem when the specified timeout expires. In case when the watchdog user chooses to initiate the `hit` event, the watchdog is disarmed. Otherwise, upon occurrence of the timeout (event `wakeup`), the watchdog will initiate the `timeout` event, invoking the warning situation.

1.3.4 Timed Interrupt Operator

The timed interrupt operator is simply a time-sensitive version of the interrupt operator. Its implementation, as depicted in Figure 5, contains the interrupt operator, and additional synchronization with a watchdog process. The watchdog is initialized, via the `start_wd` event, with the timeout value specified in the timed interrupt operator. When the guarded process (`ev1->P` in the example of Figure 5) finishes and the hit event takes place, the associated watchdog process will be disarmed. If, however, the timeout event takes place, it will cause the guarded process to be aborted, and the process specified as the second operand (process Q in the example of Figure 5) is executed.

The closed dotted line at the left-hand side of the Figure 5 encircles elements that are providing the implementation of the behaviour specified by the timed interrupt operator. The right-hand side of the Figure 5 abstracts away from those implementation details by providing a way to specify the timed interrupt operator as basic element of the SystemCSP vocabulary. In fact, a pair of blocks with timed-interrupt symbol is used to determine the scope of the operator, similarly as brackets are used in CSP expressions.

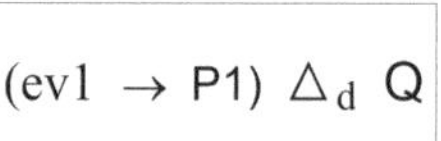

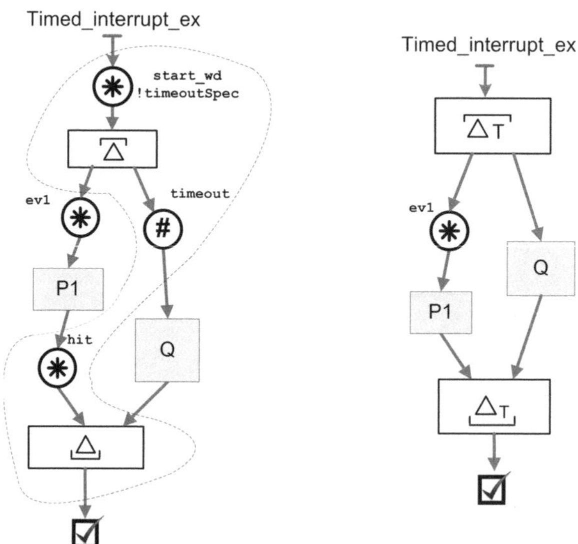

Figure 5 Timed interrupt – implementation and symbol

1.3.5 Implementation of the Timeout Operator

The timeout operator is simply a time sensitive external choice where one of the branches is a guarded process and the other one starts with a time event that will be initiated by the associated watchdog after the requested timeout expires. Following the timeout event (see Figure 6), the process specified as second operator is executed.

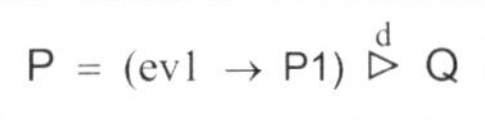

$$P = (ev1 \rightarrow P1) \overset{d}{\rhd} Q$$

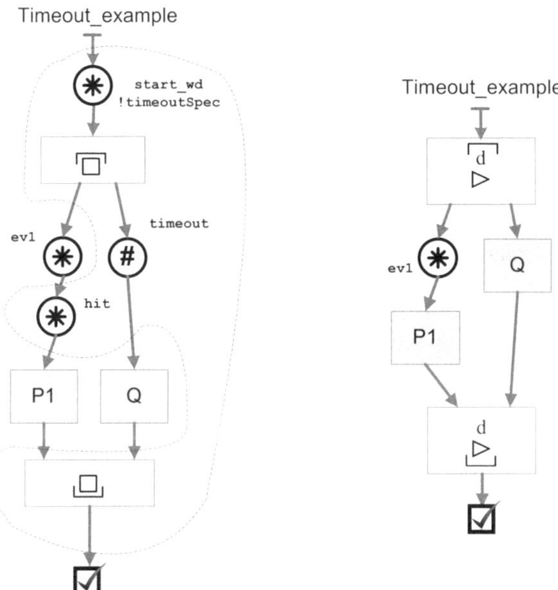

Figure 6 Timeout operator – implementation and symbol

Figure 6 depicts the implementation and a symbol of the timeout operator on a simple example and its associated visualization using the symbol for timeout operator. Instead of the letter d inside the timeout operator symbol, it is possible to use any number or variable representing time. The left hand-side of Figure 6 depicts the implementation details encircled via the dotted line, while the right-hand side introduces notation elements used to represent the timeout operator as one of the basic building blocks in SystemCSP.

2. Real-time in the Implementation of CSP-based Systems

2.1 Identifying Problems

2.1.1 Origin of Time Constraints in Implementation of Control Systems

An *embedded control system* interacts with its environment via various sensors and actuators. Sensors convert analogue physical signals to signals understandable by the embedded control system (digital quantities in case of computer-based control). Actuators (motors, valves….) perform transformation in opposite direction (note in Figure 7 the different types of arrow symbols used for digital and analogue signals).

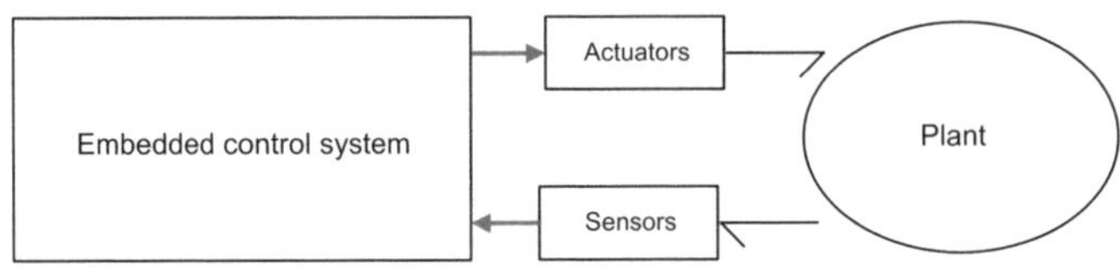

Figure 7 Typical control system

Obviously, a (control) system and its relevant environment (plant, i.e. machine to be controlled) exist concurrently. In fact, both plant and control system are often decomposed in subsystems that exist concurrently and are cooperating to achieve the desired behaviour. Thus, in the control system application area, concurrency is naturally present.

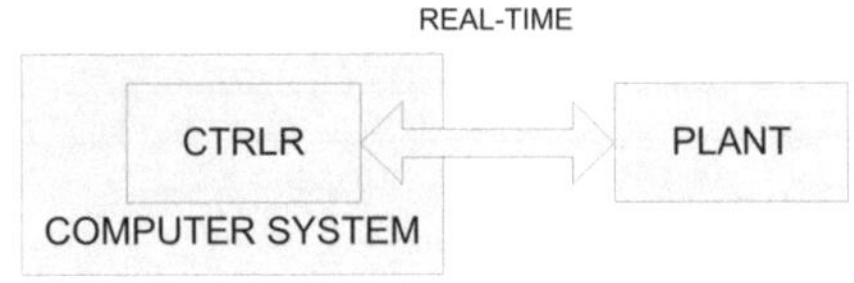

Figure 8 Implementation of computer control system

Figure 8 illustrates a computer implementation of a control system. The control algorithm (CTRLR block in Figure 8) is performed by the computer system. In general case the computer system can contain many computer nodes connected via some network. The time pattern of the interaction between the control system and its environment is based on the time constraints imposed by the underlying control theory. The computer system implementing the embedded control system must be able to guarantee that the required time properties will be met in real time. So, in real-time systems, "correctness of the system depends not only on the logical result of the computation but also on the time at which results are produced" [5]. In those systems, the response should take place in a certain time window. Real-time is not about the speed of the system, but rather about its relative speed compared to the required speed of its interaction with the environment. Rather than fast, the response of those systems should be predictable. A fast system will not be real-time if its interaction with environment requires a faster response. A slow system can work in real-time if it is faster than its interaction with environment requires.

A control loop starts with sensor data measurements and finishes with delivering command data to the actuators. The time between two subsequent measurement (sampling) points is named *sampling period* and the time between a sampling point and the related actuation action is named *control delay* [6]. Digital control theory assumes equidistant sampling and fixed control delay time. On an ideal computer system, the control loop computation is performed infinitely fast. In reality, it takes a certain time that should be bounded. This gap between ideal and real computing devices reflects itself in a design choice between two possible patterns used in practice for ordering sampling and actuation tasks.

In the Sample-Compute-Actuate approach, depicted in upper part of Figure 9, the computation time is usually assumed to be negligible, implying the computing device close to ideal. Rule of thumb is that the behaviour of a control system will still be acceptable when this computation time is kept smaller then around 20% of the sampling period. Obviously this approach does not really guarantee that system will always work as expected by control engineers. Especially in complex control systems that contain more than one control loop, or control loops closed over a network, the influence of variable control delay becomes an important factor in the resulting behaviour of the control system.

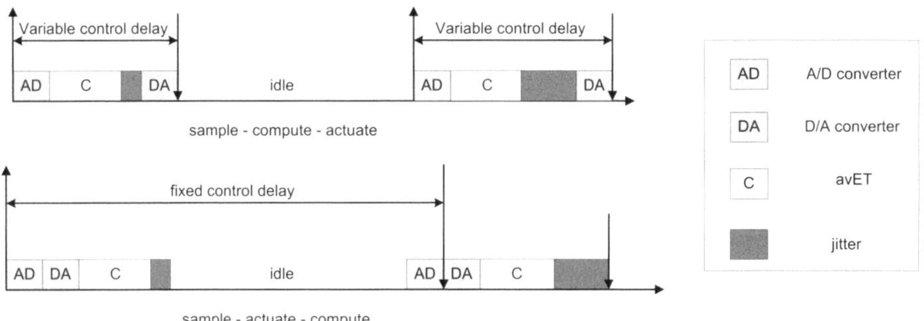

Figure 9 Sampling period and control delay (adapted from [7])

The second approach, Sample-Actuate-Compute, takes into account the non-ideal nature of the computation devices. In the approach depicted in the lower part of Figure 9, the control delay is fixed and usually set to be equal to the period. By fixing the point of actuation to be immediately after the sampling point for the next iteration, two goals are achieved: first, actuators are prevented from disturbing the next cycle of the input sampling and second, the control delay is fixed, which allows compensating for it in the control algorithm using standard digital control theory.

From these temporal requirements imposed on control theory, real-time constraints are imposed on the implementation of control systems. In both described approaches, a constant sampling frequency is achieved by performing the sampling tasks in *precisely periodic* points of time. In the first approach, the computation and actuation tasks need to get processor time as soon as possible, resulting in assigning them high priority value. The relative deadline of this task can be set using the aforementioned rule of thumb, to be 20% of the sampling period. In the second approach, as a consequence of fixing the actuation point in time, a hard real-time deadline is introduced for the computation task.

2.1.2 Classical Scheduling Theories

Real-time scheduling is nowadays a well-developed branch of computer science. It relies on the programming model where tasks communicate via shared data objects protected from the simultaneous access via a locking mechanism. Good overview of the most commonly used scheduling methods is given in [5].

Time constraints are in real-time systems met by assigning different priority levels to the involved tasks according to some scheduling policy. E.g. in Earliest Deadline First (EDF) scheduling, the task with a more stringent time requirement will get a higher priority. In Rate Monotonic (RM) scheduling, a process with higher sampling frequencies will have higher priority. Good comparison of all advantages and disadvantages of EDF and RM is given in [8]

An emerging way to schedule tasks in control systems is presented in [9]. In there, it is demonstrated that, compared to EDF and RM, better performance of a control system can be achieved when priorities are dynamically assigned according to the values of some control-system performance parameters.

Designs based on shared data objects as assumed in classical scheduling theories differ from designs based on buffered communication, or rendezvous-based communication, as assumed in CSP. In communication via shared data objects, *no* precedence constraints (set of "before"/"after" relationships between processes specifying set of relative orderings of the involved tasks) are introduced by the communication primitives.

2.1.3 Fundamental Mismatch Between CSP and Classical Scheduling

A rendezvous synchronization point introduces a pair of precedence constraint dependencies. In Figure 10, the control flow specifies that process A must be executed before process C and process B before process D. In addition, due to rendezvous synchronization on event ev1, subprocess A must be executed before subprocess D and subprocess B must be executed before subprocess D. In right-hand side part of the Figure, this is illustrated by dashed directed lines specifying precedence constraints from subprocess A to subprocess D (let us abbreviate this with A->D) and from subprocess B to subprocess C (B->C). Note that A, B, C and D are processes that can contain events and synchronize with the environment.

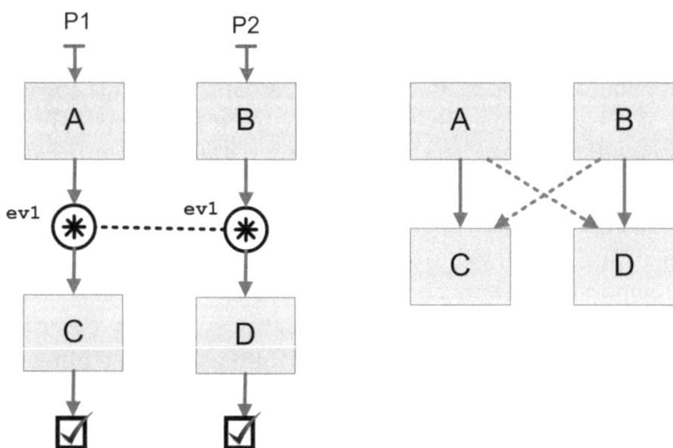

Figure 10 Rendezvous communication introduces new precedence constraints

In Figure 11, the communication from process P1 to P2 is buffered via an intermediate buffer process. Precedence relations are visualized as oriented dashed lines.

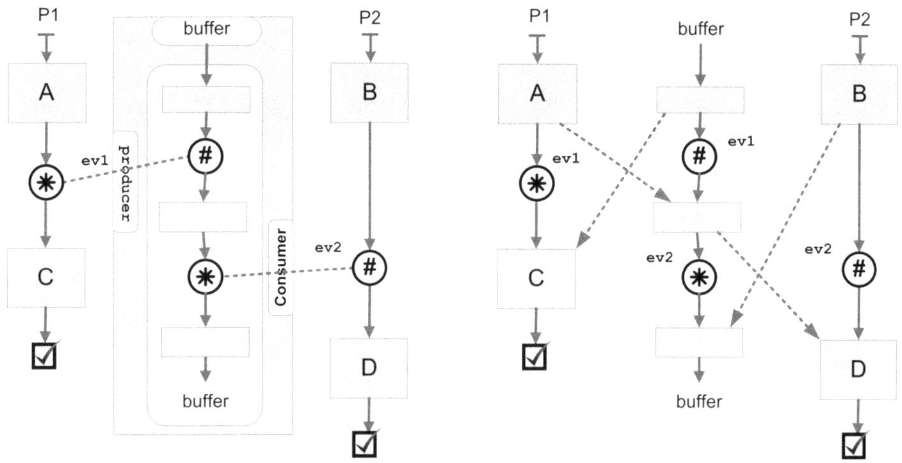

Figure 11 One-place buffer

As it can be seen in Figure 12, only the precedence dependency from A to D (A->D) exists, since the data must be present before it can be consumed. If the data flow direction for buffered asynchronous communication was from P2 to P1 then only the B->C precedence constraint would exist.

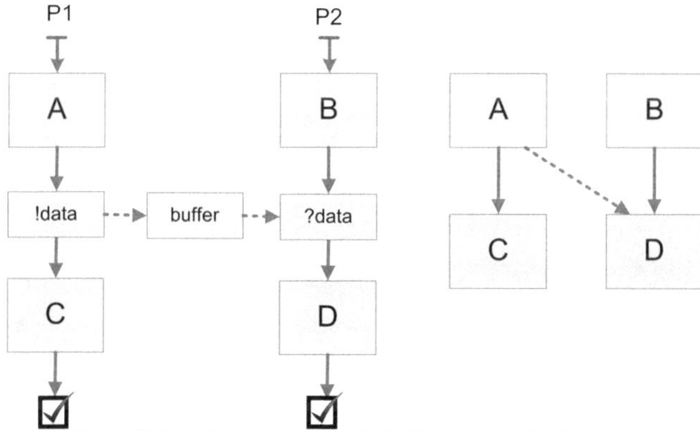

Figure 12 Precedence constraints for buffered communication

If, however, communication is via shared data objects (see Figure 13 and Figure 14), *no* precedence constraints are involved. The reason is that the shared data object has the semantics of an overwrite buffer, where a consumer always consumes the last fresh value available. In fact, in this case, it is more appropriate to use the term reader instead of the term consumer.

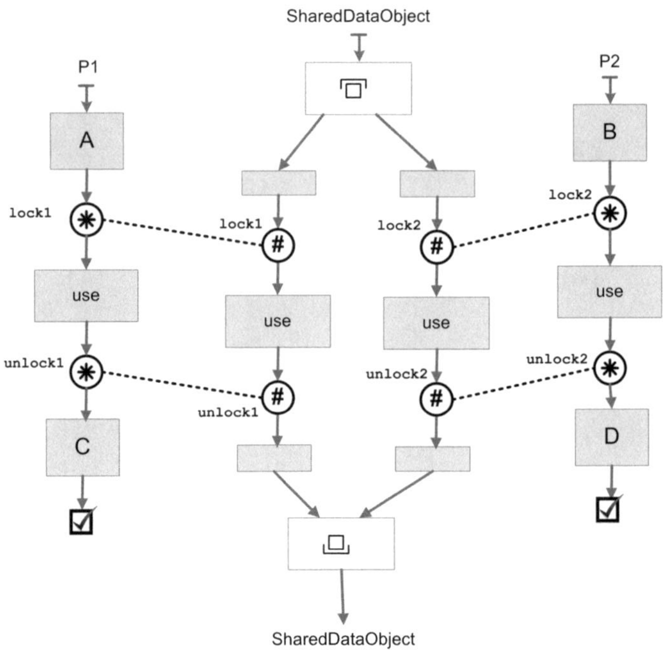

Figure 13 Shared data objects

Note that in case of shared data object communication, a process can still be blocked on waiting to access the shared data object. Scheduling theories do take into account this delay by calculating the worst-case blocking time.

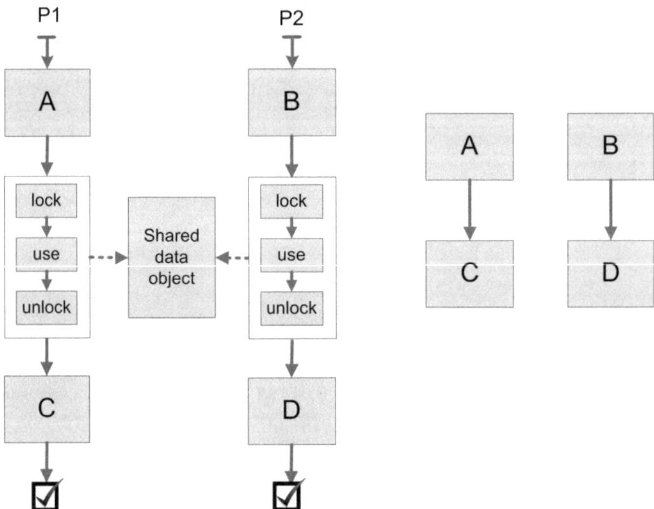

Figure 14 Precedence constraints in shared data object communication

In Figure 10, usage of rendezvous channel yields the possible orderings of subprocesses: (A || B) ->(C || D). Symbol A || B is used to abbreviate that A and B can be executed in any order, which is equivalent to composing them in parallel. When an asynchronous channel is used, it is equivalent to erasing one of the two precedence constraints (depending on the direction of data flow as explained above) and the resulting set of possible orderings is larger, allowing e.g. A->C->B->D (after A process P1 writes to the buffer and continues), which is not covered originally. When shared data objects are used, the set of possible orderings is even larger because another precedence constraint is removed. Thus, relaxation of precedence constraints, introduced by changing the type of applied communication primitive, leads to extending the set of possible behaviours.

2.1.4 Influence of Assigning Priorities on Analysis

In systems with rendezvous-based communication, using priorities reduces the set of possible traces only in pathological cases [10]. For instance, consider the system given in Figure 15, with the assumption that the highest priority level is assigned to process P1, the middle one is assigned to P2 and the lowest priority level is assigned to process P3. This priority ordering can be for instance implemented using a PriParallel construct, or alternatively, the system can rely on absolute priority settings. In any case, the relative priority ordering is from P1 to P3.

Priorities defined in this way will tend to give preference to P1i blocks compared to P2j blocks and also will give preference to P2j blocks compared to P3k blocks, but in fact the real order of execution can be any depending on the order of events accepted by the environment. Thus, the set of possible orderings of processes P1i, P2j, P3k and the set of related event traces is in general case not reduced by a priority assignment.

A PriAlternative construct is in fact giving relative priorities to event ends participating in the same PriAlternative construct. Those priorities are used only in the case when more then one event is ready. Again, the environment determines what events will be ready in run-time and thus as for PriParallel all traces are still possible.

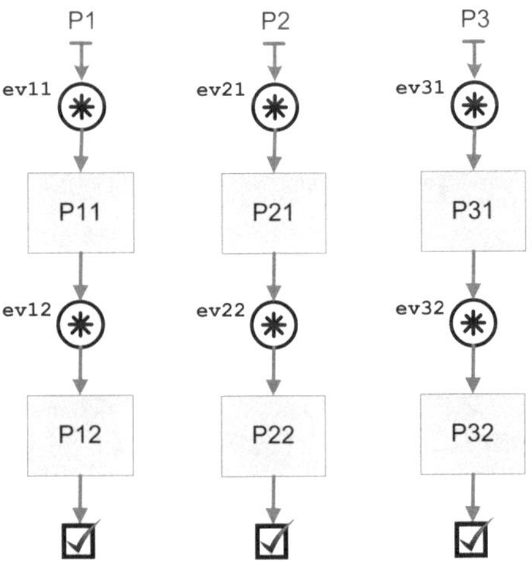

Figure 15 Some processes executed in parallel

From the discussion above, it is clear that assigning priorities to processes does not reduce the set of possible traces. Thus, to guarantee real-time, it should be verified that constraints are satisfied along any possible trace in the system. One reasonable approach for checking real-time guarantees is systematically replacing every Parallel composition with an equivalent automaton and associating execution times and deadlines with points in the control flow of that equivalent automaton. This approach is discussed in Section 2.3.2.

The conclusion is that structuring a program in the CSP way does influence schedulability analysis, because rendezvous-based communication makes processes more tightly coupled due to the additional precedence constraints stemming from the rendezvous synchronization. In rendezvous-based systems, priority of a process does not influence dominantly the order of execution. The actual execution ordering in the overall system is dominantly determined by the communication pattern encapsulated in the event-based interaction of processes. This interaction pattern inherent to the structure of the overall system, will due to the tight coupling on event synchronization points, always overrule the priorities of involved processes. Assigning a higher priority to one process, engaged in a complex interaction scheme with processes of different priorities, does not necessarily mean it will be always executed before lower-priority processes. This situation can also be seen as analogue to the priority inversion phenomenon in classical scheduling theory.

2.1.5 *Priority Inversion and Using a Buffer Process to Alleviate the Problem*

In classic scheduling, priority inversion is a situation where a higher-priority task is blocked on a resource held by a lower-priority task, which has as a consequence that tasks of intermediate priority are in the position to preempt the execution of the lower priority task and in that way prolong blocking time of the higher priority process. Let us try to view rendezvous channels in CSP-based systems as analogue to the resources shared between tasks in classic scheduling. In that light, waiting on a peer process to access the channel can be seen as analogue to the blocking time spent waiting on a peer task to free the shared resource. Viewed in this way, the system consisting of processes P1, P2 and P3 composed via a PriParallel construct depicted in Figure 16 illustrates the priority inversion problem. Process P1 has to wait for process P3 to enable occurrence of event ev3 and in meantime process P2 can preempt P3 and execute although its priority is lower than the one of P1. In this figure, the control flow of each process is given by concatenating event-ends and subrprocesses. The arrows with numbers on the left side of every process, indicate the actual order of execution that takes places due to the interaction pattern and despite the specified set of priorities.

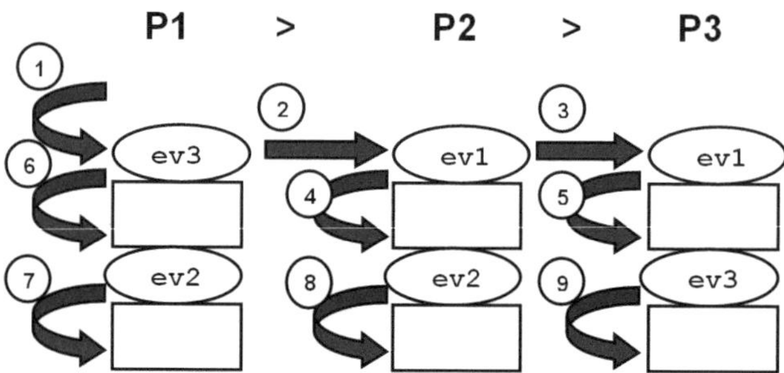

Figure 16 Process execution is dominantly determined by communication patterns rather then by process priority

Buffered channels are proposed in [4] to alleviate this problem. There, a proof of concept is given by implementing the buffered channel as a CSP process. However, a buffer process does not help when the process of higher priority (P1 in the example) is playing the role of a consumer and the process of lower priority (P3) is playing the role of a producer. In that case, the direction of the data flow introduces the precedence constraint from the event end in process P3 to the event end in process P1. The priority of the buffer process is assumed to be equal to the priority of the higher-level process for this scheme to work. Priority inversion in rendezvous-based systems is caused by precedence constraints leading from a process of lower priority to a process of higher priority. Using high-priority shared-data object communication primitives between processes of different priorities eliminates the priority inversion problem.

2.1.6 Absolute Versus Relative Specification of Priorities

Classic operating systems offer usually a fixed range of absolute priority values that can be assigned to any of the tasks/processes. In occam and the CT library, the concept of PriParallel construct is introduced that allows one to specify relative priorities instead of absolute ones. The index of a process inside a PriParallel construct determines its relative priority compared to the other subprocesses of the same construct. A program shaped as a hierarchy of nested PriParallel and Parallel constructs, results in an infinite number of possible priority levels. This approach also offers more flexibility, since new components can be added on proper places in the priority structure without the need to change priorities of already existing components.

However, while absolute priority ordering guarantees that any two processes are comparable, this is not the case in Par/Pripar hierarchies. Let's consider following example:

```
PAR
  PRIPAR
    A
    B
  PRIPAR
    C
    D
```

Where as the two PriPars define A as of having higher priority then B and C having higher priority then D, no preference is given to any when for instance B is compared to C, or A to D. They are considered to be of equal priority. So:

```
priority(C) = priority(A) > priority(B) = priority(C)
priority(C) = priority(A) > priority(B) = priority(D)
priority(D) = priority(A) > priority(B) = priority(C)
priority(B) = priority(C) > priority(D) = priority(A)
```

This looks confusing and inconsistent. If only prioritized versions of the Parallel construct (PriPar constructs) were used, there is *no* confusion. In fact, it is like collapsing a big sorted queue into smaller subqueues that can further be decomposed in subsubqueues. Only with a Par being a parent of PriPar constructs, the priority ordering problems appear.

The question is whether the relative priority ordering schemes, as the ones of occam-like hierarchies of Par and PriPar constructs, can be efficiently used in combination with the classical scheduling methods, for instance RM and EDF.

To be able to apply any priority-based scheduling method in a way that will avoid introducing priority inversion problems, as concluded in the previous section, processes of different priorities should be decoupled via consistent usage of shared data objects.

First let us consider the suitability of a PriPar construct for RM scheduling. The problem with RM scheduling is that it is not compositional. This means that if one composes two components with inner RM based schedulers, the resulting component does not preserve real-time guarantees. Thus, it would not be possible to define a Par of components on top level and to have PriPar based RM schedulers inside each of them. If, however, a hierarchy consisting only of PriPar constructs is used to implement RM priority assignment, this hierarchy can be seen as dividing one big queue into a hierarchical system of subqueues, where strict ordering is preserved. Theoretically for large queues, a hierarchical organization can significantly increase the speed of searching. In practice, however, systems usually do no need more then 8 or 16 or 32 different priority levels, which allow efficient implementation based on a single status register of size 8, 16 or 32 bits and dedicated FIFO queues for every priority level.

The conclusion is that in principle, by using only Prioritized versions of Par constructs and decoupling components with different priorities via shared data objects, it is possible to apply the RM scheme in occam-like systems. Now let us see if occam-like relative priority orderings are suitable for EDF schedulers.

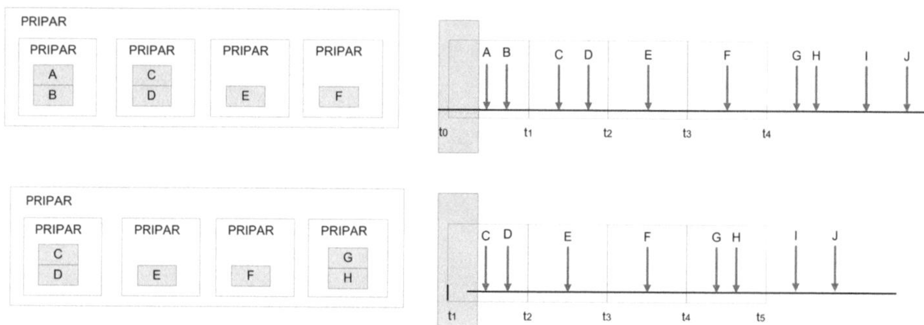

Figure 17 EDF scheduler

Implementing an EDF scheduler is tricky with fixed global priorities because the actual importance of a task is proportional to the nearness of a deadline and this nearness is a factor that keeps changing in time. Trying to assign a global priority to a process whose importance in fact changes with time is unnatural. The occam-like scheduling based on relative priorities could come as more natural solution. For instance one can divide a certain next part of the time axis into several time windows (see the right-hand side of Figure 17). Each time such a window can be associated with a single PriPar construct (compare the left-hand side and the right-hand side of Figure 17). The top level PriPar construct is used to sort the nested PriPar constructs, associated with time windows, according to their time order. Tasks with time constraints are then inserted in the PriPar construct related to the time window where their deadlines fall in. E.g. tasks C and D have deadlines falling into the interval (t1, t2) and are thus in upper part of Figure 17 mapped to the second PriPar construct.

The processes far away in the future and out of scope of any time window will be kept in a separate queue. After all tasks associated with the first time window are processed, this PriPar construct is removed from the top-level PriPar construct. The removed PriPar construct is then reused–it is associated with the first previously not mapped time interval. The non-allocated processes from the far-future queue that fall into that time window (G and H in Figure 17) are now mapped to it and the associated PriPar is now added to the top-level PriPar as the least urgent time window (lower part in Figure 17).

For this scheme to work, again, there should be no precedence constraints among tasks and thus rendezvous or buffered communication is not allowed or it should be taken into account by deriving intermediate deadlines as in EDF* (see Section 2.2.1).

The described approach for implementing EDF scheduler is based on relative priority orderings. However relative priority ordering is used for a generic implementation of the scheduler and not as a way to specify priorities of user defined processes in the application, as it was the case in occam. In order to apply this scheme in practice, the application itself should specify deadlines and not PriPar constructs or priorities.

A major problem with using relative priorities based on PriPar/Par constructs, is that it hard-codes priorities in the design, while a design of an application should be independent of priority specification. Reason is that a priority level is related to both the time requirements as specified in an application and the time properties of the underlying execution engine framework A choice is to design applications without introducing priorities and to postpone the process of assigning absolute priorities or deadlines (for EDF based scheduling) to the stage of allocation, where it is suited more naturally. Thus, the recommendation is *not* to use PriPar constructs. PriAlternative on the other hand makes sense independently of the scheduling method used.

2.2 Classic Scheduling

Most straightforward approach to making CSP designs with real-time guarantees is using CSP-based design patterns that match the programming paradigm of classical scheduling.

2.2.1 EDF*

In classical scheduling techniques, precedence constraints can be specified between tasks and special extensions exist for some scheduling theories (e.g. *modified EDF* - EDF*) to enable them to deal with the precedence constraints. EDF* takes precedence constraints into account by deriving the deadline of a task from the WCETs and deadlines of the following tasks.

When rendezvous channels are seen in the light of the introduced precedence constraints, the EDF* scheduling algorithm is applicable to rendezvous based systems. In applying EDF* to rendezvous based systems, calculation blocks can be considered to be schedulable units. A deadline of a calculation block is updated to the minimum value calculated upwards of any trace starting with some fundamental deadline and leading via the chain of precedence constraints to the current calculation block. The value is calculated starting with the time of the fundamental deadline and substracting WCET of every code block passed while going upwards the trace towards the block whose deadline is being derived in this way.

2.2.2 Design with Rendezvous Channel Communication

Regarding solving the priority inversion problem of rendezvous-based system by relaxing the used type of communication primitive, Hilderink [7] states that a deadlock-free program with rendezvous channels will still be deadlock-free when rendezvous channels are substituted with buffered channels. This is in fact intuitively explainable if we realize that deadlock is in fact a circle of precedence constraints (some being due to event prefix and sequential composition operators and some due to rendezvous communication). Since substituting rendezvous channels with buffered or shared data objects removes some of precedence constraints, this can remove some deadlock problems, but cannot introduce new ones.

Thus, a convenient design method could start with a design based on rendezvous channels. Such initial design is amenable to deadlock checking. After allocation and priority assignment, all rendezvous channels between processes of different priorities can be replaced with shared data objects allowing the usage of classic scheduling techniques, while preserving the results of deadlock checking. However, this approach might not always be feasible. Note that relaxing the precedence constraints associated with rendezvous channels results in an extended set of behaviours by including the possible behaviours that were not formally checked. As a consequence, a new implementation that was produced in this way might not anymore be a refinement of the initial specification. For conformance of such an implementation to its specification both its traces and failures must be subsets of the traces and failures defined in the specification.

2.2.3 Design Pattern for Implementation of a Typical Control System

Control systems typically function at a number of levels (see Figure 18). At the lowest level, the highest priority layer is situated. *Safety control* makes sure that functioning of the system will not endanger itself or its environment. It is especially important when embedded control systems are employed in *safety-critical* systems. *Loop control* is a hard real-time part that periodically reads inputs from sensors, calculates control signals according to the chosen control algorithms and uses the obtained values to steer the plant via actuators. *Sequence control* defines the synchronization between the involved subsystems or devices. *Supervisory control* ensures that the overall aim is achieved by using monitoring functions, safety, fault tolerance and algorithms for parameter optimization. *User interface* is an optional layer that supports the interaction of the system with an operator (user), in the form of displaying important part of the system's state to the operator and receiving the commands from the operator.

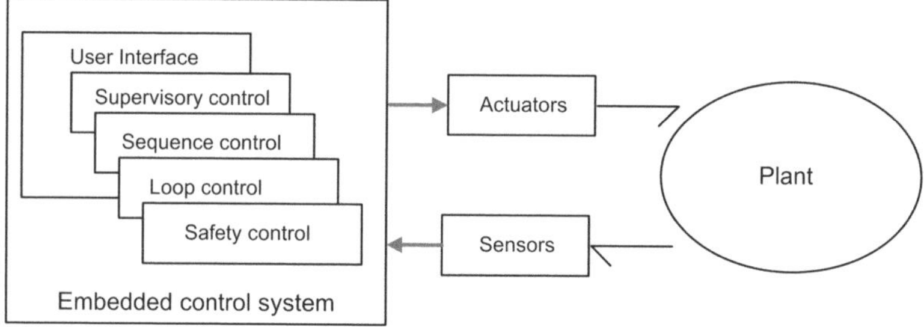

Figure 18 Typical control system

In fact, a complex control system (e.g. a production cell) typically contains several devices that need to cooperate. Every device can participate in any or all mentioned layers. The supervisory and sequence control layers are often event based and the control loop is always time-triggered and periodic, with the period in general different from one device to another. Software components in charge of devices are either situated on the same node or distributed over several nodes.

Figure 19 illustrates data/event dependencies between layers situated in the same device as well as between layers distributed over several or all participating devices. Two ways of clustering subcomponents are possible: horizontal – where a centralized supervisory layer, sequence layer, control loop layer and safety control layer exist, or vertical where parts belonging to the same device are considered to be a single component.

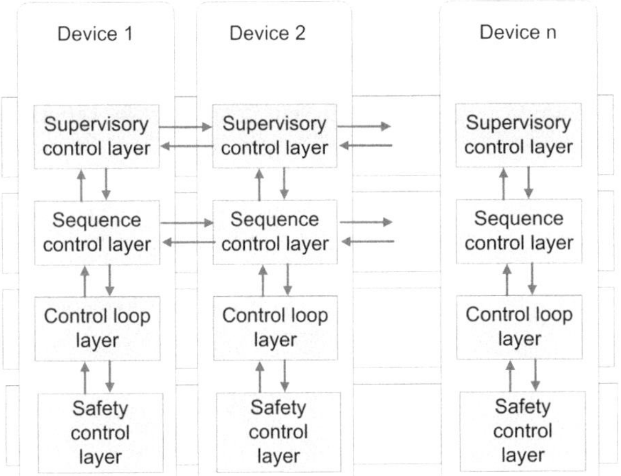

Figure 19 Typical layered structure of complex control system

In SystemCSP, design patterns can be made where either vertical or horizontal groups are structured as components and the orthogonal groupings form interaction contracts. Let us consider the case where devices are treated as components and layers as interaction contracts. Every device is in this approach a component that provides ports, which can be plugged into one of the four interaction contracts: supervision, sequence control, safety, loop control (see Figure 20).

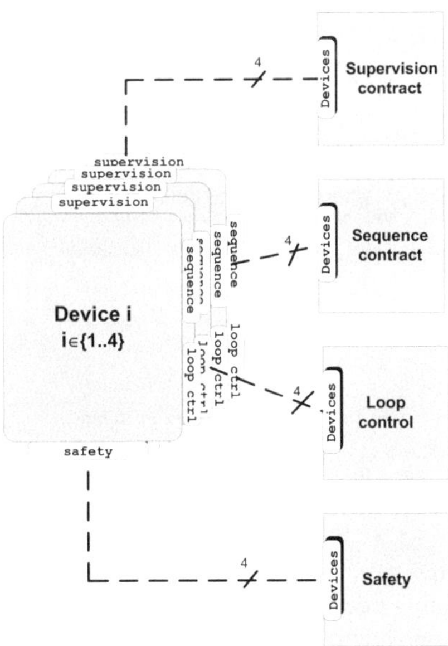

Figure 20 One SystemCSP design pattern for complex control systems

Every contract contains logic for handling several devices and managing synchronization between them. Often, it is useful to merge some of those interaction contracts into a single interaction contract (e.g. some safety measures can be in the sequence control contract or sequence and supervision layer can be merged).

A loop control contract is often implicit since there is no other dependency between control loops except for common usage of the timing and I/O subsystems to ensure precise in time execution of time-triggered periodic sampling/actuation actions. Upon performing the time triggered sampling/actuation (I/O subsystem) actions, loop-control processes of related devices are released to perform computation of control algorithms. A loop control interaction contract can for instance perform scheduling (RM, EDF) of the involved loop control processes, check whether deadlines are missed and raise alarms to the safety or supervision interaction contract when that happens. E.g. in the overload conditions, a loop control interaction contract can have a centralized policy to decrease the needed total computation time in a way that reduces performance but does not jeopardize the stability of the control system.

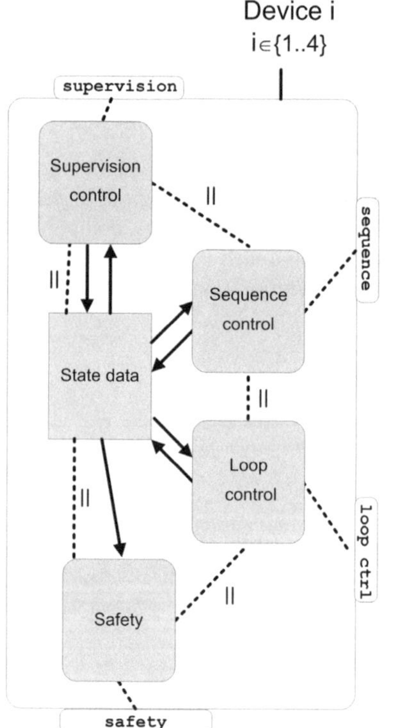

Figure 21 Device internals organization

Internally, a device component might be organized as in Figure 21, with a subcomponent dedicated to the implementation of every role that maps to one of the layers supported by the device, and a subprocess dedicated to maintaining the state data of a device. In Figure 21, in order to put emphasis on structure and data-flow and not on control flow, the GML-like interaction oriented SystemCSP diagram is used, where communication data flows and binary compositional relationships are specified. A centralized process is introduced to manage access to the data that captures the *state* variables of a component. This process is in fact the shared-data object communication pattern that allows decoupled communication between processes implementing roles of various layers. In practice, for efficiency reasons, the state data process can also be implemented as a passive object that provides the necessary synchronization. It can be, for instance, a lock-free double buffered channel. A centralized process for device's state data access is also convenient in case when the loop control process is replicated for fault tolerance reasons.

In this typical structure, subcomponents get different priorities. The safety layer is of highest priority and is activated only when it is necessary to handle alarm situations. The next range of priorities is associated with loop control subcomponents. Range of priority levels might be necessary for the implementation of the scheduling method that will guarantee their execution in real-time. Sequence control is an event based layer and thus of less importance than the time-constrained loop control layer. The Supervision layer is performing optimization and is thus of least importance.

2.3 Developing Scheduling Theory Specific for Rendezvous-based Systems

If however, the intention is to use rendezvous-based channels/events as basic primitives then a distinct scheduling theory must be developed. The topic of achieving real-time guarantees in systems with rendezvous synchronized communication is a research field that still waits for a good underlying theory.

2.3.1 Event-based Scheduling

Figure 16 illustrates that attaching priorities to processes that communicate via rendezvous channels influences the behaviour of the rendezvous-based systems much less then expected.

Instead of trying to apply classic scheduling methods, it is possible to admit the crucial role of events in CSP based systems and assign priorities to events instead of to processes. Deadlines can be seen as time requirements imposed on events or on distances between some events. Furthermore, while priority of processes is local to a node, the priority of an event is still valid throughout the whole distributed system. Such *event-based scheduling* seems to promise a way to get better insight and more control over the way the synchronization pattern influences the execution and overrules the preferred priorities.

Section 2.1.5 has introduced an analogy between the priority inversion problem in classical scheduling methods and the analogue problem in the rendezvous based systems. In classical scheduling, the standard solution to the priority inheritance problem is that the lower-priority task holding the resource needed by the higher-priority task gets a temporary priority boost until it frees the resource. If we apply this analogy to a rendezvous channel as a shared resource, then the peer process is holding the resource as long as it is not ready to engage in rendezvous. Thus to avoid priority inversions, the complete control flow of the lower-priority task, that is taking place *before* the event access point, should get a priority boost. The "before" relation is formally expressed via precedence constraint arrows. Thus, starting from some event end, the priority of all events ends upwards the precedence constraint arrows should be updated to be of equal or higher value. Keep in mind that as explained in Section 2.1.3, extra precedence constraints are introduced on every rendezvous synchronization point. Thus, the process of updating priorities propagates through rendezvous communication points to other processes. Eventually, a stable set of priorities is reached. This set of priorities is in the general case different from the initially specified set of priorities.

The user can set an initial set of preferred priorities to some subset, or to all event ends in the program. For instance, one can initially assign priorities to event ends by assigning priorities to processes, which can for instance result in automatically associating the specified process-level priority with every event end in the process. Those preferred values are used as initial values in the aforementioned procedure of systematically updating priorities of event ends. The set of priorities obtained by applying this algorithm reveal a realistic or an achievable set of values after the synchronization pattern is taken into account. In this process, priority inversions are inherently eliminated.

In the example of Figure 16, in event-based scheduling, event ends initially get priorities according to the priority specified for their parent processes. Due to precedence constraints all event ends participating in the same event need to get the same value, which is equal to the highest priority present in any of the event ends. Thus, events ev3 and ev2 would get the priority of process P1, and event ev1 the priority of process P2. However, since a precedence constraint ev1->ev2 exists, the priority of the event ev1 needs to be readjusted in order to avoid priority inversion as described above in the procedure for realigning priorities of events. Thus, although preferred priorities of process P1, P2 and P3 are different, their execution pattern results in all events ev1, ev2 and ev3 having the same priority. The event-based scheduling approach uncovers realistic, priority-inversion free, values of priority levels, achievable with the given design of synchronization pattern between processes.

The procedure is not so convenient for application in systems with lot of recursions. There are two types of recursive processes: time-triggered recursion and ordinary recursion. In ordinary recursion there is a cycle and as a result all the events in the process have the same priority. The time-triggered recursions are considered new instances of tasks with new deadline values and there is no need to perform a circular update of priorities.

2.3.2 Equivalent Automaton

From the discussion in Section 2.1.4, it is clear that assigning priorities to processes does not reduce the set of possible traces. Thus, one reasonable approach for checking real-time guarantees is treating CSP processes as automata and systematically replacing every composition of CSP processes with an equivalent automaton, and associating execution times and deadlines with points in the control flow of the equivalent automaton. In [11], it is in fact stated that timed CSP descriptions are closed-timed epsilon automata.

In order to simplify reasoning, in this paper we restrict the analyzed models to be free from the non-deterministic and too complex primitives: Systems are considered to be free from the usage of the *internal choice* operator and of those cases of the *external choice* operator that cannot be reduced to the *guarded alternative* operator. Internal choice is normally used as an abstraction vehicle and as such, it does not exist in final designs. External choice that cannot be replaced with a guarded alternative operator is another situation that is rarely used in practice and difficult to implement and also not straightforward to describe in automata representation. The decision here is to restrict final designs to be free of those two special cases. If the set of CSP operators is restricted in this way, the processes constructed using events and this restricted set of operators can be reasoned about using classic automata theory.

Automata theory [12] defines how to make a parallel composition of two automata. The example for this procedure is depicted in Figure 22. The start state of the equivalent automaton representing the composition is the combination of the initial states of the composed processes. In Figure 22, process P1 can initially engage in event a and process P2 in event b. Since event b must be accepted by both P1 and P2, only event a is initially possible. Event a will take the first automaton to state 2, while the second automaton will stay in state 1. Thus, starting from the initial state (1, 1) and following the occurrence of the event a, the composite state (2, 1) is discovered (see Figure 22).

For every reachable composite state all possible transitions are checked (taking into account when synchronization is required and when not). The resulting composite states are mapped to the equivalent automaton. After a while all transitions either lead to the already discovered composite states or to the end state (when both participating processes are in their end states) if any.

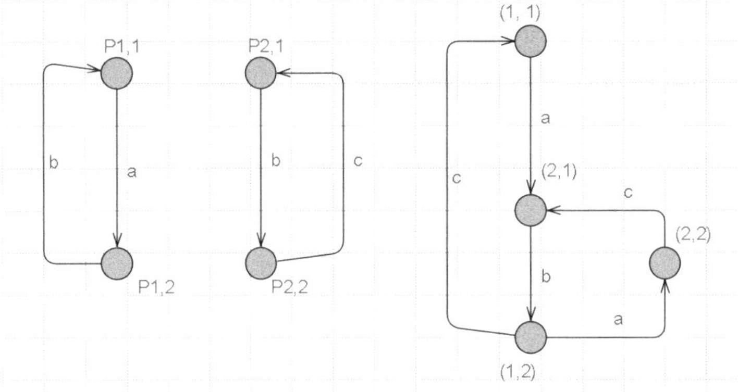

Figure 22 Construction of equivalent automaton

Some composite states are not reachable and thus not part of the equivalent automaton representing the parallel composition. In principle, composing, in parallel, a process containing 3 states with a process containing 4 states, yields a process with all 3*4 combinations possible. This is in fact the case whenever processes are composed in interleaving parallel. In non-interleaving parallel constructs, due to the involved synchronizations, the number of states is less.

A parallel composition can be seen as a way to efficiently write down complex processes that contain a number of states. Seen in that light, the introduction of the Parallel operator allows decomposing complex processes on entities that are smaller, focused on one aspect of the system at hand and simpler to understand.

The definition of the parallel operator is in CSP identical to the one in automata theory. The external choice of CSP viewed as automata is equivalent to making a composite initial state that is offering the set of initial transitions leading to the start states of the involved subprocesses and subsequently behaving as those subprocesses. The sequential composition is trivially concatenating the involved automata. If every CSP process is viewed as an automaton, creating an equivalent automaton representing a complete CSP-based application is a straightforward thing to do. The equivalent automaton defines all traces (sequences of events) possible in the system. Thus, it can be used as a model against which one can check different properties of the system – e.g. checking for deadlock/livelock freedom, checking the compliance of implementation to the related specification (refinement checking). For instance, an application has a potential for a deadlock situation if there is a state (not the end state) from which no transition is leaving. The refinement checking is about testing if a set of traces and failures defined by an automaton representing some implementation is a subset of the set of traces and failures defined by an automaton representing the related specification.

In Figure 23 SystemCSP based visualization of the equivalent automata from Figure 22 is given. The main difference of a SystemCSP representation compared to the automata way of visualizing CSP processes is that instead of on states, the focus is on events. The procedure of constructing an equivalent automaton is exactly the same. Focusing on events ensures that traces are more easily observable, especially if in SystemCSP, instead of the lines going back to the revisited states, as is always the case in automata, usage of recursion labels is enforced. Systematic usage of recursion labels will naturally separate subtraces that are repeated and thus create immediately observable trees of possible event traces.

Inspection of Figure 23 shows that possible traces in the single 'Loop' iteration of the equivalent automaton are <a,b,c> and <a,<b, a, c> ^{n},b,c>. The actual trace taken is dependent on the readiness of the environment. Recursion labels define sequences that are repeated and the IF choice and guarded alternatives divide traces into several subtraces.

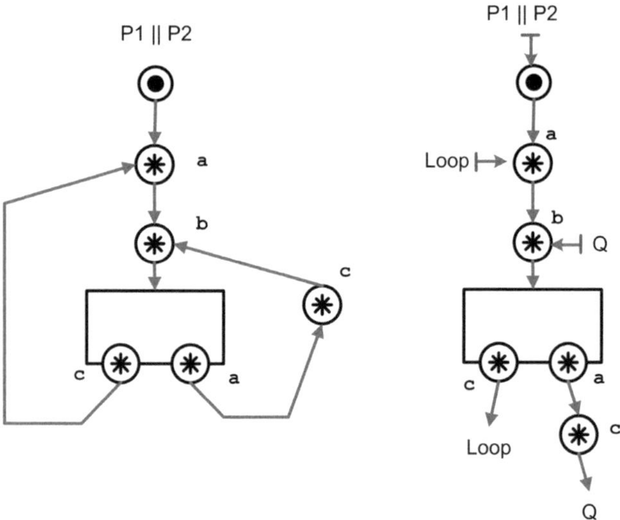

Figure 23 SystemCSP with recursion labels makes traces more obvious

2.3.3 Mapping Time Properties to Equivalent Automata

The next step is to extend the description of an equivalent automaton with time properties in a way that it will allow us to perform efficient analysis. The idea is to extend CSP descriptions with time properties in such a way that the mapping to the equivalent automaton preserves their meaning. The execution times of the calculation blocks can also be seen as related to the event ends immediately preceding them, that is, to event ends associated with events whose occurrence will allow every participating process to progress for the amount of execution time spent on the next calculation block.

The execution time of a process at a certain point of its execution is the sum of execution times along the path that brought the process to the current point. In other words, it is the sum of progress (expressed in time units) allowed by all event ends along the trace that process is following.

After specifying execution times and time constraints in the two subprocesses composed in Parallel, time properties are mapped to the equivalent automaton. If that can be done, then the analysis of time behaviour can be performed on the constructed equivalent automaton. If we are able to map the time properties from a pair of parallel composed processes to their composition, then the same can be done hierarchically in bottom-up manner yielding at the end an executable timed model of complete application.

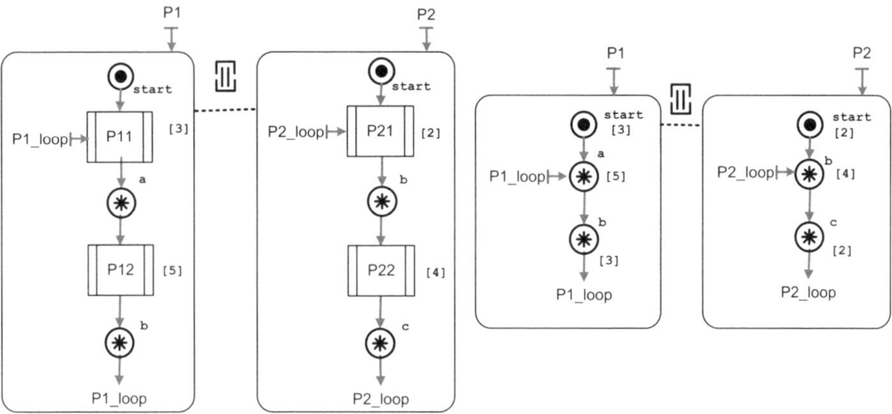

Figure 24 Specifying Execution times of code blocks

Analysis of time properties should be performed without the need to perform code block calculations. In such analysis, code blocks are substituted with their execution times and sums of execution times along all possible system traces are inspected with respect to the specified time constraints.

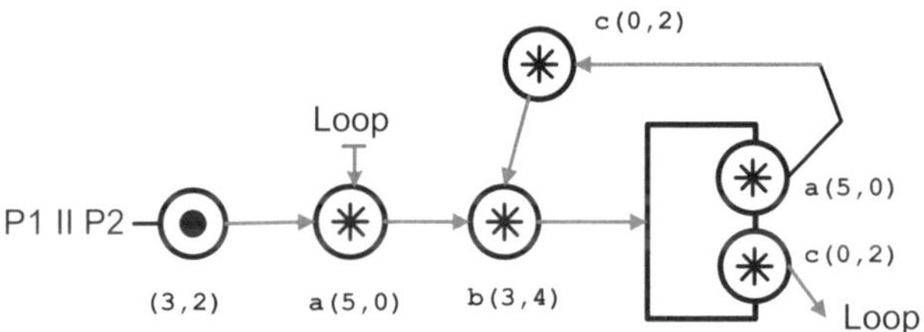

Figure 25 Equivalent Event Machine

The basic idea is that an event occurrence allows further progress of processes involved in that event occurrence. The initial event in process P1 (see Figure 24) allows process P1 to progress further in execution for 3 time units and offers event a to the environment. The initial event in P2 allows process P2 to progress in execution for 2 time units and then offer event b to environment. Thus, the composite initial event allows the involved subprocesses to progress for (3, 2) time units, where the first number maps to the event end in first subprocess and second number to the event end in the second one (see Figure 25). Event a, once it is accepted by the environment will allow progress of P1 for 5 time units and P2 for 0 time units since P2 is blocked waiting on its environment (including P1) to accept event b. Thus in the composite automaton, event a taking place following the initial event, will allow (5, 0) progress of the involved subprocesses. The subsequent occurrence of event b will allow progress of both P1 and P2, for 3 and 4 time units respectively, which is in Figure 25 expressed by associating ordered pair (3, 4) with the event b.

Note that in general, when a hierarchy of processes is resolved, it is not a good idea to capture progress of subprocesses as n-tuples. In a prospective analyzer implementation, since execution times are related to event ends, bookkeeping of allowed progress would be kept in the participating event ends and not in n-tuples, containing progress for all composed subprocesses. Different occurrences of the same event in a composite automaton can in fact have associated different progress values. Essentially, execution times are expressed as the amount of progress event ends allow.

Note that the assumption here is that the environment is always ready to accept events. In fact, when the equivalent automaton is constructed hierarchically in a bottom-up approach, it will eventually include the complete system with all events resolved internally.

Unpredictable (in sense of time) occurrences of events from the environment can of course only be analyzed for a certain chosen set of scenarios. For every scenario, the environment can also be modelled as a process with defined time properties and composed in parallel with the application to form the complete system.

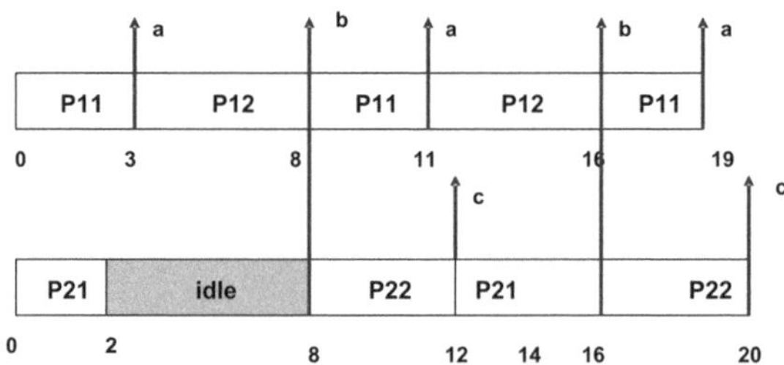

Figure 26 True parallelism

The actual time of event occurrences depends on the allocation. For the equivalent automaton of Figure 25, on Figure 26 true parallelism case is depicted, and on Figure 27 a shared CPU with P1 having higher priority and a shared CPU with P2 having higher priority. The same equivalent automaton keeps information necessary to unwrap actual timings of the involved events in all 3 cases.

1) Process 1 has higher priority

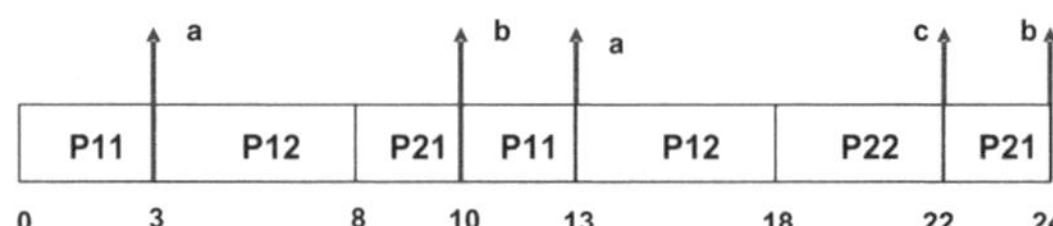

2) Process 2 has higher priority

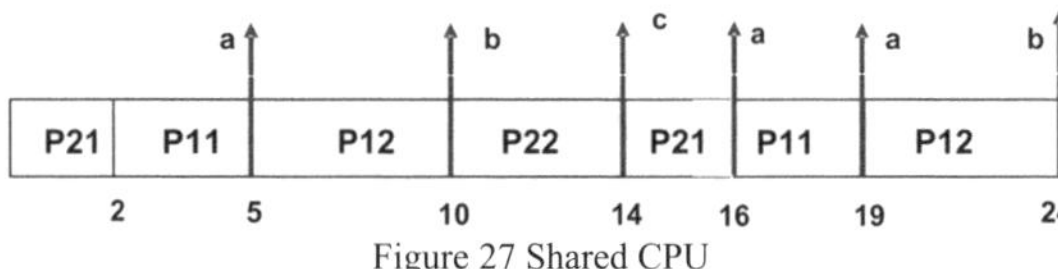

Figure 27 Shared CPU

In the case of true parallelism, components P1 and P2 are initially allowed to progress 3 and 2 time units respectively. Then event 'a' allows component P1 to progress another 5 units. Both processes synchronize on event b, meaning that their times must be same at the rendezvous point. Thus the time of this rendezvous point is max(3+5, 2+0)=8. Event b will allow components P1 and P2 to progress 3 and 4 units of time respectively. Under the assumption that the environment is always ready to accept events, event a will be accepted at time 8+3=11 and event c at time 8+4=12. This scheduling pattern is depicted in Figure 26. The scheduling pattern obtained for a shared CPU and different priorities of components is depicted in Figure 27.

Three independent timed models can be made by using in analysis either minimum, average or worst-case execution times. Using only the average execution time is a good first approximation of the system's behavior. Execution times are dependent on allocation of components to processing nodes and can in fact be measured or simulated for different targets and stored in some database. A prospective tool should be able to keep track of allocation scenarios and to simulate/analyze/compare effects of the different execution times in different allocation scenarios.

3. Conclusions

In this paper, ways to introduce time properties are defined in the scope of the SystemCSP design methodology. The specification of time properties is deduced by merging the ideas from previous work in the CSP community [1,2]. Implementation of CSP-based systems with real-time properties is then investigated. Two major directions are observed for achieving real-time: (1) introducing design patterns that can fit CSP-based systems into requirements of existing scheduling theories and (2) relying on constructing distinct scheduling theories for CSP-based systems. Comparing the two indicates is that the first proposed direction enables immediate implementation, while taking the second direction requires additional research. Thus, a recommendation for prospective tool used for editing SystemCSP designs is to use the combination of proposed design patterns and classical scheduling theories to provide real-time guarantees.

References

[1] Roscoe, A.W., The Theory and Practice of Concurrency. Prentice Hall International Series in Computer Science. 1997: Prentice Hall.

[2] Schneider, S., Concurrent and Real-Time Systems: The CSP approach. 2000: Wiley.

[3] Orlic, B. and J.F. Broenink. SystemCSP - visual notation. in CPA. 2006: IOS Press.

[4] Hilderink, G.H., Managing Complexity of Control Software through Concurrency. 2005, University of Twente.

[5] Buttazzo, G.C., Hard real-time computing systems: Predictable Scheduling Algorithms and Applications. The Kluwer international series in engineering and computer science. Real-time systems. 2002, Pisa, Italy: Kluwer Academic Publishers.

[6] Wittenmark B., N.J., Torngren M. Timing problems in Real-time control systems. in American control conference. 1995. Seattle.

[7] Boderc: Model-based design of high-tech systems, ed. M.H.E.U.o. Technology and G.M.E.S. Institute. 2006, Eindhoven, The Netherlands: Embedded Systems Institute, Eindhoven, The Netherlands.

[8] Buttazzo, G.C., Rate Monotonic vs. EDF: Judgment Day. Real-TimeSystems, 2005. 29(1): p. 5-26(22).

[9] Cervin, A. and J. Ekerz, The Control Server Model for Codesign of Real-Time Control Systems. 2006.

[10] Fidge, C.J., A formal definition of priority in CSP. ACM Trans. Program. Lang. Syst., 1993. 15: p. 681-705.

[11] Ouaknine, J. and J. Worrell, Timed CSP = closed timed epsilon-automata. Nordic Journal of Computing, 2003.

[12] Cassandras, C.G. and S. Lafortune, Introduction to discrete event systems. 1999, Dordrecht: Kluwer Academic Publishers.

Communicating Process Architectures 2007
Alistair McEwan, Steve Schneider, Wilson Ifill and Peter Welch (Eds.)
IOS Press, 2007

Testing and Sampling Parallel Systems

Jon KERRIDGE

School of Computing, Napier University, Edinburgh, EH10 5DT

Abstract The testing of systems using tools such as JUnit is well known to the sequential programming community. It is perhaps less well known to the parallel computing community because it relies on systems terminating so that system outputs can be compared with expected outputs. A highly parallel architecture is described that allows the JUnit testing of non-terminating MIMD process based parallel systems. The architecture is then extended to permit the sampling of a continuously running system. It is shown that this can be achieved using a small number of additional components that can be easily modified to suit a particular sampling situation. The system architectures are presented using a Groovy implementation of the JCSP and JUnit packages.

Keywords: JUnit Testing, Sampling, GroovyTestCase, white-box, black-box

Introduction

The concept of testing, particularly using the *white-box* and *black-box* techniques, is well known and understood by the software engineering community. White-box testing is used to ensure that the methods associated with an object oriented class definition operate in the expected manner and that their internal coding is correct. Black-box testing is used to ensure that the overall operation of the class and its methods is as expected when operating in conjunction with other classes without concern for their internal coding.

The Agile programming community [1] has developed techniques commonly referred to as unit testing. In particular, these techniques have been incorporated into an open source framework that can be used with Java, called JUnit (www.junit.org). Typically, JUnit is used to undertake white-box testing. The use of the capability has been made even easier in the Groovy scripting environment by the creation of the `GroovyTestCase` [2], by for example, ensuring that all methods starting with `test` are compiled and executed as a Groovy script. A test will normally require some form of assertion test to check that an output value is within some bound or that some invariant of the system is maintained,

An ordinary object-oriented class uses its methods to pass messages between objects and thus need to be carefully tested. Hence the JUnit test framework has been designed specifically to undertake white-box testing of these methods. An object is tested by defining it as a *fixture*, which is then subjected to a sequence of tests. After each test is completed an assertion is evaluated to determine the success of the test. An assertion can test for a true or false outcome. The testing process requires the programmer to define an input sequence of calls to one or more methods of the object and also to specify the expected outcome. The assertion tests the generated output from the object under test against the expected outcome. Thus programming becomes a process of defining inputs and expected outputs and writing the program to achieve the desired outputs. The JUnit framework automates this process further by combining sequences of tests into *testsuites*. If a change has been made to the underlying object all the tests contained in all the testsuites can be run to ensure the change has not created any unwanted side effects.

In the MIMD parallel processing environment, using JCSP (www.jcsp.org), the classes implement the interface `CSProcess` that has only one method `run()`. Any methods are private to the object and used simply to assist in the coding of the process. Hence the use of unit testing in the parallel environment can be considered more akin to black-box testing because there is only one method to test. Often processes are written in a style that runs forever rather than having a specific termination strategy. Processes can be returned to a known state using techniques such as poison [3, 4] but unless specifically required tend not to be used. Even in this situation, the process can still continue to run and may not terminate. If a network of processes does terminate then the normal testcase framework can be used. Hence a means of testing a non-terminating system has to be specially designed so the non-terminating part under test can continue to run, while the testing part terminates so that data values can be extracted for assertion testing. If the network of processes does not terminate we can never extract the values from its execution that are required to test the associated assertions. If the system has been designed to run forever then the addition of code to cause the system to terminate means the system being tested is not the one that will be used in any final deliverable. We therefore need to create a bi-partite test environment in which the process network under test is able to run forever. A terminating test part injects a finite sequence of values, with an expected outcome into the network under test. The test part also receives outputs which can be assertion tested against the expected outcome. This simple strategy is impossible to run as a single process network in a single processing node because even though the processes in the test part will terminate the network under test will not terminate and thus the complete network never terminates and thus the assertions cannot be tested. The use of the `GroovyTestCase` framework means that the testing can be even more easily automated.

Sampling a system provides a means of checking that a system remains within pre-defined bounds as it operates normally. The benefit of providing a sampling architecture that is different from the testing architecture is that it can be incorporated into the system either at design time or once it has been implemented. The primary requirement is that the processes that are used to extract the samples are as lightweight as possible. Crucially, these sampling processes must not result in any modification to the system that has already been tested.

In the next section, a generic testing architecture is presented that utilizes the capability of JCSP to place processes on different nodes of a multi-node processing network connected by means of a TCP/IP network. Section 2 then demonstrates how this architecture can be applied to a teaching example process network. Section 3 then shows how the same process network could be sampled during normal operation. Sections 4 and 5 then describe generic sampling architectures for systems that respectively communicate data by means of object data transfers and by primitive data types. Finally, some conclusions are drawn and further work identified.

1. A Generic Testing Architecture

Figure 1 shows a generic architecture in which it is presumed that the Process-Network-Under-Test (PNUT) is either a single process or a collection of processes that does not terminate. The Input-Generator process produces a finite set of inputs to the PNUT and may also create a data structure that can form one part of a test assertion. Similarly, the Output-Gatherer process collects data from the PNUT and stores it in a data structure that can be subsequently tested by Test-Network. The Assertion-Testing is only undertaken when both the Input-Generator and Output-Gatherer processes have terminated.

The goal of the architecture is to create a means by which each of the processes or sub-networks of processes can be tested and shown to operate according to the tests that have been defined for that particular process or sub-network of processes. The JCSP, due to its reliance on CSP provides a compositional semantics when processes are combined into larger networks. Other mechanisms are available, such as FDR [5] for determining the deadlock freedom of such compositions, but cannot then test the full range of values that might be applied to such a network and thus the need for a testing framework for parallel systems. The PNUT shown in Figure 1 might be a single process or a network of processes that together form a collection of testable processes that are subsequently used in a compositional manner in the final system design.

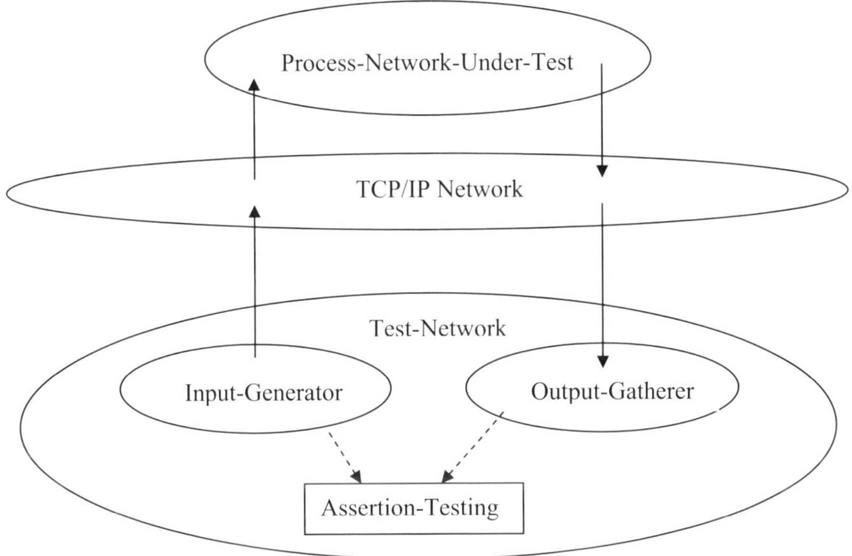

Figure 1 Generic Testing Architecture

Both the Input-Generator and Output-Gatherer processes must run as a Parallel within the process Test-Network, then terminate after which their internal data structures can be tested within Assertion-Testing. An implementation of the Test-Network process for a specific example is shown in Listing 1. It does however demonstrate the generic nature of the architecture in that the only part that has to be specifically written are the `GenerateNumbers` and `CollectNumbers` processes that implement the Input-Generator and Output-Gatherer respectively.

The class RunTestPart implements the Test-Network {1}[1] and simply extends the class `GroovyTestCase`. The method `testSomething` {3} creates the Test-Network as a process running in a node on a TCP/IP network. The node is initialized in the normal manner within the JCSP framework {5}. Two `NetChannels`, `ordinaryInput` {7} and `scaledOutput` {8} are defined and recorded within an instance of `TCPIPCNSServer` that is presumed to be running on the network prior to the invocation of both the PNUT and Test-Network. The processes are created {10, 11} using the techniques described in [6] using

---

[1] The notation {n} indicates a line number in a lisitng.

Groovy parallel helper classes. The processes are then invoked {13, 15}. Once the PAR has terminated, the properties generatedList, collectedList and scaledList can be obtained from the processes {17-20} using the Groovy dot notation for accessing class properties. In this case we know that the original generated set of values should equal the unscaled output from the collector and this is tested in an assertion {21}. In this case we also know that each modified output from the PNUT should be greater than or equal to the corresponding input value. This is implemented by a method contained in a package TestUtilities called list1GEList2, which is used in a second assertion {22}.

```
01   class RunTestPart extends GroovyTestCase {
02
03     void testSomething() {
04
05         Node.getInstance().init(new TCPIPNodeFactory ())
06
07         NetChannelOutput ordinaryInput = CNS.createOne2Net("ordinaryInput")
08         NetChannelInput scaledOutput = CNS.createNet2One("scaledOutput")
09
10         def collector = new CollectNumbers ( inChannel: scaledOutput)
11         def generator = new GenerateNumbers (outChannel: ordinaryInput)
12
13         def testList = [ collector, generator]
14
15         new PAR(testList).run()
16
17         def original = generator.generatedList
18         def unscaled = collector.collectedList
19         def scaled = collector.scaledList
20
21         assertTrue (original == unscaled)
22         assertTrue (TestUtilities.list1GEList2(scaled, original))
23
24     }
25
26   }
```

Listing 1 An Implementation of the Test-Network Process

The benefit of this approach is that we are guaranteed that the Test-Network will terminate and thus values can be tested in assertions. The fact that the PNUT continues running is made disjoint by the use of the network. This could not be achieved if all the processes were run in a single JVM as the assertions could not be tested because the PAR would never terminate. The process network comprising the PNUT and the Test-Network can be run on a single processor with each running in a separate JVM, as is the TCPIPCNSServer. RunTestPart will write its output to a console window indicating whether or not the test has passed. The console window associated with PNUT will continue to produce any outputs associated with the network being tested.

1.1 Example Generator and Gatherer Processes

Necessarily, the Generator and Gatherer processes will depend upon the PNUT. Listing 2 shows a typical formulation of a Generator process, which produces a finite sequence of numbers. The properties of the process will vary, however outChannel and generatedList will always be required. The channel is used to communicate values to the PNUT {38} and generatedList provides a means of storing the output sequence in a property {33, 39} that can be accesses once the process has terminated. The operator << {39} appends a value to a list. In this case the numbers need to be output with a delay {29,

36, 40} of one second between each number. The size of the output sequence can easily be altered by varying the value assigned to the property `iterations`, which has a default value of 20.

```
27   class GenerateNumbers implements CSProcess{
28
29     def delay = 1000
30     def iterations = 20
31
32     def ChannelOutput outChannel
33     def generatedList = []
34
35     void run() {
36       def timer = new CSTimer()
37       for (i in 1 .. iterations) {
38         outChannel.write(i)
39         generatedList << i
40         timer.sleep(delay)
41       }
42     }
43   }
```

Listing 2 An Implementation of a Generator Process

Listing 3 similarly gives the code for a Gatherer process. In this case two output lists can be collected, one which is the same as the original data stream (`collectedList`) and one which has been modified in some manner (`scaledList`). These lists will be accessible when the Test-Network process terminates as they are properties of the `CollectNumbers` process. The `results` are `read` from `inChannel` as objects of type `ScaledData`, whose properties `original` and `scaled` are appended to each of the accessible property lists.

```
44   class CollectNumbers implements CSProcess {
45
46     def ChannelInput inChannel
47     def collectedList = []
48     def scaledList = []
49
50     def iterations = 20
51
52     void run() {
53       for ( i in 1 .. iterations) {
54         def result = (ScaledData) inChannel.read()
55         collectedList << result.original
56         scaledList << result.scaled
57       }
58     }
59   }
```

Listing 3 A Gatherer Process

The basic structure of the Test-Network processes is essentially independent of the PNUT, though it does need to be specialized to its specific requirements with respect to the number and type of input channels and of its outputs. The key requirement is that some relationship between the inputs and outputs has to be testable in an assertion.

2. The Network Under Test

The Network-Under-Test used in the above example is based upon the scaling device described by Belapurkar [7]. The scaling device reads integers that appear on its input every second, hence the delay introduced in the `GenerateNumbers` process shown in Listing 2. The scaling device then outputs these inputs multiplied by a constant factor, which is initially set to 2. The constant factor is however doubled every 5 seconds. Additionally, a controlling mechanism is provided that suspends the normal operation of the scaling device. The current value of the scaling factor is read and modified as necessary. In this case, the scaling factor is incremented by 1. While the scaling device is in the suspended state any input value is output without any scaling. The scaling device is suspended after 7 seconds and remains in the suspended state for 0.7 seconds. The script that runs the scaling device is given in Listing 4.

```
60   Node.getInstance().init(new TCPIPNodeFactory ())
61
62   NetChannelInput ordinaryInput = CNS.createNet2One("ordinaryInput")
63   NetChannelOutput scaledOutput = CNS.createOne2Net("scaledOutput")
64
65   new PAR(new ScalingDevice (inChannel: ordinaryInput,
66                                outChannel: scaledOutput) ).run()
```

Listing 4 Script to Execute the ScalingDevice in its own JVM

A network node is created {60} followed by two net channels that are the ones corresponding to those created within the Test_Network {7, 8}. The single `ScalingDevice` process is then executed {65, 66}. This process does not terminate.

The `ScalingDevice` is defined by the process shown in Listing 5, which defines two channel properties for input to and output from the process {69, 70}. The `run()` method uses three channels to connect the `Scale` process to the `Controller` process. These are used to implement the suspend, reading and updating of the scale factor described above.

```
67   class ScalingDevice implements CSProcess {
68
69     def ChannelInput inChannel
70     def ChannelOutput outChannel
71
72     void run() {
73       def oldScale = Channel.createOne2One()
74       def newScale = Channel.createOne2One()
75       def pause = Channel.createOne2One()
76
77       def scaler = new Scale ( inChannel: inChannel,
78                                 outChannel: outChannel,
79                                 factor: oldScale.out(),
80                                 suspend: pause.in(),
81                                 injector: newScale.in(),
82                                 scaling: 2 )
83
84       def control =  new Controller ( testInterval: 7000,
85                                        computeInterval: 700,
86                                        factor: oldScale.in(),
87                                        suspend: pause.out(),
88                                        injector: newScale.out() )
89
90       def testList = [ scaler, control]
91
92       new PAR(testList).run()
93     }
94
95   }
```

Listing 5 The Definition of the ScalingDevice Process

The unit test described above demonstrates that the basic functionality of the scaling device is correct. However, can we be assured that the system will behave correctly over a longer period? A moment's reflection will indicate this is not the case because if you continue to double the scaling factor and add one every so often then the bound on integer values will be reached and overflow will occur. Given that we know this to be the case can we demonstrate it by means of a generalised sampling environment that is applicable in a wide variety of situations? Further, could such a sampling system be left in place permanently so that the operation of the system can be checked periodically?

3. Sampling the Scaling Device

We shall use exactly the same Scale and Controller processes as those used in the unit testing; otherwise there was no point in testing them! We shall however drive them in a slightly different manner so that we can run the system for an indeterminate period. This is shown in Figure 2.

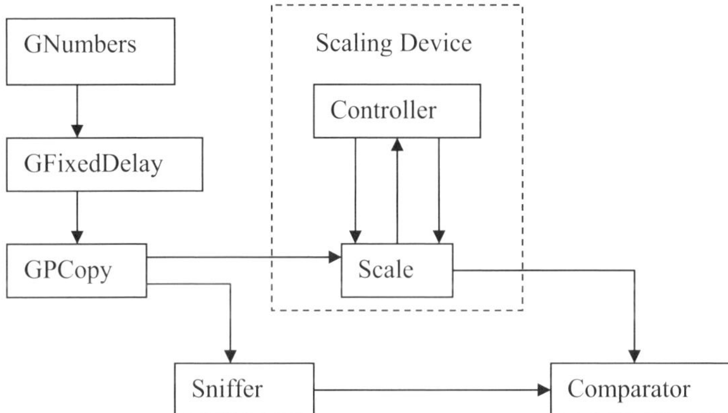

Figure 2 Sampling the Scaling Device

GNumbers, GFixedDelay are Groovy versions of the equivalent Numbers and FixedDelay processes of the package jcsp.plugNplay. GPCopy achieves the same effect as Delta2 in the same package. GNumbers generates a sequence of integers. GFixedDelay introduces a delay into the communication, in this case of 1 second. GPCopy copies any input to both its outputs in parallel.

The Sniffer process inputs all the data output to it by GPCopy. If a predefined time has elapsed since the data stream was last sniffed then the next input from GPCopy is output to the Comparator process. The Comparator process reads all the outputs from the Scale process. The Scale output comprises a data object containing both the original unscaled value and the scaled value. The definition of the scaling device is such that we know that the scaled value should either be greater than or equal to the original value. The Comparator already knows the sniffed original value and can thus check the relationship between the original value and the scaled value is correct, when it reads a record with the sniffed original value. This does presume that it takes longer for the Scale process to operate than the communication between the Sniffer and the Comparator.

On running this system we soon discover that the expected failure does occur and the scaled value goes negative and thus out of bounds. Perhaps more surprisingly a different failure mode is observed if the system is allowed to run further. We discover that the system goes into an infinite loop of a scaling factor sequence of -2, -1, -2, -1, …when we might have expected it to return to a positive number sequence as the values overflow a second time. Recall (Section 2) that the operations undertaken on the scaling factor are one of doubling and then adding one and thus this outcome is entirely reasonable but hard to predict when the processes are being defined because the operations are undertaken in different processes. The same outcome would be achieved if we quadrupled and then added three, except that the sequence would be -1, -4, -1, -4, … .

Listing 6 shows the coding of the Sniffer process. The process has two channel properties, one, fromSystemCopy, reads the outputs from the GPCopy process and the other, toComparator, writes data to the Comparator process, see Listing 7. The final property is the sampleInterval, which defaults to 10 seconds.

```
96   class Sniffer implements CSProcess{
97
98     def ChannelInput fromSystemCopy
99     def ChannelOutput toComparator
100    def sampleInterval = 10000
101
102    void run() {
103      def TIME = 0
104      def INPUT = 1
105      def timer = new CSTimer()
106      def snifferAlt = new ALT([timer, fromSystemCopy])
107      def timeout = timer.read() + sampleInterval
108      timer.setAlarm(timeout)
109      while (true) {
110        def index = snifferAlt.select()
111        switch (index) {
112          case TIME:
113            toComparator.write(fromSystemCopy.read())
114            timeout = timer.read() + sampleInterval
115            timer.setAlarm(timeout)
116            break
117          case INPUT:
118            fromSystemCopy.read()
119            break
120        }
121      }
122    }
123 }
```

Listing 6 The Sniffer Process Code

The run method defines a CSTimer {105}that is used to generate an alarm when the sampleInterval has elapsed {107, 114}. During normal INPUT {118} the data from the channel fromSystemCopy is read and ignored. When the alarm TIME has occurred the next value fromSystemCopy is read {113}and written to the channel toComparator. The next alarm time is recalculated {114,115}.

The Comparator process receives outputs from the system being sampled as well as inputs from the Sniffer process {126, 127}. The Comparator alternates over these inputs {132}. On receipt of a value from the Sniffer {137}, the process reads values from the system until the value to be evaluated is input {140-150}. It then tests the value to determine its relationship to an invariant of the system. An appropriate message is printed, which in a real system could be stored in a database.

The Sniffer and Comparator have been implemented knowing the detailed operation of the Scaling Device; in particular that it inputs a stream of integers and outputs objects containing both the original and the modified value. What happens if the input and output stream are either both objects or both streams of base types such as int, float etc?

Such requirements cannot be easily combined into a single architecture and thus in the following sections we describe approaches that enable sampling of the two types of system. Inevitably, though, the ability to create a generic sampling architecture similar to the generic testing architecture, described in Section 2, will be somewhat more difficult as the nature of the sampling system will vary with the specifics of the systems being sampled.

```
124 class Comparator implements CSProcess {
125
126   def ChannelInput fromSystemOutput
127   def ChannelInput fromSniffer
128
129   void run() {
130     def SNIFF = 0
131     def COMPARE = 1
132     def comparatorAlt = new ALT ([fromSniffer, fromSystemOutput ])
133     def running = true
134     while (running) {
135       def index = comparatorAlt.priSelect()
136       switch (index) {
137         case SNIFF:
138           def value = fromSniffer.read()
139           def comparing = true
140           while (comparing) {
141             def result = (ScaledData) fromSystemOutput.read()
142             if (result.original == value){
143               if (result.scaled >= result.original) {
144                 println "Within bounds: ${result}"
145                 comparing = false
146               }
147               else {
148                 println "Outwith Bounds: ${result}"
149                 running = false
150               }
151             }
152           }
153           break
154         case COMPARE:
155           fromSystemOutput.read()
156           break
157       }
158     }
159   }
160 }
```

Listing 7 The Comparator Process

4. An Object Based Sampling System

The basis of this sampling architecture relies on the ability of object oriented systems to extend a class such that any process that is unaware of the extension will be unable to manipulate the extended object definition. The generic architecture is shown in Figure 3.

The DataGenerator process represents a source of input objects to the Sampled Network. The Sampler process copies all inputs to its output unchanged unless it has received an input from the SamplingTimer process. The SamplingTimer process generates an output at predefined time intervals, known as the sampling period. In normal operation the Sampler process will just output the object generated by the DataGenerator. After receiving and input from the SamplingTimer the Sampler process will output an extended version of the object.

This extended version of the data object will have no effect on the SampledNetwork because it only recognizes the non-extended object. All outputs will be processed by the Gatherer process. All outputs from the Gatherer process are output to a subsequent part of the system, which in this case is a GPrint process. GPrint causes the printing of any object, provided it has a `toString()` method. Any extended data object will, in addition be communicated to the Evaluator process where its content can be evaluated against the invariants of the SampledNetwork.

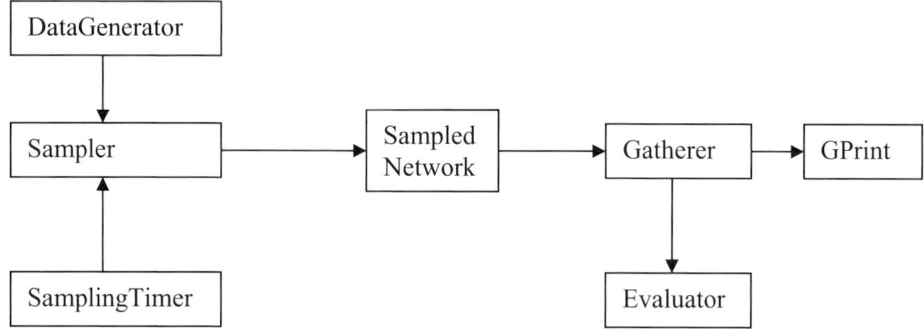

Figure 3 Generic Sampling Architecture for Object Based Input and Output

Listing 8 shows the code of the `SamplingTimer` process, which employs a simple loop that sleeps for the `sampleInterval` {169} and then outputs a signal message {170} on its `sampleRequest` channel

```
161  class SamplingTimer implements CSProcess {
162
163    def ChannelOutput sampleRequest
164    def sampleInterval
165
166    void run() {
167      def timer = new CSTimer()
168      while (true){
169        timer.sleep(sampleInterval)
170        sampleRequest.write(1)
171      }
172    }
173  }
```

Listing 8 The SamplingTimer Process

The `Sampler` process is shown at Listing 9.The channel `inChannel` receives inputs from the DataGenerator, which are output on the `outChannel` {191}. The `sampleRequest` channel is used to input requests for the generation of a sample. On receipt of the request signal {185}, the next data input is read {186} and its data values are used to create an instance of the extended object, referred to as `FlaggedSystemData` because the extension is simply a Boolean value set `true`. This extended object is then written to the system.

Necessarily, the `Evaluator` process is system dependent, an example of which is shown in Listing 10. Quite simply, values from the extended object are tested against each other {204} and a suitable message printed or saved to a database.

```
174 class Sampler implements CSProcess {
175
176   def ChannelInput inChannel
177   def ChannelOutput outChannel
178   def ChannelInput sampleRequest
179
180   void run() {
181     def sampleAlt = new ALT ([sampleRequest, inChannel])
182     while (true){
183       def index = sampleAlt.priSelect()
184       if (index == 0) {
185         sampleRequest.read()
186         def v = inChannel.read()
187         def fv = new FlaggedSystemData ( a: v.a, b:v.b, testFlag: true)
188         outChannel.write(fv)
189       }
190       else {
191         outChannel.write(inChannel.read())
192       }
193     }
194   }
195 }
```

Listing 9 The Sampler Process

```
196 class Evaluator implements CSProcess {
197
198   def ChannelInput inChannel
199
200   void run() {
201     while (true) {
202       def v = inChannel.read()
203       def ok = (v.c == (v.a +v.b))
204       println "Evaluation: ${ok} from " + v.toString()
205     }
206   }
207 }
```

Listing 10 The Evaluator Process

The `Gather` process, shown in Listing 11, repeatedly reads in objects from its `inChannel` and determines the type of the input {216, 217}.

```
208 class Gatherer implements CSProcess {
209
210   def ChannelInput inChannel
211   def ChannelOutput outChannel
212   def ChannelOutput gatheredData
213
214   void run(){
215     while (true){
216       def v = inChannel.read()
217       if ( v instanceof  FlaggedSystemData) {
218         def  s = new SystemData ( a: v.a, b: v.b, c: v.c)
219         outChannel.write(s)
220         gatheredData.write(v)
221       }
222       else {
223         outChannel.write(v)
224       }
225     }
226   }
227 }
```

Listing 11 The Gatherer Process

If the object has been extended then a non-extended version of the data is constructed and output to the rest of the system {218, 219}. The extended version of the data is also written to the Evaluator process {220}. Normally, the input data is just written to the output channel {223}.

Typical output from the sampling system is shown below where we see that the flagged data values are output twice once from `GPrint` as non-extended data and once from the `Evaluator` process where the complete `FlaggedSystemData` is printed.

```
System Data: [58, 59, 117]
System Data: [60, 61, 121]
Evaluation: true from Flagged System Data: [60, 61, 121, true]
System Data: [62, 63, 125]
System Data: [64, 65, 129]
System Data: [66, 67, 133]
System Data: [68, 69, 137]
System Data: [70, 71, 141]
System Data: [72, 73, 145]
System Data: [74, 75, 149]
System Data: [76, 77, 153]
System Data: [78, 79, 157]
System Data: [80, 81, 161]
Evaluation: true from Flagged System Data: [80, 81, 161, true]
System Data: [82, 83, 165]
System Data: [84, 85, 169]
```

5. Sampling Systems That Do Not Use Data Objects Explicitly

For systems that do not use objects explicitly, we could count the inputs to the Sampled Network. A suitable architecture is shown in Figure 4.

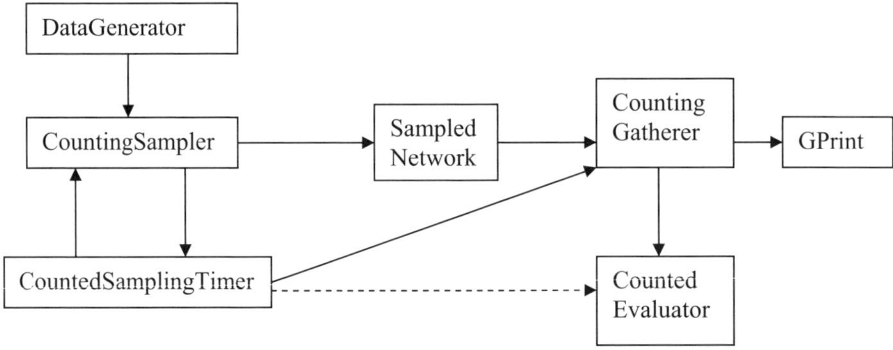

Figure 4 Generic Architecture for Sampling Networks by Counting Inputs

The CountingSampler takes inputs from the DataGenerator and copies them to the Sample Network, keeping a count of each input. At a rate determined by the sampling interval the CountedSamplingProcess will make a request to the CountingSampler, which will respond with the count value of the data input to be sampled. The CountedSamplingTimer will receive the value of the count which it sends to the CountingGatherer process. The CountingGatherer process keeps a count of the outputs from the Sample Network and on receipt of the output that corresponds to the count value it has received it outputs the count value and the output value to the CountedEvaluator process. The CountedEvaluator process can then record the sampled value and any result from a test that has been carried out. In some cases, the processes could be modified so that the CountingSampler process returns more than just the count value, for example the input data value, to the CountedSamplingTimer; in which case, this additional data may be communicated to the CountedEvaluator. The only requirement is that the time taken to undertake the two communications from the CountingSampler via the CountedSamplingTimer to the CountingGatherer must be less than the time taken to process the data in the Sampled Network. In this simple implementation we also need to ensure that every input to the Sampled Network has a corresponding output.

As a general comment we note that there is a large similarity between the architecture shown in Figures 3 and 4. This leads to the observation that some form of generic framework could be constructed that permits the easier construction of sampling architectures much in the same way as the `GroovyTestCase` framework has simplified the already relatively easy JUnit Test Case Framework.

6. Conclusions and Further Work

The paper demonstrates that is possible to use standard testing techniques commonly adopted by the software engineering community to the specialized requirements of parallel systems testing. In particular, a technique has been demonstrated that enables black-box testing using the GroovyTestCase specialization of JUnit. The paper then demonstrated that a simple set of additional processes could be easily defined that permit the sampling of running systems using a variety of approaches depending upon the nature of the data transmitted around the system.

The primary area for further work is to take the basic sampling processes and form them into a framework so they can be more easily incorporated by designers of parallel systems into their designs. In particular, the use of Groovy builders will hopefully make this a much simpler task than might be expected.

The nature of the processes being tested in this paper is somewhat limited because they were all deterministic in nature because the expected outcome was always fully determined by the input sequence. The architecture needs to be further developed to cope with non-deterministic systems where the final output is not fully determined by the input. To a certain extent the scaling system was non-deterministic in that the operation of the `Control` process was asynchronous with the `Scale` process. However, for means of explanation, it was made fully determined. Even for completely non-deterministic systems, there will probably be some processes or collection of processes that can be tested in a deterministic manner. The sampling of non-deterministic process collections will be easier to construct as the nature of the specific sampling processes is more closely tied to the underlying system.

Acknowledgements

John Savage introduced me to the delights of the GroovyTestCase specialization of the JUnit framework and convinced me that such testing could be achieved in parallel systems. He then convinced me to investigate simple ways of sampling systems. Kevin Chalmers suggested the technique of extending an object because he is a far better object oriented programmer than I. The comments of the referees were invaluable in improving the presentation of the material in this paper.

References

[1] K. Beck, Test Driven Development: By Example, Addison-Wesley, ISBN-10: 0-321-14653-0, 2003

[2] K. Barclay and J. Savage, Groovy Programming: An Introduction for Java developers, Morgan Kaufmann, San Fransisco, CA, ISBN-10: 0-12-372507-0, 2007

[3] PH Welch, Graceful Termination – graceful resetting, in A Bakkers (ed), Proceedings OUG-10: Applying Transputer Based parallel Machines, IOS Press, pp310-317, 1989.

[4] BHC Sputh, and AR Allen, JCSP-Poison: Safe Termination of CSP Process Networks, in J Broenink et al (eds), Communicating Process Architectures, 2005, IOS Press, pp 71-107, 2005

[5] FDR2 User Manual, Formal Systems Europe Ltd, http://www.fsel.com/fdr2_manual.html , accessed 11th April 2007.

[6] J.Kerridge, K. Barclay and J. Savage, Groovy Parallel: A Return to the Spirit of occam?, Proceedings of Communicating Process Architectures 2005, IOSPress, Amsterdam, 2005

[7] Belapurkar A, http://www-128.ibm.com/developerworks/java/library/j-csp2/ accessed 11th April 2007.

Communicating Process Architectures 2007
Alistair McEwan, Steve Schneider, Wilson Ifill and Peter Welch (Eds.)
IOS Press, 2007

Mobility in JCSP: New Mobile Channel and Mobile Process Models

Kevin CHALMERS, Jon KERRIDGE, and Imed ROMDHANI
School of Computing, Napier University, Edinburgh, EH10 5DT
{k.chalmers, j.kerridge, i.romdhani}@napier.ac.uk

Abstract. The original package developed for network mobility in JCSP, although useful, revealed some limitations in the underlying models permitting code mobility and channel migration. In this paper, we discuss these limitations, as well as describe the new models developed to overcome them. The models are based on two different approaches to mobility in other fields, mobile agents and Mobile IP, providing strong underlying constructs to build upon, permitting process and channel mobility in networked JCSP systems in a manner that is transparent to the user.

Keywords. JCSP Network Edition, mobile processes, mobile channels.

Introduction

The network package of JCSP [1, 2] provides a framework for distributed parallel systems development, as well a level of networked mobility of processes and channels. Originally this was complex to implement, although it was rectified by the original `jcsp.mobile` package [3]. Recent extended usage of these features has uncovered some limitations, as well as previously unknown advantages, to the original application models. The aim of this article is to examine new models for both channel and process functionality in JCSP, which have been implemented in the newest version of the `jcsp.mobile` package, as well as current plans to improve the models further.

The rest of this article is structured as follows. In Section 1 we introduce the different types of mobility, before describing process mobility further in Section 2. In Section 3 we discuss how to achieve process mobility in Java, focusing on the class loading mechanism. Section 4 discusses the original class loading model of JCSP, and Section 5 presents solutions to the problems found in the original structure. In Section 6 we provide a brief evaluation of the model developed.

Section 7 discusses channel mobility, and in particular the Mobile IP model used to define the new architecture for channel mobility in JCSP, with Section 8 presenting this new model. In Section 9, we present a more robust future implementation for channel mobility that requires some modifications to the existing JCSP Network framework, and in Section 10 we evaluate our model. Section 11 and 12 presents future work and conclusions.

1. Mobility

We are concerned with two separate types of mobility – that of channel mobility and process mobility. These can both be split further between the general headings of localized mobility (single node) and distributed mobility (between nodes). Table 1 provides a viewpoint of the complexity of these functional challenges. Also note that here process mobility is split between simple, single level processes, and complex, layered processes. The latter type of process is more difficult to migrate, and an implemented approach in JCSP has yet to be undertaken, although discussion concerning this shall be given in Section 11.

Table 1: Application Complexity of Mobility

Mobile Unit	Local Mobility	Distributed Mobility
Channel Mobility	Simple	Difficult
Simple Process Mobility	Simple	Moderate
Complex Process Mobility	Simple	Very Difficult

The argument presented in this table is that implementing mobility at a local level is simple. Although requiring additions to occam, leading to occam-π [4, 5], Java's nature meant this was possible in JCSP from conception.

These types of mobility (that we shall refer to as logical mobility [6]) allow us a new approach to developing systems, although a cost is incurred due to the level of abstraction we are dealing with. The cost the implemented models in JCSP inflict has yet to be measured. It is apparent in the pony framework of occam-π [7, 8] that overheads do exist when dealing with logical mobility at the network level. Therefore, we approached the design of these new models by examining existing applications in other fields, particularly mobile agents [9] and Mobile IP [10]. First of all we examine the concept of process mobility in the light of mobile agent approaches.

2. Process Mobility

To change the location that a program is executing requires various features, the most commonly cited being a method of code mobility [11]. This is exploited most of all in the field of mobile agents (MA). In fact, we make the argument that a MA is a specialized form of mobile process, an argument backed by a number of descriptions for agents, MAs, and their foundations.

2.1 Mobile Agents

Agents have their roots in the actor model, which are self contained, interactive, concurrently executing objects, having internal state and that respond to messages from other agents [12]. This description is similar to that of a CSP process – a concurrently executing entity, with internal state, which communicates with other processes using channels (message passing). Or another description is that an *"agent denotes something that produces or is capable of producing an effect"* [13]. Again, this is arguably similar to that of a process, which produces an effect when it communicates via a channel.

MAs are equally compelling when thinking of process mobility. Any process that can migrate to multiple hosts is a MA [13], although this may be a very broad claim, illustrated by the argument that *"the concept of mobile agent supports 'process mobility'"* from the same article. This appears to imply that mobile processes are MAs, although our argument is that this is not the case, the opposite being true.

MAs are also described as being autonomous (having their own thread of control) and reactive (receive and react to messages) [14], each of which enforce the belief that MAs and mobile processes have strong ties. Where they do differ is that process mobility is more akin to adding functionality to an existing node, enabling it to behave differently. MAs are designed to perform tasks for us, although mobile processes can be enabled to do this also. Therefore our argument that MAs are indeed mobile processes is made.

As mentioned previously, MAs exploit a feature known as code mobility [15], and it is this area we examine next.

2.2 Mobile Code

Six paradigms exist for code mobility [11]:

- *Client-server* – client executes code on the server.
- *Remote evaluation* – remote node downloads code then executes it.
- *Code on demand* – clients download code as required.
- *Process migration* – processes move from one node to another.
- *Mobile agents* – programs move based on their own logic.
- *Active networks* – packets reprogram the network infrastructure.

Each of these techniques is aimed at solving a different functional problem in distributed systems. For example, client-server code mobility involves moving a data file to the node where the program resides, executing the program and transferring the results back to the user. This allows a client to exploit functionality available on the server. MAs use an opposite initial interaction; the program moves to where the data resides, and returns to the user's node with the results.

The term process migration as given does not fully incorporate our viewpoint on process mobility. The distinction between MAs and process migration lies in where the decision to migrate takes place. MAs are autonomous, and make their own decision to migrate. In process migration, the underlying system makes the decision. In a mobile process environment, either the system or the process may make the decision to migrate.

Efficiency of each technique can be determined by the size of the mobile parts, the available bandwidth between each node, and the resources available at each node. A large data file encourages the MA paradigm, whereas a large program encourages either the client-server or remote evaluation paradigm. These are concerns when considering what method of code mobility to approach, and the distinction between data state and execution state in the size of a MA shall be covered in Section 2.4.

2.2.1 Requirements for Code Mobility

There are four fundamental requirements to support code mobility [16]:

- *Concurrency support* – entities run in tandem with currently running processes.
- *Serialization* – the functionality to be converted into data for transmission.
- *Code loading* – the ability to load code during execution.
- *Reflection* – the ability to determine the properties of a logical element.

The support for concurrency is of prime importance, and most modern development frameworks have some support for a thread based model. Threads are not the safest approach however [17], and provide us with little structure.

Serialization of logical components is also common in object-orientated platforms. A point to consider is that not all objects are serializable, and may rely on resources specific to a single node. This is a consideration when sending an object to a remote node, and how to access a required unmovable resource after migration is a problem. Mobile channels linked to the fixed resource can possibly overcome this.

Finally, code loading and reflection are common in Just-In-Time (JIT) compiled and script based platforms such as Java and the .NET Framework, and this is important to mobile code. If a system cannot load code during runtime then mobile code is not possible, unless a scripting approach is used, which is slow and sometimes impractical.

2.2.2 Reasons to Use Code Mobility

The ability to move code between computational elements promises to increase the flexibility, scalability and reliability of systems [11]. Besides these possibilities, a number of actual advantages are apparent [11, 18]. Firstly, code mobility helps to automatically reconfigure software without any physical medium or manual intervention. Secondly, thanks to moving code from a stressed machine to an idle one, code mobility can participate in load balancing for distributed systems. Finally, code mobility compensates for hardware failure, increases the opportunity for parallelism in the application, and enhances the efficiency of Remote Procedure Calls (RPC).

Other reasons to utilise a code mobility system [6] include possible reduction in bandwidth consumption, support for disconnected operations and protocol encapsulation. Disconnected operation means we do not have to rely on fixed resources on each to node to operate, allowing the logical elements to move around in a transparent manner. Protocol encapsulation involves data and the method to interpret it being sent together.

However, all these considerations must be weighed against certain attributes such as server utilisation, bandwidth usage and object size [19]. It has been noted [20] that mobile code systems are not always a better solution than traditional methods, especially if the transferred code is larger than the bandwidth saved by not sending the data direct.

2.3 Other Requirements for Mobile Agent Systems

A number of further requirements can be considered when thinking of MA systems. The list below summarizes these [13]:

- *Heterogeneity* – MAs must have the ability to move between various hardware platforms. Therefore a cross platform system is needed, and one of the key reasons why Java has a strong representation in MA frameworks [6, 12].
- *Performance* – this is a very open ended term. Performance in a mobile agent system may refer to network transference or system resource consumption. What is clear is that a definite performance benefit needs to be apparent to justify the usage of a MA system.
- *Security* – as a MA is moving code into systems for execution, certain guarantees need to be in place, or security restrictions enforced to stop malicious agents affecting a system.
- *Stand alone execution* – MAs must be able to operate on their own, without definite resources at the receiving node.
- *Independent compilation* – the MA should not require definite linkages to other components that may not be available on receiving nodes. The agent should be logically distinct in the system.

2.4 Types of Migration

Mobility of code and agent migration can be split into two groups – strong and weak mobility [15]. Weak code mobility provides data and code transference, but no state information pertaining to it. Strong code mobility transfers state also, and may be undertaken in a manner that is transparent to any communicating objects. State when referred to in this context includes execution state, and although desirable, Java lacks the ability to accomplish this. Therefore, in JCSP, only weak migration of logical elements is possible. The viability of strong migration has been questioned [11].

State itself can be split into two parts – data and execution – and it is the incorporation of execution state that dictates the type of migration. To clarify, strong mobility allows the resuming of a process directly after the migration instruction without any extra logic on the part of the developer [6]. In weak migration, the developer must create their own method of restarting the process after migration, which adds to the overall complexity of the process itself [13].

The case against strong migration is the amount of information that needs to be transferred. Both execution and data state must be transferred in this and this is time consuming and expensive.

Another consideration, in respect to how code is sent between nodes, is that of disconnected design. When migration occurs, all required code should also be sent at the same time [6]. This reflects on approaches using code repositories and the ability to accomplish this is questionable. For example, consider a mobile process that arrives at a node and then itself requires the receipt of another mobile process from its sending node. The ability to determine all code that may be required by a single mobile process is difficult, even if possible in all situations

Another viewpoint on migration is that of passive and active migration [18]. These terms are similar to those of weak and strong mobility. The difference is the amount of state transferred during the migration, with passive migration involving no state transference whatsoever. This description refers to the sending of data objects or code libraries. Active migration involves sending independently running modules.

The granularity of the code migrated is also a point to consider. Code can be thought of being either pushed to the receiving node all at once, or pulled by the receiving node as required [20]. The size of the sent component also varies depending on the type of architecture used. The approach to code mobility in Java is fine, sending only single classes at a time. A coarser grained approach is possible by utilizing JAR files. An even finer grained approach [20] has been suggested, which sends code on a per method basis instead of a per class one. The need for this in well designed software is questionable, and justification is based on unnecessary code loading occurring within Java. This has never been observed using either the original or new models for logical mobility in JCSP.

2.5 Migration Steps

There are six distinct steps involved during code migration [18]. Firstly, a decision to migrate must be made, either by the migrating unit itself or externally to it. The next step is the pre-emption of the application to avoid leaving both the application and the migrating component in an inconsistent state. The state of the component is then saved before it is sent to the next node. Any residual dependencies must then be taken into consideration, and dealt with accordingly. Finally the execution is continued when the migrating unit arrives at its new destination.

These steps represent a MA system more than just a code mobility system, and there are some issues in the form of communication used [21]. MAs use a mailbox communication system that does not guarantee instantaneous message delivery and this problem reflect upon the model for mobile channels which will be discussed in Section 7.

3. Achieving Process Mobility in Java

As was previously discussed, a key feature to achieve some form of process mobility is the usage of code mobility in a form similar to that of MAs. JCSP can take advantage of Java's code mobility mechanisms, and therefore we provide a discussion on code mobility in Java.

3.1 Java's Code Loading Model

Two objects are utilised in Java class loading – `ClassLoader` and `ObjectStream`. `ClassLoader` is responsible for loading classes for an application, and `ObjectStream` is used to serialize and deserialize objects.

Classes in Java are usually loaded by a single `ClassLoader` object – the `SystemClassLoader`. This object is responsible for loading class data for classes available on the local file system. It is also possible to develop custom `ClassLoaders` using the `ClassLoader` object as a base class.

Within the `ClassLoader` object itself, four methods are of note: `LoadClass`, `FindClass`, `DefineClass` and `ResolveClass`. Of these four methods, only `FindClass` needs to be overwritten.

3.1.1 ClassLoader Methods

The `LoadClass` method is called by any object that requires class data for an object that it uses. The `ClassLoader` being used by the object is determined by its class; whichever `ClassLoader` initially loaded the class is responsible for loading class data for that object. Consider Figure 1.

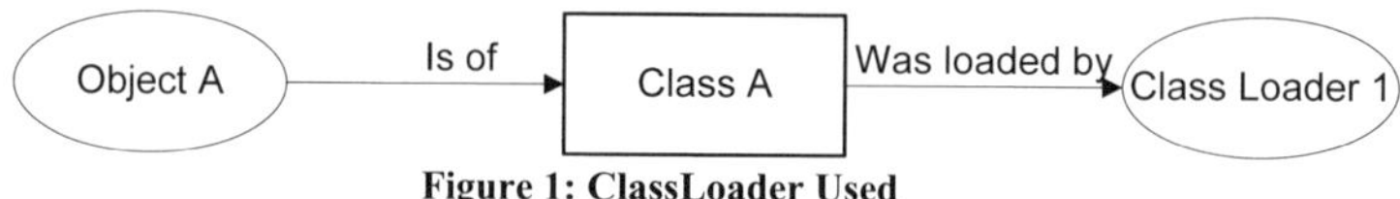

Figure 1: ClassLoader Used

If `Object` A is of `Class` A, which was loaded by `ClassLoader` 1, then any requests to load class data made by `Object` A will be carried out by `ClassLoader` 1.

`LoadClass` is called using the binary name of the class as a parameter and a `Class` object is returned. The `LoadClass` method goes through up to four stages when it is called:
1. Check if the class has already been loaded.
2. Call `FindClass` on the parent `ClassLoader` (usually the `SystemClassLoader`).
3. Call `FindClass` relevant to this `ClassLoader`.
4. Resolve (link) the `Class` if need be.

`FindClass` is used to find the data for a class from a given name. As the method used varies, the technique to find the class differs widely between implementations. What this method must do is return the `Class` of the given name and the simplest way to do this is to call `DefineClass` on an array of bytes representing the class.

3.1.2 ObjectStreams

`ObjectStreams` operate like other streams in Java. As network communications are also represented by streams, it is possible to create a object transference mechanism between nodes by plugging these into `ObjectOutputStreams` and `ObjectInputStreams`.

`ObjectInputStream` must be able to ascertain the class of the sent object and load the class data if necessary. This is performed by the `ResolveClass` method within the `ObjectInputStream`. With the `ObjectInputStream` supplied with the Java platform, the `ResolveClass` method loads the local class of the object sent using the `SystemClassLoader`. To allow a customised approach an `ObjectInputStream` is necessary that utilises a custom `ClassLoader`. The `ResolveClass` method of the new `ObjectInputStream` can then invoke the `LoadClass` method of the relevant `ClassLoader`.

With a basic description of the underlying operation of Java class loading presented, we can move on to an analysis of the operation of the original class loading mechanism within JCSP.

4. Original Class Loading System in JCSP

The original class loading mechanism for JCSP was part of the `dynamic` package provided in the JCSP.*net*. The mechanism is part of the `DynamicClassLoadingService`, which incorporates two filters, one for the serilization of objects deserialization.

4.1 Structure

Figure 2 illustrates the structure of the original `DynamicClassLoadingService` in JCSP.

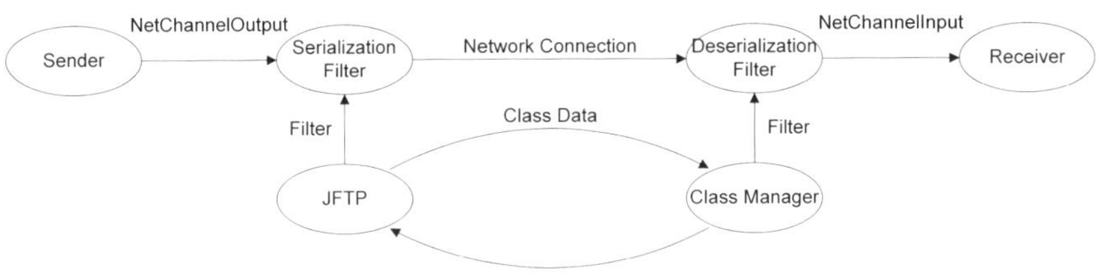

Figure 2: Original Dynamic Class Loading System

Two processes are utilised in the class loading service, `JFTP` and `ClassManager`. `JFTP` is responsible for processing requests for class data, whereas `ClassManager` requests, stores and loads class data.

4.2 Disadvantages and Problems

There are still limitations in the original `ClassLoader` itself. The most apparent is that class data requests do not propagate back through the system. This can lead to problems when trying to send an object that will traverse to another node before being utilised fully at a node on its path. For example consider the following:

```
01      public class MyClass {
02        public void run() {
03          MySecondClass obj = new MySecondClass();
04        }
05      }
```

Also consider a simple system consisting of three nodes, A, B and C. `Node A` creates an instance of `MyClass` and sends it to `Node B`. `Node B` reads in the object and then sends it onto `Node C` without calling the `run` method. `Node C` reads in the object and then invokes the `run` method.

Examining the interactions within the DynamicClassLoadingServices of these nodes we find a problem. When Node A sends the MyClass object to Node B, a request for the class data for MyClass is made from Node B to Node A. Node B now has the MyClass class file. When Node B sends the object to Node C, a request is made for the MyClass class data from Node C to Node B. When Node C invokes the run method of the MyClass object, it needs to discover what MySecondClass is. At this point it sends a request to Node B for the class data, which Node B does not have. An exception is thrown at Node C and our system fails.

The second limitation is poor handling of object properties. Consider the following:

```
06      public class MyClass {
07         private MySecondClass temp;
08      }
```

Both the MyClass and MySecondClass class representations should be loaded in one operation whenever an instance of MyClass is sent from one node to another. However, due to the method of interaction between the JFTP and ClassManager, only the MyClass data is requested and our system again fails.

The third limitation is due to the simplistic nature of the implementation of the DynamicClassLoadingService. As locally stored class data that has been received from another node is stored using the class name as a key, no two classes of the same name can be sent from two different nodes. When considering larger systems, it could occur that two classes use the same name, thereby creating an exception in the system.

One final 'fault' in the system was the handling of classes stored inside compressed JAR files. The comments inside the source code of JCSP point out that the existing method of reading in class data could not handle compressed archives. This is a trivial problem to solve, ensuring that we continue to read from the necessary stream until the number of bytes read in is the same as the number of bytes in the class file.

5. New Class Loader Model

To overcome the limitations of the current class loading mechanism, a new class loader has been developed for the jcsp.mobile package. This new model is presented as more robust, utilising a simpler method to retrieve class data, creating namespaces based on the node the class originally came from, and providing a class loading firewall as a first line of defense against malicious code attacks. The first area of development involves solving the problem of uniquely identifying classes with the same name.

5.1 Uniquely Identifying Classes

A simple solution to this problem is to link the class data with the node it originally came from. This can be achieved within JCSP by using the NodeID of the originating node that the object came from. This provides a method of creating class namespaces on nodes relevant to where the class data originated from. Consider Figure 3 for example.

Node A has sent an object to Node B and a class loading connection has been set up between them. This interaction has created a Node A class namespace within Node B. The same interaction has also occurred between Node B and Node C, creating a Node A class namespace at Node C also. Within these two separate namespaces it is quite possible to have two classes of the same name as each of the namespaces are distinct from the other.

5.2 New Class Loader Structure

The most important aspect of the new implementation is the inclusion of a method that allows class requests to propagate back through the series of nodes to the original. Essentially, the new class loader process combines both the `JFTP` process and `ClassManager` process into a single process.

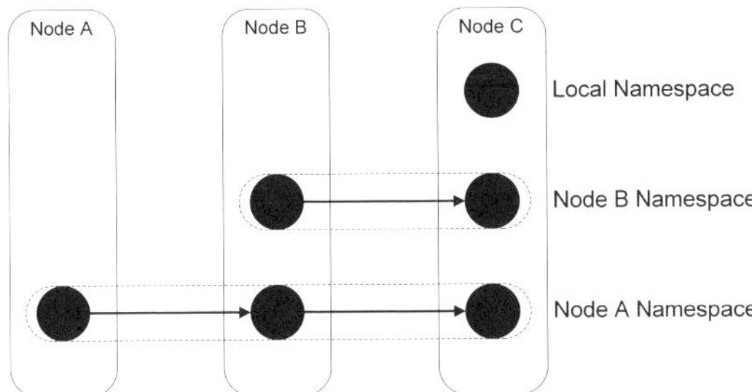

Figure 3: Node Namespaces

5.3 New Class Loader Operation

When considering the operation of the `ClassLoader`, there are four different usage scenarios that need to be examined. These scenarios are the interactions of single hop class loading between nodes for unknown internal classes and for classes that create unknown classes during their operation, and also the same two usages in a multi-hop interaction between several nodes.

5.3.1 Single Hop Operation

The first usage scenario has an object being sent from the Sender process to the Receiver process, and the sent object then creating an instance of a previously unknown class at the Receiver process's node. This is illustrated in code lines 1 to 5.

Here, the Sender first creates a new `MyClass` object and sends it to the Receiver. When the Receiver process reads the data it tries to recreate the original object using the customised `ObjectStream`. When `FindClass` is invoked, the system discovers that it cannot load the class locally and sends a request back to the node of the Sender process. This `ClassLoader` then retrieves the class for `MyClass` and sends it back to the Receiver's node. The Receiver can now invoke `run` upon the instance of `MyClass`. At this point, the `MyClass` object must create an instance of `MySecondClass`, and the same sequence of events occurs, with the same `ClassLoader` being invoked as before. The Receiver can now continue execution. Figure 4 presents a sequence diagram of this interaction.

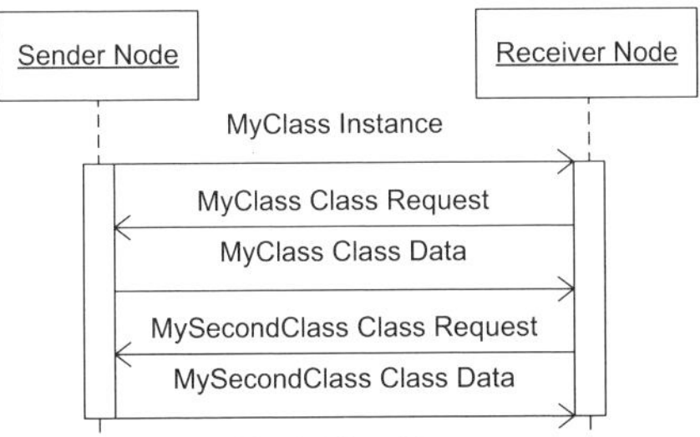

Figure 4: Single Hop Sequence

The second usage scenario for a single hop system involves an unknown object being an internal property of the sent object, as illustrated in code lines 6-8. This has the same interaction sequence, except this time `MySecondClass` is requested before a call to the `run` method is made. Our sequence diagram is the same, although the Receiver Node does not actually make a call to the Sender Node for `MySecondClass` in the latter instance until the call to `run`.

As this diagram illustrates, the interactions occurring between the Sender and Receiver are fairly simplistic, and the original model for the `DynamicClassLoadingService` could handle these, except for the internal properties of a class. The more complex interaction occurs when considering an architecture that incorporates one or more intermediary nodes.

5.3.2 Multi-hop Interaction

When class data has to pass through one or more nodes to reach its destination, we term this a multi-hop interaction. From the point of view of the Sender and Receiver, nothing has changed. The difference is that one or more remote processes sit between them, which we shall refer to as `Hop`. The purpose of the `Hop` process is to read in an object from the Sender and forward it to the Receiver.

For the usage scenario involving unknown classes the Sender creates an instance of the `MyClass` object and sends it to `Hop`. When the `Hop` process tries to recreate the object, its `ClassLoader` requests the class from the `ClassLoader` at the Sender node. When the class is received, the `Hop` node can then recreate the object and send it on to the Receiver. When the Receiver tries to recreate the object, it too requires the class for `MyClass`, and makes a request back to `Hop`. The `Hop` node can retrieve the class data it previously received and forward it onto the Receiver node. The Receiver can now invoke the `run` method upon `MyClass`. The Receiver node requires the class for `MySecondClass` at this point, and sends a request `Hop`. `Hop` does not have the required class for `MySecondClass`, and sends a request back to the Sender. It receives the necessary class back from Sender, and then stores it locally before forwarding it onto the Receiver. The Receiver can now define and resolve `MySecondClass` and allow the Receiver to continue. Figure 5 illustrates this.

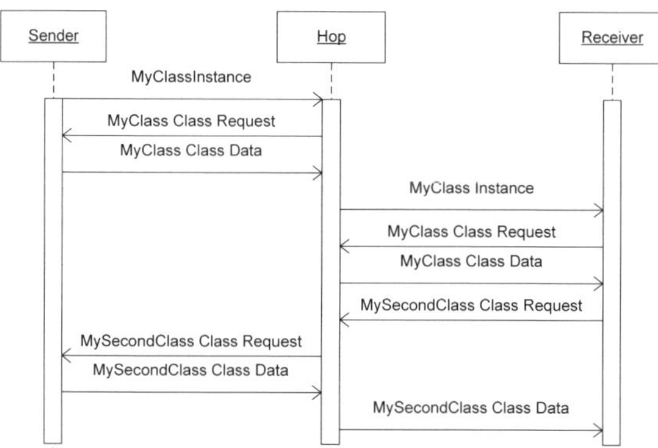

Figure 5: Multi-hop Object Creation Sequence

The interaction involving unknown classes as properties of a sent object again involves the Sender creating an instance of `MyClass` and sending it to `Hop`. When `Hop` receives the instance, it requests the class for `MyClass` from the Sender node. When this is received and an attempt to define and resolve the class is made, `Hop` discovers that `MySecondClass` is also required, and a request is sent for the `MySecondClass`. When the instance of `MyClass` is sent to the Receiver, the same requests are sent back to `Hop`. As `Hop` has both the classes, it can handle the requests itself. Figure 6 presents the sequence diagram.

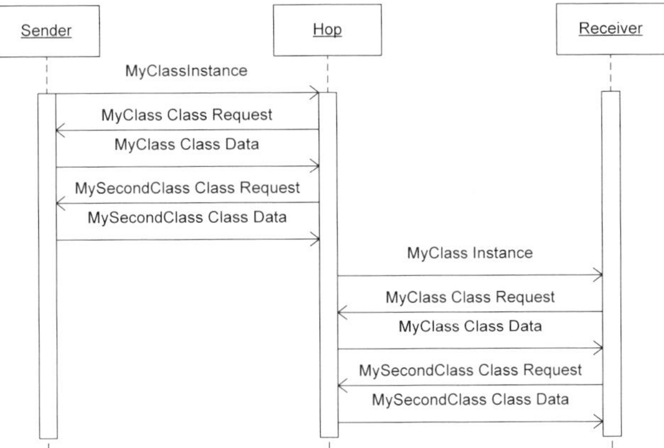

Figure 6: Multi-hop Object Property Sequence

5.4 *Class Loading Architecture*

In this section we demonstrate how `Hop` knew where to request the class data for `MySecondClass` from. Java has the ability to determine which `ClassLoader` originally defined the class of an object, and when sent from one node to another, the `DynamicClassLoadingService` determines which class loader originally defined the class. If it was the local `SystemClassLoader`, then the service retrieves the `ClassLoader` process responsible for handling requests for locally available classes. If the `ClassLoader` is a `DynamicClassLoader`, this is used instead. As each `ClassLoader` has a channel to accept requests, it's location can be provided as a property of the `DynamicClassLoaderMessage` that encapsulates a sent object.

When a node receives a `DynamicClassLoaderMessage`, it first extracts the request location from the message. It uses this location to check if it has already set up a `ClassLoader` for this object, and if not creates one. An attempt is then made to recreate the object contained within the `DynamicClassLoaderMessage`, using the relevant `ClassLoader`. This `ClassLoader` can request classes from the `ClassLoader` further down the chain, using the location provided. This allows the creation of code namespaces on each node based on where an object originally came from, as illustrated in Figure 3. As this diagram shows, each node has its own `ClassLoader` process which is relevant to its own `SystemClassLoader`, handling requests for classes stored locally at that node. Requests travel to the originating node, and responses travel forward to the destination that requested them. At a basic level, classes are always obtained from the originating node.

This structure has the further advantage of creating a trusted path from sender to receiver. Every time a request comes in, the `ClassLoader` checks its local file system first. If it can obtain the class locally, then it will send the local version. The version sent may not be the original version of the class that was sent through the node, but it does prevent malicious attempts to override trusted classes at the receiving node. In effect, this has created a code firewall between each node, with the previous node being trusted to pass on safe code only. Adding the ability to only load classes from certain channels into the node or from nodes with specific addresses enhances this level of security further.

5.5 Storing Class Data

One decision that was made in the design of this architecture was the storing of class data at each node as it is received. The contrary decision was to always allow requests to travel back to the origin node and then the class to travel back to the requesting node through all intermediaries. The latter approach would require more interactions between nodes, but fewer resources for managing classes at each node. However, the problem exists that a node may leave a system while classes it sent are still present, and if copies of classes were not retained locally, any further requests by other nodes will be unsuccessful. Also, when we consider that we wish to achieve some level of disconnected design, the approach of having classes available throughout the system is better.

The storing of code is also a major consideration for the pony framework [8]. The advantage in pony is that each system has a single master node, which could store code and distribute it to slave nodes. Although this creates a single point of failure, this is no different than if the master node went down anyway. Storing class data in a single location also builds structure and simplicity into the interactions. As pony is more controlled than the networked features of JCSP, this makes sense. The architecture for code mobility within JCSP reflects more on the fact that it is more disconnected and less controlled than pony. Therefore we make a recommendation that pony should adopt a code repository style approach for code mobility.

6. Evaluation of the Code Mobility Model

The architecture described for code mobility, and thereby mobile processes, has been developed with Java, JCSP and mobile agents in mind, and this does lead the limitation of having weak mobility only. However, the architecture developed is sufficiently disconnected to allow systems where the mobility of code between any nodes is almost at a very high level.

There are some other considerations. Our approach does allow mobile processes to be sent to any node that we are directly connected to, and code to be loaded between nodes as long as a connection back to the origin exists, or one of the nodes has the code required. If this is not the case then a system will fail. Although we have developed an architecture that exhibits a level of disconnected design, this serious vulnerability is still in place.

Consider a comparison between the JCSP approach and the recommendation for pony given above. We shall use the idea of locations and abstract locations [14], a location being a single node in which a mobile process can execute, and an abstract location being a collection of nodes within which a mobile process can execute. Within JCSP, only the former viewpoint is possible, as our nodes and architecture have left the system loosely coupled. Within pony, a mobile process can be considered at either the single node level or multiple nodes in the form of a mobile process within an application. The more rigid control offered by pony allows this. However, a mobile process wishing to move outside of the application has more difficulty in pony than in JCSP. This difference is inherent in the structures developed for JCSP and occam.

Mobile Agents and mobile processes also provide us with a very advanced abstraction, allowing a different viewpoint for systems development and understanding. Although it has been stated that there is no 'killer application' for mobile agent systems [6], as anything achievable with a MA system is also achievable using traditional methods, the simplicity of the abstraction is very powerful.

Finally, handling connection to, and management, of resources is an issue [6]. The proposal we make for this is the use of mobile channels that allow both our process and channel ends to move together, allowing the process to stay connected to a resource. This is also true for mobile processes connected to other mobile processes. Communication between agents is seen as a problem in MA systems, and sent messages may end up chasing a mobile element around a system. This has been taken into consideration when designing the mobile channel model.

7. Channel Mobility

The model we have developed to allow channel mobility in JCSP is partially based on that of Mobile IP [10]. As this model was developed to allow mobility of devices in standard IP based networks, it was determined that it was a worthwhile model to replicate for mobile channels. We will first describe the shortcomings of the original model for migratable channels in JCSP.

7.1 Migratable Channels

Migratable cannels allowed a form of channel mobility in a rigid manner. The `MigratableChannelInput` had to be prepared prior to an actual move occurring. This was not a major issue, as this can be done by the movement function. The major flaw was data waiting to be read on the channel being rejected, and therefore lost. This is a dangerous system; losing messages is not safe. Also, the exception thrown by the JCSP system when data is rejected is not catchable, but handled by the underlying framework. So, although we can move a networked channel, the likelihood is the system will break during the process.

This problem leaves us with the need to create a new model to achieve channel mobility within JCSP. Therefore, the existing model for Mobile IP was examined and then replicated in a manner to allow channel mobility within JCSP.

7.2 Mobile IP

For Mobile IP to operate, three new functional entities are introduced [10]:

- *Mobile node* – the mobile device, which has a constant IP address.
- *Home agent* – resides on the mobile node's home network and forwards messages onto the mobile node's actual location.
- *Foreign agent* – resides on the network the mobile node is currently located, and receives messages from the home agent for the mobile node.

Also of import are the care-of-address (COA), which is where the mobile node is currently situated, and the home address, which is where the mobile node originated from. The Mobile IP model is presented in Figure 7.

When a mobile node is started, it is registered with a normal IP address on its home network. When it leaves its home network, the home agent is informed and is prepared to intercept any messages sent to the mobile node's address. The node then moves and when it arrives at its new location, informs the foreign agent, which registers the mobile node. The home agent is informed of the address of the foreign agent (the COA) and forwards messages onto this address.

Let us consider more fully the model in Figure 7. The communicating host is initially communicating with the mobile node using the home address. When the node moves, it informs the home agent to preempt messages sent to its home address and prepare them for forwarding to a new address. When the mobile node arrives at the foreign network, the

COA is determined and sent to the home agent, which forwards all messages onto this new address. From the point of view of the communicating host, nothing has changed, and messages are still sent to the same home address. When the mobile node wishes to communicate directly to the communicating host, it does so through the foreign agent.

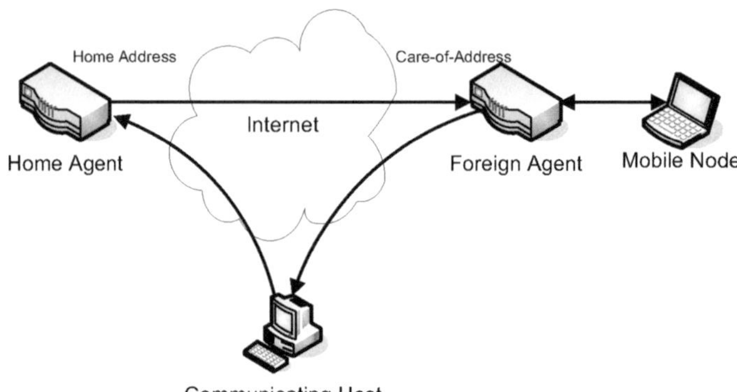

Figure 7: Mobile IP Model

Within JCSP we have the advantage that channels are one way only, meaning we can simplify our mobile channel model from that of Mobile IP. However, first we provide a basic overview of how JCSP networking operates.

7.3 JCSP Networking

A number of functional elements provide us with networked channel capabilities in JCSP. When we describe a networked channel in this respect, we mean an input end and output end connected together across a communication medium, and these are two distinct objects usually located on different nodes. We have:

- *Node* – a JVM where a JCSP application resides. Multiple nodes may make up an entire JCSP system, and multiple nodes may reside on a single machine.
- *Node address* – the unique identifier for a node.
- *Link manager* – listens for new connections on the node address. It is also responsible for creating new connections to other nodes.
- *Link* – a connection between two nodes. A node may have multiple links. The link is a process that listens for messages on its incoming connection and forwards them onto the net channel input.
- *Net channel input* – an input end of a networked channel.
- *Net channel location* – the unique location associated with a net channel input. The location takes the node address and a channel number as unique identification.
- *Net channel input process* – receives messages sent by links and passes them onto the net channel input. This process accepts messages from all links, which must determine which net channel input to pass the message onto by the unique channel number passed in the original address.

When a networked JCSP system is started, the node is initiated and allocated a unique address. The `LinkManager` on the node is started and listens for new connections from other `Nodes`. When a connection occurs, or the application requests that a new connection to a remote `Node` be made, a new `Link` process is started. When a new `NetChannelInput` is created, a new `NetChannelInputProcess` is started, and a unique number allocated to the channel. Now when a `Link` receives a message from its connected `Node`, it extracts the unique channel number destination, and uses this to forward the body of the message to the `NetChannelInputProcess` which reads in the message and immediately writes to the `NetChannelInput`. The `NetChannelInput` is buffered so the `NetChannelInputProcess` does not block. These parts are placed together to provide the model presented in Figure 8.

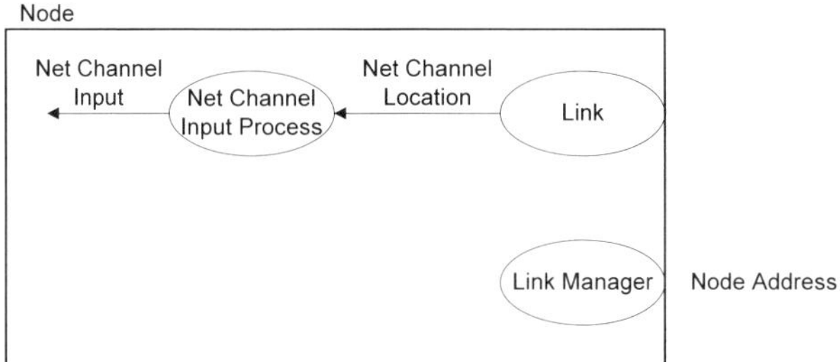

Figure 8: JCSP Networking

8. Mobile Channel Model

If we compare the two models of Mobile IP and JCSP Networking, we can see that there are similarities between them. For example the `Link` process acts as a router for messages received from a connection to a `NetChannelInputProcess`, similar to the foreign agent in Mobile IP, which forwards messages onto the mobile node on a foreign network. Therefore we can remove the foreign agent from our interpretation of the Mobile IP model, and implement the home agent to provide the functionality we need. The home address is a normal `NetChannelLocation`, and the COA likewise. This gives us the model presented in Figure 9.

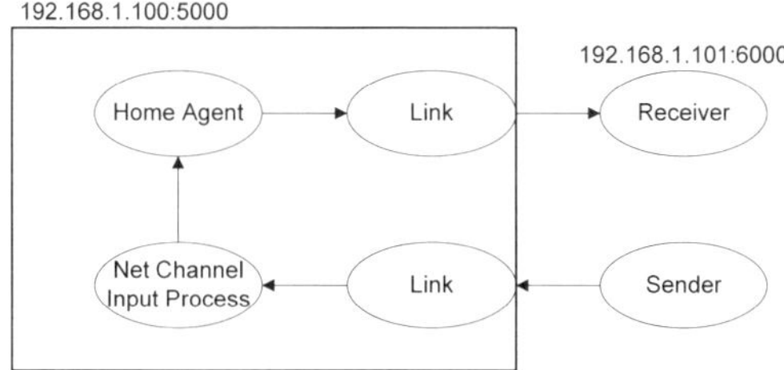

Figure 9: Mobile Channel Model

We have condensed the Sender and Receiver nodes down to single processes to keep the diagram simple. When a `NetChannelInput` migrates, it leaves a `HomeAgent` process that receives messages from the `NetChannelInputProcess`, and forwards them on accordingly. The sequence of interactions is:

1. Sender sends a message to the home address of the `NetChannelInput`. In Figure 9, the Sender would send to 192.168.1.100:5000/7 – 7 being the unique number associated with the channel.
2. The `Link` connecting to the Sender's `Node` reads the message, extracts the location, and forwards the message to the `NetChannelInputProcess`.

3. The `NetChannelInputProcess` reads in the message and sends it to the `HomeAgent`.

4. The `HomeAgent` reads the message and sends it to the new location of the Receiver, via the `Link` connecting to the `Node` that the Receiver is located. In Figure 9, the `HomeAgent` would forward the message to 192.168.1.101:6000/12.

5. The Receiver can now read the message sent by the Sender.

This model permits channels to move from one `Node` to another transparently, allowing multiple senders to send messages to a single address, instead of the system trying to inform them of the new location. As networked channels are Any-2-One, this is a simpler solution. If channels were One-2-One, we could inform a single sender. As this is not so, we implement a system allowing all senders to remain virtually connected to the receiver.

8.1 Shortcomings

Although this model does allow mobile channels, we still have a number of problems to overcome. Firstly, we still run the risk of losing a message whenever a channel moves to another location. Secondly, we lose synchronization between sender and receiver due to passing through the `HomeAgent`. Thirdly, we are creating a chain of `HomeAgent` processes between the mobile channel's home location, and all the locations it has visited during its lifetime. This problem also causes difficulties for the second shortcoming.

To overcome this, we introduce another channel to the `HomeAgent`, a configuration channel, which is responsible for receiving movement notifications from the mobile channel end, as well as being responsible for requesting messages from the `HomeAgent` when the mobile channel end is ready to read. The `HomeAgent` process waits for a request from the mobile channel end before forwarding the next message from its input channel. The configuration channel can also be used to send updates for locations; resulting in only one `HomeAgent` for a mobile channel. The operation of the `HomeAgent` is now:

```
09      Message msg = (Message)config.read();
10      if (msg instanceof MoveMessage) {
11        NetChannelLocation newLoc = (NetChannelLocation)config.read();
12        toMobile = NetChannelEnd.createOne2Net(newLoc);
13      }
14      else {
15        toMobile.write(in.read());
16      }
```

`toMobile` is connected to the mobile channel end, and `in` is the original channel set up to receive incoming messages to the mobile channel.

Using this extra channel overcomes the three problems mentioned above. We will no longer lose messages as no messages are stored at the mobile channel end; they are forwarded when a read occurs. Synchronization between reader and writer has been regained, as the writer is not released until the reader has committed to a read, and we no longer need to have multiple `HomeAgent` processes, as the original one is kept informed of the location of the mobile channel.

8.2 Comparison to pony

The mobile channel model in pony is implemented differently to the one we have described here for JCSP. The major difference is that pony takes a controlled approach to channel mobility, with a process controlling access to a networked channel, and ensuring messages are received by monitoring the location of channels. The model we have proposed for

JCSP has less centralised control, and acts in a peer-to-peer manner. Our approach therefore has certain advantages over pony. For example, pony has a single point of failure in its controlling process, although this makes channel mobility easier to control. The failure of a single `HomeAgent` process will only bring down a single mobile channel, but it does mean that the control of channel mobility is distributed, and harder to maintain.

There are still two major problems with the approach taken in JCSP. Our model does not allow selection of a mobile channel in an alternative, and each mobile channel requires an extra process to operate. As Java cannot handle a large number of processes (around 7000 threads is the maximum on a Windows platform [22, 23]), this can lead to a problem when large number of mobile channels are in operation. Our model requires three processes, a `NetChannelInputProcess` at the home location and at the new location, and a `HomeAgent` process at the home location, to provide the functionality of a single channel. Therefore we need to refine our model to provide more functionality, and use fewer resources at the same time. We therefore also propose a, as yet unimplemented, further model for channel mobility.

9. Future Mobile Channel Model

To implement a `MobileAltingChannelInput`, we can take advantage of recent additions to JCSP in the form of Extended Rendezvous [24]. This allows us to read a single message from a `NetChannelInput` into the `HomeAgent` and not release the writer until the forwarded message has been successfully received at the mobile channel's actual location. The code to achieve such an operation would be:

```
17      public void run() {
18        Object message = in.beginExtRead();
19        toMobile.write(message);
20        in.endExtRead();
21      }
```

This blocks the writer until the full forwarding operation has been completed, but does not take into account rejection of messages at the mobile end that occurs when a channel moves. Our `HomeAgent` process therefore becomes:

```
22      public void run() {
23        Guard[] guards = {config, in};
24        Alternative alt = new Alternative(guards);
25        while (true) {
26          switch (alt.priSelect()) {
27            case 0:
28              NetChannelLocation newLoc = (NetChannelLocation)config.read();
29              toMobile.recreate(newLoc);
30              break;
31            case 1:
32              boolean writing = true;
33              Object message = in.beginExtRead();
34              while (writing) {
35                try {
36                  toMobile.write(message);
37                  writing = false;
38                catch (ChannelDataRejectedException cdr) {
39                  NetChannelLocation loc =(NetChannelLocation)config.read();
40                  toMobile.recreate(newLoc);
41              }
42              in.endExtRead();
43              break;
44      }}}
```

This code also defines how to handle the location of the channel changing. First we check if the configuration channel or the `in` channel into the `HomeAgent` is ready (lines 24-26). If the configuration channel is ready, we read in the new channel location and recreate the channel to the mobile channel end, allowing the `HomeAgent` to communicate with the new location (lines 27-30).

If the `in` channel has data ready (line 31), we begin reading from the channel and try to write to the mobile channel. There are two possible outcomes. Either, the data is successfully read by the mobile channel and we can continue (lines 36-37), or the mobile channel end could move before the data is read and therefore the data is rejected (line 38). In this case, we read the new channel location and recreate the channel connected to the mobile end (lines 39-40) and try to write again. Once the data has been successfully written, the writing process can be released (line 42) and the process continued.

This proposed `HomeAgent` process can also accommodate the functionality of the `NetChannelInputProcess`, thereby removing the need for two processes on the home node and reducing the amount of required resources. Thus, we have solved both the problem of selection and resource usage. The mobile end can alternate as it knows if it has received a message on its buffer.

10. Evaluation of Mobile Channel Model

The model for mobile channels presented coincides with the distributed nature of JCSP, and is therefore deemed a suitable approach for providing mobility of channels. It differs from that of pony, although this is due to the different approach taken for distributed systems. Although each has their good and bad points, it is left to the individual to choose whether the controlled approach of pony or distributed approach of JCSP is better for the system they wish to implement. As both systems have fairly similar performance in their networked channels, this is justified.

The JCSP approach as currently implemented does have its shortcomings however. Selection of a mobile channel and the number of resources used are an issue. A plan for the future approach to channel mobility does overcome these shortcomings, but does require some changes to the underlying functionality of JCSP as it stands.

There is still a question of performance of mobile channels in JCSP, as each sent message has to be forwarded onto the actual location of the channel end. This doubles the amount of time taken to send a message, as it must make a hop before arriving at its destination. The performance of the mobile channels has yet to be evaluated, and this is left for future work.

As the model developed is based on an already implemented model, that of Mobile IP, we can argue that it is sufficiently sound as an approach for channel mobility. Although we have not described how output channel mobility is achieved, this is trivial. Networked channels are Any-2-One, so when an output channel end moves we need only send the location of the input end and reconnect at the new location. No data is kept at the output end, and the channel blocks until the write operation is completed, so the output end cannot move. Therefore, we have both input and output channel end mobility across the network, implemented in a manner that is transparent to both reader and writer.

11. Future Work

Although we have presented models for both channel mobility and process mobility, we have not discussed how more complex processes can be moved from one node to another due to the complexity of suspending a running network of processes. This has been discussed at length in terms of graceful resetting [25] and how to place processes into a state so that they can be migrated safely [5]. These ideas have been presented for occam, although it is also possible to implement them in JCSP with the ideas presented for JCSP poison [26]. These principles have been implemented in the new edition of JCSP [24].

Also, what we can exploit specifically in Java to allow complex process migration has not been investigated. As Java uses object serialization to send objects to remote machines, the mechanism can be taken advantage of to allow signaling of a process when it should start preparing itself to move. This has also been taken advantage of in the channel mobility model to a certain degree. Any serializable object in Java can override methods to allow the dictation of how the object behaves when it is being written to or read from an object stream. The write method can be used to allow the signaling of the process that it should prepare itself to move. However, this will require a great deal of extra logic built into existing processes. This can be directly linked back to Java's lack of support for strong code mobility.

Other future work includes the evaluation of the mobile models against other approaches to code mobility, especially mobile agent platforms. It is hoped that this will allow JCSP to be put forward as a possible framework for mobile agent systems in the future, with the advantage of taking a more formal and structured approach than the standard remote procedure call and object-orientated one.

12. Conclusions

By utilising existing features of both Java and JCSP, we have shown that it is possible to develop transparent methods of both process and channel mobility between distinct networked nodes. By developing our models from those of mobile agents and Mobile IP, we have taken into account some of the pitfalls already discovered for other logical mobility systems. Our models still need to be thoroughly tested and examined, and for mobile channels in particular additions need to be made to JCSP itself to allow the best use of the resources available, as well as allow selection of mobile channels in an alternative.

References

[1] P. H. Welch and J. M. R. Martin, "A CSP Model for Java Multithreading," in Proceedings International Symposium on, Software Engineering for Parallel and Distributed Systems, 2000, pp. 114-122, 2000.
[2] P. H. Welch, J. R. Aldous, and J. Foster, "CSP Networking for Java (JCSP .net)," in P. M. A. Sloot, C. J. K. Tan, J. J. Dongarra, and A. G. Hoekstra (Eds.), Proceedings International Conference Computational Science - ICCS 2002, Part II, Lecture Notes in Computer Science 2330, p. 695, Springer Berlin / Heidelberg, 2002.
[3] K. Chalmers and J. M. Kerridge, "jcsp.mobile: A Package Enabling Mobile Processes and Channels," in J. Broenink, H. Roebbers, J. Sunter, P. Welch, and D. Wood (Eds.), Communicating Process Architectures 2005, pp. 109-127, IOS Press, 2005.
[4] F. R. M. Barnes and P. H. Welch, "Prioritised Dynamic Communicating Processes: Part 1," in J. Pascoe, P. H. Welch, R. Loader, and V. Sunderam (Eds.), Communicating Process Architectures 2002, pp. 321-352, IOS Press, 2002.
[5] F. Barnes and P. H. Welch, "Communicating Mobile Processes," in I. R. East, D. Duce, M. Green, J. M. R. Martin, and P. H. Welch (Eds.), Communicating Process Architectures 2004, pp. 201-218, IOS Press, 2004.
[6] G. P. Picco, "Mobile Agents: an Introduction," Microprocessors and Microsystems, 25(2), pp. 65-74, 2001.
[7] M. Schweigler, A Unified Model for Inter- and Intra-Processor Concurrency. PhD Thesis, The University of Kent, Canterbury, UK, 2006.

[8] M. Schweigler and A. Sampson, "pony - The occam-π Network Environment," in P. H. Welch, J. M. Kerridge, and F. R. M. Barnes (Eds.), Communicating Process Architectures 2006, IOS Press, pp. 77-108 2006.

[9] J. White, "Mobile Agents White Paper," General Magic, 1994. Available from: http://citeseer.ist.psu.edu/white96mobile.html

[10] C. E. Perkins, "IP Mobility Support for IPv4," IETF, Technical Report RFC 3334, 2002.

[11] R. R. Brooks, "Mobile Code Paradigms and Security Issues," IEEE Internet Computing, 8(3), pp. 54-59, 2004.

[12] H. S. Nwana, "Software Agents: An Overview," Knowledge Engineering Review, 11(3), pp. 205-244, 1996.

[13] K. Rothermel and M. Schwehm, "Mobile Agents," in A. Kent and J. G. Williams (Eds.), Encyclopedia for Computer Science and Technology, New York, USA: M. Dekker Inc., 1998.

[14] V. A. Pham and A. Karmouch, "Mobile Software Agents: an Overview," IEEE Communications Magazine, 36(7), pp. 26-37, 1998.

[15] A. Fuggetta, G. P. Picco, and G. Vigna, "Understanding Code Mobility," IEEE Transactions on Software Engineering, 24(5), pp. 342-361, 1998.

[16] M. Delamaro and G. P. Picco, "Mobile Code in .NET: A Porting Experience," in N. Suri (Ed.), Proceedings of Mobile Agents: 6th International Conference, MA 2002, Lecture Notes in Computer Science 2535, pp. 16-31, Springer Berlin / Heidelberg, 2002.

[17] P. B. Hansen, "Java's Insecure Parallelism," SIGPLAN Notices, 34(4), pp. 38-45, 1999.

[18] P. Troger and A. Polze, "Object and Process Migration in .NET," in Eighth IEEE Workshop on Object-Orientated Real-Time Dependable Systems (WORDS '03), p. 139, IEEE Computer Society, 2003.

[19] K. A. Hummel, S. Póta, and C. Schusterreiter, "Supporting Terminal Mobility by Means of Self-adaptive Communication Object Migration," in WMASH '05: Proceedings of the 3rd ACM International Workshop on Wireless Mobile Applications and Services on WLAN Hotspots, pp. 88-91, ACM Press, 2005.

[20] P. Braun, S. Kern, I. Müller, and R. Kowalczyk, "Attacking the Migration Bottleneck of Mobile Agents," in AAMAS '05: Proceedings of the Fourth International Joint Conference on Autonomous Agents and Multiagent Systems, pp. 1239-1240, ACM Press, 2005.

[21] X. Zhong, C.-Z. Xu, and H. Shen, "A Reliable and Secure Connection Migration Mechanism for Mobile Agents," in Proceedings 24th International Conference on Distributed Computing Systems Workshops - W4: MDC (ICDCSW'04), pp. 548-553, IEEE Computer Society, 2004.

[22] N. C. Brown and P. H. Welch, "An Introduction to the Kent C++CSP Library," in J. F. Broenink and G. H. Hilderink (Eds.), Communicating Process Architectures 2003, pp. 139-156, IOS Press, 2003.

[23] K. Chalmers and S. Clayton, "CSP for .NET Based on JCSP," in P. H. Welch, J. M. Kerridge, and F. R. M. Barnes (Eds.), Communicating Process Architectures 2006, pp. 59-76, IOS Press, 2006.

[24] P. H. Welch, N. Brown, J. Moores, K. Chalmers, and B. Sputh, "Integrating and Extending JCSP," in A. McEwan, S. Schneider, W. Ifill, and P. Welch (Eds.), Communicating Process Architectures 2007, IOS Press, 2007.

[25] P. H. Welch, "Graceful Termination - Graceful Resetting," in A. W. P. Bakkers (Ed.), oUG-10: Applying Transputer Based Parallel Machines, pp. 310-317, 1989.

[26] B. Sputh and A. Allen, "JCSP Poison," in J. Broenink, H. Roebbers, J. Sunter, P. Welch, and D. Wood (Eds.), Communicating Process Architectures 2005, IOS Press, 2005.

Communicating Process Architectures 2007
Alistair A. McEwan, Steve Schneider, Wilson Ifill, and Peter Welch
IOS Press, 2007

C++CSP2: A Many-to-Many Threading Model for Multicore Architectures

Neil BROWN

Computing Laboratory, University of Kent, Canterbury, Kent, CT2 7NF, UK.
neil@twistedsquare.com/nccb2@kent.ac.uk

Abstract. The advent of mass-market multicore processors provides exciting new opportunities for parallelism on the desktop. The original C++CSP – a library providing concurrency in C++ – used only user-threads, which would have prevented it taking advantage of this parallelism. This paper details the development of C++CSP2, which has been built around a many-to-many threading model that mixes user-threads and kernel-threads, providing maximum flexibility in taking advantage of multicore and multi-processor machines. New and existing algorithms are described for dealing with the run-queue and implementing channels, barriers and mutexes. The latter two are benchmarked to motivate the choice of algorithm. Most of these algorithms are based on the use of atomic instructions, to gain maximal speed and efficiency. Other issues related to the new design and related to implementing concurrency in a language like C++ that has no direct support for it, are also described. The C++CSP2 library will be publicly released under the LGPL before CPA 2007.

Keywords. C++CSP, C++, Threading, Atomic Instructions, Multicore

Introduction

The most significant recent trend in mass-market processor sales has been the growth of multicore processors. Intel expected that the majority of processors they sell this year will be multicore [1]. Concurrent programming solutions are required to take advantage of this parallel processing capability. There exist various languages that can easily express concurrency (such as occam-π [2]), but the programming mainstream is slow to change languages; even more so to change between programming 'paradigms'. Sequential, imperative, procedural/object-oriented languages remain dominant. It is for this reason that libraries such as JCSP [3], C++CSP [4,5], CTC++ [6] and CSP for the .NET languages [7,8] have been developed; to offer the ideas of occam-π and its ilk to programmers who choose (or are constrained) to use C++, Java and the various .NET languages.

C++CSP has previously been primarily based on user-threads, which simplify the algorithms for the primitives (such as channels and barriers) as well as being faster than kernel-threads, but cannot take advantage of multicore processors. A review of the latter problem has given rise to the development of C++CSP2.

This report details the design decisions and implementation of C++CSP2 [9]. Section 1 explains the new threading model that has been chosen. Section 2 briefly clarifies the terminology used in the remainder of the paper. Section 3 provides more details on the implementation of the run-queue and timeout-queue. Section 4 describes how to run processes in a variety of configurations in C++CSP2's threading model. Section 5 details a new barrier algorithm specifically tailored for the chosen threading model. Section 6 then presents benchmarks and discussion on various mutexes that could be used as the underpinning for many of C++CSP2's algorithms. Section 7 contains details on the modified channel-ends and section

8 discusses the channel algorithms. Section 9 highlights some issues concerning the addition of concurrency to C++. Finally, section 10 provides a brief discussion of networking support in C++CSP2, and section 11 concludes the paper.

Notes

All the benchmarks in this report are run on the same machine; an Intel Core 2 Duo 6300 (1.9 Ghz) processor with 2GB DDR2 RAM. The 'Windows' benchmark was run under Windows XP 64-bit edition, the 'Linux' benchmark was run under Ubuntu GNU/Linux 64-bit edition.

C++CSP2 is currently targeted at x86 and (x86-64) compatible processors on Windows (2000 SP4 and newer) and Linux (referring to the Linux kernel with GNU's glibc).

Windows is a registered trademark of Microsoft Corporation. Java is a trademark of Sun Microsystems. Linux is a registered trademark of Linus Torvalds. Intel Core 2 Duo is a trademark of Intel Corporation. Ubuntu is a registered trademark of Canonical Ltd. occam is a trademark of STMicroelectronics.

1. Threading

1.1. Background

Process-oriented programming, based on CSP (Communicating Sequential Processes) [10] principles, aims to make concurrency easy for developers. In order to provide this concurrency, developers of CSP implementations (such as JCSP [11], CTJ [12], KRoC [13]) must use (or implement) a threading mechanism. When running on top of an Operating System (OS), such threading mechanisms fall into three main categories: user-threads, kernel-threads and hybrid threading models.

User-threads (also known as user-space or user-level threads, or M:1 threading) are implemented in user-space and are invisible to the OS kernel. They are co-operatively scheduled and provide fast context switches. Intelligent scheduling can allow them to be more efficient than kernel-threads by reducing unnecessary context switches and unnecessary spinning. However, they do not usually support preemption, and one blocking call blocks all the user-threads contained in a kernel-thread. Therefore blocking calls have to be avoided, or run in a separate kernel-thread. Only one user-thread in a kernel-thread can be running at any time, even on multi-processor/multicore systems. C++CSP v1.3 and KRoC use user-threads.

Kernel-threads (also known as kernel-space or kernel-level threads, or 1:1 threading) are implemented in the OS kernel. They usually rely on preemption to perform scheduling. Due to the crossing of the user-space/kernel-space divide and other overheads, a kernel-thread context switch is slower than a user-space context switch. However, blocking calls do not cause any problems like they do with user-threads, and different kernel-threads can run on different processors simultaneously. JCSP (on Sun's Java Virtual Machine on most Operating Systems) uses kernel-threads.

Hybrid models (also known as many-to-many threading or M:N threading) mix kernel-threads and user-threads. For example, SunOS contained (adapting their terminology to that used here) multiple kernel-threads, each possibly containing multiple user-threads, which would dynamically choose a process to run from a pool, and run that process until it could no longer be run. Much research on hybrid models has involved the user-thread and kernel-thread schedulers sharing information [14,15].

In recent years hybrid models have faded from use and research. The predominant approach became to increase the speed of kernel-threading (thereby removing its primary drawback) rather than introduce complexity with hybrid models and/or ways to circumvent user-threading's limitations. This was most obvious in the development of the NGPT (Next Generation POSIX Threads) library for Linux alongside the NPTL (Native POSIX Thread Library). NGPT was a complex hybrid threading library, whereas NPTL was primarily centred on a speed-up of kernel-threading. NPTL is now the default threading library for Linux, while development of NGPT has been quietly abandoned.

1.2. C++CSP2 Threading Options

This section explores the three different threading models that could be used for C++CSP2.

1.2.1. User Threads

The previous version of C++CSP used only user-threads. With the mass-market arrival of multicore processors, the ability to only run on one processor simultaneously became an obvious limitation that needed to be removed. Therefore, continuing to use only user-threads is not an option.

1.2.2. Kernel Threads

The most obvious move would have been to change to using only kernel-threads. In future, it is likely that the prevalence of multicore processors will continue to motivate OS developers to improve the speed of kernel-threads, so C++CSP would get faster without the need for any further development.

Channels and barriers between kernel-threads could be implemented based on native OS facilities, such as OS mutexes. The alternative would be to use atomic instructions – but when a process needed to wait, it would either have to spin (repeatedly poll – which is usually very wasteful in a multi-threaded system) or block, which would involve using an OS facility anyway.

1.2.3. Hybrid Model

The other option would be to combine kernel-threads and user-threads. The proposed model would be to have a C++CSP-kernel in each kernel-thread. This C++CSP-kernel would maintain the run queue and timeout queue (just as the C++CSP kernel always has). Each process would use exactly one user-thread, and each user-thread would always live in the same kernel-thread.

It is possible on Windows (and should be on Linux, with some assembly register hacking) to allow user-threads to move between kernel-threads. However, in C++ this could cause confusion. Consider a process as follows:

```
void run()
{
    //section A
    out << data;
    //section B
}
```

With moving user-threads, the code could be in a different thread in section A to section B, without the change being obvious. If this was a language like occam-π, where the code in these sections would also be occam-π code, this may not matter. But C++CSP applications will almost certainly be interacting with other C++ libraries and functions that may not handle concurrency as well as occam-π. These libraries may use thread-local storage or otherwise depend on which thread they are used from. In these cases it becomes important to the programmer to always use the library from the same thread. So allowing user-threads to move would cause confusion, while not solving any particular problem.

Moving user-threads would allow fewer kernel-threads to be created (by having a small pool of kernel-threads to run many user-threads) but the overheads of threading are primarily based on the memory for stacks, so there would not be any saving in terms of memory.

Another reason that hybrid models are considered inferior is that using priority can be difficult. A high priority kernel-thread running a low-priority user-thread would run in preference to a low-priority kernel-thread running a high-priority user-thread. However, C++CSP has never had priority so this is not yet a concern. I believe that the benefits of a hybrid threading model outweigh the drawback of making it difficult to add priority to the library in the future.

1.2.4. Benchmarks

Benchmarking threading models directly is difficult. While context-switches between user-threads can be measured explicitly, this is not so with kernel-threads; we cannot assume that our two switching threads are the only threads running in the system, so there may be any number of other threads scheduled in and out between the two threads we are testing. Therefore the best test is to test a simple producer-consumer program (one writer communicating to one reader), which involves context-switching, rather than trying to test the context switches explicitly.

Four benchmark tests were performed. One test used user-threads on their own (without any kernel-threads or mutexes) and another test tested kernel-threads on their own (without any user-threads or C++CSP2 kernel-interactions). These 'plain' benchmarks reflect a choice of a pure user-threads or pure kernel-threads model. The two hybrid benchmarks show the result of using user-threads or kernel-threads in a hybrid framework (that would allow the use of either or both). See Table 1.

Table 1. Single communication times (including associated context-switches) in microseconds.

Threading	Windows Time	Linux Time
Plain user-threads	0.19	0.19
Hybrid user-threads	0.27	0.28
Plain kernel-threads	2.7	2.4
Hybrid kernel-threads	3.1	2.8

1.2.5. Analysis

It is apparent that kernel-threads are at least ten times slower than user-threads. The nature of the benchmark means that most of the time, only one kernel-thread will really be able to run at once. However, these times are taken on a dual-core processor which means that there may be times when less context-switching is needed because the threads can stay on different processors. For comparison, the ratio between user-threads and kernel-threads on single core Linux and Windows machines were both also almost exactly a factor of ten.

For simulation tasks and other high-performance uses of C++CSP2, the speed-up (of using user-threads rather than kernel-threads) would be a worthwhile gain. However, running all processes solely as user-threads on a multi-processor/core system would waste all but one of the available CPUs.

A hybrid approach would allow fast user-threads to be used for tightly-coupled processes, with looser couplings across thread boundaries. Using a hybrid model with kernel-threads would seem to be around 15% slower than 'plain' kernel-threads. If only kernel-threads were used by a user of the C++CSP2 library, the hybrid model would be this much slower than if C++CSP2 had been implemented using only kernel-threads. In fact, this deficit is because no additional user-threads are used; much of the time in the hybrid model benchmark is spent blocking on an OS primitive, because the user-thread run-queue is empty (as explained in section 3.1). This cost would be reduced (or potentially eliminated) if the kernel-thread contained multiple user-threads, because the run-queue would be empty less often or perhaps never.

A hybrid approach flexibly offers the benefits of user-threads and of kernel-threads. It can be used (albeit with slightly reduced speed) as a pure user-thread system (only using one kernel-thread), or a pure kernel-thread system (where each kernel-thread contains a single user-thread, again with slightly reduced speed), as well as a combination of the two. Therefore I believe that the hybrid-threading model is the best choice for C++CSP2. The remainder of this paper details the design and implementation of algorithms for the hybrid-threading model of C++CSP2.

2. Terminology

Hereafter, the unqualified term 'thread' should be taken to mean kernel-thread. An unqualified use of 'kernel' should be taken to mean a C++CSP2-kernel. An unqualified 'process' is a C++CSP process. A 'reschedule' is a (C++CSP2-)kernel call that makes the kernel schedule a new process, without adding the process to the end of the run-queue; i.e. it is a blocking context-switch, rather than a yield (unless explicitly stated). An alting process refers to a process that is currently choosing between many guards in an Alternative. The verb 'alt' is used as a short-hand for making this choice; a process is said to alt over multiple guards when using an Alternative.

3. Process and Kernel Implementation

3.1. Run Queues

Each thread in C++CSP2 has a kernel, and each kernel has a run-queue of user-threads (each process is represented by exactly one user-thread). Any thread may add to any thread-kernel's run-queue. The run-queue is built around a monitor concept; reads and writes to the run-queue are protected by the mutual exclusion of the monitor. When a kernel has no extra processes to run, it waits on a condition variable associated with the monitor. Correspondingly, when a thread adds to another kernel's empty run-queue (to free a process that resides in a different thread), it signals the condition variable.

In Brinch Hansen [16] and Hoare's [17] original monitor concepts, signalling a condition variable in a monitor effectively passed the ownership of the mutex directly to the signalled process. This had the advantage of preventing any third process obtaining the monitor's mutex in between. In the case of the run-queue, using this style of monitor would be a bad idea. A kernel, X, adding to the run-queue of a kernel, K, would signal the condition variable and then return, without releasing the mutex (which is effectively granted to K). Any other threads that are scheduled after X but before K would not be able to add to K's run-queue because the mutex would still be locked. These processes would have to spin or yield, which would be very inefficient and unnecessary.

Instead, the monitor style used is that described by Lampson and Redell [18]. In their system, the signalling kernel (X) does not grant the mutex directly to the signalled kernel K. Instead, it merely releases the mutex. K will be scheduled to run at some future time, at which point it will contend for the mutex as would any other kernel. This allows other kernels to add to K's run-queue before K has been scheduled – but only X (which changes the run-queue from empty to non-empty) will signal the condition variable.

There is one further modification from Lampson and Redell's model. They replace Hoare's "IF NOT (OK to proceed) THEN WAIT C" with "WHILE NOT (OK to proceed) DO WAIT C"; because the monitor can be obtained by another process in between the signaller and the signallee, the other process could have changed the condition ("OK to proceed") to false again. In the case of C++CSP2, kernels may only remove processes from their *own* run-queue. Therefore the condition (the run-queue being non-empty) can never be invalidated by another kernel (because they cannot remove processes from the queue).

Hoare has shown that monitors can be implemented using semaphores [17] (in this case only two would be needed – one to act as a mutex, one for the condition variable). Therefore one implementation option would be to use an OS semaphore/mutex with an OS semaphore. In section 6 we will see that our own mutexes are considerably faster than OS mutexes, therefore a faster implementation is to use our own mutex with an OS semaphore. This is how it is implemented on Linux – on Windows, there are 'events' that are more naturally suited to the purpose fulfilled by the semaphore on the Linux.

There were other implementation options that were discounted. The semaphore could have been implemented using atomic instructions in the same way most of the mutexes in section 6 are. This would inevitably have involved spinning and yielding. The process will be blocked for an indefinite amount of time, which makes spinning and yielding inefficient. The advantage of an OS semaphore/event is that it blocks rather than spinning, which will usually be more efficient for our purposes. The other discarded option is that the POSIX threads standard supports monitors directly (in the form of a combination of a mutex and condition variable). Benchmarks revealed this option to be at least twice as slow as the mutex/semaphore combination that C++CSP2 actually uses. The newly-released Windows Vista also provides such support for monitors, but I have not yet been able to benchmark this.

3.2. Timeouts

Similar to previous versions of C++CSP, a timeout queue is maintained by the kernel. It is actually stored as two queues (in ascending order of timeout expiry) – one for non-alting processes and one for alting processes. Before the kernel tries to take the next process from the run queue (which may involve waiting, as described in the previous section), it first checks the timeouts to see if any have expired. If any non-alting timeouts have expired, the processes are unconditionally added back to the run queue. If any alting timeouts have expired, an attempt is made to add the processes back to the run-queue using the `freeAltingProcess` algorithm described in the next section.

The previous section described how a kernel with an empty run-queue will wait on a condition variable. If there any timeouts (alting or non-alting) that have not expired, the wait is given a timeout equal to the earliest expiring timeout. If no timeouts exist, the wait on the condition variable is indefinite (i.e. no timeout value is supplied).

3.3. Alting

Processes that are alting pose a challenge. Alting is implemented in C++CSP2 in a similar way to JCSP and KRoC. First, the guards are enabled in order of priority (highest priority first). Enabling can be seen as a 'registration of interest' in an event, such as indicating that we may (conditionally) want to communicate on a particular channel. If no guards are ready (none of the events are yet pending) then the process must wait until at least one event is ready to take place. As soon as a ready guard is found, either during the enable sequence or after a wait, all guards that were enabled are disabled in reverse order. Disabling is simply the reverse of enabling – revoking our interest in an event. At the end of this process, the highest priority ready guard is chosen as the result of the alt.

Consider a process that alts over two channels and a timeout. It may be that a process in another thread writes to one of the channels at around the same time that the process's kernel finds that its timeout has expired. If the process is waiting, *exactly one* of these two threads should add the process back to the run queue. If the process is still enabling, the process should not be added back to the run queue.

This problem has already been solved (and proven [19]) in JCSP, so C++CSP2's algorithm is an adaptation of JCSP's algorithm, that changes the monitor-protected state variable into a variable operated on by atomic instructions. The skeleton of JCSP's algorithm is as follows:

```
class Alternative
{
  private int state; //can be inactive, waiting, enabling, ready

  public final int priSelect ()
  {
    state = enabling;
    enableGuards ();
    synchronized (altMonitor) {
      if (state == enabling) {
        state = waiting;
        altMonitor.wait (delay);
        state = ready;
      }
    }
    disableGuards ();
    state = inactive;
    return selected;
  }

  //Any guard that becomes ready calls schedule:
  void schedule () {
    synchronized (altMonitor) {
      switch (state) {
        case enabling:
          state = ready;
          break;
        case waiting:
          state = ready;
          altMonitor.notify ();
          break;
        // case ready: case inactive:
        // break
      }
    }
  }
}
```

C++CSP2's algorithm is as follows. Note that unlike JCSP, it is possible that the `freeAltingProcess` function might be called on a process that is not alting – hence the case for dealing with `ALTING_INACTIVE`.

```cpp
unsigned int csp::Alternative::priSelect()
{
    int selected, i;
    AtomicPut(&(thisProcess->altingState),ALTING_ENABLE);

    //Check all the guards to see if any are ready already:
    for (i = 0;i < guards.size();i++)
    {
        if (guards[i]->enable(thisProcess))
            goto FoundAReadyGuard;
    }
    i -= 1;

    if (ALTING_ENABLE == AtomicCompareAndSwap(&(thisProcess->altingState),
        /*compare:*/ ALTING_ENABLE, /*swap:*/ ALTING_WAITING))
    {
        reschedule(); //wait
    }

FoundAReadyGuard: //A guard N (0 <= N <= i) is now ready:
    for (;i >= 0;i--)
    {
        if (guards[i]->disable(thisProcess))
            selected = i;
    }

    AtomicPut(&(thisProcess->altingState),ALTING_INACTIVE);
    return selected;
}

void freeAltingProcess(Process* proc)
{
    usign32 state = AtomicCompareAndSwap(&(proc->altingState),
        /*compare:*/ ALTING_ENABLE, /*swap:*/ ALTING_READY);

    //if (ALTING_ENABLE == state)
        //They were enabling, we changed the state. No need to wake them.
    //if (ALTING_READY == state)
        //They have already been alerted that one or more guards are ready.
        //No need to wake them.

    if (ALTING_INACTIVE == state)
    {
        freeProcess(proc); //Not alting; free as normal
    }
    else if (ALTING_WAITING == state)
    {
        //They were waiting.  Try to atomically cmp-swap the state to ready.
        if (ALTING_WAITING == AtomicCompareAndSwap(&(proc->altingState),
            /*compare:*/ ALTING_WAITING, /*swap:*/ ALTING_READY))
        {
            freeProcess(proc); //We made the change, so we should wake them.
        }
        //Otherwise, someone else must have changed the state from
        //waiting to ready. Therefore we don't need to wake them.
    }
}
```

Thus, the above algorithm does not involve claiming any mutexes, except the mutexes protecting the process's run-queue – and this mutex is only claimed by a maximum of one process during each alt. This makes the algorithm faster, and avoids many of the problems caused by an 'unlucky' preemption (the preemption of a thread that holds a lock, which will cause other processes to spin while waiting for the lock).

JCSP's algorithm has a "`state = ready`" assignment after its wait, without a corresponding line in C++CSP2. This is because the wait in JCSP may finish because the specified timeout has expired – in which case the assignment would be needed. In C++CSP2 timeouts are handled differently (see section 3.2), so the process is always woken up by a call to `freeAltingProcess`, and therefore the state will always have been changed before the reschedule function returns. With the addition of atomic variables in Java 1.5, it is possible that in future ideas from this new algorithm could be used by JCSP itself.

4. Running Processes

The vast majority of processes are derived from the `CSProcess` class. The choice of where to run them (either in the current kernel-thread or in a new kernel-thread) is made when they are run; the process itself does not need to take any account of this choice. The one exception to this rule is described in section 4.1.

For example, the following code runs each process in a separate kernel-thread[1]:

```
Run( InParallel
    (processA)
    (processB)
    (InSequence
        (processC)
        (processD)
    )
);
```

To run processes C and D in the same kernel-thread, the call `InSequenceOneThread` would be used in place of `InSequence` in the previous code. To instead run processes A and B in one kernel-thread, and C and D in another kernel-thread, the code would look as follows:

```
Run( InParallel
    ( InParallelOneThread
        (processA) (processB)
    )
    ( InSequenceOneThread
        (processC) (processD)
    )
);
```

To run them all in the current kernel-thread:

```
RunInThisThread( InParallelOneThread
    (processA)
    (processB)
    (InSequenceOneThread
        (processC)
        (processD)
    )
);
```

---

[1]The syntax, which may seem unusual for a C++ program, is inspired by techniques used in the Boost 'Assignment' library [20] and is valid C++ code

In occam-π terminology, we effectively have PAR and SEQ calls (that run the processes in new kernel-threads) as well as PAR.ONE.THREAD and SEQ.ONE.THREAD calls. Notice that the shorter, more obvious method (InParallel and InSequence) uses kernel-threads. Novice users of the library usually assume that, being a concurrent library, each process is in its own kernel-thread. They make blocking calls to the OS in separate processes, and do not understand why (in previous versions, that used only user-threads) this blocked the other processes/user-threads. Therefore it is wise to make the more obvious functions start everything in a separate kernel-thread, unless the programmer explicitly states not to (usually for performance reasons, done by advanced users of the library) by using the InParallelOneThread/InSequenceOneThread calls.

The reader may notice that there is very little difference from the user's point of view between InSequence and InSequenceOneThread. The two are primarily included for completeness; they are used much less than the corresponding parallel calls, because sequence is already present in the C++ language. A call to "Run(InSequence(A)(B));" is equivalent to "Run(A);Run(B);".

4.1. Blocking Processes

As stated in the previous section, most processes can be run as a user-thread in the current kernel-thread, or in a new kernel-thread – decided by the programmer using the process, not the programmer that wrote the process. Some processes, for example a file-reading process, will make many blocking calls to the OS. If they are placed in a kernel-thread with other user-threads, this would block the other user-threads repeatedly. Therefore the programmer writing the file-reading process would want to make sure that the process being run will always be started in a new kernel-thread. Only the sub-processes of the file-reading process can occupy the same kernel-thread, otherwise it will be the only process in the kernel-thread.

This is done in C++CSP2 by inheriting from ThreadCSProcess instead of CSProcess. The type system ensures that the process can only be run in a new kernel-thread. This will not be necessary for most processes, but will be applicable for those processes repeatedly interacting with OS or similar libraries, especially if the call will block indefinitely (such as waiting for a GUI event, or similar).

5. Barrier Algorithm

Like JCSP, C++CSP2 offers a barrier synchronisation primitive. Unlike most implementations of barriers, dynamic enrollment and resignation is allowed. That is, the number of processes enrolled on a barrier is not constant. The implementation of barriers in JCSP (from which the original C++CSP barrier algorithm was taken) has a 'leftToSync' count (protected by a mutex) that is decremented by each process that synchronises. The process that decrements the count to zero then signals all the other waiting threads and sets the leftToSync count back up to the number of processes enrolled (ready for the next sync). This section details a new replacement barrier algorithm for use in C++CSP2.

The idea of using software-combining trees to implement a barrier on a multi-processor system is described by Mellor-Crummey and Scott in [21]. The processors are divided into hierarchical groups. Each processor-group synchronises on its own shared counter, to reduce hot-spot contention (due to shared-cache issues, reducing the number of processors spinning on each shared 'hot-spot' is desirable). The last ('winning') processor to synchronise in the group goes forward into the higher-level group (which has a further shared counter) and so on until the top group synchronises. At this point the method is reversed and the processors go back down the tree, signalling all the shared counters to free all the blocked processes in the lower groups that the processor had previously 'won'.

This idea can easily be transferred to multi-threaded systems, with each thread blocking rather than spinning. A spinning thread is usually wasteful in a system with few processors but many threads. In order for it to finish spinning it will likely need to be scheduled out, and the other thread scheduled in to finish the synchronisation. Therefore, yielding or blocking is usually more efficient than spinning in this situation.

C++CSP2 uses a many-to-many threading approach. The software-combining tree approach can be adapted into this threading model by making all the user-threads in a given kernel-thread into one group, and then having another (higher-tier) group for all the kernel-threads. This forms a two-tier tree. This tree allows for optimisations to be made as follows.

Consider a group for all the user-threads in a kernel-thread. In C++CSP2 each user-thread is bound to a specific kernel-thread for the life-time of the user-thread. The user-threads of a particular kernel-thread can never be simultaneously executing. This means that a group shared among user-threads does not need to be protected by a mutex during the initial stages of the synchronisation, nor do the operations on it have to be atomic. This allows speed-up over the traditional barrier implementation where all the user-threads (in every kernel-thread) would always need to claim the mutex individually.

The code for this optimised version would look roughly as follows:

```
struct UserThreadGroup
{
    int leftToSync;
    int enrolled;
    ProcessQueue queue;
};

//Returns true if it was the last process to sync
bool syncUserThreadGroup(UserThreadGroup* group)
{
    addToQueue(group->queue,currentProcess);
    return (--(group->leftToSync) == 0);
}

void sync(UserThreadGroup* group)
{
    if (syncUserThreadGroup(group))
        syncKernelThread();
    else
        reschedule();
}
```

The `reschedule()` method makes the C++CSP2-kernel pick the next user-thread from the run-queue and run it. It does not automatically add the current user-thread back to the run-queue – it effectively blocks the current process.

Only the higher-tier group (that is shared among kernel-threads) needs to consider synchronisation. This group *could* be mutex-protected as follows:

```
int threadsLeftToSync;
map<KernelThreadId, UserThreadGroup > userThreadGroups;
Mutex mutex;

void syncKernelThread()
{
    mutex.claim();
    if (--(threadsLeftToSync) == 0)
    {
        int groupsLeft = userThreadGroups.size();
```

```
        for each group in userThreadGroups
        {
            group->leftToSync = group->enrolled;
            if (group->enrolled == 0)
            {
                remove group from userThreadGroups;
                groupsLeft -= 1;
            }
            freeAllProcesses(group->queue);
        }
        threadsLeftToSync = groupsLeft;
    }
    mutex.release();
}
```

The code only finishes the synchronisation if all the user-thread groups have now synchronised (that is, `threadsLeftToSync` is zero). The user-thread groups are iterated through. Each one has its `leftToSync` count reset. If no processes in that group remain enrolled, the group is removed. Finally, the `threadsLeftToSync` count is reset to be the number of kernel-threads (user-thread groups) that remain enrolled.

During this synchronisation, we modify the `UserTheadGroups` of other kernel-threads, even though they are not mutex-protected. This is possible because for us to be performing this operation, all currently enrolled processes must have already synchronised (and hence blocked) on the barrier, so they cannot be running at the same time *until after the freeAllProcesses call* (which is why that call is made last in the for-loop). If a process tries to enroll on the barrier, it must claim the mutex first. Since we hold the mutex for the duration of the function, this is not a potential race-hazard. The resign code would look as follows:

```
void resign(UserThreadGroup* group)
{
    group->enrolled -= 1;
    if (--(group->leftToSync) == 0)
        syncKernelThread();
}
```

The enrolled count is decremented, as is the `leftToSync` count. If this means that all the user-threads in the group have now synchronised (or resigned), we must perform the higher-tier synchronisation. The mutex does not need to be claimed unless as part of the `syncKernelThread()` function. The enroll code is longer:

```
UserThreadGroup* enroll()
{
    UserThreadGroup* group;
    mutex.claim();
    group = find(userThreadGroups,currentThreadId);
    if (group == NULL)
    {   //Group did not already exist, create it:
        group = create(userThreadGroups,currentThreadId);
        group->enrolled = group->leftToSync = 1;
        threadsLeftToSync += 1; //Increment the count of threads left to sync
    } else
    {   //Group already existed:
        group->enrolled += 1;
        group->leftToSync += 1;
    }
    mutex.release();  return group;
}
```

There is one further (major) optimisation of the algorithm possible. All but the final thread to call `syncKernelThread()` will merely claim the mutex, decrement a counter and release the mutex. This can be simplified into an atomic decrement, with an attempt only being made to claim the mutex if the count is decremented to zero:

```
int threadsLeftToSync;
map<KernelThreadId, UserThreadGroup > userThreadGroups;
Mutex mutex;

void syncKernelThread()
{
    if (AtomicDecrement(&threadsLeftToSync) == 0)
    {
        mutex.claim();
        // Must check again:
        if (AtomicGet(&threadsLeftToSync) == 0)
        {
            int groupsLeft = 0;

            for each group in userThreadGroups
            {
                if (group->enrolled != 0)
                    groupsLeft += 1;
            }

            AtomicPut(&threadsLeftToSync,groupsLeft);

            for each group in userThreadGroups
            {
                group->leftToSync = group->enrolled;
                if (group->enrolled == 0)
                    remove group from userThreadGroups;

                freeAllProcesses(group->queue);
            }
        }
        mutex.release();
    }
}
```

There are some subtle but important features in the above code. The `threadsLeftToSync` count is first reset. This is important because as soon as any processes are released, they may alter this count (from another kernel-thread) without having claimed the mutex. Therefore the groups must be counted and the `threadsLeftToSync` variable set before freeing any processes. This could be rearranged to set the `threadsLeftToSync` count to the size of the `userThreadGroups` map at the start, and performing an atomic decrement on the `threadsLeftToSync` variable each time we find a new empty group. However, it is considered that the above method, with a single atomic write and two iterations through the map, is preferable to repeated (potentially-contested) atomic decrements and a single iteration through the map.

The other feature is that the `threadsLeftToSync` count is checked before and after the mutex claim. Even if our atomic decrement sets the variable to zero, it is possible for an enrolling process to then claim the mutex and enroll before we can claim the mutex. Therefore, once we have claimed the mutex, we must check again that the count is zero. If it is not zero (because another process has enrolled) we cannot finish the synchronisation.

5.1. Benchmarks

The proposed new algorithm is more complicated than a 'standard' barrier algorithm. This complexity impacts maintenance of the code and reduces confidence in its correctness; it has not been formally verified. In order to determine if the new algorithm is worthwhile, its speed must be examined. Barrier synchronisations were timed, the results of which are given in Table 2.

Table 2. The column headings are *(Number of kernel-threads)*x*(Number of processes in **each** kernel-thread).* Each time is per single barrier-sync of all the processes (in microseconds).

OS	Barrier	1x100	1x1000	1x10000	2x1	2x5000	100x1	100x100
Windows	New	20	370	7,500	3.5	5,900	170	6,400
	Standard	24	490	8,600	3.4	7,700	300	9,500
Linux	New	19	200	5,700	2.4	4,400	180	5,100
	Standard	21	400	6,400	2.9	5,600	240	7,100

The new algorithm is at least as fast as the standard algorithm in all cases bar one. As would be expected, the performance difference is most noticeable with many user-threads in each of many kernel-threads. The new algorithm eliminates use of the mutex among sibling user-threads, where the standard algorithm must claim the mutex each time – with competition for claiming from many other threads. The expectation is that with more cores (and hence more of these contesting threads running in parallel), the new algorithm would continue to scale better than the standard algorithm.

6. Mutexes

Most C++CSP2 algorithms (such as channels and barriers) use mutexes. Therefore fast mutexes are important to a fast implementation. As well as mutexes provided by the operating system (referred to here as OS mutexes) there are a number of mutexes based on atomic instructions that could be used. This section describes various mutex algorithms and goes on to provide benchmarks and analysis of their performance.

6.1. Spin Mutex

The simplest mutex is the spin mutex. A designated location in shared memory holds the value 0 when unclaimed, and 1 when claimed. An attempt at claiming is made by doing an atomic compare-and-swap on the value. If it was previously 0, it will be set to 1 (and therefore the mutex was claimed successfully). If it is 1, nothing is changed – the process must re-attempt the claim (known as spinning). Spinning endlessly on a system that has fewer processors/cores than threads is often counter-productive; the current thread may need to be scheduled out for the thread holding the mutex before a claim will be successful. Therefore C++CSP2 spins an arbitrary number of times before either scheduling in another process in the same thread or telling the OS to schedule another thread in place of the spinning thread (i.e. yielding its time-slice). For the purposes of this benchmark, the latter option was implemented.

6.2. Spin Mutex Test-and-Test-and-Set (TTS)

The TTS mutex was developed for multi-processor machines where an attempted atomic compare-and-swap would cause a global cache refresh. Multiple attempted claims on a much-contested location would cause what is known as the 'thundering herd' problem, where multiple caches in the system have to be updated with each claim. The TTS mutex spins on a read-only operation, only attempting a claim if the read indicates it would succeed. Although the thundering herd problem should not occur on the benchmark system, the TTS mutex is included for completeness.

6.3. Queued Mutex

The Mellor-Crummey Scott (MCS) algorithm is an atomic-based mutex with strict FIFO (first-in first-out) queueing. It is explained in greater detail in [21], but briefly: it maintains a queue of processes, where the head is deemed to own the mutex. New claimers add themselves to the tail of the current list and spin (in the original MCS algorithm). When the mutex is released, the next process in the queue notices, implicitly passing it the mutex.

The MCS algorithm has been adapted to C++CSP2 by removing the spinning. Instead of spinning, the process immediately blocks after inserting itself into the queue. Instead of a process noticing the mutex is free by spinning, the releasing process adds the next process in the queue back to the appropriate run-queue. When it runs again, it implicitly knows that it must have been granted the mutex.

This mutex has the benefit of being strictly-FIFO (and hence avoids starvation) as well as having no spinning (except in a corner-case with unfortunate timing). The memory allocation for the queue is done entirely on the stack, which will be quicker than using the heap.

6.4. OS Mutex

Both Windows and Linux provide native OS mutexes. In fact, Windows provides two (a 'mutex' and a 'critical section'). They can be used as blocking or non-blocking, as described in the following sections.

6.4.1. Blocking Mutexes

Blocking mutexes cannot be used with C++CSP2. One user-thread cannot block *with the OS* on a mutex, because this would block the entire kernel-thread. Instead, processes (user-threads) must block with the C++CSP2-kernel, or not block at all (spinning or yielding). Therefore blocking OS mutexes are not a candidate for use with C++CSP2. The performance figures are given only for comparison, had C++CSP2 been purely kernel-threaded – in which case it could have used such mutexes.

6.4.2. Non-Blocking Mutexes

In contrast to the blocking mutexes, non-blocking OS mutexes are real candidates for use in C++CSP2.

6.5. Benchmarks

Benchmarks for each of the four mutexes are given Table 3 (five in the case of Windows). 'Uncontested' means that the mutex is claimed repeatedly in sequence by a single process – i.e. there is no parallel contention. '2x1' is two concurrent kernel-threads (each with one user-thread) repeatedly claiming the mutex in sequence. '10x10' is ten concurrent kernel-threads (each with ten concurrent user-threads) repeatedly claiming the mutex in sequence – a total of one hundred concurrent claimers.

Table 3. The column headings are *(Number of kernel-threads)x(Number of processes in **each** kernel-thread)*. B = Blocking, NB = Non-Blocking. All figures in nanoseconds (to 2 significant places).

OS	Mutex	Uncontested	2x1	10x10
Windows	Spin	30	86	6,100
	Spin TTS	33	140	4,100
	Queued	53	6,000	180,000
	OS (Mutex), B	1,000	5,500	280,000
	OS (Mutex), NB	1,100	2,800	230,000
	OS (Crit), B	53	360	19,000
	OS (Crit), NB	56	310	17,000
Linux	Spin	35	85	6,700
	Spin TTS	35	84	6,400
	Queued	53	3,500	180,000
	OS, B	62	150	13,000
	OS, NB	58	120	7,200

6.6. Analysis

It is clear that the Windows 'mutex' is much slower than the alternatives, especially when uncontested.

Performance of the queued mutex is of the same order of magnitude as the other mutexes when uncontested, but scales badly. This is because of the continued interaction with the C++CSP run-queues. Consider what will happen if a process is preempted while holding a mutex in the 10x10 case. The next thread will be run, and each of the ten user-threads will probably queue up on the mutex. Then each of the further eight threads will run, and each of the ten user-threads in each will probably queue up on the mutex. So 90 user-threads in total may be scheduled. Compare this to the spin mutexes, where only 10 user-threads would be scheduled (each performing a thread-yield).

The reason for the queued mutex's appalling performance in the 2x1 case is not as immediately clear. A clue can be found in the performance on a single-core system, which is only a factor of two behind the fastest mutexes, rather than a factor of over 40. Consider the two threads running simultaneously (one on each core), repeatedly claiming and releasing. Each time a claim is attempted, it is reasonably likely that the other thread will hold the mutex. The second process will queue up, and if the release does not happen soon enough, the run-queue mutex will be claimed, and the condition variable waited upon. Thus, a wait on a condition variable is reasonably likely to happen *on each and every claim*. Therefore the performance is particularly bad for repeated claims and releases by kernel-threads with no other processes to run.

The Linux OS (now 'futex'-based [22]) mutex and Windows critical section work in a similar manner to each other. They first attempt to claim the mutex using atomic instructions. If that does not immediately succeed (potentially after spinning for a short time), a call is made to the OS kernel that resolves the contention, blocking the thread if necessary. Therefore when there is no or little contention the performance is very close to the spin mutexes, and only becomes slower when there is more competition and hence more calls need to be made to the OS kernel to resolve the contention.

The benchmarks were carried out with no action taking place while the mutex was held. For the channel mutexes, this is fairly accurate. Only a couple of assignments are performed while the mutex is held, and a maximum of two processes compete for the mutex. Therefore the best mutex for channels is clearly the spin mutex, which has the best performance with little or no contention.

The mutex for a barrier (under the new algorithm) is only claimed by an enrolling process or by the last process to sync (that is, it is only claimed once per barrier-sync, barring any enrollments). It is not contended if no processes are enrolling. Therefore the spin mutex is also the best choice for the barrier algorithm. The best mutex for the run-queues (explained in section 3.1) is similarly the spin mutex.

The other major use of a mutex is for shared channel-ends. Unlike all the other uses of a mutex, in this case the mutex will be held indefinitely (until the channel communication has completed). Therefore spinning is not advisable. The queued mutex is ideally suited for this case. While it does not perform as well as the other mutexes for quick claim-release cycles, it offers no spinning and strict-FIFO ordering, which suits shared channel-ends (to prevent starvation).

7. Channel Class Design

Like all the other CSP systems mentioned in this paper, C++CSP has the important concept of channels. Channels are typed, unidirectional communication mechanisms that are fully synchronised. In C++CSP, channels are templated objects that are used via their channel-ends (a reading end and a writing end).

C++CSP v1 had two channel-end types: `Chanin` and `Chanout` [4]. The former supplied methods for both alting and extended rendezvous, and threw an exception if an operation was attempted on a channel that did not support it (for example, channels with a shared reading-end do not support alting). This was bad design, and has now been rectified. There are now two channel reading ends (`Chanout` remains the only writing-end); `Chanin` and `AltChanin`. The former does not provide methods to support alting, whereas the latter does. In line with the latest JCSP developments [23], they both support extended rendezvous on all channels (including buffered channels).

In JCSP the `AltingChannelInput` channel-end is a sub-class of `ChannelInput`. However, in C++CSP2 `AltChanin` is not a sub-class of `Chanin`. This is because channel-ends in C++CSP2 are rarely held by pointer or reference, so sub-classing would be of no advantage (and indeed would suffer additional virtual function call overheads) – except when passing parameters to constructors; specifically, an `AltChanin` could be passed in place of a parameter of type `Chanin`. To facilitate this latter use, implicit conversions are supplied from `AltChanin` to `Chanin` – but not, of course, in the opposite direction.

8. Channel Algorithms

In [24] Vella describes algorithms for implementing CSP channels based on atomic instructions, for use in multi-processor systems. C++CSP2 even has an advantage over the constraints that Vella had to work with. Vella is careful to not re-add a process to the run-queue before it has blocked, in case another thread takes it off the run-queue and starts running it simultaneously. In C++CSP2, this is not possible, because processes cannot move between threads (so it will only be re-added to the run-queue for its own thread).

C++CSP2 does not use Vella's algorithms however, because the complications that are added by supporting poisoning have not yet been resolved with the difficult atomic algorithms. Instead, a mutex is used to wrap around the channel algorithms (one mutex per channel). There are two other changes from the original C++CSP channel algorithms (described in [4]), which are motivated in the following two sub-sections on poison and destruction.

8.1. Poison

C++CSP has always offered poisonable channels. Poisoning a channel is used to signal to other processes using that channel that they should terminate. Either end of a channel can be used to poison it, and both ends will 'see' the poison (a poison exception will be thrown) when they subsequently try to use the channel.

The channel algorithms in C++CSP v1 had a curious behaviour with regards to poison. Imagine, for example, that a reader was waiting for input on a channel. A writer arrives, provides the data and completes the communication successfully. As its next action the writer poisons the channel. When the reader wakes up, it sees the poison straight away and throws a poison exception. The data that the writer thought had 'successfully' been written is lost. This could be further obscured if on a shared channel, one writer completed the communication and another writer did the poisoning.

Sputh treats this as a fault in his JCSP algorithm, and corrects it [25]. I think that his decision is correct, and the consequent implication that C++CSP's original semantics (with regards to poison) were flawed is also correct. This problem is solved by introducing an additional state flag into the channel, which indicates whether the last communication completed successfully (before the poison) or not (it was aborted due to poison).

Another area in which poison semantics have been corrected are buffered channels. Previously, when a writer poisoned a buffered channel, the reader would see the poison immediately, even if there was unread data in the buffer. This caused a similar problem to the one above – data that the writer viewed as successfully sent would be lost. The new effects of poisoning buffered channels are summarised below:

Writer poisons the channel: The channel is flagged as poisoned, the buffer is not modified.
Reader poisons the channel: The channel is flagged as poisoned, and the buffer is emptied.
Writer attempts to use the channel: Poison is always noticed immediately.
Reader attempts to use the channel: Poison is noticed only when the buffer is empty.

The semantics are asymmetric. The simplest rationale behind their choice is that poisoning a channel that uses a first-in first-out buffer of size N now has a similar effect to poisoning a chain of N identity processes.

8.2. Destruction

There are often situations in which the user of C++CSP2 will want to have a single server process serving many client processes. Assuming the communication between the two is a simple request-reply, the server needs some way of replying specifically to the client who made the request. One of the easiest ways of doing this is for the client to send a reply channel-end with its request (channel-ends being inherently mobile in C++CSP2):

```
//requestOut is of type Chanout< pair< int,Chanout<int> > >
//reply is of type int
{
    One2OneChannel<int> replyChannel;
    requestOut << make_pair(7,replyChannel.writer());
    replyChannel.reader() >> reply;
}
```

The corresponding server code could be as follows:

```
//requestIn is of type Chanin< pair< int,Chanout<int> > >
pair< int, Chanout<int> > request;
requestIn >> request;
request.second << (request.first * 2);
```

For this trivial example, requests and replies are integers, and the server's answer is simply double the value of the request.

If the old algorithms were used, this code would be potentially unsafe. The following trace would have been possible:

1. Client sends request, server receives it.
2. Server attempts to send reply, must block (waiting for the client).
3. Client reads reply, adds server back to the run-queue.
4. Client continues executing, destroying `replyChannel`.
5. Server wakes up and needs to determine whether it woke because it was poisoned or because the communication completed successfully. The server checks the poison flag; a member variable of `replyChannel`.

This situation thus leads to the server checking a flag in a destroyed channel. To help avoid this problem, the first party to the channel (the one who must wait) creates a local stack variable that will be used to indicate whether the communication completed successfully, and puts a pointer to it in a channel variable. The second party uses the pointer to modify the variable. When the first party wakes up, it can then check its local stack variable successfully, even if the channel has been destroyed.

9. Scoped Forking

Scope is a useful part of structured programming. In most languages, variable storage is allocated when variables come into scope and de-allocated when variables go out of scope. In C++ classes this concept is built on to execute a constructor when an object variable comes into scope, and a destructor to be called when an object variable goes out of scope. This feature, which is not present in Java, can be both useful and dangerous in the context of C++CSP2. Both aspects are examined in this section.

C++CSP2 takes advantage of the scope of objects to offer a `ScopedForking` object that behaves in a similar manner to the FORKING mechanism of occam-π [26]. In occam-π, one might write:

```
FORKING
  FORK some.widget()
```

In C++CSP2 the equivalent is:

```
{
    ScopedForking forking;
    forking.fork(new SomeWidget);
} //end of block
```

The name of the `ScopedForking` object is arbitrary (`forking` is as good a name as any). At the end of the scope of the `ScopedForking` object (the end of the block in the above code), the destructor waits for the forked processes to terminate – the same behaviour as at the end of the FORKING block in occam-π.

The destructor of a stack object in C++ is called when the variable goes out of scope – this could be because the end of the block has been reached normally, or because the function was returned from, or an exception was thrown. In these latter two cases the destructor will still be executed.

For example:

```
{
    ScopedForking forking;
    forking.fork(new SomeWidget);

    if (something == 6)
        return 5;

    if (somethingElse == 7)
        throw AnException();
}
```

Regardless of whether the block is left because of the return, the throw, or normally, the code will only proceed once the SomeWidget process has terminated. Using such behaviour in the destructor allows us to emulate some language features of occam-π in C++, and even take account of C++ features (such as exceptions) that are not present in occam-π. However, there is one crucial difference – the occam-π compiler understands the deeper meaning behind the concepts, and can perform appropriate safety checks. In contrast, the C++ compiler knows nothing of what we are doing. In section 8.2, one potential problem of using objects concurrently was demonstrated. There are two further mistakes that can be made using the new ScopedForking concept, which are explained in the following two sub-sections.

9.1. Exception Deadlock

Consider the following code:

```
One2OneChannel<int> c,d;
try
{
    ScopedForking forking;
    forking.fork(new Widget(c.reader()));
    forking.fork(new Widget(d.reader()));
    c.writer() << 8;
    d.writer() << 9;
}
catch (PoisonException)
{
    c.writer().poison();
    d.writer().poison();
}
```

At first glance this code may seem sensible. The try/catch block deals with the poison properly, and the useful ScopedForking process makes sure that the sub-processes are waited for whether poison is encountered or not. Consider what will happen if the first Widget process poisons its channel before the example code tries to write to that channel. As part of the exception being thrown, the program will destroy the ScopedForking object *before* the catch block is executed. This means that the program will wait for both Widgets to terminate before poisoning the channels. If the second Widget is waiting to communicate on its channel, then deadlock will ensue.

This problem can be avoided by moving the declaration of the ScopedForking object to outside the try block. The general point, however, is that the C++ compiler can offer no protection against this mistake. In a language such as Rain [27], which offers both concurrency and poison exceptions, the compiler could avoid such problems by detecting them at compiletime in the first place, or by ensuring that all catch blocks for poison are executed before the wait for sub-processes.

9.2. Order of Destruction

Consider the following code:

```
{
    ScopedForking forking;
    One2OneChannel<int> c,d;
    forking.fork(new WidgetA(c.reader(),d.writer()));
    forking.fork(new WidgetB(d.reader(),c.writer()));
}
```

This code creates two processes, connected together by channels, and then waits for them to complete[2]. This code is very unsafe. In C++, objects are constructed in order of their declaration. At the end of the block, the objects are destroyed in reverse order of their declaration. This means that at the end of the block in the above code, the channels will be destroyed, and then the ScopedForking object will be destroyed. So the processes will be started, the channels they are using will be destroyed, and then the parent code will wait for the processes to finish, while they try to communiate using destroyed channels.

Again, this problem can be avoided by re-ordering the declarations. This code is dangerous (in the context of our example):

```
    ScopedForking forking;
    One2OneChannel<int> c,d;
```

This code is perfectly safe:

```
    One2OneChannel<int> c,d;
    ScopedForking forking;
```

The subtle difference between the two orderings, the non-obvious relation between the two lines, and the ramifications of the mistake (in all likelihood, a program crash) make for a subtle error that again cannot be detected by the C++ compiler. In languages such as occam-π or Rain, this mistake can be easily detected at compile-time (variables must remain in scope until the processes that use them have definitely terminated) and thus avoided.

The documentation for the C++CSP2 library explains these pitfalls, and offers design rules for avoiding the problems in the first place (for example, always declare all channels and barriers outside the block containing the ScopedForking object). The wider issue here is that adding concurrency to existing languages that have no real concept of it can be a dangerous business. Compile-time checks are the only real defence against such problems as those described here.

10. Networking

C++CSP v1 had in-built support for sockets and networked channels, as detailed in [5]. The network support was integrated into the C++CSP kernel; every time the kernel was invoked for a context switch, it checked the network for new data, and attempted to send out pending transmissions. Happe has benchmarked this model against other models (with different threading arrangements) [28]. His results showed that using separate threads (one for waiting on the network and one for processing requests) produced the best performance. C++CSP2's network support (which has not yet been implemented) will be considered in light of these results, and with consideration for facilities available only on Linux (such as `epoll`) or Windows (such as I/O completion ports).

---

[2]The same effect could have easily been achieved *safely* using the `Run` and `InParallel` constructs demonstrated in section 4.

11. Conclusions

C++CSP2 now supports true parallelism on multicore processors and multi-processor systems. This makes it well-positioned as a way for C++ programmers to take advantage of this parallelism, either by wrapping the process-oriented methodology that C++CSP2 offers around existing code, or by developing their programs on top of C++CSP2 from the outset.

This paper has presented benchmarks of various mutexes and selected the fastest for C++CSP2's algorithms. Where possible these algorithms have actually used atomically-updated variables, avoiding the use of mutexes, in order to reduce contention for mutexes and minimise the chance of processes being scheduled out while holding a mutex. The effect of this work is to make C++CSP2 as fast as possible on multicore and multi-processor machines by reducing spinning and blocking to a minimum.

This work should prove relevant to the efforts to take advantage of multicore processors in other CSP implementations. The atomic alting algorithm described in section 3.3 could prove useful in JCSP, while the barrier algorithm and mutex benchmarks may be applicable to the implementation of occam-π.

The major opportunities for future work are implementing the network support (mentioned briefly in section 10) and formally proving some of the new algorithms presented in this paper. There are also new features being added to JCSP, such as alting barriers, output guards and broadcast channels [23] that would be advantageous to add to C++CSP2.

The C++CSP2 library will have been released before this paper is published, and can be found at [9]. In this new version, particular effort has been put into improving and expanding the documentation to make the library accessible to both novice and advanced users.

11.1. Final Thought

"I don't know what the next major conceptual shift will be, but I bet that it will somehow be related to the management of concurrency." – Bjarne Stroustrup, MIT Technology Review, December 7, 2006.

References

[1] Anton Shilov.   Single-core and multi-core processor shipments to cross-over in 2006 – Intel. `http://www.xbitlabs.com/news/cpu/display/20051201235525.html`, 10 February 2007.

[2] Fred Barnes. occam-pi: blending the best of CSP and the pi-calculus. `http://www.occam-pi.org/`, 10 February 2007.

[3] Peter H. Welch. Java Threads in Light of occam/CSP (Tutorial). In Andrè W. P. Bakkers, editor, *Proceedings of WoTUG-20: Parallel Programming and Java*, pages 282–282, 1997.

[4] Neil C. C. Brown and Peter H. Welch. An Introduction to the Kent C++CSP Library. In Jan F. Broenink and Gerald H. Hilderink, editors, *Communicating Process Architectures 2003*, pages 139–156, 2003.

[5] Neil C. C. Brown. C++CSP Networked. In Ian R. East, David Duce, Mark Green, Jeremy M. R. Martin, and Peter H. Welch, editors, *Communicating Process Architectures 2004*, pages 185–200, 2004.

[6] B. Orlic. and J.F. Broenink. Redesign of the C++ Communicating Threads Library for Embedded Control Systems. In F. Karelse STW, editor, *5th Progress Symposium on Embedded Systems*, pages 141–156, 2004.

[7] Alex Lehmberg and Martin N. Olsen. An Introduction to CSP.NET. In Frederick R. M. Barnes, Jon M. Kerridge, and Peter H. Welch, editors, *Communicating Process Architectures 2006*, pages 13–30, 2006.

[8] Kevin Chalmers and Sarah Clayton. CSP for .NET Based on JCSP. In Frederick R. M. Barnes, Jon M. Kerridge, and Peter H. Welch, editors, *Communicating Process Architectures 2006*, pages 59–76, 2006.

[9] Neil Brown. C++CSP2. `http://www.cppcsp.net/`, 10 February 2007.

[10] C.A.R. Hoare. *Communicating Sequential Processes*. Prentice-Hall, 1985.

[11] Peter Welch.   Communicating Sequential Processes for Java.   `http://www.cs.kent.ac.uk/projects/ofa/jcsp/`, 10 February 2007.

[12] Gerald Hilderink. Communicating Threads for Java. `http://www.ce.utwente.nl/JavaPP/`, 10 February 2007.

[13] Fred Barnes. Kent Retargetable occam Compiler. `http://www.cs.kent.ac.uk/projects/ofa/kroc`, 10 February 2007.

[14] Thomas E. Anderson, Brian N. Bershad, Edward D. Lazowska, and Henry M. Levy. Scheduler Activations: Effective Kernel Support for the User-Level Management of Parallelism. *ACM Transactions on Computer Systems*, 10(1):53–79, 1992.

[15] Brian D. Marsh, Michael L. Scott, Thomas J. LeBlanc, and Evangelos P. Markatos. First-class user-level threads. In *SOSP '91: Proceedings of the thirteenth ACM symposium on Operating systems principles*, pages 110–121, New York, NY, USA, 1991. ACM Press.

[16] Per Brinch Hansen. *Operating System Principles*. Prentice-Hall, 1973.

[17] C. A. R. Hoare. Monitors: an operating system structuring concept. *Communications of the ACM*, 17(10):549–557, 1974.

[18] Butler W. Lampson and David D. Redell. Experience with processes and monitors in Mesa. *Commun. ACM*, 23(2):105–117, 1980.

[19] Peter H. Welch and Jeremy M. R. Martin. Formal Analysis of Concurrent Java Systems. In Peter H. Welch and Andrè W. P. Bakkers, editors, *Communicating Process Architectures 2000*, pages 275–301, 2000.

[20] Thorsten Ottosen. Boost.Assignment Documentation. `http://www.boost.org/libs/assign/doc/`, 10 February 2007.

[21] John M Mellor-Crummey and Michael L. Scott. Algorithms for scalable synchronization on shared-memory multiprocessors. *ACM Transactions on Computer Systems*, 9(1):21–65, February 1991.

[22] Ulrich Drepper. Futexes are tricky. Technical Report 1.3, Red Hat, December 2005.

[23] Peter Welch, Neil Brown, Bernhard Sputh, Kevin Chalmers, and James Moores. Integrating and Extending JCSP. In Alistair A. McEwan, Steve Schneider, Wilson Ifill, and Peter Welch, editors, *Communicating Process Architectures 2007*, pages –, 2007.

[24] Kevin Vella. *Seamless Parallel Computing On Heterogeneous Networks Of Multiprocessor Workstations*. PhD thesis, University of Kent, 1998.

[25] Bernhard Herbert Carl Sputh. *Software Defined Process Networks*. PhD thesis, University of Aberdeen, August 2006. Initial submission.

[26] Fred Barnes and Peter Welch. Prioritised Dynamic Communicating Processes - Part I. In James Pascoe, Roger Loader, and Vaidy Sunderam, editors, *Communicating Process Architectures 2002*, pages 321–352, 2002.

[27] Neil C. C. Brown. Rain: A New Concurrent Process-Oriented Programming Language. In Frederick R. M. Barnes, Jon M. Kerridge, and Peter H. Welch, editors, *Communicating Process Architectures 2006*, pages 237–252, 2006.

[28] Hans Henrik Happe. TCP Input Threading in High Performance Distributed Systems. In Frederick R. M. Barnes, Jon M. Kerridge, and Peter H. Welch, editors, *Communicating Process Architectures 2006*, pages 203–214, 2006.

Communicating Process Architectures 2007
Alistair McEwan, Steve Schneider, Wilson Ifill and Peter Welch (Eds.)
IOS Press, 2007

Design Principles of the SystemCSP Software Framework

Bojan ORLIC, Jan F. BROENINK
Control Engineering,
Faculty of EE-Math-CS, University of Twente
P.O.Box 217, 7500AE Enschede, the Netherlands
{B.Orlic, J.F.Broenink}@utwente.nl

Abstract. SystemCSP is a graphical design specification language aimed to serve as a basis for the specification of formally verifiable component-based designs. This paper defines a mapping from SystemCSP designs to a software implementation. The possibility to reuse existing practical implementations was analyzed. Comparison is given for different types of execution engines usable in implementing concurrent systems. The main part of the text introduces and explains the design principles behind the software implementation. A synchronization mechanism is introduced that can handle CSP kind of events with event ends possibly scattered on different nodes and OS threads, and with any number of participating event ends, possibly guarded by alternative constructs.

Keywords. Concurrency, CSP, SystemCSP, code generation

Introduction

Concurrency is one of the most essential properties of the reality as we know it. In every complex system, it can be perceived that many activities are taking place simultaneously. Better control over concurrency structure should automatically reduce the problem of complexity handling. Thus, a structured way to deal with concurrency is needed.

SystemCSP [1] is a graphical design specification language aimed to serve as a basis for the specification of formally verifiable component-based designs of distributed real-time systems. It aims to cover various aspects needed for the design of distributed real-time systems. SystemCSP is based on principles of both component-based design and CSP process algebra. According to [2] "CSP was designed to be a notation and theory for describing and analyzing systems whose primary interest arises from the ways in which different components interact". CSP is a relevant parallel programming model and the SystemCSP design specification method aims to foster its utilization in the practice of component-based design.

Occam was a programming language loosely based on CSP. Nowadays, occam-like libraries exist for modern programming languages. JCSP [4] developed in Kent, and CT libraries [5, 6] developed in our lab, are examples of occam-like libraries. Both approaches rely on OOP principles to implement an API that mimics the syntax of occam.

This paper defines the architecture of a framework for the software implementation of SystemCSP designs. As illustrated in Figure 1, software implementation is one of the possible target domains for a model specified in the SystemCSP design domain. This paper does focus on the infrastructure needed in the target domain to support the implementation of a model specified in SystemCSP (e.g. the one on Figure 2 or Figure 3).

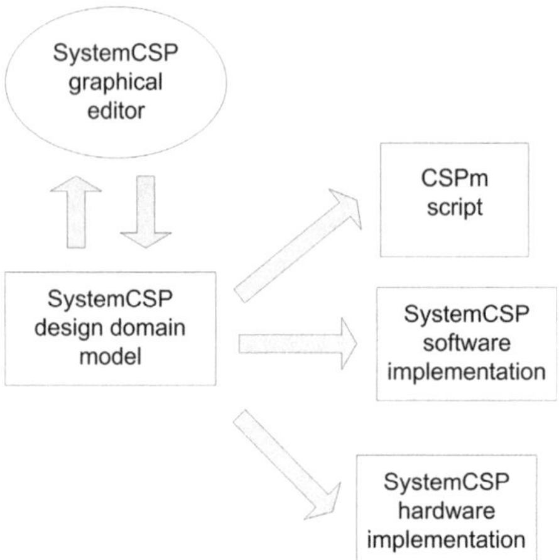

Figure 1 SystemCSP source and target domains

The SystemCSP notation has a control flow oriented part that is more or less a direct visualization of CSP primitives, and an interaction oriented part based on binary compositional relationships. In addition, the primitives for component-based software engineering are introduced.

The following CSP expression

$$P1= \text{ev}1\rightarrow(P;(T;(Q \parallel R)))$$

is in Figure 2 represented using the control flow oriented part of SystemCSP.

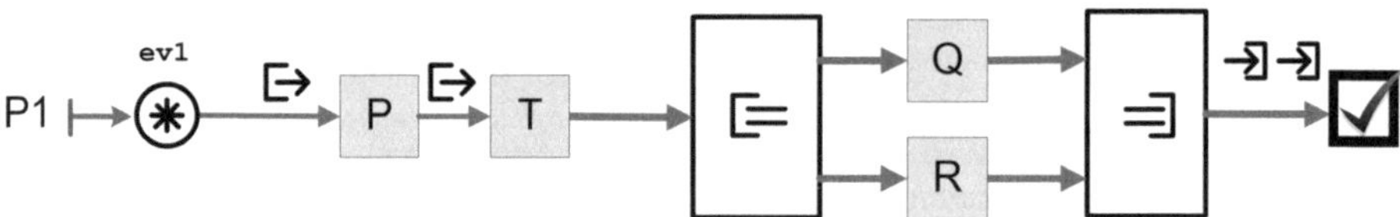

Figure 2 Example of control flow oriented SystemCSP design

In Figure 3, on the right-hand side, a control flow oriented design is visualized, and on the left-hand side two views are shown, each focusing on a part of the interaction between the involved components. Note also that instead of process symbols as used in Figure 2, in Figure 3 symbols for components and interaction contracts are used.

A detailed introduction of SystemCSP elements is out of the scope of this paper. For more details about SystemCSP design domain notation, the reader is referred to [6].

In this paper, in the Section 1, the discussion is focused on the possibility to reuse the CT library, developed at our lab, as a target domain framework for code generation.

After discarding the possibility to reuse the CT library, the discussion about the basic design principles for a new library starts in Section 2 with investigating practical possibilities for implementing concurrency. Possible types of execution engines are listed in Section 2.1. In Section 2.2, a flexible architecture is proposed that allows a designer to make trade-offs regarding the used structure of execution engines. In Section 2.3, a design of component internals is introduced, that allows subprocesses to access variables defined in parent components and offers a way to reuse processes in same way as components.

Section 2.4 explains the way in which function-call based concurrency is applied to structure concurrency inside components. An example is given illustrating how this mechanism actually works.

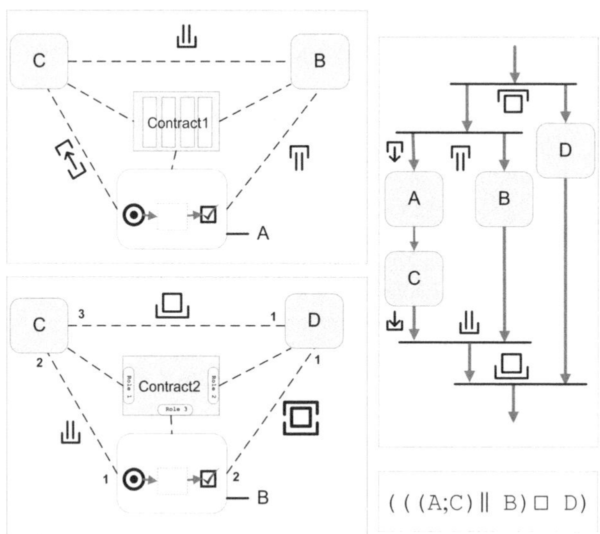

Figure 3 Example illustrating the relation between an interaction-oriented part and a control-based part

Section 3 explains a synchronization mechanism designed to handle CSP kind of events with any number of participants and with some of them possibly participating in several guarded alternative constructs. A special problem that was solved related to this was achieving mutual exclusion when event ends and the associated synchronization points are potentially scattered in different operating system threads or on different nodes.

Section 4 introduces design of a mechanism that implements exception handling and of mechanisms that provide support for logging and tracing.

1. Why Yet Another CSP Library?

In this section we focus on a possibility to reuse the CT library, the occam-like library developed in our lab, as a framework for the software implementation of SystemCSP models. The CT library follows the occam model as far as possible. SystemCSP builds upon the CSP legacy. It does in addition introduce new elements related to the area of component-based engineering. However, those newly introduced elements are: 1) components and interaction contracts that both map to CSP processes and 2) ports that are just event-ends exported by such CSP processes.

In fact, SystemCSP defines auxiliary design time operators like the fork and join control flow elements and binary compositional relationships of FORK, JOIN, WEAK and STRONG types. Those auxiliary operators do exist only during the design process and are therefore after grouping, in mapping to CSPm target domain substituted with CSP operators, and in mapping to software implementations with constructs like the ones existing in occam and CT library.

Basic SystemCSP control flow elements and binary relationships do map to the constructs as it is the case in the CT library. However, since SystemCSP aims to correspond exactly to CSP, it cannot be implemented completely by occam-like approaches that do put only restricted part of CSP into practical use. Following text will explore those differences in more details.

In CT library, like in its role-model occam, a `Parallel` construct spawns separate user-level threads for every subprocess. Synchronization points are defined by channel interconnections. The SystemCSP design domain allows both the CSP way of event synchronization (through a hierarchy of processes), and the occam-way with direct channel interconnections. Thus, a software implementation of SystemCSP designs needs mechanism for the hierarchical CSP-like CSP event synchronization.

In SystemCSP, as in CSP, data communication over a channel can be multidirectional involving any number of data flows. The CT library, as occam, has only unidirectional channels. In addition those channels are strongly typed using the template mechanism of the C++ language and as a consequence, they are not flexible enough to be reused in constructing the support for multidirectional communication. Thus, the channel framework of the CT library is not reusable.

The CT library implements the Alternative construct as a class whose behavior is based on the ideas of the occam ALT construct. The implementation of the Alternative construct [5] allows several different working modes (preference alting, PriAlternative, fair, FIFO), introduced to enable an alternative way to make a deterministic choice in case when more then one alternatives are ready for execution at the same time. The alting in CT library assumes that a channel can be guarded by some alternative construct only from one of the exactly two event-end sides (there can be either an input or an output guard associated with a channel). A guarded channel is just a channel with an associated guard. A guard is an object inside an alternative construct associated with a channel and a process. When a guarded channel is accessed by the peer process, then the guard becomes ready and is added to the alting queue. The way in which guards are ordered in this queue, determines the working mode (preference alting, PriAlternative, fair, FIFO) of the alternative construct. An alternative construct is thus a single point where the decision of a choice is made.

The SystemCSP design domain makes a difference between an external choice and a guarded alternative operators and in that sense adhere strictly to CSP. Thus, an implementation is needed that can support both. Event-ends contained by a guarded alternative or the ones resolving the parent external choice operator need to delegate their roles in the process of CSP event synchronization to the related guarded alternative or external choice operator. In case when, in an event occurrence, any number of guarded event-ends can participate, the whole alting mechanism must be completely different then the one applied in CT library. This means that in fact for CSP event synchronization mechanism completely different implementation of alting needs to be implemented. Thus again in this respect too, the CT library is not useful.

Simple CSP processes, made out of only event synchronization points connected via the prefix and the guarded alternative operator, are often visualized using a *Finite State Machine* (FSM). With the guarded alternative of CSP, no join of branches is assumed, and the branches can lead to any other state. The occam/CT library choice (ALT construct) requires that all alternatives are eventually joined. Thus a natural FSM interpretation is not possible anymore.For SystemCSP, the ability to implement FSM-like designs in a native way is especially important. Thus, implementation of the guarded alternative operator should not assume the join of branches.

In addition, it should be possible to use process labels to mark process entry points and allow recursions other then repetitions as in the SystemCSP design domain. Since in occam and the CT library, processes are structural units like components in SystemCSP, the use of recursion different then a loop is not natural there. A strict tree hierarchy of processes and constructs as basic architecture design pattern of occam and CT library is a misfit for our purpose. Thus, again the CT library does not meet the requirements imposed by SystemCSP.

In fact, instead of processes as structural units arranged in strict tree hierarchy, flexibility can be introduced by using classes for implementation of some processes and functions and labels for other processes. For instance, a single FSM-like design can contain many named processes that in fact do name the relevant states. Certainly, those processes cannot map to the occam notion of process. They are more convenient to be implemented as labels, while the whole finite state machine is convenient to be a single function.

In addition, SystemCSP is intended to be used as a design methodology for design and implementation of component-based systems. This needs to be supported by introducing appropriate abstractions and also possibilities for easy reconfiguration, interface checking, and so on.

To conclude, the mismatch between the CT library and the needs of SystemCSP is to big to allow reusing the CT library as a framework for the software implementation of SystemCSP designs.

2. Execution Engine Framework

2.1 Brief Overview of Execution Engines

Concurrency in a particular application assumes the potential of parallel existence and parallel progress of the involved processes. If processes are implemented in hardware, or if each of the processes is deployed on a dedicated node, these processes can truly progress concurrently. In practice, multiple processes often share the same processing unit.

Operating systems provide users with the possibility to run multiple OS processes (programs). Every OS process has its own isolated memory space and its own set of allocated resources. Within OS processes it is possible to create multiple OS threads that have their own dedicated workspaces (stack), but share other resources with all threads belonging to the same process. Synchronization in accessing those resources is left to the programmer. OS synchronization and communication primitives (semaphores, locks, mutexes, signals, mailboxes, pipes…)[7] are not safe from concurrency related hazards caused by bad design [4]. OS thread context switch is heavyweight, due to allowing preemption to take place at any moment of time.

User-level threading is an alternative approach that relies on creating a set of own threads in the scope of a single OS thread. Those threads are invisible to the underlying OS-level scheduler and their scheduling is under the control of the application. The main advantages compared to OS threads are the speed of context switching and gaining control over scheduling. The use of Operating System calls from inside any user-level thread is blocking the complete OS thread with all nested user-level threads (operating system call problem).

Another approach is to implement concurrency via function-calls, where the concurrent progress of parallel processes is achieved by dividing every process into little atomic steps. After every atomic step, the scheduler gets back control and executes the function that performs the next atomic step in one of the processes. There is no need to dedicate a separate stack for every process. Steps are executed atomically and cannot be preempted. A function-calls based approach is often used to mimic concurrency in simulation engines. There is even an operating system (Portos [8]) that is based on scheduling prioritized function calls.

2.1.1 Discussion

SystemCSP [1] structures concurrency, communication and synchronization using primitives directly coupled to appropriate CSP operators. To implement concurrent behaviour, it is possible to use any of the approaches described in Section 2.1.

The CT library is based on user-level threading. Every process in the CT library that can be run concurrently (i.e. every subprocess of the (Pri)Parallel construct) has a dedicated user-level thread. A scheduler exists that can choose the next process to execute according to the hierarchy of Parallel/PriParallel constructs. As in occam, rendezvous channels are the basic communication and synchronization primitives. Possible context switching points are hidden in every access to local channels.

The first important issue related to the SystemCSP framework is what type of execution engine is best to choose. Actually, the optimal choice depends on the application at hand and is a compromise between the level of concurrency, the communication overhead and other factors. The best solution is, therefore, to let the designer choose the type(s) of execution engines on which the application will execute. A way to do this is to separate the application from the execution engines, and to let the designer map the components of his application to the underlying architecture of execution engines.

2.2 Four Layer Execution Engine Architecture

An application is in SystemCSP organized as a containment hierarchy of components and processes. A component is the basic unit of composition, allocation, scheduling and reconfiguration. Inside every component, contained components, processes and event-ends are related via CSP control flow elements (sequential, parallel, choice …). While a subprocess is inseparable part of its parent component, a subcomponent is independent and can for example be located on some other node.

As a result of the previous discussion, flexible execution engine architecture is proposed, that allows the user to adjust the level of concurrency to the needs of the application at hand. The execution engine architecture is hierarchical, based on four layers: node/OS Thread/UL thread/component managers. Any component can be assigned to any execution engine on any level in such a hierarchy.

The class diagram given in Figure 4 defines the hierarchy of the execution engines. In the general case, inside an operating-system thread, a user-level scheduler exists, which can switch context between its nested user-level threads. Inside a user-level thread is, in the general case, a component manager that can switch between the contained components. Every component has an internal scheduler that will use a function-call based concurrency approach to schedule nested subprocesses.

Internalizing the scheduler inside every component allows more flexibility in the sense that some levels in the 4-layer architecture can be skipped. The concurrency of the node execution engine can be delegated to operating system threads or to user level threads or to component managers or it can execute a single component directly without providing support for lower-level execution engines. It is even possible to have a single component per node. Similarly operating system threads can execute a set of user level threads, or a component manager or a single component. A user-level thread is able execute just a single component or a set of components via the component manager. The possibility to choose any of those combinations is actually reflected in Figure 4.

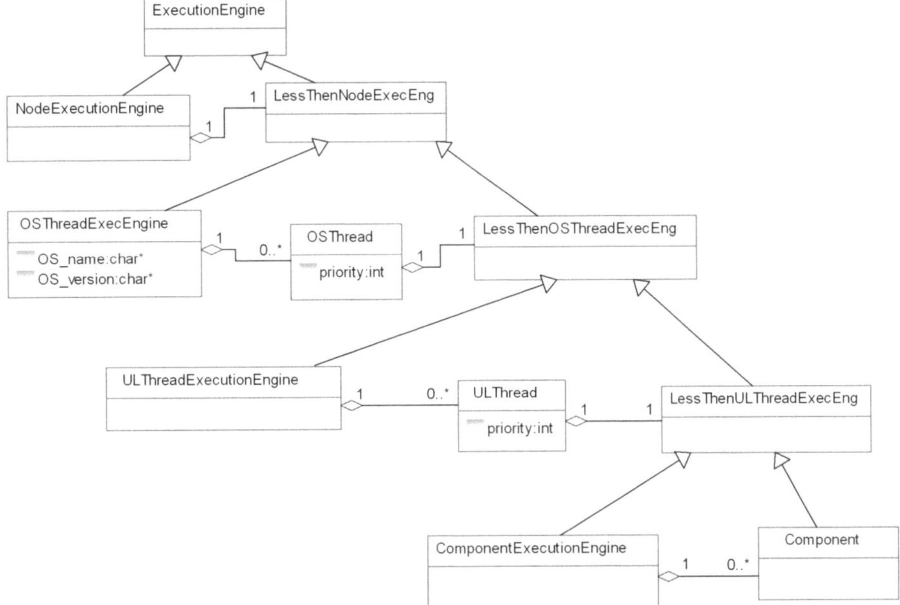

Figure 4 Class diagram of the 4-layer execution engine framework

The OS thread execution engine is in fact representing the scheduling mechanism of the underlying operating system. Therefore, in the design domain this class contains the name and version number of the used operating system as attributes. In software implementation, there is no matching class since implementation is provided by the underlying operating system. The OS thread class in the software implementation domain does have a dedicated subclass for every supported operating system. In that way, the portability is enhanced by isolating platform-specific details in the implementation of subclasses. Auxiliary abstract classes LessThenodeExecEng, LessThenOSThreadExecEng and LessThenUL-ThreadExecEng are.introduced to enable the described flexibility in structuring the hierarchy of execution engines.

2.2.1 Allocation

An allocation procedure as the one depicted in Figure 5 (below here), is a process of mapping components from the application hierarchy of components to the hierarchy of execution engines. The criteria for the choice of the execution framework and for the allocation, is setting the proper level of concurrency while optimizing performance by minimizing overhead. Two components residing on different nodes can execute simultaneously. Two components allocated to the same node, but to different operating system threads can be executed simultaneously only in the case of multi-core or hyper-threading nodes. Communication overhead between two components is directly proportional to the distance between the execution engines that execute them.

Figure 5 Allocation = mapping components from application hierarchy to hierarchy of execution engines

Control flow (as specified by parallel, sequential and alternative constructs) is decoupled from its execution engines. As a result, components can be reconfigured more easily. A component can be moved from one (node, operating-system thread, user-level thread) execution engine to another. Components can be dynamically created, moved around and connected to interaction contracts. On dynamical reconfiguration, checking compatibility of the interface required by the interaction contract with the interface supported by the component is done.

2.2.2 Priority Assignment

CSP is ignorant of the way concurrency is implemented. Concurrency phenomena involving parallel processes interacting via rendezvous synchronizations are the same regardless whether concurrent processes are executed on dedicated nodes, or sharing CPU time of the same node is done according to some scheduling algorithm. However, temporal characteristics are different in these two cases. The most commonly applied scheduling schemes are based on associating priorities with processes. In real-time systems, achieving proper temporal behavior is of utmost interest. Therefore, in real-time systems priorities are attached to schedulable units according to some scheduling algorithm that can guarantee meeting time requirements.

In addition to the PAR (parallel) construct, in occam a prioritized version of the parallel construct, the PRIPAR construct, was introduced. It specifies parallel execution with priorities assigned according to the order of adding subprocesses to the construct. However, on transputer platforms only two priority levels were supported. Additional priority levels were sometimes implemented in software [9].

Following occam, the CT library introduces a PriParallel construct with the difference that inside one PriParallel up to 8 subprocesses can be placed. While all subprocesses of a Parallel construct have the same priority, priorities of processes inside a PriParallel are based on the order in which they are added to the construct. This allows for a user-friendly priority assignment based on the notion of the, more or less intuitive, relative importance of a process compared to the other processes. The `PriParallel` construct is as any other construct also a kind of process, and as such it can be further nested in a hierarchy of constructs. This leads to the possibility to use a hierarchy of `PriParallel` and `Parallel` constructs to create a program with an unbounded number of different priority levels. Note however, that priority ordering, of all processes in a system, if defined in this way is not necessarily a strict ordering, but rather a set of partial orderings. If only PriParallel constructs were used, a set of partial orderings results in global strict priority ordering.

As in execution-engine architecture issues, where the conclusion was that flexibility can be achieved by separating hierarchy of components belonging to the application domain, from the hierarchy of execution engines, the similar reasoning applies to specifying priorities. The PriPar construct of the occam-like approaches is hard-coding priorities in the design, where a intuitively priority assignment is related to the execution of processes on the real target architecture. Priority values are in fact the result of a trade-off due to temporal requirements that belong to the application domain and processing time that belongs to the domain of underlying architecture engines. Therefore, the choice is *not* to follow the occam-like approach. Priorities belong to the execution engine framework and not to the application framework. Instead of relative priorities in each Par construct, a component from application hierarchy of components can be mapped to the execution engine of appropriate priority.

Every operating-system thread has a priority level used by the underlying operating-system scheduler to schedule it. Every user-level thread has its own priority level which defines its importance compared to the other user-level threads belonging to the same operating-system thread. In this way, a 2-level priority system exists and any component can be assigned to the pair of operating-system thread and user-level thread with appropriate priority levels

Note that the priorities specified on higher levels in an execution engine hierarchy overrule the ones specified on lower levels. This is the case because a higher-level execution engine (an operating-system execution engine) is not aware of the lower-level schedulable units (e.g. a user-level thread).

A problematic situation occurs when two components of different user-level thread priorities are allocated to two different operating-system threads of the same operating-system thread priority. In that case, it can happen that advantage is given to the component that has a lower user-level thread priority. In case when such a scenario should be avoided, two components with the same operating-system thread priority should always be in the same operating-system thread. In other words, this problem is avoided when there are no operating-system threads of the same priority on one node.

An additional issue is priority inversion that happens when a component of higher priority interacts with one of lower priority via rendezvous channels. For more details about this problem and possible solutions, the reader is referred to the related paper[10].

2.3 Components, Processes and Variables

The UML class diagram in Figure 6 illustrates the hierarchy of classes related to the internal organization of components. Every component has an internal scheduler that can handle various schedulable units (construct, processes, guarded alternative operators and event ends).

Variables are in SystemCSP defined in the scope of the component they reside in, and should be easily accessible from subprocesses of that component. A subprocess is allowed to access the variables defined in its parent component, but subcomponent cannot – because a subcomponent can be executed in a different operating-system thread or even on a different node. Instead of defining actual variables, the process class does define references to these variables (see Figure 6). Those references are in the constructor of the process associated with real variables defined in the scope of the component. In this way, subprocesses can access variables defined in components without restrictions; Component definitions are divided into smaller parts that are easier to understand and processes become as reusable as components are.

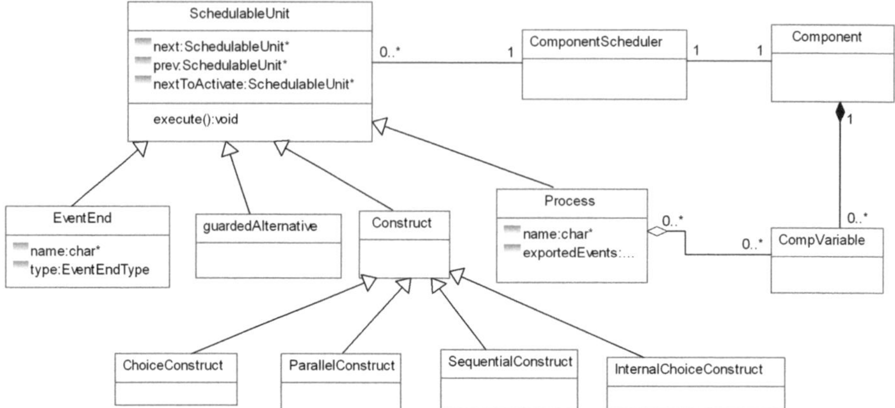

Figure 6 UML class diagram illustrating the relations between components and processes

Subcomponents that are executed in different execution engines do have associated proxy subporocess in their parent component (see Figure 7). In that way, the synchronization between the remote subcomponent and its parent component is done indirectly via that proxy process. The Proxy process and remote subcomponent synchronize on start events and termination events via regular channels.

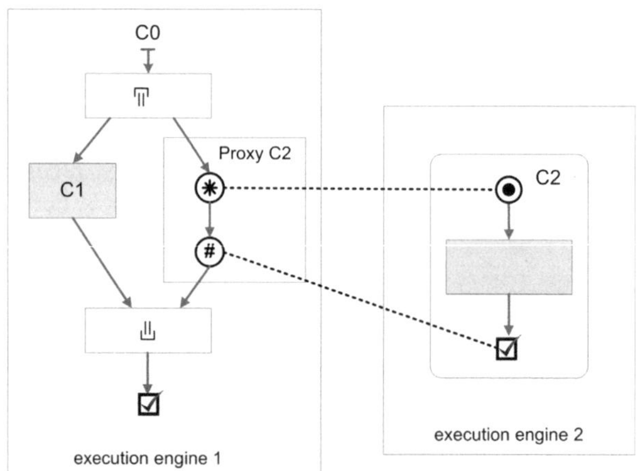

Figure 7 Using proxy processes to relate remote subcomponents to parent constructs

2.4 Function Call Based Concurrency Inside Components

The class diagram in Figure 6 defines that each component contains an internal scheduler. The dispatcher of a component is in its execute() function. It will use a scheduling queue

(FIFO or sorted queue) to obtain the pointer to the next schedulable unit ready to be executed.

Every schedulable unit inside a component is implemented as a finite-state machine that performs one synchronization and computation step per each function call, and subsequently returns control back to the component scheduler. The current place where the schedulable unit stopped with its execution is remembered in its internal state variable. When the schedulable unit is activated a next time, it will use this value to continue from where it had stopped. Every schedulable unit does have associated a pointer to the next schedulable unit to activate when its execution is finished. This is either its parent construct or the next schedulable unit in sequence (if the parent is a sequential construct).

Every construct exists inside some parent component. Constructs (Parallel, Alternative and Sequential) as well as channel/event ends are designed as predefined state-machines that implement behavior expected from them.

For instance, a simplified finite state machine implementing the Parallel construct would have two states: one with forking subprocesses (the FORK state in code snippet bellow), and one waiting for all subprocesses to finish (JOIN state in the code snippet). In reality a mechanism for handling errors and exceptional situations requires one or two additional states.

```
Parallel::run(){
  switch(state){
        case FORK:
                parentComponent->scheduler->add(subprocesses);
                state = JOIN;
                result =0;
                break;
        case JOIN:
                if(finishedCount == size)
                {
                        state = FORK;
                        finishedCount=0;
                        parentComponent->scheduler->add(next);
                        result =1;
                }
                break;
    }
    return result;
}


Parallel::exit() {
  finishedCount++;
  if(finishedCount ==size) parentComponent->scheduler->add(this);
}
```

The subprocesses use the exit() function to notify the Parallel construct that they have finished their execution. Since all subprocesses are in the same component and executed in atomic parts in function-call based concurrency manner, there are no mutual exclusion hazards involved.

When a construct finalizes successfully its execution, it returns a status flag equal to 1 or higher. For its parent it is a sign that it can move to the next phase in its execution by updating its state variable. In case of a guarded alternative, the returned number is in the parent process understood as the index of the branch to be followed and it is used to determine the next value of the state variable.

Thus, the system works by jumping in a state-machine, making one step (e.g. executing a code block or attempting event synchronization or forking subprocesses), and then jumping out. This might seem inefficient, but actually also in the user-level thread situation, a similar thing is done: testing the need for a context switch is hidden in every event attempt. Only performance testing can show which way is actually more efficient under what conditions. Recursions that are used to define auxiliary, named, process entry points are not implemented in a separate class. Instead they are naturally implemented using labels.

Let us use the example given in SystemCSP (Figure 8), and also in CSPm code above the figure to display how its software implementation would look like in this framework.

```
Restricted_program_use = Program \ {install, uninstall}
Program = install  → StartMenu
StartMenu = openProg  → UseProg | uninstall  → Program
UseProg = closeProg  → StartMenu | openDoc  → Work
Work = UpdateDoc  → Work | saveDoc  → Work
                | closeDoc  → UseProg | closeProg  → StartMenu
```

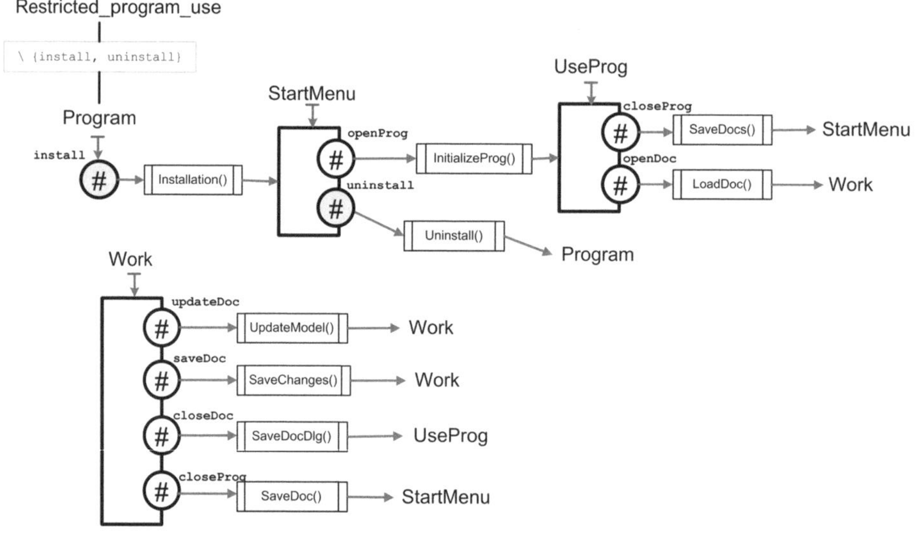

Figure 8 SystemCSP design used as an example for software implementation

The code is as follows:

```
Program(){
 switch (state){
  case START:
     status = install->sync();
     if(status == 0) return;
     elseif(status == 1){
        Installation();
        state = START_MENU;
```

```
    }
     else state = ERROR;
    break;
 case START_MENU:
       status = guardedAlt_StartMenu->select();
       if(status == 0) return;
       elseif(status == 1) {
           InitializeProg();
           state = USE_PROG;
       }
   else if(status== 2) {
       UninstallProg();
       state = START;
   }
   else state = ERROR;
   break;
case USE_PROG:
   status = guardedAlt_UseProg ->select();
   if(status == 0) return;
   elseif (status == 1) {
       SaveDocs();
       state = START_MENU:
   }
   else if (status == 2) {
       LoadModel();
       state = WORK;
   }
   else state = ERROR;
   break;
case WORK:
   status = guardedAlt_Work->select();
   if(status == 0) return;
   elseif(status == 1) {
       UpdateModel();
       state = WORK;
   }
   elseif(status == 2) {
       SaveChanges();
       state = WORK;
   }
   elseif(status == 3) {
       SaveDocDlg();
       state = USE_PROG;
   }
   elseif(status == 4) {
       SaveDocs();
       state = USE_PROG;
   }
   else state= ERROR;
   break;
```

```
case ERROR:
    printf(" process P got invalid status ");
  break;
}
```

In the constructor of the class defining this process, objects for the contained event ends and constructs are instantiated. For instance, the guarded alternative named StartMenu is on creation initiated using the offered event ends (openProg and uninstall) as arguments:

```
guardedAlt* StartMenu = new guardedAlt(openProg, uninstall);

EventEnd* openProg = new EventEnd(parentESP);
```

Code blocks are defined as member functions of a class that represent the process in which they are used. Code blocks that are used in more then one subprocess are usually defined as functions on the level of the component. Note that all code blocks (even a fairly complex sequential OOP subsystem that contains no channels, events and constructs) will be executed without interruption. Their execution can only be preempted by the operating-system thread of higher priority. As explained, user-level scheduling and function-call based execution engines are not fully preemptive. Thus, the events that need immediate reaction should be handled by operating-system threads of higher priorities.

3. Implementing CSP Events and Channels

Event ends are schedulable units implemented as state machines. They participate in the synchronization related to the occurrence of the associated event. This includes communicating their readiness to upper layers and waiting till the event is accepted by all participating event ends. This section describes in more detail how precisely this synchronization is performed.

3.1 Event synchronization mechanism

CSP events use the hierarchy of constructs for synchronization. An event end can be nested in any construct and it has to notify its parent construct of its activation.

In Figure 9, component C0 contains a parallel composition of components C1, C2 and C3 that synchronize on events a and b. Component C2 contains a parallel composition of C11 and C12 that synchronize on event a. The guarded alternative located in component C21 offers to its environment both events a and b.

Every process needs to export not-hidden events further to its environment, that is to a higher level synchronization mechanism. Every construct in the hierarchy must provide support for synchronizing events specified in its synchronization alphabet. This synchronization is done by dedicated objects – instances of the ESP (EventSynchronizationPoint) class (see Figure 10). The event-end will actually notify the ESP object of its parent construct about its readiness. A guarded alternative offers a set of possible event ends and thus instead of signaling its readiness to its parent construct, it can only signal conditional readiness.

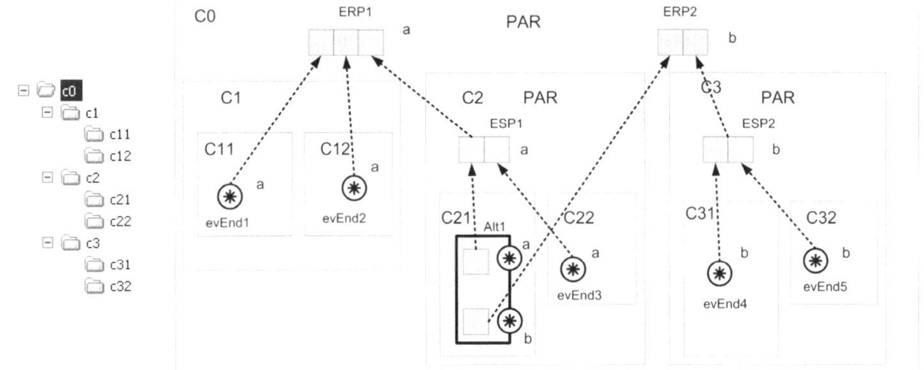

Figure 9 Hierarchical synchronization of CSP events

An ESP will, when all branches under its control are ready (conditionally or unconditionally) to synchronize on the related event, forward the readiness signal further to its parent ESP. When an event is not exported further, that construct is the level where the event occurrence is resolved. In that case, instead of an ordinary ESP object, a special kind of it exist (Event Resolution Point or ERP class) that performs the event resolution process. If some event ends are only conditionally ready, the ERP object will initiate a process of negotiation with the nested guarded alternative elements willing to participate in that event. When all event ends agree on accepting the event, ERP will notify all of them about the event occurrence.

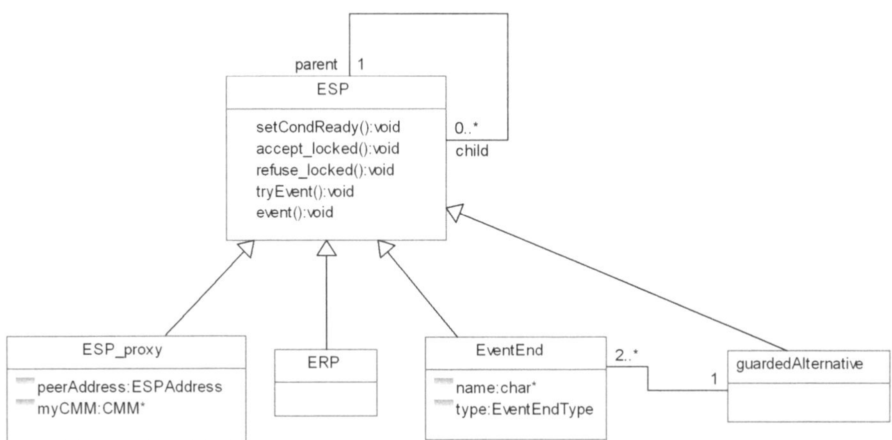

Figure 10 Event synchronization point classes

When on the top-level, in ERP, all fields, representing readiness of the associated branches, are ready or conditionally ready, a procedure of negotiation with sources of conditional readiness starts. This action results in every participating guarded alternative being asked to accept the event. If not previously locked by accepting negotiation with some other ERP, the queried guarded alternative will respond by accepting the event conditionally and locking till the end of the negotiation process. The attempt to start negotiation with already locked guarded alternative results in a rejection. In that case, the conditional readiness of the guarded alternative is canceled for that event and the negotiation process stops. When all guarded alternative constructs participating in the negotiation process have accepted the event (and are locked - rejecting other relevant events attempts), the ERP declares that the event is accepted by notifying all participating event ends (including the guarded alternatives) about the event occurrence. However, after one of the involved guarded alternatives has rejected the event acceptance, the event attempt did not succeed and all involved guarded alternatives are unlocked. Guarded alternatives unlocked in this way do again state conditional readiness for those event ends for which it might have been canceled during the negotiation procedure.

The class hierarchy defining types and relationships between event synchronization points is illustrated in Figure 10. For every type of the negotiation message, the ESP class declares a dedicated function. In case of local synchronization, a parent and the related children ESPs communicate via function calls. In case that synchronizing parent/child ESPs are residing in different OS threads or nodes, the ESP_proxy abstraction is used.

In the table below, the list of exchanged messages is specified as an illustration of an attempt to synchronize participating event-ends in a scenario based upon the example from Figure 9.

Table 1 One synchronization scenario

source	destination	message
evEnd1, evEnd2	ERP1	Ready
ALT1	ESP1	Conditionally Ready
ALT1	ERP2	Conditionally Ready
evEnd3	ESP1	Ready
ESP1	ERP1	Conditionally Ready
evEnd4	ESP2	Ready
ERP1	ESP1	Try event
ESP1	ALT1	Try event
evEnd5	ESP2	Ready
ALT1	ESP1	Accept_locked
ESP2	ERP2	Ready
ERP2	Alt1	Try event
ALT1	ERP2	Refuse_locked
ESP1	ERP1	Accept_locked
ERP1	ESP1, evEnd1, evEnd2	event
ESP1	ALT1, evEnd3	event

3.2 Solving the Mutual Exclusion Problem

Let us assume that allocation of the application hierarchy from Figure 9 to the hierarchy of execution engines is performed as in Figure 11. Clearly, simultaneous access to variables, which is possible in the case of distributed systems and operating-system thread based concurrency, must be prevented while implementing the previously explained event synchronization mechanism.

Event synchronization is more or less a generalization of the synchronization process used for channels. Let us therefore use channel synchronization as an example to show where the simultaneous access can cause problems.

In CT, a channel is a passive object. The process that first accesses the rendezvous channel will be blocked (taken out of the scheduler) and the pointer to that process thread is preserved in the channel. The process thread that arrives secondly will then copy the data and add the blocked process (one that has arrived first) to the scheduler. In CT, there is no problem of simultaneous access because the whole application is located in single OS thread.

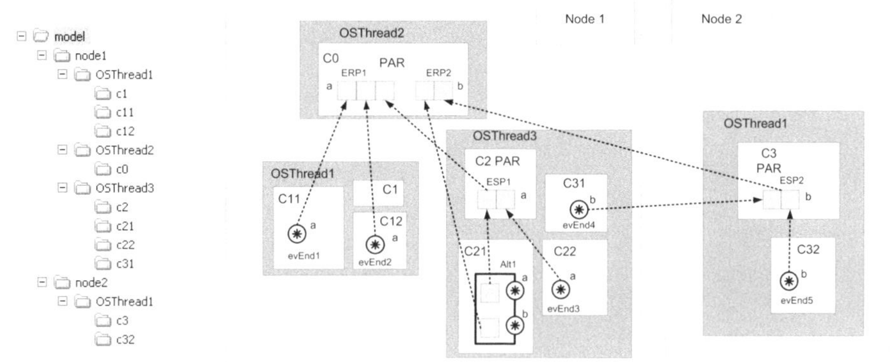

Figure 11 Synchronization of event ends allocated to different execution engines

In the SystemCSP framework, due to the possibility of using several OS threads as execution engines, protection from simultaneous access needs to be taken into account in order to make safe design.

Problematic points for channel communication when truly simultaneous access is possible are: (1) making the decision who arrived first to the channel and (2) adding the blocked process/component/user-level thread to its parent scheduler that can be accessed simultaneously from many OS threads.

Constructing a custom synchronization mechanism using flag variables is complex and error-prone. Besides, it is highly likely that such mechanism will fail to be adequate in case of hyperthreading and multi-core processors.

Using blocking synchronization primitives provided by the underlying operating systems causes the earlier mentioned problem of blocking all components nested in an operating-system thread that makes the blocking call. Besides unpredictable delay, this introduces additional dependency that can result in unexpected deadlock situations. It also does not provide a solution for an event synchronization procedure in case the participating components are located on different nodes.

If non-blocking calls, to test whether critical sections can be entered, are used, the operating-system thread that comes first can do other things and poll occasionally whether a critical section is unlocked. However, this approach makes things really complicated. For instance, the higher priority operating-system thread needs to be blocked so that the lower priority one can get access to the CPU and be able to access the channel. To block only the component, which accessed the channel and not the whole operating-system thread, one needs later to be able to reschedule it. For safe access to the scheduler from the context of another operating-system thread, another critical section is needed.

The previously discussed attempts to solve the mutual exclusion problem do apply only for processes located in different OS threads, but on the same node. In essence, from the point of view of the mutual exclusion problem, an operating system thread is equally problematic as synchronization with parts of a program on another node. Thus, it is convenient if the solution for both problems relies on the same mechanism.

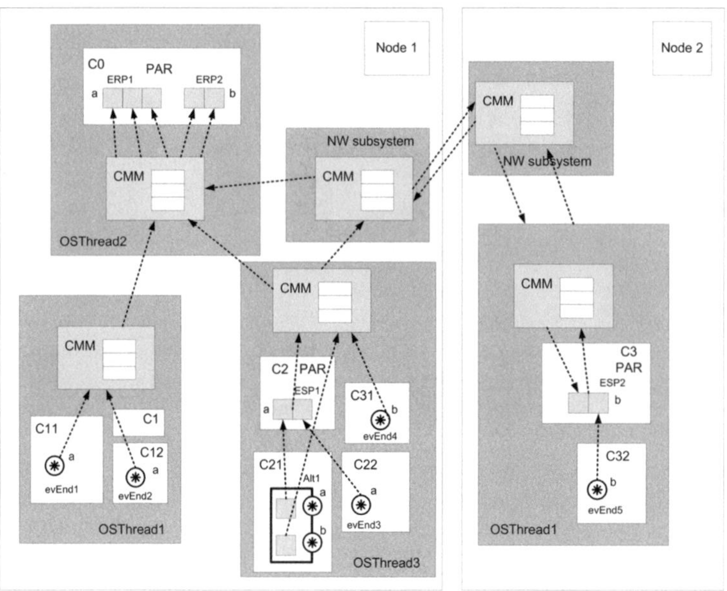

Figure 12 Using message queue based CMM to provide safe usage of concurrency

We propose that every operating-system thread has an associated message queue (operating systems provide message queues as a way to have non-blocking communication between operating-system threads). Thus, every OS thread, that interacts with other OS threads, will contain a control message manager (CMM) component that dispatches control messages (like event ready, event conditionally ready, try event, event accepted and similar) to message queues of other operating-system threads and transforms the received control messages to the appropriate function calls. For synchronization between nodes, networking subsystem can be located in a dedicated operating system thread that has a similar CMM component. This CMM will use the networking system to dispatch control messages to other nodes and will dispatch control messages received from other nodes to the message queues associated with CMMs of appropriate operating-system threads.

ESP_proxy (see Figure 10) communicates messages and addresses to local CMM, which further transfer it to the peer's CMM. The peer's CMM will then deliver the message by invoking direct function calls of appropriate ESP objects.

3.3 Channels Capable of Multidirectional Communication

Channels are special types of events where only two sides participate and in addition data communication is performed. As such, channels can be implemented in a more optimized way then events by avoiding the synchronization through hierarchy. Similar optimizations can be done for barriers with always fixed participating event ends, shared channels (any2One, One2Any) and simple guarded alternatives where all participating events are channels that are guarded only on one side.

One of the requirements (imposed by CSP as opposed to occam) for channels is that data communication can contain a sequence of several communications in either direction. A design choice made here is to separate synchronization from communication. To achieve flexible multidirectional communication, the part dealing with communication is further decomposed to pairs of sender and receiver communication objects (TxBuffer and RxBuffer) instead of using the template C++ language mechanism to parameterize complete channels with parameters specifying transferred data types, only RxBuffers and TxBuffers are parameterized. In this way flexibility is enhanced. Every channel end will contain an array consisting of one or more TX/RxBuffer objects connected to their pairs in the other end of the channel.

Since TxBuffers and RxBuffers contain pointers to the peer TxBuffer<T>/RxBuffer<T> objects, checking type compatibility of connected channel ends is done automatically at the moment of making the channel connection. This is convenient in case when connections between components are made dynamically during run-time. Otherwise, design time checks would be sufficient. Decoupling communication and synchronization via Tx./RxBuffers is also convenient for distribution.

3.4 Distribution/Networking

The CMM based design with control messages is straightforwardly extendable to distributed systems. In a distributed system, compared to operating-system thread based concurrency, besides control messages, also data messages are sent. Every node has a network subsystem with a role to exchange data and control messages with other nodes. The network subsystem takes control over RxBuffer and TxBuffer objects of a channel-end from the moment when the event is attempted, and returns control to the OS thread where the channel end is located after the data transfer is finished. This is done by exchanging (via the CMM mechanism) control messages related to location, locking and unlocking of data.

Of course, distributed event resolution comes with a price of increased communication overhead due to network layer usage. But, the task of the execution framework is to create conditions for this distribution to take place and the task of the designer of a concrete application is to optimize its performance by choosing to distribute on different nodes only those events whose time constraints allow for this imposed overhead.

4. Other Relevant Parts of the Software Implementation

4.1 Exception Handling

In SystemCSP, exception handling is specified by the take-over operator related to the interrupt operator of CSP. The take-over operator specifies that when an event offered to the environment by the process specified as second operand (exception handler) is accepted, the further execution of the process specified as the first operand (interrupted process) is aborted.

Upon the abort event (see Figure 13), the exception handler process is added to the scheduling queue of its parent component. Since the exception handler is a special kind of process recognizable as such by the scheduler, it is not added to the end of FIFO queue as other, 'normal' processes, but at its head. The preempt flag of the component manager is set to initiate preemption of the currently executing process. In that way, the situation where the exception handler needs to wait, while the interrupted process might continue executing, is avoided as much as possible.

As illustrated in Figure 13, the preempted process is appended to the end of FIFO queue of the component scheduler. If the preempted process is in fact the interrupted one then it will be taken out from the FIFO queue later during the abort procedure.

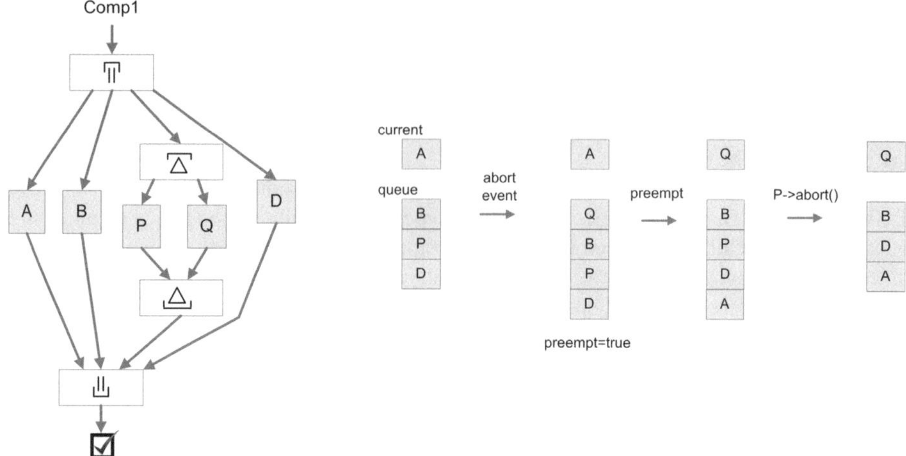

Figure 13 Example used to explain the implementation of take-over operator

The first step in the interrupt handler process is calling the abort() function of the interrupted process. The default version of abort() will cancel the readiness of all event ends for which the aborted process has declared readiness or conditional readiness. If the process is in the scheduling queue, it will be removed from there. Further, if the process is a construct, abort() will be invoked for all its subprocesses.

This exception handling mechanism does not influence the execution of other components that might have higher priority than the component where interrupted process resides.

4.2 Support for Development and Run Time Supervision

4.2.1 Logging

Logging is the activity of collecting data about the changes in values of certain chosen set of variables during some time interval. Not every change needs to be logged, but one should be able to use the obtained values to get insight in what was/is going on in some process/component. In this framework, the design choice is to allow logging only for the variables defined on the component-level. The main reason is obtaining a very structured and flexible way of logging that allows on-line reconfiguration of logging parameters. Thus all data constituting the state of the component should be maintained in the shape of component level variable. Every component can have a bit field identifying which of its variables are currently chosen for logging. The interface is defined that allows human operators to update this bit field at any time and thus change the set of logged variables.

Logging points are predetermined in design. In control flow diagram of SystemCSP, symbol used for logging point (a circle with big L inside) is associated with a prefix arrow as its property. The reason for this is a choice to treat a set of logging points as an optionally visualized layer added on top of the design. In implementation however prefix arrows do not exist, while logging points are inserted to the appropriate location in execution flow, as defined by the position of prefix arrow in the design.

Any logging point, either uses set of variables set for logging on component level using the described bit field mechanism, or defines its own bit field with set of variables to log. The operator is via the NodeManager allowed to inspect logging points and update their bit fields. Every logging point has a tag (or ID) unique in scope of its parent component, that is used to uniquely identify it. On the target side of the application, this tag can be a pointer to the object implementing the logging point. On the operator side of the application this tag is mapped to the unique ID of the logging point as specified in the system design.

The reason to opt for this kind of logging is predictability. The logging activity is considered to be part of the design and all the needed resources (e.g. CPU time, memory, network bandwidth and storage capacity) can be preallocated. Logging points can in design be inserted in such a way that it is possible to reconstruct change of every variable during the time. This approach to logging is considered here to be more structured and predictable then tracking every change for a chosen set of variables.

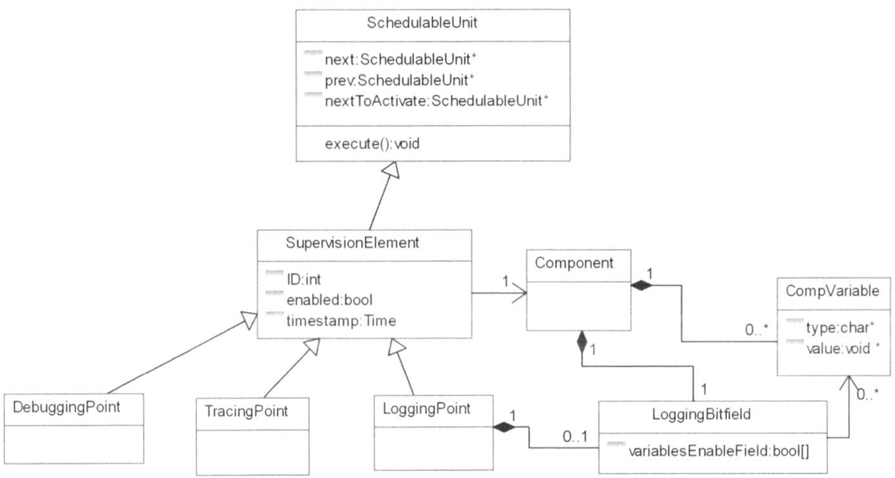

Figure 14 Supervision elements

4.2.2 Tracing

Tracing is an activity similar to logging. The difference is that instead of data, the information communicated to the human operator is the current position in execution flow of the application. Control flows leading to error states are always traced. Errors that are not fatal for the functionality of the system are logged as warnings. Other tracing points can be used for debugging or for supervising control. As it is the case for logging, the tracing is here considered to be part of the design and as such performed in predefined points of the execution flow.

SystemCSP defines a circle with a big T inside as a symbol of tracing point. Again it is associated with prefix arrow element, defining in that way the precise position of a tracing point. Every tracing point has a tag (or ID) that is unique per component and communicated to the operator to notify the occurrence of control flow passing over a tracing point. In addition, every function entry/exit is a potential tracing point.

5. Conclusions

This paper introduces design principles for the implementation of a software architecture that will support SystemCSP designs. The paper starts with explaining the reasons to discard the possibility to reuse the CT library as a framework for software implementation of SystemCSP models. The rest of the paper introduces the design principles for the implementation of the framework infrastructure needed in the software domain to support the implementation of a models specified in SystemCSP.

One of the main contributions of this paper is the decoupling application domain hierarchy of the components (related via CSP control flow elements and parent-children relationship) from the execution engine framework. In addition, this framework is constructed to allow maximal flexibility in choosing and combining execution engines of different types. In this way, flexible and reconfigurable component-based system is obtained. The priority specification is related to the hierarchy of execution engines and has thus become part of the deployment and not application design process.

Another significant contribution is solving the problem of implementing the mechanism for synchronizing CSP events in a way that is safe from mutual exclusion problems and is naturally suited for distribution. Besides that, the paper describes and documents the most important design choices in the architecture of the SystemCSP software framework.

Recommendation for future work is to fully implement everything presented in this paper. Furthermore, a graphical development tool is needed that will be capable to generate code. The described software framework would be used as a basic infrastructure that supports the proper execution of generated code.

References

[1]   Orlic, B. and J.F. Broenink. SystemCSP - visual notation. in CPA. 2006: IOS Press.
[2]   Roscoe, A.W., The Theory and Practice of Concurrency. Prentice Hall International Series in Computer Science. 1997: Prentice Hall.
[3]   Welch, P.H. and D.C. Wood, The Kent Retargetable occam Compiler, in Parallel Processing Developments -- Proceedings of WoTUG 19. 1996, IOS Press: Nottingham, UK. p. 143 -166.
[4]   Welch, P.H. The JCSP Homepage. 2007, http://www.cs.kent.ac.uk/projects/ofa/jcsp/.
[5]   Hilderink, G.H., Managing Complexity of Control Software through Concurrency. 2005, University of Twente.
[6]   Orlic, B. and J.F. Broenink, Redesign of the C++ Communicating Threads Library for Embedded Control Systems, in 5th PROGRESS Symposium on Embedded Systems, F. Karelse, Editor. 2004, STW: Nieuwegein, NL. p. 141-156.
[7]   Tanenbaum, A., Modern Operating Systems. 2001.
[8]   Chrabieh, R., Operating System with Priority Functions and Priority Objects. 2005.
[9]   Sunter, J.P.E., Allocation, Scheduling and Interfacing in Real-time Parallel Control Systems, in Faculty of Electrical Engineering. 1994, University of Twente: Enschede, Netherlands.
[10] Orlic, B. and J.F.Broenink. CSP and real-time – reality or an illusion? in CPA. 2007: IOS Press.

Communicating Process Architectures 2007
Alistair A. McEwan, Steve Schneider, Wilson Ifill, and Peter Welch
IOS Press, 2007

229

PyCSP - Communicating Sequential Processes for Python

John Markus BJØRNDALEN [a,1], Brian VINTER [b] and Otto ANSHUS [a]

[a] *Department of Computer Science, University of Tromsø*
[b] *Department of Computer Science, University of Copenhagen*

Abstract. The Python programming language is effective for rapidly specifying programs and experimenting with them. It is increasingly being used in computational sciences, and in teaching computer science. CSP is effective for describing concurrency. It has become especially relevant with the emergence of commodity multi-core architectures. We are interested in exploring how a combination of Python and CSP can benefit both the computational sciences and the hands-on teaching of distributed and parallel computing in computer science. To make this possible, we have developed PyCSP, a CSP library for Python. PyCSP presently supports the core CSP abstractions. We introduce the PyCSP library, its implementation, a few performance benchmarks, and show example code using PyCSP. An early prototype of PyCSP has been used in this year's Extreme Multiprogramming Class at the CS department, university of Copenhagen with promising results.

Keywords. CSP, Python, eScience, Computational Science, Teaching, Parallel, Concurrency, Clusters

Introduction

Python [1] has become a popular programming language in many fields. One of these fields is scientific programming, where efforts such as SciPy (Scientific Tools for Python) [2] has provided programmers with tools for developing new simulation models, as well as tools for scripting, managing and using existing codes and applications written in C, C++ and Fortran in new applications.

For many scientific applications, the time-consuming operations can be executed in libraries written in lower-level languages that provide faster execution, while automation, analysis and control of information flow and communication may be more easily expressed in Python. To see some examples of current uses and projects, we refer to the 2007 May/June issue of IEEE Computer in Science & Engineering, which is devoted to Python[1]. A study of the performance when using Python for scientific computing tasks is available in [3]. Langtangen's book [4] provides an introduction and many examples of how Python can be used for Scientific Computing. More information can also be found at the SciPy homepage [2].

There are several libraries for Python supporting many communication paradigms, allowing programmers to take advantage of clusters and distributed computing. However, to the best of our knowledge, there is no implementation of the basic abstractions of CSP (Communicating Sequential Processes) [5,6] for Python. This is the situation that we are trying to remedy with our implementation of CSP for Python: PyCSP.

---

[1]Corresponding Author: *John Markus Bjørndalen, Department of Computer Science, University of Tromsø,*
N-9037 Tromsø, Norway. Tel.: +47 7764 5252; Fax: +47 7764 4580; E-mail: `jmb@cs.uit.no`.
[1]Available on line at `http://www.computer.org/cise`.

PyCSP is under development at the University of Tromsø, Norway, and University of Copenhagen, Denmark. It is intended both as a research tool and as a compact library used to introduce CSP to Computer Science and eScience students. Students may already be familiar with the Python programming language from other courses and projects, and with the support for CSP they get better abstractions for expressing concurrency.

Early experiences with PyCSP are promising: PyCSP was offered as an option along with occam, C++CSP [7] and JCSP [8,9,10,11] in this year's Extreme Multiprogramming Class at the CS department, university of Copenhagen. Several students opted for PyCSP even with the warning that it was early prototype software. No students experienced problems related with the stability of the code however. An informal look-over seems to indicate that the solutions that uses PyCSP were shorter and easier to understand than solutions using statically typed languages.

PyCSP can be downloaded from [12].

1. Background Ideas

This paper assumes familiarity with the basic abstractions of CSP, as well as some familiarity of other recent implementations of CSP, such as CSP for Java, JCSP.

1.1. eScience

eScience refers to the application of computational methods in natural science. While no formal definition exists, common components in eScience are mathematical modeling, data acquisition and handling, scientific visualization and high performance computing. eScience thus expects no formal computer science training but rather a strong scientific background in general. A consequence of this is that applications for eScience often lack the most basic computer science techniques for ensuring correctness and performance.

Computational methods are becoming pervasive in science, and a growing number of students will need increasing knowledge of sequential, concurrent, parallel, and distributed computing to be efficient in their studies and research. We observe that the knowledge in many areas are lacking, including choice of language, methods for reuse and techniques for parallelization and distribution.

In our experience, the choice by students of which languages to use is typically made based on which languages they already know. The less they know, the worse match there will be between tools and problem. Scientific communities can end up using a programming language because of dependency of older programs they are using and enhancing. The tendency to stay with previously used languages limits how practical it is to use a language better suited to solve the problems at hand.

Java is used in many sciences. The availability of text-books for Java may have contributed to this, but we do not think it is because of its (relatively low) performance or (large) memory footprint [13]. We see extensive use of Perl in biology (see [14,15,16] for some pointers) and C++ in physics, though both languages require an in-depth knowledge of their implementation for efficient, let alone correct, use.

We prefer the use of Python for scientific computing because it is easy to adapt to the problem at hand: it requires little knowledge of the language to use it correctly, the source code is usually relatively short and readable, and through efforts such as SciPy, it seamlessly supports integration of high performance implementations of common scientific libraries.

Finally, multi-core architectures are now becoming the standard and as eScience is insatiable for performance, using multiple cores will soon be the norm in scientific computing. Multi-core architectures, shared memory multi-processors, cluster-computers and even meta-computing systems may easily be utilized by PyCSP simply by changing the run-time environment to match the architectures, without requiring the programmer to rewrite applications or consider the underlying architecture.

1.2. Computer Science

Concurrency, distribution and parallelism is often considered to be a hard subject. Educating future computer scientists and programmers, now that increased parallelism appears to become the norm, is a challenge that needs to be met even if we do not yet know which programming models will prevail. Clearly, much research remains to be done [17].

One approach is to educate students by guiding them through hands-on use of the models, and experimentally comparing them. This should provide students with a set of tools and models, aiding them in handling legacy systems on existing architectures, porting legacy systems to new architectures, and creating new systems. Python is a promising language to help us do this.

For more advanced students, Python can be used for the introduction and comparison of concepts, while more specialized languages that focus on given programming models can be used to study the respective models in greater detail. This would give us an opportunity to discuss trade-offs between using specialized languages that may have less library support vs. general purpose languages that need to support the models through libraries.

Systems that are candidates to use include: MPI (Message Passing Interface) [18], which can be taught using systems such as Pypar[19] or pyMPI[20]. To cover Tuple Spaces[21], we have SimpleTS[22]. RMI (Remote Method Invocation) and similar approaches can be taught using Fnorb [23] or Pyro (Python Remote Objects). PATHS [24,25] uses Python internally, and has a Python interface. Multi-threading and shared memory with various approaches to synchronization can be taught using libraries and modules that come with Python.

Most of these approaches and systems have implementations that we can use from Python, but we lack a Python CSP implementation. This is the situation that we are trying to remedy with PyCSP.

1.3. Terminology and Conventions

We will refer to CSP processes as *processes*, while we refer to user level processes scheduled by the operating system as *OS processes*. Many CSP processes will run inside a (Python) user level OS process.

To reduce the size of code listings in the paper, we have chosen to remove documentation strings and some of the comments in the code listings. The code is instead explained in the text of the paper.

1.4. Organization

The paper is organized as follows: Section 2 provides a short introduction to and mini-tutorial of PyCSP. Section 3 describes the implementation of PyCSP and some of the design choices we have made. Section 4 describes some of the eScience applications we are working on. Section 5 present the ever-present commstime benchmark, while future work and conclusions are presented in Sections 6 and 7 respectively.

2. A Short Introduction to PyCSP

Two central abstractions of CSP and PyCSP are the process and the channel. A running PyCSP program typically comprise several CSP processes communicating by sending messages over channels.

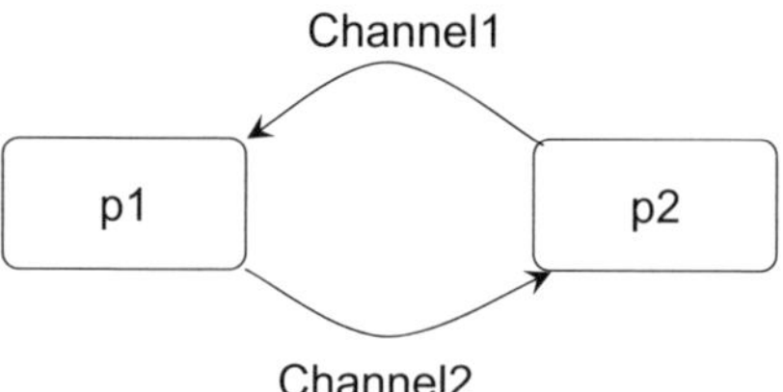

Figure 1 Basic PyCSP process network, with two processes: P1 and P2. The processes are connected and communicate over two channels: Channel1 and Channel2.

Listing 1: Complete PyCSP program with two processes

```
1  import time
2  from pycsp import *
3
4  def P1(cin, cout):
5      while True:
6          v = cin()
7          print "P1, read from input channel:", v
8          time.sleep(1)
9          cout(v)
10
11 def P2(cin, cout):
12     i = 0
13     while True:
14         cout(i)
15         v = cin()
16         print "P2, read from input channel:", v
17         i += 1
18
19 chan1 = One2OneChannel()
20 chan2 = One2OneChannel()
21
22 Parallel(Process(P1, chan1.read, chan2.write),
23          Process(P2, chan2.read, chan1.write))
```

Figure 1 shows an example, where two processes, P1 and P2, communicate over two channels: Channel1 and Channel2.

Listing 1 shows a complete PyCSP program implementing the process network in Figure 1. A PyCSP process is created by instantiating the Python *Process* class, passing as the first parameter a function that implements the functionality of the CSP process. The rest of the parameters to the Process constructor are passed directly to the function when the function starts.

A PyCSP channel is created by instantiating one of the channel classes. In the example, we create two One2OneChannels, which are PyCSP channels that can only have one reader and one writer attached[2].

Processes are usually connected in a network by passing channel ends to each of the processes when they are created. In the example, the reading end of channel 1 is passed to process P1, while the writing end of channel 1 is passed to process P2. To the functions implementing the processes, the channel ends appears as ordinary functions passed as parameters (`cin` and `cout` in the example).

---

[2]This design is inherited from JCSP. Other channel variants exist, and will be described later.

Accidentally accessing the wrong end of a channel can cause deadlocks, or at the very least lead to incorrect behaviour. To ensure that processes only use the correct ends of channels, PyCSP uses a simple approach: by passing the read and write methods of channels to the corresponding processes, we are certain that processes do not accidentally use the wrong methods. Python makes this trick simple since an object's methods are bound to the object. If a PyCSP process only has a reference to the read() method of a channel (and not to the channel object), the read() method is still able to access the internals of the channel.

This is similar to the way channel ends are used in JCSP Network Edition (from Quickstone) and the recently integrated JCSP core and network edition [26]. The main difference is that we don't need to define and implement separate interfaces and functions for returning the channel ends (like JCSP's in() and out()) since we use functionality built into Python.

Our simple program also contains another central CSP abstraction: the Parallel construct. To allow processes to execute in parallel, we have to specify which processes should be executed, and initialize execution within the Parallel construct. In PyCSP, this is done by writing Parallel, and listing all the processes that should be executed in parallel.

The Parallel construct initiates execution of the provided processes, and then waits for the completion of all of them before it finishes.

The output of the program in Listing 1 is as follows[3]:

```
~/pycsp/pycsp-0-1/test> python2.5 simple.py
P1, read from input channel: 0
P2, read from input channel: 0
P1, read from input channel: 1
P2, read from input channel: 1
P1, read from input channel: 2
P2, read from input channel: 2
P1, read from input channel: 3
P2, read from input channel: 3
...
```

2.1. Alternative

Another central abstraction in PyCSP is the Alternative command. One of the basic examples of the use of an Alternative construct is a process that needs to select from a number of input channels depending on which one has data already available for reading. This is done in CSP using a set of *guards*, which have two states: ready and unready. If a guard is ready, the expression associated with that guard can be executed. If several guards are ready in an Alternative statement, one of them is (non-deterministically) selected and the statements being guarded by the selected guard are executed.

PyCSP uses a similar approach to JCSP: a channel's *read()* method can act as a guard. When a number of channel *read()* operations are registered with an Alternative command, the Alternative's *select()* method can be used to detect which of the channels have available input. The following listing is an example:

Listing 2: Alternative example

```
1 ch1 = One2OneChannel ()
2 ch2 = One2OneChannel ()
3 ...
4 alt = Alternative(ch1.read, ch2.read)
5 ...
6 ret = alt.select()
7 print "Reading from the selected channel:", ret()
```

---

[3]Note that PyCSP does not provide any guarantees about output interleaving when using the standard Python 'print' statement.

Two channels are created, and the input methods of the channels are passed to the Alternative construct. An Alternative object is created (`alt`), which can be used to select from the input guards.

In PyCSP, the selected guard is returned from the select() call. Thus, since we select from two input guards, it is possible to read directly from the returned object.

2.2. *Library Contents*

The PyCSP library currently contains the following constructs:

- Channels: One2One, One2Any, Any2One, Any2Any, BlackHole
- Channel Poison
- Alternative
- Guards: Guard, Skip, and input channels
- Parallel and Sequence constructs
- Processes
- Some components based on the JCSP.plugNplay library

3. PyCSP Implementation

PyCSP is implemented as pure Python code, aiming for a portable implementation. This enables us to run PyCSP on devices ranging from mobile phones and embedded devices up to most desktop and high performance computing architectures.

Another goal is to aim for compact and readable code that can be taught to students. We intend to walk students through the implementation without having to spend time on many of the problems created by, for instance, statically typed languages which sometimes tend to be rather verbose and hide some of the abstractions that we try to teach.

We are currently using Python version 2.5, which provides us with some new language features that help us write more compact understandable code. An example is the new *with* statement, which, among other things, simplifies some of the code using locks. Listing 3 shows an example where scaffolding with *try/finally* is replaced with a single *with* statement.

Listing 3: Simplifying lock usage by using the *with* statement

```
 1 # old-style lock usage, ensuring that locks are always
 2 # released when returning from the code
 3 somelock.acquire()
 4 try:
 5     dosomething
 6 finally:
 7     somelock.release()
 8
 9 # alternative Python code using the 'with' statement
10 with somelock:
11     dosomething
```

The *with* statement takes care of acquiring the lock before executing the following code block ("dosomething"). It also takes care of releasing the lock when leaving the code block, even if exceptions are thrown in the code block.

The PyCSP implementation mainly borrows ideas from the JCSP [8] implementation, but also uses ides from C++CSP [7], and two independent CSP implementations for Microsoft's .NET platform (Chalmers [27] and Lehmberg [28]).

3.1. Threads, CSP Processes and OS Processes

The current implementation uses the Python *threading.Thread* class to implement CSP Processes. Python uses kernel threads to implement multi-threading, which should allow us to draw advantage of multi-core and multi-processor architectures.

The main drawback with this is that Python has a single interpreter lock, restricting the execution of Python byte-code to a single thread at a time. This may not be a major problem for our intended use, since we expect most of the execution time of compute-intensive applications to be spent in C or Fortran libraries, which can release the interpreter lock when executing the library code, allowing multiple threads to execute concurrently.

Another limitation is that we use up a kernel thread for every CSP Process, limiting the number of CSP Processes we can run concurrently since operating systems usually have an upper limit on the number of threads and processes a user can run concurrently.

Both problems can be solved by introducing network channels, which allow us to use PyCSP to communicate between multiple OS Processes on the same host or on multiple hosts in clusters. Network channels is on our agenda (see Section 6).

Kernel threads introduce extra scheduling and synchronization overhead compared to CSP implementations that use fibers or user threads to implement CSP Processes. For the intended applications that are expected to use C or Fortran libraries for compute-intensive and time-consuming tasks, we do not expect the difference to cause any major performance problems. User threads are also likely to introduce extra complexity compared to the current implementation when we try to avoid stalling the rest of the process network when one CSP Process calls a blocking system call or a time-consuming library call.

3.1.1. Synchronization and Python Decorators

Python has no *synchronized* keyword, but with recent versions of Python, *decorators* allow us to implement similar functionality. Code listing 4 shows an implementation of a synchronized decorator and its usage in a class. The Python decorators are essentially wrappers around a given method, allowing us to wrap code around existing functions.

We started the project using Python 2.4, where the @synchronized decorator took care of the necessary framework of lock acquiring and releasing as well as exception handling with try/finally (see the first part of listing 3). When we decided to use Python 2.5, with its *with* statement, the @synchronized decorator was simplified to the code shown in listing 4. The self._cond attribute is a standard condition variable from the Python *threading* module.

Listing 4: Python decorator for monitor/synchronized

```
1 def synchronized(func):
2     "Decorator to help create monitors"
3     def _call(self, *args, **kwargs):
4         with self._cond:
5             return func(self, *args, **kwargs)
6     return _call
```

A decorator is applied to a function by prefixing the function with @decoratorname, as in the following example where the decorator wraps a method in a class:

Listing 5: Example decorator use

```
1 class Foo:
2     def __init__(self):
3         ....
4     @synchronized
5     def somemethod(self, arg1, arg2):
6         dosomething...
```

Calling *somemethod* in this example, will result in the call being redirected through the *synchronized* decorator function, which handles the lock and then forwards the call to the original function.

Compared to Java, which has *synchronized* built-in, this adds extra code, although not by much. Decorators, however, allow us to use similar techniques to simplify other tasks in the CSP library, such as managing channel poison propagation (see Section 3.4.1).

We are currently evaluating whether the *@synchronized* decorator should be removed in future version of PyCSP. The advantage of keeping it there is that it clearly labels the intention of the programmer, but the drawback is that decorators can only be applied to functions, while the *with* statement can be applied to any block of code.

Another reason for keeping the decorator is that we can insert *@synchronized* before other method decorators, ensuring that the lock is acquired before executing other decorators. This is currently used for channel poisoning.

3.2. Processes

PyCSP processes are encapsulated using the `Process` class, which is a subclass of the Python `threading.Thread` class. Listing 6 shows an implementation of the `Process` class (the full implementation uses the new `run()` function necessary for handling channel poisoning, shown in Section 3.4.1, Listing 11).

Listing 6: PyCSP process implementation.

```
1 class Process(threading.Thread):
2     def __init__(self, fn, *args, **kwargs):
3         threading.Thread.__init__(self)
4         self.fn = fn
5         self.args = args
6         self.kwargs = kwargs
7     def run(self):
8         self.fn(*self.args, **self.kwargs)
```

Rather than creating a new class for each type of process, we have chosen to use the Process class directly as the Process construct in PyCSP. Programmers create a PyCSP process by creating an instance of the Process object, passing as the first argument a Python function that implements the process. The rest of the arguments to the Process object are passed to the function as arguments and keyword arguments. This is similar to one of the methods of creating threads in Python: passing a function to the constructor of the `threading.Thread` class.

The advantage of this is that source code tends to be shorter and clearer than source code where classes have to be made for every type of process. Listing 7 shows an example, where we first define the *Successor* process used in the *commstime* benchmark, and then create a successor process, passing two channel ends to the process.

Listing 7: Process example

```
1 def Successor(cin, cout):
2     while True:
3         cout(cin()+1)
4
5 p = Process(Successor, chan1.read, chan2.write)
```

Although we believe this method to be easier than creating classes for most uses, users may still want to create a new class for some process types. This can be supported either by sub-classing *Process*, or by taking advantage of the fact that Python objects can act as functions: any Python object can behave as a function if it has a `__call__()` method. Any object with a call method can be passed to *Process* in the same way as the Successor function was in listing 7.

A process object does not start automatically after creation. Instead, it exists as a container for a potential execution. To start the execution, a Parallel or Sequence construct is needed[4].

3.3. Parallel and Sequence

Parallel and Sequence have the following straight-forward implementations (see listing 8). Parallel is implemented as a class where the constructor takes a list of Process objects, calls start() on each of the processes (initiating execution of the processes), and then calls join() on each of the processes to synchronize with the termination of all of the processes. The constructor of the Parallel container object returns when the processes in the Parallel construct have terminated.

Sequence is similar, but instead of starting the threads and joining with them, the Sequence constructor calls the run() method directly on each of the processes in the sequence specified by the programmer.

Listing 8: Implementation of Parallel and Sequence

```
 1 class Parallel:
 2     def __init__(self, *processes):
 3         self.procs = processes
 4         # run, then sync with them.
 5         for p in self.procs:
 6             p.start()
 7         for p in self.procs:
 8             p.join()
 9
10 class Sequence:
11     def __init__(self, *processes):
12         self.procs = processes
13         for p in self.procs:
14             p.run()
```

3.4. Channels

Similar to the Chalmers et. als CSP for .NET implementation [27], we protect the user from accidentally using the wrong end of a channel. We do this by passing write and read methods of the channel objects directly to processes. The necessary bits for doing this already exist in the language, so in PyCSP, channel ends are passed to the processes as in listing 9.

Listing 9: Passing channel ends to processes. The read end of a channel is passed to a new process (bar).

```
 1 def bar(cin):
 2     val = cin()
 3     print 'Got a value from the channel:', val
 4
 5 ch = Channel('foo')
 6 Parallel(Process(bar, ch.read),
 7         ...
```

---

[4]In practice, a user can abuse the fact that the process object is a Python thread, and start it manually with p.start() or p.run(), but this is not the intended use in PyCSP.

In Python, the methods of an instantiated object are already bound to the object the methods belongs to. Thus, a function that only has a reference to one of the channel methods can still call the method by treating the reference as an ordinary function. In listing 9, bar uses the passed channel input (cin = read) directly by calling the cin function.

The PyCSP channels allow any object to be passed over the channel, including channel ends and CSP processes. This may not be the case for the future network channels as some objects, such as CSP Processes, will be more difficult to pass across a network connection (see Section 6).

PyCSP channels also take a name as an optional argument to the constructor, as in listing 9. Channel names are currently only used for debugging purposes.

The current PyCSP version implements the following channels from JCSP: One2One, Any2One, One2Any, Any2Any, and BlackHole. The One2One and Any2One channels can be used as input guards (see Section 3.5).

3.4.1. Channel Poisoning

PyCSP channels supports *Channel Poison* [29] to aid in terminating a process network. Any process that tries to read or write to a poisoned channel is terminated, and the channels passed to that process upon creation are also poisoned. There is currently no support for automatic poisoning of channels created inside the process, or mobile channels passed between processes.

Poisoning and poison propagation is implemented by adding a poisoncheck decorator around the channel methods (Listing 10). The poisoncheck decorator checks whether a channel is poisoned before and after calls to the channel and throws a ChannelPoisonException if poison is detected. The exception is caught in the `Process` class (specifically in the run() method). The process object then examines the parameters to the process and poisons any channels passed to the process.

Listing 10: Channel poison check decorator

```python
def chan_poisoncheck(func):
    "Decorator for making sure that poisoned channels raise exceptions"
    def _call(self, *args, **kwargs):
        if self.poisoned:
            raise ChannelPoisonException(self)
        try:
            return func(self, *args, **kwargs)
        finally:
            if self.poisoned:
                raise ChannelPoisonException(self)
    return _call
```

Listing 11: PyCSP process - adding support for poison propagation.

```python
class Process(threading.Thread):
    ....
    def run(self):
        try:
            self.fn(*self.args, **self.kwargs)
        except ChannelPoisonException, e:
            # look for any chan. end methods (only inspect method objects)
            InstMethType = type(self.run)
            for chfn in [x for x in self.args if type(x) == InstMethType]:
                ch = chfn.im_self # get channel object from channel end
                if isinstance(ch, Channel):
                    # Only call poison() on proper channels
                    ch.poison()
```

3.5. Alternative

Alternative in PyCSP follows the implementation in JCSP in principle, but with a few alterations to allow for a more Python-style implementation.

Alternative is a Python class, where the constructor takes a list of guards (see example in listing 12). When the priSelect() operation is called, each of the guards are enabled in turn, as in the JCSP implementation.

Listing 12: Alternative example

```
1 # assuming that we already have two channel inputs: in1, and in2
2 sg = Skip()
3 alt = Alternative(in1, in2, sg)
4 ret = alt.priSelect()
5 if ret != sg:
6     # Alt did not return the skip guard
7     print "Reading from the selected channel:", ret()
```

Contrary to JCSP, the PyCSP Alternative returns a reference to the selected guard, which allows the program to use the guard directly. In the above example, we check the returned object. If it is the skip guard, we ignore the results. Otherwise, we attempt to read from the returned channel.

Listing 13: JCSP Alt example, modified from Regulate demo

```
1  ...
2  final Skip sg = new Skip();
3  final Guard[] guards = {in1, in2, sg};    // prioritised order
4  final int IN1 = 0, IN2 = 1, SG = 2;       // index into guards
5
6  final Alternative alt = new Alternative (guards);
7
8  switch (alt.priSelect()) {
9    case IN1:
10       x1 = in1.read();
11       break;
12    case IN2:
13       x2 = in2.read();
14       break;
15    case SG:
16       break;
17 }
```

The advantage of returning the guard directly, compared to the JCSP example in listing 13, is that the programmer can not mix up the indexes into the provided guard array, and we do not need a switch when the returned guard can be called directly as a function. The latter should be common when selecting from multiple inputs.

When we need to check the identity of the returned guard, the PyCSP code needs to use a series of if- and elif-statements comparing the identity of the returned guard with the guards provided to the Alternative construct. We do not consider this a drawback compared to the JCSP method: Python does not have a switch statement, and a pattern similar to the provided JCSP example would normally be implemented using a series of if- and elif-statements.

PyCSP currently only supports priSelect() which mimics the behaviour of priSelect() in JCSP. Select() is using priSelect() as an implementation of select(). As soon as fairSelect() is implemented, select() will be set to use fairSelect() to mimic the behaviour in JCSP.

3.6. Guards

The guards in PyCSP follows the implementation of JCSP guards. Since our current examples and test-applications have not demanded many of the JCSP guards yet, the current implementation only has two guards apart from the channel input guards: Skip and the Guard base class. Other guards will be added in the future to handle cases such as timeouts.

The One2One and Any2One channels can be used as input guards.

4. Applications

4.1. Radiation Planning

The first eScience application targets planning of stereotactic radiation of brain-tumours. The challenge in the problem is to set up a number of radiation sources in such a manner that it minimizes the amount of energy that is absorbed by healthy brain-tissue while still depositing enough energy within the tumour. The modeling of the radiation is a simple Monte-Carlo simulation of a radiation source where the probability of a ray depositing its energy in any point is proportional to the density of the tissue, or inversely proportional to the light on the CT scan of the brain. The images in Figure 2 are before and after the simulation.

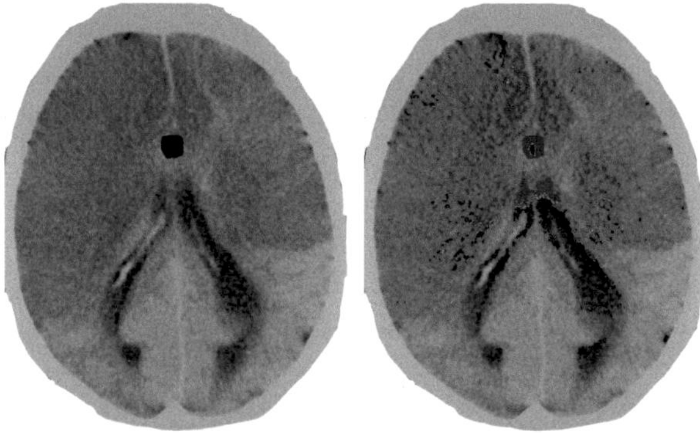

Figure 2 Radiation Planning. CT brain scan Before (left) and after (right) radiation simulation.

Since we have a number of radiation sources, parallelization of the application through CSP has a trivial solution by allowing each radiation source to be simulated in parallel. Unfortunately that approach limits the potential parallelism in the application to the number of radiation sources, thus a more scalable solution was chosen where the radiation sources produces vectors of particles and a number of processes receive these vectors through an any2any channel and traces the particles in the vector through the brain tissue. Applying this approach allows us to reach very high degrees of parallelism, in principle hundreds of millions of ray-tracing processes, since the number of rays that are simulated in real-world scenarios are in the billions.

Figure 3 shows the CSP network used in the application. The code for this setup, including termination process, is shown in listing 14. In the listing, the names of the processes have been shortened to fit in the listing: `source` is the Radiation Source and `trace` is the Ray Tracer.

When a radiation source has finished creating all its particles, it sends a "finished" message on its termination channel (the "c" channel in listing 14). The terminator process waits for all radiation sources to finish, then poisons the channel used to transmit particle vectors to the ray tracers (the "ec" channel). This terminates all ray tracer processes when they attempt to read a new particle vector.

Terminating the network this way is safe, since: a) a radiation source will not terminate until it has safely transmitted all its particles to a ray tracer, and thus, the Terminator will

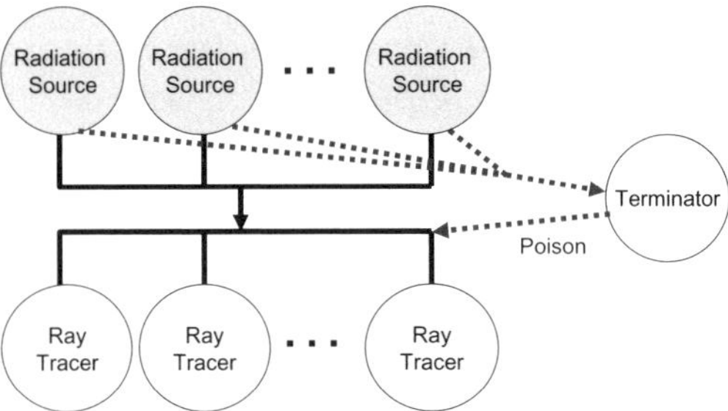

Figure 3 CSP network for parallelizing the brain-tumour radiation simulation. Note that there are usually more Ray-tracers than there are radiation sources.

not poison the channel before all radiation has been transmitted, and b) a Ray tracer process will not be poisoned and terminated until it goes back to read from its input channel after processing the final radiation.

Note that the source code contains unnecessary replication of code. The main reason for this is to provide a simple example. Larger networks of processes could use standard Python list comprehensions to create similar networks with fewer lines of code than this listing.

Listing 14: PyCSP raytrace network

```
 1 def terminator(n, cin, cout):
 2     for i in range(n):
 3         cin()
 4     poisonChannel(cout)
 5
 6 c = Any2AnyChannel()
 7 ec = Any2OneChannel()
 8
 9 Parallel(Process(source,   (85.0,75.0),   (1.0,0.8),  50000, c.write, ec.write)
10          Process(source, (10.0,230.0),   (1.0,0.0),  50000, c.write, ec.write)
11          Process(source,(550.0,230.0),  (-1.0,0.0),  50000, c.write, ec.write)
12          Process(source,  (475.0,90.0), (-1.0,.75),  50000, c.write, ec.write)
13          Process(source,   (280.0,0.0),  (0.0,1.0),  50000, c.write, ec.write)
14          Process(terminator, 5, ec.read, c.write),
15          Process(trace, c.read),
16          Process(trace, c.read),
17          Process(trace, c.read),
18          Process(trace, c.read),
19          Process(trace, c.read),
20          Process(trace, c.read))
```

4.2. Circuit Design

As an exercise in designing simulated experiments we have another example where digital circuits are build as a networks of CSP processes each functioning as a trivial small Boolean logic gate These gates may by grouped to form more complex components, adders and multiplexers, etc. Even simple circuits includes tens of processes and easily hundreds or even thousands[5].

The circuit design code is straightforward except for wire-junctions, which electrically are trivial, but in a CSP model needs to be handled explicitly by a junction. Thus a full adder need to be set up as in listing 15.

---

[5]Welch [30] provides examples and a more detailed discussion about emulating digital logic using CSP and occam.

Listing 15: PyCSP circuit design code

```
 1 def Adder(A, B, Cin, S, Cout):
 2     Aa = One2OneChannel()
 3     Ab = One2OneChannel()
 4     Ba = One2OneChannel()
 5     Bb = One2OneChannel()
 6     Ca = One2OneChannel()
 7     Cb = One2OneChannel()
 8     i1 = One2OneChannel()
 9     i1a = One2OneChannel()
10     i1b = One2OneChannel()
11     i2 = One2OneChannel()
12     i3 = One2OneChannel()
13
14     Parallel(Process(delta, A.read, Aa.write, Ab.write),
15              Process(delta, B.read, Ba.write, Bb.write),
16              Process(delta, Cin.read, Ca.write, Cb.write),
17              Process(delta, i1.read, i1a.write, i1b.write),
18              Process(XOR, Aa.read, Ba.read, i1.write),
19              Process(XOR, i1a.read, Ca.read, S.write),
20              Process(AND, Ab.read, Bb.read, i2.write),
21              Process(AND, i1b.read, Cb.read, i3.write),
22              Process(OR, i2.read, i3.read, Cout.write))
```

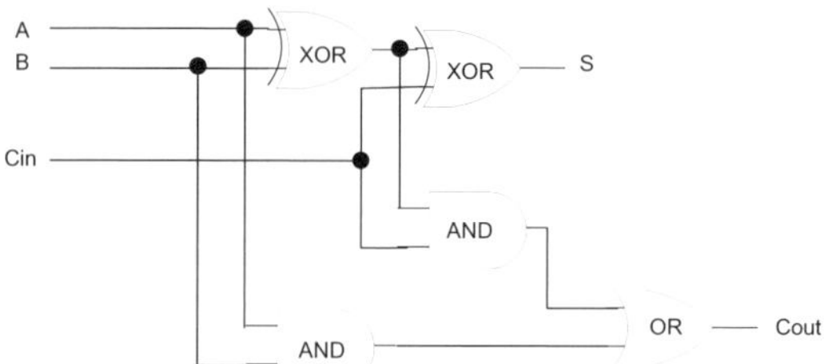

Figure 4 Full-adder diagram, the CSP implementation has 9 processes in it.

4.3. Protein Folding

Protein folding is an extremely hot topic in medical research these days, unfortunately protein folding is extremely computationally demanding and requires a huge supercomputer to fold even the simplest proteins. Luckily the task of calculating protein-foldings is quite well suited for parallel processing.

Proteins are made up of amino-acids, of which there are 20 types. Thus a protein can be viewed as a sequence of amino-acids and folding such a sequence means that the sequence "curls up" until there is a minimum of unbound energy present in the protein. For teaching purpose we need not concern ourselves with the chemistry behind the protein-foldings. Instead we can play with a simplified version of proteins called prototeins – proto-type proteins.

Our simplified prototeins are folded in only two dimensions and only in 90 degree angles. This is much simpler than real three dimensional foldings with angles depending on the amino-acids that are present at the fold, but as a model it is quite sufficient. Our amino-acids are also reduced to two types; Hydrophobic (H) and Hydrophilic (P). When our prototein is folded it will seek the minimal unbound energy, modeled by the highest number of H-H neighborships.

Each folding results in an residual energy level and the solution to a protein-folding problem is to find the folding that has the minimum residual energy level. The actual folding

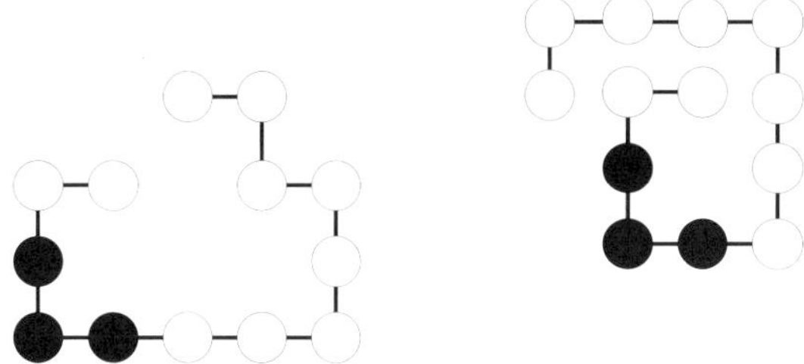

Figure 5 A non-optimal (left) and an improved (right) prototein folding of 13 amino-acids.

is performed as a search-tree of the potential solutions, much like the Travelling-Salesman-Problem, but without the option for branch-and-bound. Thus the CSP solution is well known and implemented as a producer-consumer algorithm.

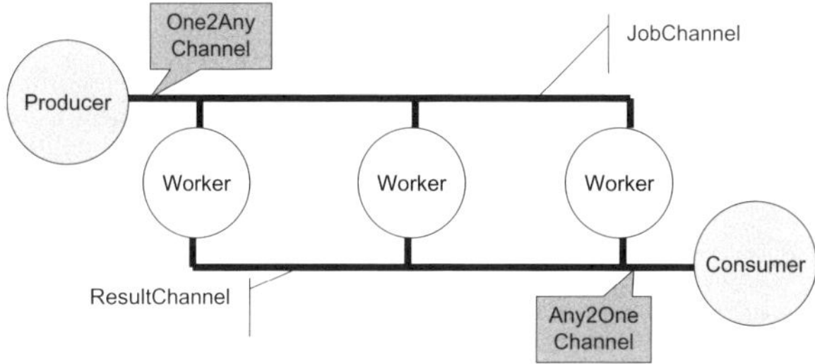

Figure 6 CSP network for handling the prototein folding example.

The code for this model, including a termination process that is not shown in the figure looks as:

Listing 16: PyCSP prototein network

```
1  feeder = One2AnyChannel()
2  collector = Any2OneChannel()
3  done = Any2OneChannel()
4
5  Parallel( Process(producer, protein, map, place, feeder.write),
6            Process(worker, feeder.read, collector.write, done.write),
7            Process(worker, feeder.read, collector.write, done.write),
8            Process(worker, feeder.read, collector.write, done.write),
9            Process(worker, feeder.read, collector.write, done.write),
10           Process(worker, feeder.read, collector.write, done.write),
11           Process(barrier, 5, done.read, collector.write),
12           Process(consumer, collector.read) )
```

4.4. Commstime

The classic *commstime* benchmark [31] is used in many of the recent CSP papers. The source code for the consumer process, written as a PyCSP process, is shown in listing 17. Listing 18 shows the source code for setting up and running the network of processes in the benchmark. The network uses a Delta2 process that is similar to the JCSP SeqDelta2Int process: the process forwards its input to the two output channels in sequence.

The output from the Consumer process is the execution time per communication, computed as time per loop divided by 4, which is reported as "microseconds/communication" in JCSP.

The final line in listing 17 shows an example usage of channel poison, terminating the commstime process network when the consumer process finishes.

Performance results of commstime are shown in Section 5.2

Listing 17: Consumer process

```
1 def Consumer(cin):
2     "Commstime consumer process"
3     N = 5000
4     ts = time.time     # get a reference to the time function
5     t1 = ts()
6     cin()              # warmup, only one loop
7     t1 = ts()
8     for i in range(N):
9         cin()
10    t2 = ts()
11    dt = t2-t1
12    tchan = dt / (4 * N)
13    print "DT = %f.\nTime per comm : %f/(4*%d) = %f s = %f us" % \
14            (dt, dt, N, tchan, tchan * 1000000)
15    print "consumer done, poisoning channel"
16    poisonChannel(cin)
```

Listing 18: Commstime benchmark

```
1 def CommsTimeBM():
2     # Create channels
3     a = One2OneChannel()
4     b = One2OneChannel()
5     c = One2OneChannel()
6     d = One2OneChannel()
7
8     print "Running commstime"
9     Parallel(Process(Prefix, c.read, a.write, prefixItem=0),
10             Process(Delta2, a.read, b.write, d.write),
11             Process(Successor, b.read, c.write),
12             Process(Consumer, d.read))
```

5. Experiments

Since we do not yet have network support, and since the execution of Python byte-code is limited to one thread at the time, the potential for parallelism is not very large for our application examples. Thus, we currently only have performance numbers for the commstime benchmark.

The benchmarks were executed on the following hosts, all using Python 2.5:

AMD AMD Athlon 64 X2 Dual-Core 4400+, 2.2GHz, 2GB RAM, Running Ubuntu Linux 6.10 in 32-bit mode. Both cores were enabled.

R360 Dell Precision Workstation 360, Intel P4 Prescott, 3.2GHz, 2GB RAM, with Hypterthreading enabled. Running Rocks cluster distribution of Linux.

R370 Dell Precision Workstation 370, Intel P4 Prescott, 3.2GHz, 2GB RAM, with Hypterthreading enabled. Running Rocks cluster distribution of Linux, in 64-bit mode.

Qtek Qtek 9100 mobile phone, 195MHz TI OMAP 850 processor, 64MB RAM, Windows Mobile 5 operating system.

5.1. Optimization

Python compiles the source code to byte-code and runs the byte-code in an interpreter. Further optimizations of the byte-code can be made with the Psyco [32] Python module, which works similarly to a just-in-time compiler.

Enabling Psyco optimization is as easy as importing the Psyco module and calling one of the optimizer functions in the module:

Listing 19: Using Psyco byte-code optimization

```
1 import psyco
2 psyco.full()
```

According to the Psyco documentation, users often experience speed-ups of 2 to 100 times on Python programs, with a typical result being a 4x speed-up of the execution time.

The benchmarks below are presented with and without Psyco optimizations for the two machines running in 32-bit mode. There is no Psyco-support for 64-bit Linux or for the Qtek mobile phone, so Psyco-optimization experiments were not tried on these machines.

5.2. Commstime

The commstime benchmark was executed with N set to 5000 on all hosts (see Listing 17), with the exception of the Qtek mobile phone, where it was set to 500 due to the slower CPU in the phone. The reported numbers in Table 1 are the minimum, maximum and average of 10 runs of the commstime benchmark.

In addition, we ran the JCSP benchmark on the AMD machine, using JCSP 1.0rc7 with Sun JDK 1.5-06. For the JCSP experiments, we specified that we wanted sequential output from the Delta process (using SeqDelta2Int) rather than parallel output. The reported results are the minimum, maximum and average of the "microseconds / communication" output from 20 runs of the benchmark. No errors or spurious wakeups were reported by commstime.

Table 1 Commstime results

Implementation	Optimization	min	max	avg
AMD, PyCSP		$74.78\mu s$	$88.40\mu s$	$84.81\mu s$
AMD, PyCSP	Psyco	$48.15\mu s$	$54.91\mu s$	$52.67\mu s$
R360, PyCSP		$141.67\mu s$	$142.51\mu s$	$142.09\mu s$
R360, PyCSP	Psyco	$89.50\mu s$	$91.57\mu s$	$90.37\mu s$
R370, PyCSP		$128.14\mu s$	$129.12\mu s$	$128.61\mu s$
Qtek mobile phone, PyCSP		$6500\mu s$	$6500\mu s$	$6500\mu s$
AMD, JCSP, w/SeqDelta		$6\mu s$	$9\mu s$	$8.1\mu s$

There is clearly an advantage in running Psyco to optimize the Python byte code: a factor 1.6 improvement in the AMD case, and a factor 1.57 in the R360 case. This is lower than the improvement that the Psyco developers claim is common, but some of the explanation for this may be that a large fraction of the commstime execution time is outside the reach of Psyco: C library code for locks, system calls and operating system code.

There is also a significant difference in execution time favoring the AMD multi-core processor compared to the Intel Hyperthreading processors. We do not know the reason for this yet, as there are several factors changing between the processors: Hyperthreading vs. multi-core, AMD vs. Intel P4 Prescott implementation of instruction sets, memory buses, and Linux distributions with different kernel versions.

Comparing the "average" column for PyCSP and JCSP, we see that PyCSP without Psyco is about an order of magnitude slower than JCSP, and PyCSP with Psyco is about 6.5 times slower than JCSP. This is within the range that we expected. PyCSP is not intended for fine granularity CSP networks where a significant part of the time is spent communicating. It is intended for reasonable CSP performance in applications where most of the computation time is spent in C or Fortran library code. In that sense, commstime is the worst-case benchmark: it stresses the code that we expect to spend the least amount of time to check that PyCSP does not introduce unreasonable communication overhead.

The experiments show that PyCSPs channel communication overhead is not prohibitive for scientific applications.

6. Future Work

Network support is an important addition to PyCSP since this will allow us to make use of clusters. It may also improve utilization of multi-core architectures, remedying some of the problems with Pythons Global Interpreter Lock (GIL)[6] since we can run multiple OS processes on the same host, each hosting a set of PyCSP processes.

Initial prototyping of network support is likely to use Pyro (Python Remote Objects) to speed up development efforts, keep the code small, and to allow us to identify issues and potential implementation techniques. We do not expect to follow the JCSP or C++CSP implementations too closely, since we are hoping that Python will allow us to express some of the ideas in a more compact way.

One of the problems we are likely to encounter is how to handle mobile processes and mobile channels, or whether we should allow them in the first place. Passing processes and channels over channels within the same Python process is not a problem, since channels can essentially pass any object, and passing a PyCSP process across a channel would not influence the execution of the PyCSP process.

Moving PyCSP processes across network channels is not as simple though. Migration could be handled by suspending the PyCSP Process, passing the state across a network channel, and restarting the state in another OS Process. There are two complicating factors however: the first is that we expect users to make use of C and Fortran libraries, and we have no control of pointers and references used by those libraries. The second factor is that we are using kernel threads to implement PyCSP processes. Suspending a PyCSP Process by suspending the Python kernel thread executing it and handling potential problems with locks held by the thread, open files and other objects may prove difficult, save in the most trivial cases. Thus, it may in fact be impossible to migrate PyCSP processes to another address space in a safe way.

The same might end up being a problem with channels and channel ends: waiting queues for locks are difficult to migrate in a safe way if they are maintained by the operating system. Migrating a channel reference across to another address space, however, should be safe if we ensure that any access to the referenced object is forwarded back to the home node of the channel.

An alternative approach is to introduce remote evaluators and code execution. With Python and Pyro, we can pass expressions (as text strings), functions (as objects), classes and even entire Python modules across the network to remote Python processes, and have the remote Python process evaluate and execute the provided code. We have used this in other projects, and it may be a viable alternative to moving processes across address spaces.

---

[6]We have seen several questions and discussions about removing the GIL over the years, but it appears that the GIL is here to stay for the foreseeable future. For more information, please see the Python Frequently Asked Questions on the Python homepage.

Further development to support more constructs from the core JCSP and plugNplay libraries are also underway.

7. Conclusions

In this paper we have presented the preliminary results from working on integrating CSP in the standard Python model. PyCSP does not seek to be a high-performance CSP implementation but, like Python itself, seeks to provide an easy and transparent environment for scientists and students to work on models and algorithms. If high performance is needed, it may be achieved through the use of native-code libraries, and we do not envision CSP used at that level.

We believe that we have shown how scientists may easily model experiments similar to physical setups by using CSP enabled processes as black-boxes for more complex experiments. The advantage of CSP in this context becomes the isolation of private data and thus elimination of race-conditions and legacy dependencies that may come from using an object oriented model for scientific computing. While performance is not key to this work, we have shown that with commstime round-trip as low as $50\mu s$, the overhead of using PyCSP will not become prohibiting to using the model for real scientific computations.

The presented version of PyCSP is still work in progress and significant changes may still be applied. However, future developments will be directed towards portability, scalability, network support, and usability rather than performance and "feature-explosion".

Early experiences with PyCSP are promising: PyCSP was offered as an option along with occam, C++CSP [7] and JCSP [8,9,10] in this year's Extreme Multiprogramming Class at the CS department, university of Copenhagen. Several students opted for PyCSP even with the warning that it was early prototype software. No students experienced problems related with the stability of the code however. An informal look-over seems to indicate that the solutions that uses PyCSP were shorter and easier to understand than solutions using statically typed languages.

PyCSP can be downloaded from [12].

References

[1] Python programming language home page. `http://www.python.org/`.

[2] Scientific tools for Python (SciPy) homepage. `http://www.scipy.org/`.

[3] Xing Cai, Hans Petter Langtangen, and Halvard Moe. On the performance of the python programming language for serial and parallel scientific computations. *Scientific Programming, Vol 13, Issue 1, IOS Press*, pages 31–56, 2005.

[4] Hans Petter Langtangen. *Python Scripting for Computational Science, 2nd Ed.* Springer-Verlag Berlin and Heidelberg Gmbh & Co., 2005.

[5] C.A.R. Hoare. Communicating sequential processes. *Communications of the ACM, 21(8):666-677*, pages 666–677, August 1978.

[6] C.A.R. Hoare. *Communicating sequential processes.* Prentice-Hall, 1985.

[7] Neil Brown and Peter Welch. An introduction to the Kent C++CSP Library. *CPA, Communicating Process Architectures*, September 2003.

[8] JCSP - Communicating Sequential Processes for Java. `http://www.cs.kent.ac.uk/projects/ofa/jcsp/`.

[9] J.Moores. Native JCSP: the CSP-for-java library with a Low-Overhead CPS Kernel. In P.H.Welch and A.W.P.Bakkers, editors, *Communicating Process Architectures 2000*, volume 58 of *Concurrent Systems Engineering*, pages 263–273. WoTUG, IOS Press (Amsterdam), September 2000.

[10] P.H.Welch. Process Oriented Design for Java: Concurrency for All. In H.R.Arabnia, editor, *Proceedings of the International Conference on Parallel and Distributed Processing Techniques and Applications (PDPTA'2000)*, volume 1, pages 51–57. CSREA, CSREA Press, June 2000.

[11] P.H.Welch, J.R.Aldous, and J.Foster. CSP networking for java (JCSP.net). In P.M.A.Sloot, C.J.K.Tan, J.J.Dongarra, and A.G.Hoekstra, editors, *Computational Science - ICCS 2002*, volume 2330 of *Lecture Notes in Computer Science*, pages 695–708. Springer-Verlag, April 2002.

[12] PyCSP distribution. `http://www.cs.uit.no/~johnm/code/PyCSP/`.

[13] Ronald F. Boisvert, J. Moreira, M. Philippsen, and R. Pozo. Java and numerical computing. *Computing in Science and Engineering, Volume 3, Issue 2*, pages 18–24, 2001.

[14] BioPerl. http://www.bioperl.org/.

[15] Perl for Bioinformatics and Internet. http://biptest.weizmann.ac.il/course/prog/.

[16] James Tisdall. *Beginning Perl for Bioinformatics*. O'Reilly, 2001. ISBN 0-596-00080-4. Also see http://www.perl.com/pub/a/2002/01/02/bioinf.html.

[17] Krste Asanovic et. al. The Landscape of Parallel Computing Research: A View from Berkeley, 2006. Technical Report UCB/EECS-2006-183, Electrical Engineering and Computer Sciences University of Californai at Berkeley.

[18] MPI: A Message-Passing Interface Standard. *Message Passing Interface Forum*, March 1994.

[19] Pypar software package. `http://datamining.anu.edu.au/~ole/pypar/`.

[20] pyMPI software package. `http://pympi.sourceforge.net/`.

[21] N. Carriero and D. Gelernter. Linda in Context. *Commun. ACM*, 32(4):pp. 444–458, April 1989.

[22] SimpleTS - Tuple Spaces implementation in Python, John Markus Bjørndalen, unpublished. Source code available at `http://www.cs.uit.no/~johnm/code/`.

[23] Fnorb software package. `http://fnorb.sourceforge.net/`.

[24] John Markus Bjørndalen, Otto Anshus, Tore Larsen, and Brian Vinter. PATHS - Integrating the Principles of Method-Combination and Remote Procedure Calls for Run-Time Configuration and Tuning of High-Performance Distributed Application. In *Norsk Informatikk Konferanse*, pages 164–175, November 2001.

[25] John Markus Bjørndalen, Otto Anshus, Tore Larsen, Lars Ailo Bongo, and Brian Vinter. Scalable Processing and Communication Performance in a Multi-Media Related Context. *Euromicro 2002, Dortmund, Germany*, September 2002.

[26] Peter H. Welch, Neil Brown, James Moores, Kevin Chalmers, and Bernhard Sputh. Integrating and Extending JCSP. In A.A.McEwan, S.Schneider, W.Ifill, and P.Welch, editors, *Communicating Process Architectures 2007*, jul 2007.

[27] Kevin Chalmers and Sarah Clayton. CSP for .NET Based on JCSP. *CPA, Communicating Process Architectures*, September 2006.

[28] Alex A. Lehmberg and Martin N. Olsen. An Introduction to CSP.NET. *CPA, Communicating Process Architectures*, September 2006.

[29] Bernhard H.C. Sputh and Alastair R. Allen. JCSP-Poison: Safe Termination of CSP Process Networks. *CPA, Communicating Process Architectures*, September 2005.

[30] P.H. Welch. Emulating Digital Logic using Transputer Networks (Very High Level Parallelism = Simplicity = Performance). In *Proceedings of the Parallel Architectures and Languages Europe International Conference*, volume 258 of *Springer-Verlag Lecture Notes in Computer Science*, pages 357–373, Eindhoven, Netherlands, June 1987. Springer-Verlag. sponsored by the CEC ESPRIT Programme.

[31] Fred Barnes and Peter H. Welch. Prioritised Dynamic Communicating Processes - Part I. In James Pascoe, Roger Loader, and Vaidy Sunderam, editors, *Communicating Process Architectures 2002*, pages 321–352, sep 2002.

[32] Psyco optimizer for Python. `http://psyco.sourceforge.net/`.

Communicating Process Architectures 2007
Alistair A. McEwan, Steve Schneider, Wilson Ifill, and Peter Welch
IOS Press, 2007

A Process-Oriented Architecture for Complex System Modelling

Carl G. RITSON and Peter H. WELCH

Computing Laboratory, University of Kent, Canterbury, Kent, CT2 7NF, England.
{cgr,phw}@kent.ac.uk

Abstract. A fine-grained massively-parallel process-oriented model of platelets (potentially artificial) within a blood vessel is presented. This is a CSP inspired design, expressed and implemented using the occam-pi language. It is part of the TUNA pilot study on nanite assemblers at the universities of York, Surrey and Kent. The aim for this model is to engineer emergent behaviour from the platelets, such that they respond to a wound in the blood vessel wall in a way similar to that found in the human body – i.e. the formation of clots to stem blood flow from the wound and facilitate healing. An architecture for a three dimensional model (relying strongly on the dynamic and mobile capabilities of occam-pi) is given, along with mechanisms for visualisation and interaction. The biological accuracy of the current model is very approximate. However, its process-oriented nature enables simple refinement (through the addition of processes modelling different stimulants/inhibitors of the clotting reaction, different platelet types and other participating organelles) to greater and greater realism. Even with the current system, simple experiments are possible and have scientific interest (e.g. the effect of platelet density on the success of the clotting mechanism in stemming blood flow: too high or too low and the process fails). General principles for the design of large and complex system models are drawn. The described case study runs to millions of processes engaged in ever-changing communication topologies. It is free from deadlock, livelock, race hazards and starvation *by design*, employing a small set of synchronisation patterns for which we have proven safety theorems.

Keywords. occam-pi, concurrency, CSP, complex systems

Introduction

In this paper, a process-oriented architecture for simulating a complex environment and mobile agents is described. The environment is modelled by a fixed topology of stateful processes, one for each unit of space. State held includes the strength of specific environmental factors (e.g. chemicals), local forces and the presence of agents. Agents are mobile processes interacting directly with the space processes in their immediate neighbourhood and, when they sense their presence, other agents. Mechanisms for dynamically structuring hierarchies among agents are also introduced, allowing them to display complex group behaviours. The architecture combines deadlock free communications patterns with (phased barrier controlled) shared state, maintaining freedom from race hazards and high efficiency. We have used occam-π [1,2] as our implementation language.

This research is part of the TUNA project [3,4,5,6,7,8,9] at the universities of York, Surrey and Kent, which seeks to explore simple and formal models of emergent behaviour. Medium term applications are for the safe construction of massive numbers of nanotechnology robots (*nanites*) and their employment in a range of fields such as the dispersion of pollution and human medicine. With this goal in mind, this paper introduces our generic simulation architecture through specific details of how it has been used to simulate platelets in the human blood stream and the clotting response to injury.

1. Architecture

1.1. Dynamic Client-Servers

The simulation architecture is constructed in layers. At the bottom lie the *site* processes, representing distinct points (or regions) in the simulated space and managing information associated with that locality. Each site is a pure *server* process, handling requests on the server-end of a channel bundle (unique for each site). It will have a dynamically changing set of *client* processes (mobile agents), competing with each other to access the client-end of its channel bundle. Each channel bundle contains two channels used in opposite directions: one from a client to the server (*request*) and one from the server to a client (*response*). All communication is initiated by one of the clients successfully laying claim to its end of the channel bundle and making a request. Once accepted, the server and this client engage in a bounded conversation over the channel bundle, honouring some pre-agreed protocol. So long as no closed cycle of such *client-server* relationships exists across the whole process network, such communication patterns have been proven to be deadlock free [10,11].

1.2. Space Modelling

To model *connected* space, each site has reference to the client-ends of the channel bundles serviced by its immediate *neighbours*. These references are only used for forwarding to visiting clients – so that they can explore their neighbourhood and, possibly, move. Sites must never directly communicate with other sites, since that could introduce client-server cycles and run the risk of deadlock. The inter-site references define the *topology* of the simulation world. For standard Euclidean space, these neighbourhood connections are fixed. For example, each site in a 3D *cubic* world might have access to the sites that are immediately above/below, left/right or in-front/behind it. In a more fully connected world, each site might have access to all 26 neighbours in the 3x3x3 cube of which it forms the centre. Other interesting worlds might allow dynamic topologies – for example, the creation of *worm-holes*.

1.3. Mobile Channels and Processes

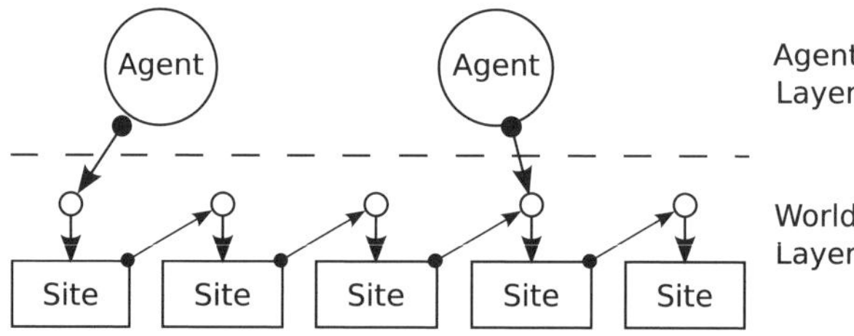

Figure 1. A simplified representation of sites and agents. Each site services an exclusive channel bundle for communicating with visiting agents. Agents obtain connections to their next site from references held by their current site.

The world layer (Figure 1) is homogeneous – only sites. The (first) agent layer is hetrogeneous. There can be many kinds of agent process, visiting and engaging with sites as they move around their world. Agent-site protocols fall into three categories: querying and modifying the current site state, obtaining access to neighbouring sites, and moving between sites. Agents move through the simulated world registering and de-registering their presence in sites (commonly by depositing free channel-ends through which they may be contacted), using environmental information (held in the sites) to make decisions as they go and, possibly, modifying some environmental factors. An agent only needs to hold the channel-end of its current site and, when relevant, the next site it wishes to enter. For all this the concept of channel-end mobility [12], a feature of occam-π based on items from the π-calculus [13], is essential.

Figure 1 shows a one-dimensional world where each site has access only to the neighbour immediately to its right. In this world, agents can only move in one direction. The arrows with circles on their bases represent client-server relations (pointing to the server). The client-ends of these connections are *shared* between other sites and agents (shown by the arrows with solid disc bases). Recall that these connections do provide two-way communications.

1.4. Barriers and Phases

Agents use *barriers* [14,15] to coordinate access to the sites into time-distinct *phases*. An occam-π BARRIER is (almost) the same as a multiway synchronisation event in CSP: *all* enrolled processes must reach (synchronise upon) the barrier in order for *all* of them to pass. The resulting phases ensure that they maintain a consistent view of their environment, and keep to the same simulation step rate. To prevent agents viewing the world while it is in flux, at least two phases are required:

discovery: where agents observe the world and make decisions;

modify: where agents change the world by implenting those decisions (e.g. by moving and/or updating environmental parameters).

The basic agent logic is:

```
WHILE alive
  SEQ
    SYNC discovery
    ... observe my neighbourhood
    SYNC modify
    ... change my neighbourhood
```

where `discovery` and `modify` are the coordinating barriers.

1.5. Site Occupancy and Agent Movement

In a typical simulation, only one agent will be allowed to occupy a given site at any point in time. Within our architecture, sites enforce this constraint. If two agents attempt to enter a site in the same simulation cycle, the decision can be left to chance (and the first agent to arrive enters), or made using an election algorithm (the *best* candidate is picked). In the case of an election algorithm, the modify phase should be sub-divided:

first modify sub-phase: agents request to enter the site providing some sort of candidacy information (e.g. mass, aggressiveness, or unique ID). When the site receives a new candidate, it compares it to the exiting one and overwrites that if the new candidate is *better*.

second modify sub-phase: all agents query the site(s) they attempted to enter again, asking who *won*? On receiving the first of these queries, the site installs its current *best* candidate as the new occupier and passes those details back to the asker and to any subsequent queries.

However, an optimisation can be made by including the first *modify* sub-phase in the *discovery* phase! Only offers to move are made – no world state change is detectable by the agents in this phase. The second *modify* sub-phase simply goes into the *modify* phase. This optimisation saves a whole barrier synchronisation and we employ it (section 2.5).

1.6. Agent-Agent Interaction

Some agents in the same locality may need to communicate with each other. To enable this, they deposit in their current site the client-end of a channel bundle that they will service. This client-end will be visible to other agents (observing from a neighbouring site). However, agents must take care how they communicate with each other in order to avoid client-server cycles and deadlock. A simple way to achieve this is to compose each agent from at least two sub-processes: a server to deal with inter-agent transactions and a client to deal with site processes and initiate inter-agent calls.

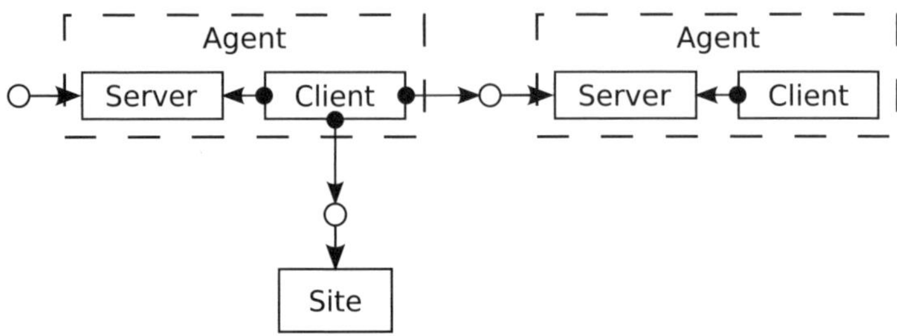

Figure 2. Agents are composed from client and server sub-processes to prevent client-server loops and maintain deadlock freedom.

In Figure 2, the agent *server* process manages agent state: its clients are the *client* processes of its own and other agents. The agent *client* process drives all communication between the agent and the rest of its environment (the sites over which it roams, other agents in the neighbourhood and higher level agents to which it reports – section 1.7). Technically, it would be safe for the agent *server* also to communicate with the sites.

1.7. Layers of Agents

So far, agents have occupied a single site. Complex agents (e.g. a blood clot) may grow larger than the region represented by a single site and would need to span many, registering with all it occupies. This may be done from a single agent process (as above) or by composing it from many sub-processes (one *client* part per site). We view the latter approach as building up a *super-agent* (with more complex behaviour) from many lower level agents (with simpler behaviour and responsibilities). It introduces a third layer of processes.

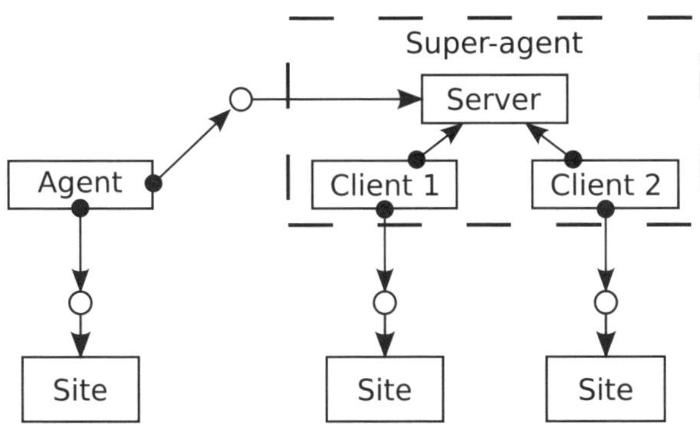

Figure 3. Super-agents as a layered composition of processes.

In figure 3, clients 1 and 2 share a higher level server process, holding information from both that enables them to act in a coordinated manner. Agents outside the super-agent just see a single server off a single agent. Such sharing of higher level servers allows us to create groups of arbitrarily large coordinated agents. The approach can be continued hierarchically to create ever more complex groups, while keeping the complexity of each process manageable – see figure 4. Note that some processes are pure servers (the sites and mega-agents), some are pure clients (the lowest level agents) and some are servers that sometimes act as clients to fulfil their promised service (the super-agents). Note that there are no client-server cycles and that the pure clients (the lowest level agents) are the initiators of all activity.

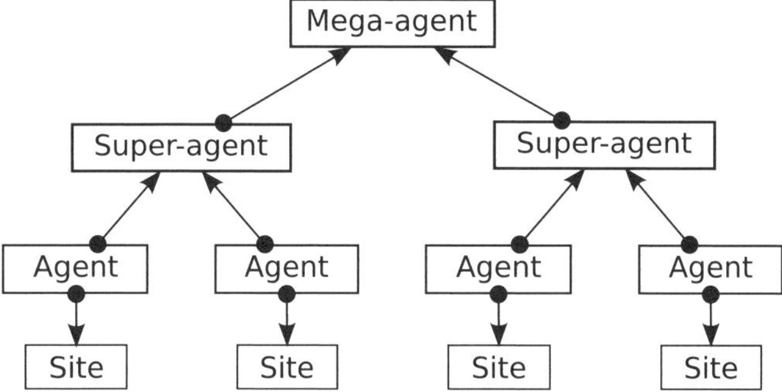

Figure 4. A hierarchy can be constructed among agents to give increasingly complex group behaviours.

2. Human Blood Clotting Simulation

We have introduced the principle components of the simulation architecture: a hierarchical client-server network of sites, agents and super-agents. We now look at how this has been applied to simulate the clotting of platelets in the human blood stream [8].

Haemostasis is the response to blood vessel damage, whereby platelets are stimulated to become *sticky* and aggregate to form blood clots that seal small wounds, stemming blood loss and allowing healing. Platelets are non-living agents present in certain concentrations in blood; they are continually formed in bone marrow and have a half-life of around 10 days. Normally, they are inactive. They are triggered into becoming sticky by a complex range of chemical stimuli, moderated by a similarly complex range of inhibitors to prevent a lethal chain reaction. When sticky, they combine with each other (and proteins like fibrin) to form physically entangled clots. Summaries can be found in [16,17,18], with extensive details in [19].

The work present in this paper employs a highly simplified model of haemostasis. We model the smooth and sticky states of platelets, with transition triggered by encountering a sufficient amount of a single chemical *factor* released by a simulated wound to the blood vessel wall. We model no inhibition of clotting, instead focusing only on the initial reaction to a wound, and relying on a sufficient rate of blood flow to prevent a chain reaction until it is observed.

Clots form when sticky platelets bump together and, with some degree of probability, become permanently entangled. The velocity of an individual clot decreases with respect to the rate of blood flow as its size increases. We are not modelling other factors for the clotting material (such as fibrin). Nevertheless, even with this very simple model, we have reached the stage where emergent behaviours (the formation of blood clots and the sealing of wounds) are observed and simple experiments are possible that have scientific interest.

2.1. Sites

Sites define the space of the simulated environment. Our sites are arranged into cubic three-dimensional space (giving each site 26 neighbours). Sites are pure server processes, responding to agent (client) offers of, or requests for, information. They operate independently, engaging in no barrier synchronisations.

Interacting with the sites, the lowest level agents are blood *platelets* and chemical *factors* (which, when accumulated in the sites above a certain threshold, can switch passing platelets into their *sticky* state). Blood clots are super-agents, composed of many stuck-together platelets.

The sites allow one platelet to be resident at a time and store a unique ID number, stickiness, size (of the blood clot, if any, of which it is a part) and transaction channel-end (for later agent-agent communications). Sites use the (clot) size and unique ID to pick the best candidate during the entry elections described in section 1.5.

In addition to platelet/clot information, the sites also store a clotting chemical factor level (obtained from passing factor processes), a unit vector (indicating the direction of blood flow) and a *blocking* flag (indicating whether the site is part of the blood vessel wall – in which case agents are denied entry).

Although using agents to simulate the wall would also be possible, we choose to implement it as a feature of space to save the memory overhead of having more agents (with very trivial behaviour).

Finally, each site has access to a *voxel* (a byte from a shared 3D-array), which it is responsible for maintaining. Whenever the site changes, it computes a transfer function over its state to set this voxel. The voxel itself is used to visualise the simulation via volume rendering techniques.

2.2. Platelets (Agents)

Our simulation agents model individual platelets in the blood. As in figures 3 and 4, platelets are pure clients and do not communicate directly with each other. However, they are clients to their clot super-agent and it is this that keeps them together. A platelet may be in one of two states:

> *non-sticky:* the platelet queries its local site and reports the blood-flow direction and clotting factor level to its super-agent. It then initiates any movement as instructed by the super-agent. The clot's size and unique ID are used to register presence in the sites.
>
> *sticky:* in addition to the above non-sticky behaviour, the platelet searches neighbouring sites for other sticky platelets, and passes their details to its super-agent.

Platelets, along with the chemical factor processes (section 2.3), move and update their environment. Together with the processes generating them and the processes controlling visualisation, they are enrolled and synchronise on the *discovery* and *modify* barriers – dividing the timeline into those respective phases (sections 1.4 and 2.5.1).

Note: for programming simplicity, *all* platelets in our current model have a clot process – even when they are not sticky or part of any clot. We may optimise those clot processes away later, introducing them only when a platelet becomes sticky. Most platelets in most simulations will not be sticky!

2.3. Clots (Super-agents)

Clots coordinate groups of platelets. They accumulate the blood-flow vectors from their platelets' sites and make a decision on the direction of movement. That decision also depends on the size of clots, with larger clots moving more slowly. They also change platelets from non-sticky to sticky if sufficient levels of clotting factor are encountered (these accumulate over many simulation steps).

When two or more clots encounter each other, if they contain sticky platelets they *may* become stuck together and merge. One of the clots takes over as super-agent for all sets of platelets in the bump group – the other clots terminate.

In [15], a clotting model in a one-dimensional blood stream was presented (as an illustration of mobile channels and barriers). In that system, deciding which clot process takes over is simple. Only two clots can ever be involved in a collision so, arbitrarilly, the one further upstream wins.

Stepping this model up to two dimensions, multiway collisions are possible since clots can be shaped with many leading edges in the direction of movement – for example, an "E"-shaped clot moving rightwards. Furthermore, those multiple collisions may be with just a single or many other clots. Fortunately, stepping this up to three dimensions does not introduce any further difficulties.

To resolve the decision as to which clot survives the collision, another *election* takes place involving direct communication between the clot super-agents. This is outside the client-server architecture shown in figure 3 (for whose reasoning this election is deemed to be a bounded internal computation). The clot processes must engage in nothing else during this election and that must terminate without deadlock. Reasoning about this can then be independent from reasoning about all other synchronisations in the system.

The trick is to order all the communications in a sequence that all parties know about in advance. Each clot has an ID number which is registered in all sites currently occupied by its constituent platelets. Each clot has had reported back to it, by its platelets, the clot IDs of all clots in the collision.

The platelets also place the client-end of a server channel to their clot in the site they are occupying. They report to their clot the client-ends of the other clots in the collision. Thus, each clot now has communication channels to all the other clots in its collision.

High number clots now initiate communication to low number clots. The lowest numbered clot is the winner and communicates back the election result, with communication now from low number clots to high. The choice that *low* numbered clots should win was not arbitrary. Clots are introduced into the world with increasing ID numbers, so having low number clots win means that low number clots will tend to amass platelets. In turn, this reduces the number of times those platelets need to change super-agent after collision. Although our algorithm for ordering communication (not fully outlined here) has yet to undergo formal proof, it has so far in practice proven reliable.

Platelets communicate with their clot using the shared client-end of a server bundle. By keeping track of the number of platelet processes it contains, a clot knows how many communications to expect in each phase (and, so, does not have to be enrolled in the barriers used by the platelets to define those phases). See section 2.5 for more details of clot and platelet communications.

2.4. Factors (Agents)

The second and final type of agent in our simulation is one that models the chemical factors released into the blood by a wounded (damaged) blood vessel. Since they move and modify their environment (the sites), they must engage on the same *discovery* and *modify* barriers as the platelets.

Factors are launched (forked) into the simulation with an initial vector pointing away from the wound and into the blood vessel. Every simulation step, the factor integrates a proportion of its current site's blood flow vector with its own vector and uses the result to determine its next move. The effect is cumulative so that eventually the factor is drawn along with the blood flow. At each site it enters, the factor increases the factor strength field, and modifies the site's blood flow vector to point back to the wound. The second of these two actions simulates both the slight pressure drop from an open wound and other biological mechanisms which draw platelets to open wounds.

Finally, it should be noted that factors are not considered to take up any space – being tiny molecules as opposed to full cells. Hence, many are allowed to occupy individual sites.

2.5. *Simulation Logic*

To provide more detail, here is some pseudo-code (loosely based on **occam-π** [1,2]) for the platelet and clot processes.

2.5.1. *Platelet Process*

Initially, a platelet is attached to its launch site, is not `sticky`, has a clot process to which only it belongs and has no knowledge of its neighbourhood (which it assumes is empty of platelets/clots). Platelets decide whether they want to move in the *discovery* phase; however, the movement is election based (section 1.5), and the result of the election is not queried until the *modify* phase. This means that although movement offers are made in the *discovery* phase, actual movement does not happen until the *modify* phase.

The *"channels"* `site`, `new.site` and `clot/clot.b`, used (illegally) in both directions below, represent SHARED client ends of channel bundles containing request and reply channels (flowing in opposite directions and carrying rich protocols). For further simplicity, the necessary CLAIM operations have also been omitted. They connect, respectively, to the current and (possible) future *site* locations of the platelet and the *clot* process of which it forms a part.

```
SEQ

  WHILE still in the modelled blood vessel
    SEQ

      SYNC discovery                   -- all platelets and factors wait here for each other

      site ! ask for local chemical factor level and motion vector
      site ? receive above information
      clot ! factor.vector.data; forward above information

      IF
        sticky
          SEQ
            site ! get clot presence on neighbour sites (in directions that were previously empty)
            site ? receive above information
            clot ! forward information only on clots different to our own (i.e. on clot collisions)
        TRUE
          SKIP

      -- clot decides either on transition to sticky state or merger of bumped clots

      clot.b ? CASE
        update; clot; clot.b      -- our clot has bumped and merged with others
          SKIP                    -- we may now belong to a different clot process
        become.sticky
          sticky := TRUE          -- accumulated chemical factors over threshold
        no.change
          SKIP

      -- clot decides which way, if any, to try and move
```

```
clot ? CASE

  no.move
    SYNC modify                        -- empty phase for us, in this case

  move; target
    SEQ
      site ! get.neighbour; target    -- get the channel end of the new site
      site ? new.site
      new.site ! enter; clot          -- offer to enter new site, giving our clot reference

      SYNC modify                     -- wait for all other offers to be made

      new.site ! did.we.enter; clot   -- ask if we were successful
      new.site ? CASE
        yes
          SEQ
            clot ! ok                  -- report ability to move
            clot.b ? CASE
              ok                       -- all platelets in clot can move
                SEQ
                  site ! leave         -- leave present site
                  site := new.site     -- commit to new site
              fail
                new.site ! leave       -- give up attempted move
        no
          SEQ
            clot ! fail                -- report failure to move
            clot.b ? CASE fail         -- clot cannot move as this platelet failed

SEQ                                    -- we have exited the modelled region of space
  SYNC discovery                       -- must get into the right phase for last report
  clot ! terminated
```

2.5.2. Clot Process

Initially, a clot is not `sticky` and starts with a platelet count (`n.platelets`) of 1. A clot runs
for as long as it has platelets. It does not need to engage in the *discovery* and *modify* barriers,
deducing those phases from the messages received from its component platelets. At the start
of each phase, a clot is `sticky` *if and only if* all its component platelets are `sticky`.

The *"channels"* `platelets`/`platelets.b` used (illegally) in both directions, represent the
server ends of two channel bundles containing request and reply channels (flowing in opposite
directions and protocol rich). They service communications from and to all its component
platelets (and are the opposite ends to the `clot`/`clot.b` channels shared by those platelets).

```
WHILE n.platelets > 0
  SEQ

    -- nothing will happen till the discovery phase starts
    -- we just wait for the reports from our platelets to arrive

    SEQ i = 0 FOR n.platelets
      platelets ? CASE
        factor.vector.data; local chemical factor level and motion vector
          ... accumulate chemical factor level and motion vector
        terminated
          n.platelets := n.platelets - 1
```

```
IF
  sticky
    SEQ
      SEQ i = 0 FOR n.platelets
        platelets ? report on any bumped clots
      IF
        sufficiently hard collision anywhere
          SEQ
            ... run clotting election to decide which clot takes over the merger
            SEQ i = 0 FOR n.platelets
              platelets.b ! update; winner; winner.b
            IF
              this.clot = winner
                ... update number of platelets to new size of clot
              TRUE
                n.platelets := 0      -- i.e. terminate
        TRUE
          SEQ i = 0 FOR n.platelets
            platelets.b ! no.change

  accumulated.chemical.factor > sticky.trigger.theshold
    SEQ
      sticky := TRUE
      SEQ i = 0 FOR n.platelets
        platelets.b ! become.sticky

  TRUE
    SEQ i = 0 FOR n.platelets
      platelets.b ! no.change

target := pick.best.move.if.any (n.platelets, motion.vector)

IF
  target = no.move
    SEQ
      SEQ i = 0 FOR n.platelets
        platelets ! no.move
      -- platelets synchronise on modify barrier

  TRUE
    SEQ
      SEQ i = 0 FOR n.platelets
        platelets ! move; target
      -- platelets synchronise on modify barrier

      all.confirm := TRUE
      SEQ i = 0 FOR n.platelets
        platelets ? CASE
          ok
            SKIP
          fail
            all.confirm := FALSE
      IF
        all.confirm
          SEQ i = 0 FOR n.platelets
            platelets.b ! ok
        TRUE
          SEQ i = 0 FOR n.platelets
            platelets.b ! fail
```

2.6. Spatial Initialisation

The simulated environment must be initialised before platelets are introduced. It needs to contain some form of bounding structure to represent the walls of the blood vessel and the vectors in the sites must direct platelets along the direction of blood flow.

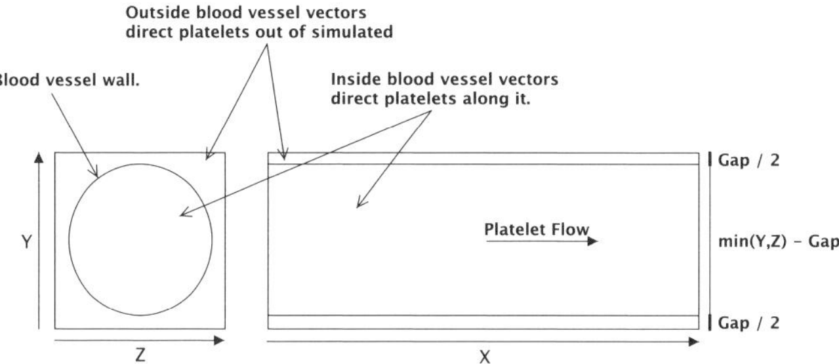

Figure 5. Layout of the simulated space in relation to blood vessel.

The blood vessel wall is placed so that it runs parallel to an axis in simulated space – the X-axis in our simulations (see figure 5). Our simulated blood vessel is simple: a cylinder with wall thickness of approximately two sites. The wall is simulated by setting the sites to which it belongs to *blocking*.

Force vectors inside the blood vessel are initialised so that there is a 55%[1] chance of moving forward along the blood vessel, an 6% chance of moving left or right, and an 8% chance of moving up or down. A given site vector can only point in one direction per axis, so the vectors point either left or right, and either up or down, e.g. left and down. The directions are select randomly per site, with an even distribution between each. Changing the initialisation of these vectors can give subtle changes in simulation behaviour – something left largely unexplored at this time.

The vectors outside the blood vessels are programmed to draw platelets to the edges of the simulated space and beyond. This enhances the blood loss effect when the vessel wall is broken. If this were not done, platelets would continue along much the same path just outside the blood vessel.

2.7. Optimisations

A few optimisations to our implementation were made to improve performance and memory usage.

Instead of giving each site an array of client-ends to neighbours, a single global array was used. This array is shared by all sites and significantly reduces memory requirement. This is safe as this connectivity information is static – we are not dealing with *worm-holes* and dynamic space topologies yet! occam-π does not yet have a language mechanism to enforce this read-only behaviour (of the connectivity) at compile time; but manual checking of our code is simple and deemed sufficient for our purposes here.

For performance enhancement, our implementation was designed so that platelets (agents) need only query their current site to discover the state of their local neighbourhood. This is accomplished in two stages. Firstly, site state data is placed into an array shared by all sites. This allows sites to retrieve data from their neighbours on behalf of an agent just by accessing (shared) memory. This is safe in our simulation because agent query and modification

---

[1] These are experimental values (not reflecting any biology).

are separated by barriers and individual updates to a site's state are serialised through that site's server interface. Secondly, agents now query their neighbourhood through their current site, passing it a *mobile* array of unit vectors and a *mobile* record. The site copies from the shared site state array the data for the specified vectors into the mobile memory, which it then returns along with its own state. Use of mobile data passed back and forth is very efficient and removes the need for dynamic memory allocation during normal agent run-time.

Our final optimisations were to reduce the neighbourhood searched by the agents. The previous optimisation reduced an individual search mainly to memory copies. As a first step, search is limited to (the obvious) six directions from the 26 available – although movement is permitted in any direction. When a platelet is part of a clot with other platelets, each platelet remembers the relative position of other platelets discovered around it and does not search those directions again. Futhermore, if a platelet becomes completely surrounded by platelets of the same clot, it terminates. For our simulation purposes, only the outline of clots need be maintained.

3. Support Processes

A small number of other processes complete the simulation and provide interaction and (3D) visualisation.

3.1. Platelet Generator

The platelet generator is a process that injects platelets at the *upstream* end of the blood vessel. It is enrolled on the *discovery* and *modify* barriers and restricts the injection (i.e. forking) of platelets to the *modify* phase (so that each platelet starts correctly synchronised, waiting for the *discovery* barrier). The platelet generator is programmed with a rate that can be varied at runtime. This rate (together with the cross-sectional area of the blood vessel) determines platelet density in the bloodstream. It sets a forward velocity (slightly randomised around an average of a 55% probability of movement).

At each simulation step, the number of platelets to be introduced is added to a running count; the truncated integer value of this count used to calculate the number of actual platelets to be forked. For each new platelet, two random numbers are generated: a Y and Z offset from the centre of the blood vessel. So long as these lie within the blood vessel, the platelet is injected at that position.

3.2. Wound Process

The wound process allows a user to punch a hole in the blood vessel wall. The *wound tool* is rendered as a sphere in the user interface and the user attacks the blood vessel with it. It creates a hole where there is an intersection between the sphere and the blood vessel walls. To do this, it uses the position of the sphere and its radius. If a point lies within the sphere, the corresponding site is tested to see if it is *blocking* (i.e. part of the blood vessel wall). If so, it is set to *unblocking* and four chemical factor processes are forked at its location (as a reaction to the damage). The initial movement vector of each factor process is initialised (with slight randomised jitter) so that it travels into the blood vessel.

3.3. Drawing Process

The drawing process has the task of informing the user interface when it is safe to render the voxel volume. It does this by signaling the user interface after the *discovery* barrier and before the *modify* barrier. When the user interface finishes rendering the volume, this process synchronises on the *modify* barrier. Using this sequence, the voxels are only rendered during the stable *discovery* phase, and the user interface stays in step synchronisation with the simulation. Rendering of only *one-in-n* simulation steps is implemented by a simple counter in this process.

3.4. User Interface and Visualisation

Our simulation architecture is not tied to any specific form of visualisation or interface. We have built simulations using 2D text and graphical interfaces; however, for our 3D blood clotting simulations we choose to employ the open source Visualisation Toolkit (VTK) from Kitware [20]. Binding foreign language routines into occam-π is straightforward [21].

VTK is an open source library written in C++, with Python, Tcl/Tk and Java wrappers. It has several hundred different classes and a selection of examples illustrating their use. However, the focus of this toolkit is on loading static content from files, not the visualisation of realtime simulations (known as *tracking*).

For our visualisations, VTK is employed as a *volume renderer*. This means we can directly visualise what is in effect a 3D array of pixels. Internally, the vtkVolumeTextureMapper2D class is used, which turns slices of the 3D volume into 2D textures that are rendered using OpenGL. This approach is much faster than ray tracing. Two transfer functions map the byte voxel data into colour and opacity before it is rendered. In theory, and there is evidence of its use in the field, modern 3D hardware could be programmed to do this mapping in real time, reducing CPU load and improving rendering times.

Also provided by VTK is a wealth of 3D interaction tools. In practice this means that VTK handles mouse input to manipulate the camera, and the user-controllable sphere used to project wounds onto the blood vessel. Input event handlers are registered so that interaction events, including key strokes, are recorded in an overwriting ring buffer from which the occam-π user interface process can access them.

4. Results and Further Work

4.1. Emergent Behaviour

Using the architecture and simple processes and behaviours described, we have been able to achieve results surprisingly similar to those in the human body. Given the *right* concentration of platelets (figure 6), wounds to our simulated blood vessel (figures 7 and 8) triggers the formation of clots (figure 9) that eventually form a *plug* covering the wound and preventing further blood loss (figure 10). Too low a concentration and the clotting response is too weak to let sufficiently large clots form. Too high a concentration and a clot forms too early, gets stuck in the blood vessel *before* the wound and fails to seal it. The clot also gets bigger and bigger until it completely blocks all blood flow – which cannot be too the good!

The concentration boundaries within which successful sealing of a wound is observed are artifacts of the current simulation model, i.e. they do not necessarily correspond with the biology. However, the fact that this region exists for our models gives us encouragement that they are beginning to reflect some reality.

In the human blood stream, clotting stimulation (and inhibition, which we have not yet modelled but is certainly needed) involves many different chemical factors, cell types (there are different types of platelet) and proteins (e.g. fibrinogen). It is encouraging that our modelling techniques have achieved some realistic results from such a simple model.

The clotting response we observe from our model has been engineered, but not explicitly programmed. The platelets are not programmed to spot wounds and act accordingly. They are programmed only to move with the flow of blood, become sticky on encountering certain

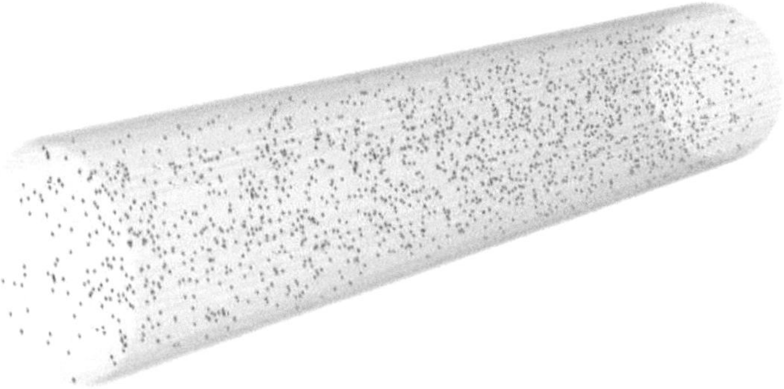

Figure 6. Simulated blood vessel represented by the cylinder, dots are platelets.

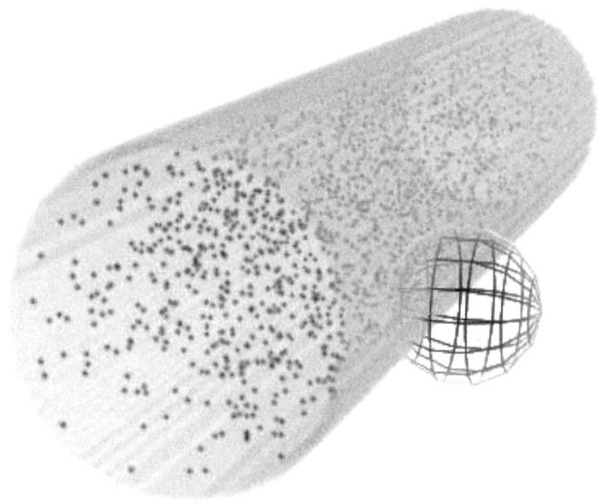

Figure 7. Simulation viewed from different angle, with wound placement tool on right.

levels of chemical and, then, clump together when they bump. Refining this so that greater and greater levels of realism emerge should be possible through the addition of processes modelling different stimulators and inhibitors of the clotting reaction, along with different platelet types and other participating agents. Because of the compositional semantics of CSP and occam-π, such refinement will not intefere with existing behaviours in ways that surprise – but should evolve to increase the stability, speed, accuracy and safety of the platelets' response to injury.

4.2. Performance

Our process oriented model implemented in occam-π has proved stable and scalable. Simulations have been run with with more than 3,000,000 processes on commodity desktop hardware (P4, 3.0Ghz, 1GB RAM). Memory places a limit on the size of our simulations. However, as our site processes only become scheduled when directly involved in the simulation, the available processing power only limits the number of active agents. Bloodstream platelet densities of up to 2% (an upper limit in healthy humans) imply up an average of around 60,000 agents – actual numbers will be changing all the time. Cycling each with an average processing time of 2 microseconds (including barrier synchronisation, channel communica-

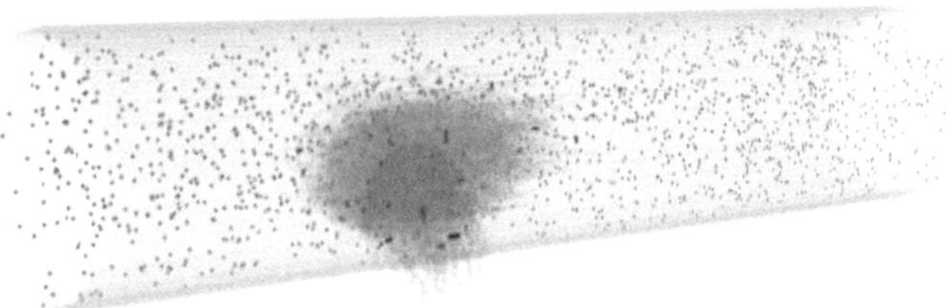

Figure 8. Having placed a wound, platelets "fall" out of the blood vessel, and chemical factors can be visualised by the darkened area.

Figure 9. Given time, chemical factors flow down the blood vessel and (small) clots can be seen forming as dark blobs.

tion and cache miss overheads) still enables around 8 simuations steps per second, which is very useable.

Figure 11 shows performance for simulations on a world of size 256x96x96 (2.3M+ sites). The different curves are for different levels of platelet concentration (0.5%, 1.0% and 2.0%). The x-axis shows simulation step numbers (*generations*), starting from an (unrealistic) bloodstream devoid of any platelets – but with them starting to arrive from upstream. Performance does not stablise until the blood vessel is filled with platelets, which takes 500 generations. This is as expected, given a volume 256 sites in length and with a roughly even chance of any platelet moving forwards. At 0.5% platelet concentration (an average of approximately 5,000 agents), we are achieving around 13 simulation/steps a second. All these results have visualisation disabled; in practice, most commodity graphics hardware has difficult rendering simulations this size at rates greater than 10 frames per second. As the number of agents doubles to 1.0%, and then 2.0%, performance degrades linearly. Again, this is expected, given that the computation load has doubled and that occam-π process management overheads are independent of the number of processes being managed.

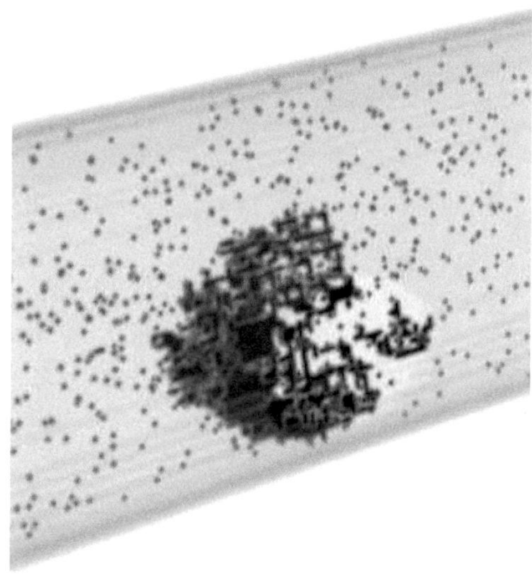

Figure 10. With sufficient time and a high enough platelet concentration a clot forms over the wound.

For the simulations whose results are shown in Figure 12, the platelets and their associated clots are initialised sticky. This is the worst case (and unrealistic) scenario where clots will form whenever two platelets collide. As expected, performance is lower than that in Figure 11, because the there are more agents. As clots form, they slow down. This means that platelets leave the simulation at a lower rate than they are entering and numbers rise. Even then, performance rates stabilise given sufficient time and the relationship between the levels of platelets is consistent.

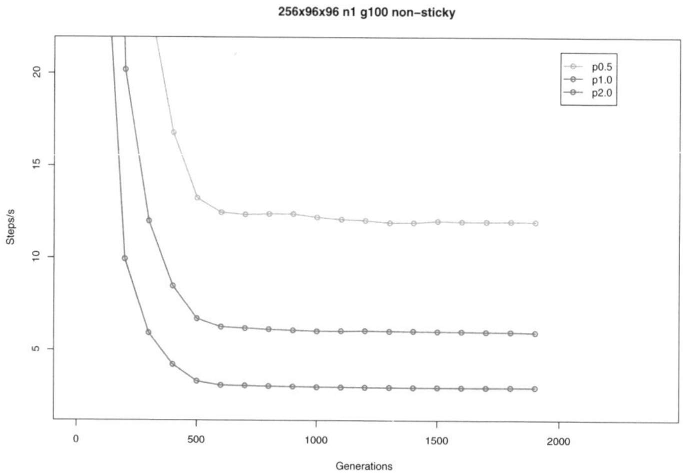

Figure 11. 256x96x96 simulations with non-sticky platelets.

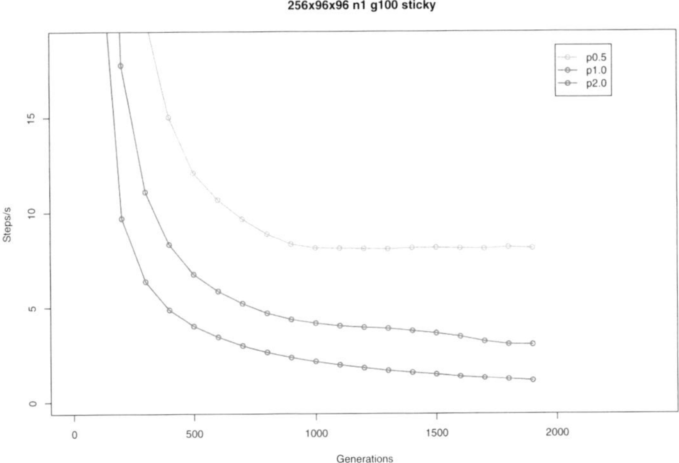

Figure 12. 256x96x96 simulations with sticky platelets.

4.3. Future Work

The next steps in our research are to expand and refine our simulations. For the former, we need to use either more powerful single machines or, more sensibly, clusters of machines. The later will be possible using pOny [22], an networking environment for the occam-π runtime system. We have begun tesing a cluster-based implementation of these simulation models and initial results, not published here, are quite promising.

For refining the accuracy of the model, we would like to achieve the return of our simulated blood vessel to a *normal* state once blood loss through a wound has been stemmed. We need to introduce factors that inhibit the production of further clots and *bust* existing ones (e.g. all those little ones that were washed away by the bloodstream before they could clump to the wound). So long as the wound is open, chenical factors would continue to be released, gradually lowering as the wound is closed. Inhibitor agents would also reduce clotting factor levels and correct blood flow vectors. The blood vessel wall also needs to be able to reform under the protective clot. Eventually, with the wound healed, the clot would dissipate and the factors that caused it would disappear.

Further refinement could be explored by integrating aspects of other research, both physical and simulated, into the flow of platelets within the blood stream [23]. In order to model these properties we will need to introduce aspects of fluid dynamics into our model, and allow our simulated clots to roll and sheer. By removing the rigid movement constraints on platelets within a clot and giving them a degree of individual freedom, the introduction of these new behaviours should be attainable. For example, by adding and appropriate vector (changing with time) to each of the platelets within a clot, the clot as a whole could be made to roll or tumble as it moves through the blood vessel.

Finally, we believe that the massively concurrent process-oriented architecture, outlined in this paper for this simulation framework, can be applied generically to many (or most) kinds of complex system modelling. We believe that the ideas and mechanisms are natural, easy to apply and reason about, maintainable through refinement (where the cost of change is proportional to the size of that change, not the size of the system being changed) and can be targetted efficiently to modern hardware platforms. We invite others to try.

References

[1] P.H. Welch and F.R.M. Barnes. Communicating mobile processes: introducing occam-pi. In A.E. Abdallah, C.B. Jones, and J.W. Sanders, editors, *25 Years of CSP*, volume 3525 of *Lecture Notes in Computer Science*, pages 175–210. Springer Verlag, April 2005.

[2] The occam-pi programming language, June 2006. Available at: http://www.occam-pi.org/.

[3] S. Stepney, P.H. Welch, F.A.C. Pollack, J.C.P. Woodcock, S. Schneider, H.E. Treharne, and A.L.C. Cavalcanti. TUNA: Theory Underpinning Nanotech Assemblers (Feasibility Study), January 2005. EPSRC grant EP/C516966/1. Available from: http://www.cs.york.ac.uk/nature/tuna/index.htm.

[4] P.H. Welch, F.R.M. Barnes, and F.A.C. Polack. Communicating Complex Systems. In Michael G. Hinchey, editor, *Proceedings of the 11th. IEEE International Conference on Engineering of Complex Computer Systems (ICECCS-2006)*, pages 107–117, Stanford, California, August 2006. IEEE. ISBN: 0-7695-2530-X.

[5] S. Schneider, A. Cavalcanti, H. Treharne, and J. Woodcock. A Layered Behavioural Model of Platelets. In Michael G. Hinchey, editor, *ICECCS-2006*, pages 98–106, Stanford, California, August 2006. IEEE.

[6] S. Stepney, H.R. Turner, and F.A.C. Polack. Engineering Emergence *(Keynote Talk)*. In Michael G. Hinchey, editor, *ICECCS-2006*, pages 89–97, Stanford, California, August 2006. IEEE.

[7] F. Polack, S. Stepney, H. Turner, P.H. Welch, and F.R.M.Barnes. An Architecture for Modelling Emergence in CA-Like Systems. In Mathieu S. Capcarrère, Alex Alves Freitas, Peter J. Bentley, Colin G. Johnson, and Jon Timmis, editors, *Advances in Artificial Life, 8th European Conference on Artificial Life (ECAL 2005)*, volume 3630 of *Lecture Notes in Computer Science*, pages 433–442, Canterbury, UK, September 2005. Springer. ISBN: 3-540-28848-1.

[8] C. Ritson and P.H.Welch. TUNA: 3D Blood Clotting, 2006. https://www.cs.kent.ac.uk/research/groups/sys/wiki/3D_Blood_Clotting/.

[9] A.T. Sampson. TUNA Demos, January 2005. Available at: https://www.cs.kent.ac.uk/research/groups/sys/wiki/TUNADemos/.

[10] P.H. Welch, G.R.R. Justo, and C.J. Willcock. Higher-Level Paradigms for Deadlock-Free High-Performance Systems. In R. Grebe, J. Hektor, S.C. Hilton, M.R. Jane, and P.H. Welch, editors, *Transputer Applications and Systems '93*, volume 2, pages 981–1004, Aachen, Germany, September 1993. IOS Press, Netherlands. ISBN 90-5199-140-1. See also: http://www.cs.kent.ac.uk/pubs/1993/279.

[11] J.M.R. Martin and P.H.Welch. A Design Strategy for Deadlock-free Concurrent Systems. *Transputer Communications*, 3(4):215–232, October 1996.

[12] F.R.M. Barnes and P.H. Welch. Prioritised dynamic communicating and mobile processes. *IEE Proceedings – Software*, 150(2):121–136, April 2003.

[13] R. Milner, J. Parrow, and D. Walker. A Calculus of Mobile Processes – parts I and II. *Journal of Information and Computation*, 100:1–77, 1992. Available as technical report: ECS-LFCS-89-85/86, University of Edinburgh, UK.

[14] F.R.M. Barnes, P.H. Welch, and A.T. Sampson. Barrier synchronisations for occam-pi. In Hamid R. Arabnia, editor, *Parallel and Distributed Processing Techniques and Applications – 2005*, pages 173–179, Las Vegas, Nevada, USA, June 2005. CSREA press. ISBN: 1-932415-58-0.

[15] P.H. Welch and F.R.M. Barnes. Mobile Barriers for occam-pi: Semantics, Implementation and Application. In J.F. Broenink, H.W. Roebbers, J.P.E. Sunter, P.H. Welch, and D.C. Wood, editors, *Communicating Process Architectures 2005*, volume 63 of *Concurrent Systems Engineering Series*, pages 289–316. IOS Press, September 2005. ISBN: 1-58603-561-4.

[16] Hemostatis. URL: http://en.wikipedia.org/wiki/Haemostatis.

[17] Fibrin. URL: http://en.wikipedia.org/wiki/Fibrin.

[18] Disorders of Coagulation and Haemostasis. Available at: http://www.surgical-tutor.org.uk/default-home.htm?core/preop2/clotting.%htm.

[19] J. Griffin, S. Arif, and A. Mufti. *Immunology and Haematology (Crash Course) 2nd Edition*. C.V. Mosby, July 2003. ISBN: 0-7234-3292-9.

[20] W. Schroeder, K. Martin, and B. Lorensen. *The Visualisation ToolKit*. Kitware, 2002.

[21] D.J. Dimmich and C.L. Jacobsen. A Foreign Function Interface Generator for occam-pi. In J.F. Broenink et al., editor, *Communicating Process Architectures 2005*, volume 63 of *Concurrent Systems Engineering Series*, pages 235–248. IOS Press, September 2005. ISBN: 1-58603-561-4.

[22] M. Schweigler and A.T. Sampson. pony - The occam-pi Network Environment. In *Communicating Process Architectures 2006*, Amsterdam, The Netherlands, September 2006. IOS Press.

[23] I.V. Pivkin, P.D. Richardson, and G. Karniadakis. Blood flow velocity effects and role of activation delay time on growth and form of platelet thrombi. *Proceedings of the National Academy of Science*, 103(46):17164–17169, October 2006.

Communicating Process Architectures 2007
Alistair McEwan, Steve Schneider, Wilson Ifill and Peter Welch (Eds.)
IOS Press, 2007

267

Concurrency Control and Recovery Management for Open e-Business Transactions

Amir R. RAZAVI, Sotiris K. MOSCHOYIANNIS and Paul J. KRAUSE
Department of Computing, School of Electronics and Physical Sciences,
University of Surrey, Guildford, Surrey, GU2 7XH, UK.
{a.razavi, s.moschoyiannis, p.krause}@surrey.ac.uk

Abstract. Concurrency control mechanisms such as turn-taking, locking, serialization, transactional locking mechanism, and operational transformation try to provide data consistency when concurrent activities are permitted in a reactive system. Locks are typically used in transactional models for assurance of data consistency and integrity in a concurrent environment. In addition, recovery management is used to preserve atomicity and durability in transaction models. Unfortunately, conventional lock mechanisms severely (and intentionally) limit concurrency in a transactional environment. Such lock mechanisms also limit recovery capabilities. Finally, existing recovery mechanisms themselves afford a considerable overhead to concurrency. This paper describes a new transaction model that supports release of early results inside and outside of a transaction, decreasing the severe limitations of conventional lock mechanisms, yet still warranties consistency and recoverability of released resources (results). This is achieved through use of a more flexible locking mechanism and by using two types of consistency graph. This provides an integrated solution for transaction management, recovery management and concurrency control. We argue that these are necessary features for management of long-term transactions within "digital ecosystems" of small to medium enterprises.

Keywords. concurrency control, recovery management, lock mechanism, compensation, long-term transactions, service-oriented architecture, consistency, recoverability, partial results, data dependency, conditional-commit, local coordination, business transactions.

Introduction

This paper focuses on support for long-term transactions involving collaborations of small enterprises within a Digital Business Ecosystem [1]. Although there is significant current work on support for business transactions, we argue that this all rely on central coordination that provides unnecessary (and possibly threatening) governance over a community of collaborating enterprises. To address this, we offer an alternative transaction model that respects the local autonomy of the participants. This paper focuses on the basic transactional model in order to highlight the concurrency issues that are inherent in these kinds of reactive systems. Formal analysis of this model is in hand, and first results are reported in [2].

The conventional definition of a transaction [3] ACID properties: *Atomicity* – either all tasks in a transaction are performed, or none of them are; *Consistency* – data is in a consistent state when the transaction begins, and when it ends; *Isolation* – all operations in a transaction are isolated from operations outside the transaction; *Durability* – upon successful completion, the result of the transaction will persist.

Several concurrency control mechanisms are available for maintaining consistency of data items such as: turn-taking [4], locking [5], serialization [6], transactional locking mechanism [7-8], and operational transformation [9]. Lock mechanisms, as a widely used method for concurrency control in transaction models [8], provide enough isolation on modified data items (Exclusive lock) to ensure there is no access to any of these data items before a transaction that is accessing or updating them commits [8]. The constraint of atomicity requires that a transaction either fully succeeds, or some recovery management process is in place to ensure that all the data items being operated on in the transaction return to their original state should the transaction fail at any point prior to commitment. Recovery management must use an original copy of the unmodified data to ensure the possibility of recovering the system to a consistent check point (before running the faulty transaction). Recovery management may also use a log system (which works in parallel with the lock mechanism of the concurrency control), to support reversing, or rolling back, the actions of a transaction following failure. However, as we will discuss, if these properties are strictly adhered to in business transactions, they can present unacceptable limitations and reduce performance [10]. In order to design a transaction model suitable for Digital Business Ecosystems, we will focus on three specific requirements which cause problems for conventional transaction models [10-13] *long-term transactions* (also called long-running or long-life transactions); *lack of partial results*; and *omitted results*.

Within the Digital Business Ecosystem (DBE) project [1], the term "Digital Business Ecosystem" is used at a variety of levels. It can refer to the run-time environment that supports deployment and execution of e-services. It can include the "factory" for developing and evolving services. But most importantly it can be expanded to include the enterprises and community that uses the ecosystem for publishing and consuming services. It is this last that is the most important driver for the underlying technology, since it is the ability to support a healthy and diverse socio-economic ecosystem that is the primary "business goal" of the project. From that comes a specific focus on supporting and enabling e-commerce with Small and Medium-sized Enterprises – contributors of over 50% of the EU GDP. The DBE environment, as a service oriented business environment, tries to facilitate business activities for SMEs in a loosely coupled manner without relying on a centralized provider. In this way, SMEs can provide services and initiate business transactions directly with each other. The environment is highly dynamic and service relatively frequent unavailability and/or change of SME providers is to be expected. Therefore we can anticipate these necessary attributes in such environment:

- *Long-term transactions:* a high range of B2B transactions (business activities [14], [15] or business transactions), has a long execution time period. Strictly adhering to ACID properties for such transactions can be highly problematic and can reduce concurrency dramatically. The application of a traditional lock system (as the concurrency control mechanism [3], [8]) for ensuring Isolation (or capturing some version of serializability[3], [8]), reduces concurrency and the general performance of the whole system (many transactions may have to wait for a long-term transaction to commit and release its resources or results). As a side effect, the probability for deadlock is also increased since long-term holding of locks directly increases the possibility for deadlock. Furthermore, the lack of centralized control in a distributed transactional environment such as DBE hinders the effective application of a deadlock correction algorithm.

- *Partial results:* according to business semantics; releasing results from one transaction to another before commitment is another challenge in the DBE environment. According to conventional transaction models, releasing results before transactions commit is not legal, as it can misdirect the system to an inconsistent state should the transaction be aborted before final commit. Allowing for partial results and (ensuring) consistency are two aspects that clearly cannot fit within a conventional lock mechanism for concurrency control and log- or shadow-based recovery management [3], [15]. A wide range of business scenarios, however, demand partial results in specific circumstances. Therefore we need to reconsider this primary limitation but at the same time also provide consistency for the system.

- *Recoverability and failures:* in the dynamic environment of distributed business transactions, there is a high probability for failure due to the temporary unavailability of a particular service. Thus, recoverability of transactions is important. Recovering the system in the event of failure or abortion of a transaction needs to be addressed in a way that takes into account the loosely-coupled manner of connections. This makes a recoverability mechanism in this context even more challenging. As we can not interfere with the local state of the underlying services, the recovery has to be done at the deployment level [16],[17] and service realization (which includes the state of a service) has to be hidden during recovery. This is a point which current transactional models often fail to address as will be further discusses in the sequel.

- *Diversity and alternative scenarios:* by integrating SMEs, the DBE provides a rather diverse environment for business transactions. The provision for diversity has been discussed in the literature for service composition [16], [17], [2], [18]. When considered at the transaction model and/or business processes, it can provide a unique opportunity in not only covering a wider range of business processes but also in designing a corresponding recovery system [2], [18]. In conventional concurrency control and recovery management there is no technical consideration of using diversity for improving performance and reliability of the transactions.

- *Omitted results:* one point of criticism for recovery systems often has to do with wasting intermediate results during the restart of a transaction after accruing a failure. The open question here is how much of these results can be saved (i.e. not being rolled back) and how. In other words, how we can preserve as much progress–to-date as possible. Raising to this challenge within a highly dynamic environment such as DBE can have significant direct benefits for SMEs in terms of saving time and resources.

Similar Approaches for Business Environment:

In 2001, a consortium of companies including Oracle, Sun Microsystems, Choreology Ltd, Hewlett-Packard Co., IPNet, SeeBeyond Inc. Sybase, Interwoven Inc., Systinet and BEA System, began work on the Organization for Advance Structured Information Systems (OASIS) Business Transaction Protocol (BTP), which was aimed at B2B transactions in loosely-coupled domains such as Web services. By April 2002 it had reached the point of a committee specification [19].

At the same time, others in the industry, including Microsoft, Hitachi, IBM, IONA, Arjuna Technologies and BEA Systems, released their own specifications: Web Services Coordination (WS-Coordination) and Web Services Transactions (WS-AtomicTransactions and WS-BusinessActivity) [20], [14]. Recently, Choreology Ltd. has started to make a joint protocol which tries to cover both models and this effort has highlighted the caveats of each as mentioned in [15].

The coordination mechanism of these well-known transaction models for web-services, namely BTP and WS-BusinessActivity, is based on WS-Coordination [21]. A study of this coordination framework however, reported in [22], shows it to suffer from some critical decisions about the internal build-up of the communicating parties; a view also supported in [23].

The Coordinator and Initiator roles are tightly-coupled and the Participant contains both business and transaction logic. These presumptions are against the primary requirements of SOA, particularly loose-coupling of services and local autonomy, and thus are not suitable for a digital business ecosystem, especially when SMEs are involved.

A further concern has to do with the compensation mechanism. Behavioural patterns such as "validate-do" and "provisional-final" [23], [2], [15] are not supported while the "do-compensate" pattern, which is supported, results in a violation of local autonomy, since access to the service realisation level is required (see [22] for further details). Prescribing internal behaviour at the realisation level raises barriers for SMEs as it inevitably leads to their tight-coupling with the Coordinator.

In previous work [2], [18], [15] we have been concerned with a distributed transaction model for digital business ecosystems. We have shown how a thorough understanding of the transaction behaviour, before run-time, can ease the adoption of behaviour patterns and compensation routines necessary to prevent unexpected behaviour (but without breaking local autonomy). In this paper, we are present a lock system that provides concurrency control, for data items within and between DBE transactions. Further, with the local autonomy of the coordinators in mind, we introduce two additional locks, an internal and a conditional-commit lock, which allow for exchange of data both inside and across transactions. We show how the lock system together with the logs generated by the transaction model can provide full consistency and ultimately lead to automation in this model.

In the next section, we provide an overview of our primary log system, which has been introduced in [2]. In Section 3 we describe a mechanism for releasing uncommitted results between subtransactions of a transaction. Section 4 is concerned with the issue of releasing partial results between transactions (to the outside world). Section 5 recapitulates our concurrency model for a full recovery mechanism. The issue of omitted results is addressed in Section 6 which also describes a forward recovery mechanism. The paper finishes with some concluding remarks and a discussion on future extensions of this work.

1. Log System and Provided Graphs for Recoverability

We have seen that in our approach [2] transactions are understood as pertaining to SOC [17] for B2B interactions. Hence, a transaction has structure, comprising a number of subtransactions which need to be coordinated accordingly (and locally), and execution is long-term in nature.

In order to relax the ACID properties, particularly Atomicity and Isolation without compromising Consistency, we need to consider some additional structure that will warranty the consistency of the transaction model. Maintaining consistency is critically important within a highly dynamic and purely distributed environment of a Digital

Ecosystem. To reach this aim, we categorize the solution in two stages: providing recoverability and consistency by introducing a transaction model.

In our approach, a transaction is represented by a tree structure. Each node is either a coordinator (a composition type) or a basic service (a leaf). Five different coordinator types are considered, drawing on [16], [2], [18], [15] that allow for various forms of service composition to be expressed in our model.

1.1 Local Coordinators

At the heart of this transactional model are the local coordinators. They have to handle the complexities of the model and control/generate all logs. At the same time, they should have enough flexibility for handling the low bandwidth (and low processing power) limitations from some nodes in the network.

Based on different types of compositions [16], we use different type of coordinators. Therefore a transaction will split to a nested group of sub-transactions with a tree structure (nested transaction model). The root of this tree is the main composition, which is a coordinator and each sub-transaction is either a coordinator or a simple service (in the leaf). There are five different coordinator types plus delegation coordination for handling delegation:

- *Data oriented coordinator:* This coordinator is specifically working on data oriented service composition; including fully atomic and simple service oriented which is dealing with released data item inside of a transaction or using partial results, released by other transactions.

- *Sequential process oriented coordinator:* This coordinator is invoking its sub-transactions (services) sequentially. The execution of a sub-transaction is dependent on its previous service, i.e., one cannot begin unless the previous sub-transaction commits. In fact this coordinator handles Sequential process oriented service composition by covering both Sequential with commitment dependency (SCD) and Sequential with data dependency (SDD).

- *Parallel process oriented coordinator:* In the Parallel oriented coordinator all the sub-transaction (component services) can be executed in parallel but different scenarios can be considered which can make different situations (implementations) in the transactional outlook which covers; Parallel with data dependency (PDD), Parallel with commit dependency (PCD) and Parallel without dependency (PND).

- *Sequential alternative coordinator:* This coordinator indicates that there are alternative sub-transactions (services) to be combined, and they are ordered based on some criterion (e.g., cost, time, etc). They will be attempted in succession until one sub-transaction (service) produces the desired outcome. In fact it is for supporting Sequential alternative composition (SAt) and it may use dynamically for forward recovery.

- *Parallel alternative coordinator:* Unlike the previous coordinator, alternative sub-transactions (services) are pursued in parallel. As soon as any one of the sub-transaction (service) succeeds the other parallel sub-transactions are aborted (as it has clear, this coordinator rely on reliable compensation mechanism). Actually the Parallel alternative coordinator handles Parallel alternative composition (PAt).

- *Delegation coordinator:* The whole transaction or a sub transaction can be delegated to another platform; delegation can be by sending request specification or service(s) description. Figure 1 shows the DBE transaction model structure [13],[2],[16].

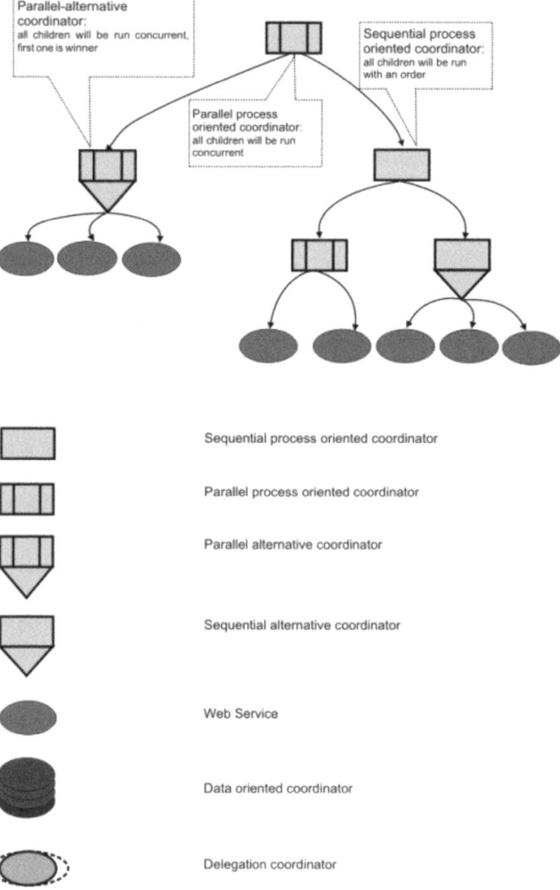

Figure 1. Transaction model structure

1.2 Internal Dependency Graph (IDG)

Two different graphs are introduced for: keeping track of data (value) dependencies; providing support for reversing actions; supporting a deadlock control mechanism; and, transparency during delegation. These graphs provide important system logs, which are stored locally on a coordinator and will be effected locally (in terms of local faults, forward recovery and contingencies plan) and globally (abortion, restart etc).

The *Internal Dependency Graph* (IDG) is a directed graph in which each node represents a coordinator and the direction shows the dependency between two nodes. Its purpose is to keep logs on value dependencies in a transaction tree. In further explanation, when a coordinator wants to use a data item belonging to another coordinator, two nodes have to be created in the IDG (if they do not already exist) and an edge generated between them (the direction of which shows the dependency between the two coordinators). Figure 2 shows an example of a SDD coordinator when IDSi releases data item(s) to IDSi+1 and IDSi+1 releases data items to IDSi+2. This means that IDSi+1 is dependent on IDSi+1 and IDSi+1 is dependent on IDSi (on the other hand if some failure happen for IDSi, the coordinator by traversing the graph knows who used the results from IDSi which are not consistent anymore).

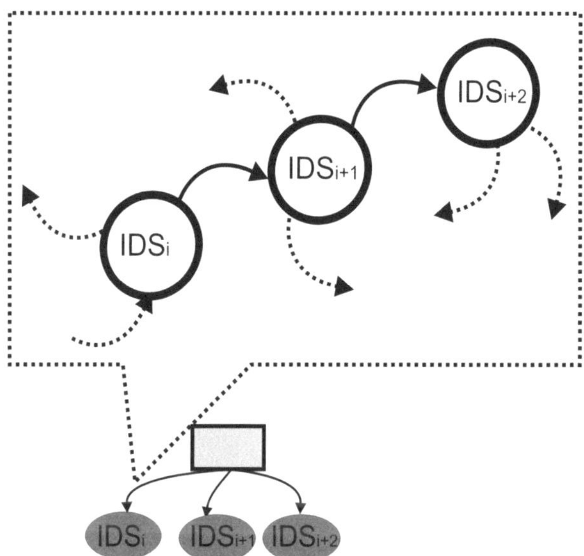

Figure 2. Sequential Data Dependency Coordinator and Associated Internal Dependency Graph

1.3 Conditional Commit, External Dependency Graph (EDG)

When a subtransaction needs to access a released data item which belongs to another DBE transaction this dependency is shown by creating a directed link between these two nodes from the owner to the user of that data item. As an example, Figure 3 shows the release of partial results from two subtransactions of IDHC1 to IDHC2. As shown in the figure the two nodes appear linked in the corresponding EDG – notice the direction is towards the consumer of data thus indicating this data item usage.

If each of these nodes are absent in EDG, they must be added and if nodes and a connection between them already exist, there is no need for repetition. The most important usage of this graph is in the creation of compensatory transactions during a failure.

By using the IDG and EDG, we have provided a framework which shows dependencies between coordinators and the order of execution in the transaction tree. This gives a foundation for recoverability. But the internal structure of the local coordinator (local coordination framework) still is not explained and the feasibility of the model relies on it. The IDG and EDG can support provision of a routine for recovering the system in a global view but they show neither the internal behaviour of a coordinator, nor the automated routines of each coordinator for avoiding the propagation of failure or clarifying support for loosely coupled behaviour patterns.

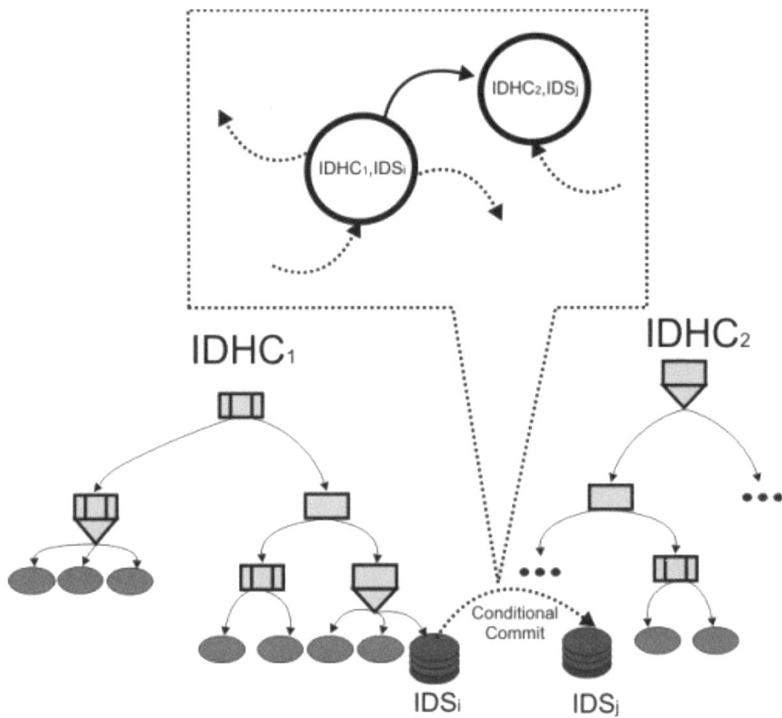

Figure 3. EDG for releasing partial results

We may ask these questions: how can a coordinator release deployed data items to other coordinators of the same transaction, and which safeguards/procedures should be considered in concurrency control (for example on SDD or PDD coordinators)? How, when and based on which safeguards (which lock mechanism on the concurrency control), can deployed data items be released to a coordinator of another transaction (Partial results) and which internal structure will support this procedure? How will failure and abortion of a transaction be managed internally in a coordinator and how can complete failure be minimized and recovery automated? The next section provides answers to these questions.

2. Releasing Data Inside a Transaction

Implementing locks as a conventional mechanism in concurrency control provides a practical mechanism for preserving consistency while allowing for (restricted) concurrency in the transactional system. However, the traditional two-phase S/X lock model does not give permission for releasing data items before a transaction commits. Based on this model [3], [8], once a Shared lock (S_Lock) is converted to an Exclusive lock (X_Lock), the respective data item can only be accessed by the owner of the lock (who triggered the transition from S_Lock to X_Lock). In this way subtransactions cannot share their modified data items between (as they have been locked by Exclusive lock and can not be released before the transaction commits). In contrast, in our approach deployed data items are made available to other subtransactions of the same transaction (by using the corresponding IDG). We introduce an Internal lock (I_Lock) which in combination with the IDG can provide a convenient practical mechanism for releasing data inside a transaction.

When a subtransaction needs to release some results before commitment, it will use the I_Lock as a relaxed version of X_Lock. This has the effect that other subtransactions can use these results by adding an entry to the IDG. For example, in a parallel coordinator each child not only can use S_Lock and X_Lock but also it can convert an X_Lock to I_Lock and release that data item for the other children of the coordinator (applying data dependency). This means that the other subtransactions can (provisionally) read/modify this data item, as well as the owner/generator of the data item. In comparison with the conventional usage of X_Lock, which decreases concurrency dramatically since it isolates deployed data items, I_lock not only supports a higher level of collaboration inside a transaction, but also allows more concurrent subtransactions to be executed and their results shared. It also provides a proper structure for any compensation during possible failures, as will be discussed in Section 4.

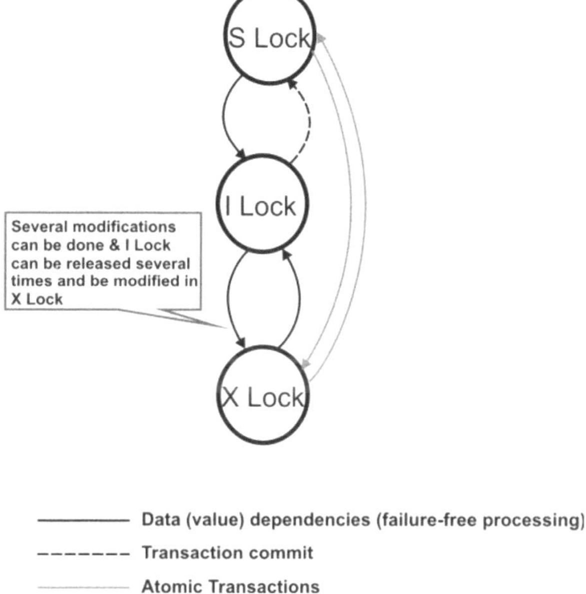

Figure 4. Internal lock (I_Lock) schema

2.1 I_Lock as Mechanism for Releasing the Un-committed Results Inside a Transaction

The use of I_Lock (Figure 4.) allows for the generation of new logs, which could be used in creating/updating the corresponding IDG [2], [15]. The necessary information from the owner of each I_Lock is the unique identifier of the main transaction (IDT), the identifier of the parent (parallel coordinator), IDSh, and the identifier of the subtransaction (IDS). When another subtransaction needs to access a data item, a validation process will compare IDSh with the parallel scheduler of the subtransaction.

In the sequential coordinator with data dependency (SDD), I_Lock is again used for data access with a similar method. When each child modifies any data item, it uses X_Lock on it and after committing the child (subtransaction), X_Lock will be converted to I_Lock. Remember that only subtransactions (children) with the same parent id can access that data item and modify it.

In the case of a value dependency, a data item will be released when converting the X_Lock to I_Lock. This means the other children of the same parent (the I_Lock owner) can use the data item. The combination of I_Lock and the IDG shows the chain of dependencies between different coordinators. In the final section of this paper, we will discuss a possible algorithm that this enables us to design for the combination for deadlock detection and correction. In Figure 4 we show the schema for I_Lock conversion from/to X_Lock and final commit of the transaction. Converting the lock to S_lock provides the possibility to share the result or return it to the initiator of the transaction.

3. Partial Results

One of the novel aspects of our transaction model [2] for DBEs has to do with the release of partial results. These are results which are released (to some other transactions) during a long-term transaction before the transaction commits (conditional-commit). This requires a mechanism for concurrency control and recovery management to be designed to maintain the integrity and consistency of all data.

3.1 Conditional Commit by Using C_Lock (after 1st Phase of Commit)

As we have seen in the previous chapter, I_Lock in collaboration with the IDG provides for the possibility of releasing data items to the other subtransactions of the same transaction. But another important problem concerns releasing results to other transactions. The inability to do this not only stops transactions from being executed concurrently, but also, according to the nature of business activities which may have a long duration or life time, can stop a wide range of transactions from reaching their targets.

Using a similar approach to that of introducing the internal lock I_Lock, we introduce a *conditional-commit* lock (C_Lock) which in collaboration with the EDG can provide a safe mechanism for releasing partial results to a subtransaction within another transaction. It works as follows. In the first step, a transaction can release its data item by using C_Lock on them (before commit). When a data item has C_Lock, that data item is available but some logs must be written during any usage of data (in the corresponding EDG). The released data item is from a data-oriented coordinator to the other data-oriented coordinator of another transaction. If a failure occurs, the compensating mechanism must be run. In this mechanism, transactions that used the released data item must run the same mechanism (rollback/abort).

In the process of conditional-commit a data item with X_Lock can be converted to C_Lock (or I_Lock for Internal data release) for releasing partial results. The necessary information from the owner of each C_Lock is the unique identification of a transaction (IDT) and the identification of the compensatory subtransaction (IDS). The combination of C_Lock lifecycle and EDG (and IDG) provide a practical mechanism for compensation (recoverability) which warranties the consistency of the system and will be discussed in the next chapter.

Figure 5, shows the lifecycle of C_Lock (without considering any failure. This is covered in the following section). In these circumstances, the final commit will be the final stage of C_Lock which triggers conversion of the C_Lock to an S_Lock. At that point, results can be returned to the initiator of the transaction and a signal can be sent to the other dependent transactions for permission to proceed with their commit.

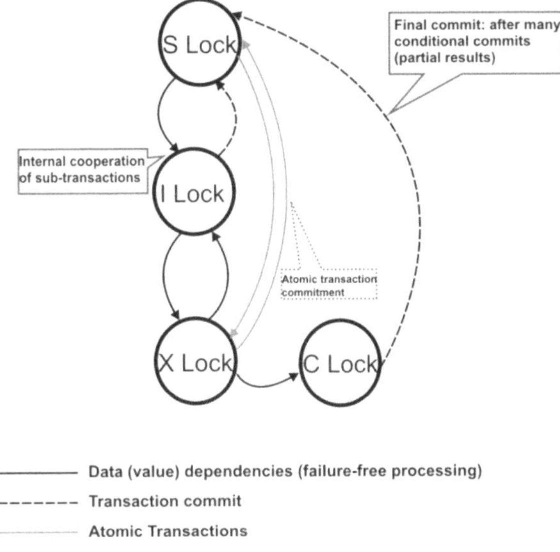

Figure 5. Conditional commit lock (C_Lock) schema

4. Recovery Management

Recovery management in a Digital Business Ecosystem has to deal with specific challenges which other similar models do not have to handle. One of the most important differences is the purely distributed nature of a DBE and the participation of SMEs. The (necessary) lack of a strong central point for managing the recovery procedure, forces the model towards a distributed algorithm which is supposed to not only handle but also predict failures. Localising recovery, guides us to delegate full responsibility to local coordinators. We start by considering loss of communication between two coordinators as the first sign of possible failure. Based on this presumption, we provide a mechanism for applying an effective policy for a local coordinator to rolling back the transaction effects on the environment.

The other challenge is the high speed of failure propagation, which can lead the system towards a completely inconsistent situation. By using an oracle from nature (cf Sections 4.2 and 4.3), we introduce a mechanism for limiting the side effects of a failure and apply the recovery procedure in two phases. Using distributed (and probably replicated) logs (provided by the IDG and EDG), gives more opportunity for generalising our mechanism.

The cost of the recovery and the range of waste during the procedure, was another motivation for applying an optimization mechanism (5) and trying to avoid a full system rollback, during the recovery procedure. There are two established methods for designing recovery management [3]: Shadow paging; and, Log-based. As the proposed model is a fully distributed model which can be widely generalised, shadow paging can not be considered as a suitable method because of the global overheads [24]. The structure of our model has a similarity with a log-based system with several features that make the method feasible for such a complex environment.

Two types of information are released before final commitment, which provide certain complexities for our recovery management. The first type is the release of results between subtransactions *within* a transaction. The second type is release of partial results between *different* transactions before their commitment. In order to support release of results within a transaction we have introduced an internal log with a graph structure that records the internal dependencies for a recovery routine when a failure occurs (drawn from the IDG). To support release of partial results (release of information *between* DBE transactions) we can use the other dependency log which is recording external dependencies (EDG).

The graph creation, the order of the recovery manager execution, and the routines for the release of results (both within and between transactions) has been analysed so far and, conventionally, it is considered to be the responsibility of the concurrency control mechanism. In contrast with the conventional methodology, one of the DBE necessities (given the dynamicity and unpredictable nature of the environment) was to merge these two and as we explain in this section, our design reflects this fact.

4.1 Fully Isolated Recovery and Using R-lock

The nature of business activities and long-term transactions infers that considering the recovery system as a practical mechanism directly attached to the transaction, leads to an unacceptable long time period of the recovery system. Accruing a fault in a DBE transaction does not necessarily mean full abortion of the transaction (because of the nature of a distributed network and the diversity of the DBE environment there is possibility to perform a task in different ways). Rather, it could necessitate the restart of some subtransaction or repair and/or choosing of some alternative scenario. Additionally, it is important to note that restart/repair mechanisms can become part of an abort/restart chain (in a different transaction). This is why Recovery Management is one of the most crucial and important parts of the transaction model.

In order to design this part, we drew analogies from the biochemical diagnosis of an infectious disease; the isolation of enzymes from infected tissue can also provide the basis of a biochemical diagnosis of an infectious disease [25]. Common strategies of public health authorities for containing the spread of a contagious illness rely on isolation and quarantine [26]. This provides further inspiration for designing our recovery model. Overall, Recovery Management in combination with the concurrency control procedure runs in two phases:

1. *Preparation phase:* by sending a message (abort/restart) to all subtransactions that puts them (and their data) to an isolated mode (preparing for recovery). This helps avoid any propagation of inconsistent data before rollback.

2. *Atomic Recovery Transaction routine:* recovery routine will be run as an atomic-isolation (quarantine) procedure that can rollback or just pass (without applying any changes) a subtransaction.

It can be seen that the first task of Recovery Management in the transaction model is to isolate the failed transaction and related transactions (those using its partial results directly or indirectly), then to determine the damaged part (where the failure occurred), and finally to rollback to a consistent virtual check point (our system does not work based on determining actual check points, but virtually by using the logs and structural definitions of coordinators - we can rollback the system to the part of transaction tree in which the corresponding coordinator is working well and then that specific coordinator can lead the transaction to the next step). The compensable nature of our model can help on what could be done by compensating transactions (after applying the preparation phase).

Another benefit of a two phase recovery management is the possibility for saving valuable results provided by safe subtransactions until the transaction is restarted.

4.2 Two Phase Recovery Routine

In the first phase, Recovery Management tries just to isolate the damaged (or failed) part of the system by distributing a message that can isolate all worked data-items of those subtransactions. In the transaction model, we have seen that modified data items can be locked by two different locks, I_Lock and C_Lock.

As it was shown, data items that are locked by I_Lock, can be used just internally (IDG). Therefore, when the transaction is aborted (or restarted) there is no danger of misuse of these data items by the other transactions (because they do not have access to these items). These data items naturally can be considered atomic. They will be rolled back (if necessary) by using the IDG. The only issue is whether we need to rollback all data items. Only the damaged part (and related data items) of a transaction must be rolled back (and all related parts as determined by IDG).

The other modified data items are locked by the C_Lock and so are available for all other transactions. Meanwhile by following the EDG, the other transactions which used these partial results are in danger of abortion (or a restart), at least in some parts of a transaction. Therefore they must be identified as soon as possible. In fact, this must be done in the preparation phase, because the procedure of rollback for C_Lock can result in chains of rollback operations which can take time to commit.

4.3 Solution for Isolation in Recovery

For the critical part of the problem (C_Lock), the lock must be converted to R_Lock (Recovery Lock) by using the EDG and without any processing on data. The R_Lock must restrict access to data purely to Recovery Management in a transaction. This stops problem (failure) propagation until the Recovery routine is finished.

For the I_Lock optimization, we define T_Lock (Time-out lock) by some key abilities in a DBE transaction. The T_Lock is rather like giving a time-out before rollback of a data item. In addition, access to the data item will be limited to Recovery routines (avoiding failure propagation). Before finally considering a time-out, Recovery Management has the opportunity for reconverting a T_Lock to I_Lock (if rollback is not necessary). However, after finishing the time-out the data item will be rolled back automatically (figure 6 shows the effect of Recovery on the locking system).

5. Omitted Results and Forward Recovery

The probability for failure (for example because of a disconnection between different coordinators) can activate recovery and the preparation phase of recovery can be started. As we have seen, in the preparation phase, C_Locked (in all related transactions) data items were converted to R_Lock by using the EDG and all I_Locked data items were converted to T_Lock by using the IDG. Therefore Recovery Management in phase two behaves like a full ACID transaction in that it is fully isolated during the lifetime of a transaction.

However, using a suitable data structure the recovery manager transaction is optimized by providing not only special concurrent (by introducing the isolated T_Lock structure) operations, but also enables the possibility for saving key results of some sub-transactions even when the transaction has failed and been restarted.

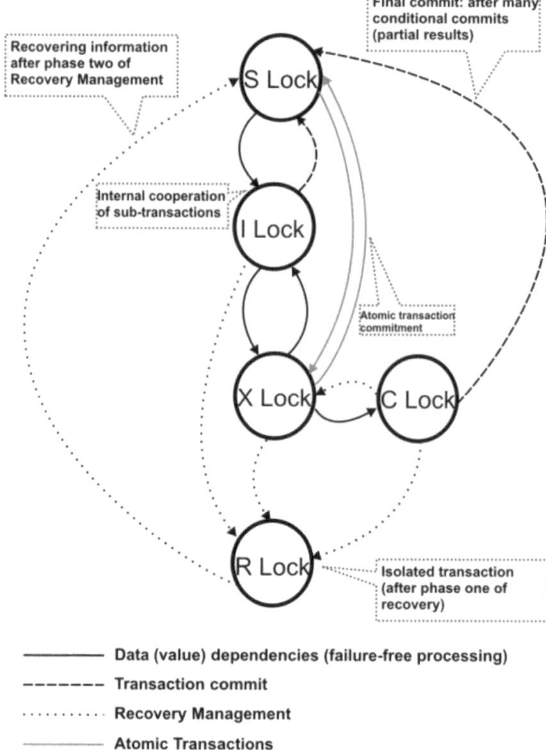

Figure 6. Recovery lock (R_Lock) schema

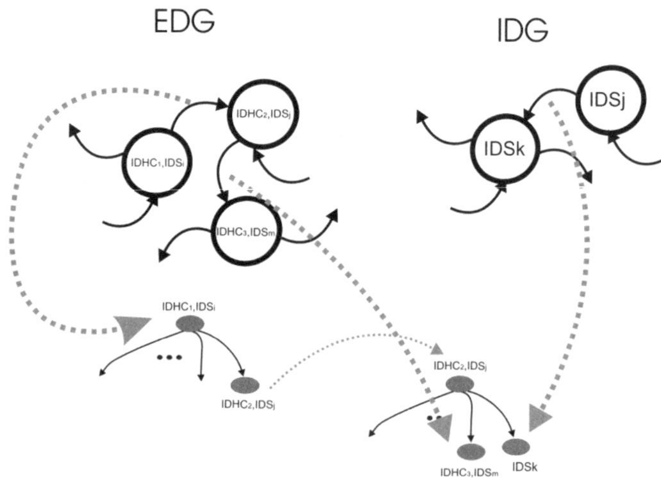

Figure 7. Creating compensating routines using EDG and IDG

The normal procedure of phase two of recovery of the transaction will be done by traversing EDG and IDG. For rollback of partial results, traversing the EDG will help to create/execute compensatory transactions (figure 7); the T_Lock can provide for an automatic rollback operation (after passing the time-out). However, for revalidating the correct data-items before time-out, the recovery manager transaction traverses the IDG and recalculates the data items. Then for unchecked data items it reconverts T_Lock to I_Lock, which can be useful in forward recovery and/or restart of aborted transactions. In this way, recalculating a specific data item is unnecessary. This will be happen only if a data item of T_lock has not been dependant to some inconsistent data item in the IDG graph.

5.1 Forward Recovery

Within a Digital Business Ecosystem, a number of long-running and multi-service transactions take place. Each comprises an aggregation of sub-transactions. There is an increased likelihood that at some point a subtransaction might fail. This may be due to a platform failure or its coordinator not responding or, simply, because it is a child of a Parallel Alternative coordinator and some alternative sub-transaction has already met the pre-set condition. There must be a way to compensate for such occasions and defer from aborting or even restarting the whole transaction.

Forward recovery is reliant on alternative coordinators (SAt and PAt at section 1.1) and the compensation operation in recovery management (section 4). By failing one sub-transaction of an alternative coordinator, that specific sub-transaction should be fully rolled back (by some compensation mechanism) and then the alternative coordinator tries to commit the transaction with its other sub-transaction(s).

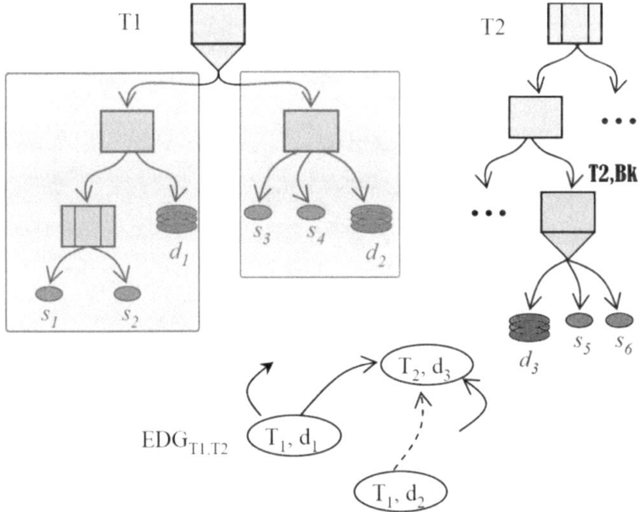

Figure 8. Forward recovery in the transaction model

Figure 8, shows an example in which transaction T1 is using a sequential alternative coordinator on top of a transaction tree; naturally T1 tries to run first sub-transaction ('T1,B1' on the figure 8). If a failure happens (for example failure on s1), T1,B1 must be compensated (in this scenario some partial results has been released to d3 from transaction T2 and this means by using EDG, those results should be rolled back too. This will be reflected in the compensation tree). After this compensation, the alternative sequential coordinator of T1 tries to run the second sub-transaction; T1,B2 and partial results will be released to transaction T2 from this sub-transaction (it will be reflected on the corresponding EDG). If T1,B2 is not successful too, transaction T1 will be fully aborted (recovered). But the interesting part is on transaction T2, which even after abortion of T1, needs to compensate any dependant results to d3 and then will try s5, which means the whole transaction will not fail even when T2 uses partial results from the aborted transaction (T1). Only dependant sub-transactions will be rolled back, and T2 will try to continue the execution and successfully commit.

This example (Figure 8), shows forward recovery in two different levels; firstly when we have an alternative coordinator and all dependencies are internal (T1,B1 to T1,B2), and secondly when a transaction (T2) uses a partial result of another transaction (T1) and it is dependant on that transaction (still T2 tries to avoid full recovery and will only modify its affected part).

6. Full Lock Schema

In total, there are 6 different locks for Concurrency Control in our transaction model. Two locks (R_Lock and T_Lock) are related to maintaining atomicity, and optimization during recovery. The S_Lock and X_Lock (eXclusive Lock) have similar behaviour to a conventional two-phase commit transaction model. However, value dependency and conditional commitment (partial results) can change the S_Lock /X_Lock behaviour (Figure 9 shows the full life cycle of locking system).

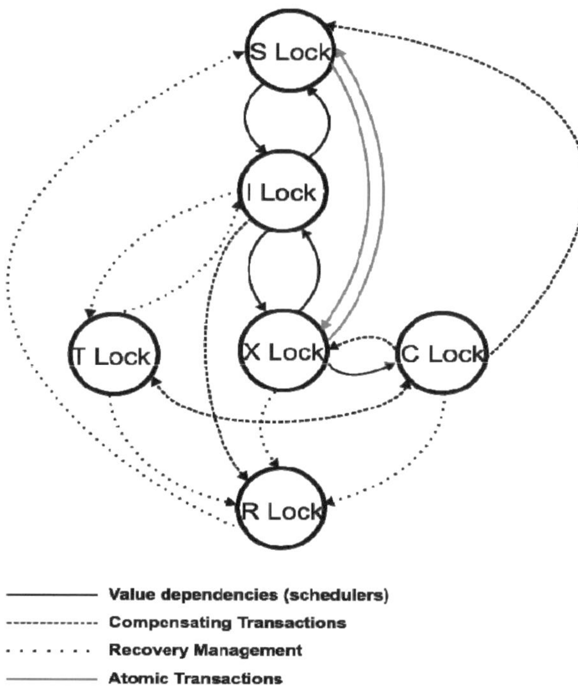

Figure 9. Full life cycle of the locking system

By using I_Lock, we relax the X_Lock and increase the support for concurrency inside a long-term transaction. Using C_Lock ables us to provide concurrency even when there are data dependencies between transactions (conventionally this was not possible, as with X_Lock there is not any permission for sharing data items before a transaction commits). IDG and EDG, as a two types of dependency graph have a complementary role of providing full recoverability for the transaction model.

7. Further Work and Conclusion

The nature of the transactions that take place in a highly dynamic and distributed environment such as that of a Digital Business Ecosystem (DBE) raises a number of non-trivial issues with respect to defining a consistent transaction model. In this paper, we have presented the fundamentals of a concurrency control mechanism based on an extended lock implementation for DBE transaction model that satisfies a number of issues that arise in providing a collaborative distributed software environment for SMEs.

The long-term nature of business transactions frames the concept of a transaction in Digital Business Ecosystems. Conceptually support for recoverability and data consistency, cause considerable limitations on concurrency, which are reflected in the limitations of conventional concurrency control mechanisms as applied in a transactional environment [8]. We have described an extended locking mechanism that supports the DBE transaction model. This is done in a way that ensures data consistency and transaction recoverability; at the same time it maximizes concurrency by relaxing the concurrency control limitations and introduces a flexible structure to support that. More specifically, we described the use of two locks, namely I-Lock and C-Lock, for ensuring consistency between the distributed logs as provided by the IDG and EDG and the local concurrency model. We also introduced a lock, the so-called T-Lock, for covering omitted results in common distributed events. Finally, we described a lock for recovery, named R-Lock, which facilitates an isolated two-phase recovery routine.

These different locking schemes, as apart of the concurrency control, can provide mechanisms to support compensation and forward recovery in a way that ensures local progress-to-date is preserved as much as possible. The locking mechanism is set up in such a way that it allows us to introduce a customised three-phase commit (3PC) communication mechanism, where the intermediate phase is used for addressing unexpected failures in the commit state.

7.1 Further Approaches and Future Work

Apart from increasing concurrency, another benefit of our work is that by relaxing the lock system and relying on logs for consistency and recoverability, the average duration of locks is reduced (comparing with conventional model in which a simple X_Lock could have the same time duration as its transaction and only after transaction commit could release data items to other transactions), it is possible to claim potential for a dramatic reduction on the probability of deadlock.

Our interest for future work is not just measuring this reduction, but also designing deadlock detection/prevention algorithms. In the case of deadlock correction, we are interested to reduce the probability for a transaction blocking and starvation (abortion of a transaction for avoiding and/or correcting a deadlock scenario).

Our preliminary approaches show that by detecting loops in the IDG and EDG, and a combined graph of both, it is possible to find all possibilities of deadlock. On the other hand, the primary proposed method for avoiding starvation is relying on alternative scenarios and forward recovery during prevention of deadlock, instead of restarting the whole transaction. In this way, that specific transaction which causes the loop on an EDG, can abort one its subtransactions (coordinators) and use an alternative subtransaction for avoiding the creation of a loop on the graph (deadlock scenario).

As is clear, checking against deadlock and other pathological properties has potential for further integration on this model. Meanwhile connection of the model with the semantic of particular business processes of SMEs is another area that sponsors of the Digital Business Ecosystem would like to have a solution for. The minimum requirements for the structural infrastructure of DBE network that supports this model is another issue of discussion and research in the wider scope of Digital Business Ecosystems.

Acknowledgements

This work was supported by the EU FP6-IST funded projects DBE (Contract No. 507953) and OPAALS (Contract No. 034824).

References

[1]   Digital Business Ecosystems (DBE) EU IST Integrated Project No 507953. Available http://www.digital-ecosystem.org [19 Sep 2006].

[2]   A. Razavi, S.Moschoyiannis, P.Krause. A Coordination Model for Distributed Transactions in Digital Business Ecosystems. In Proc. IEEE Int'l Conf on Digital Ecosystems and Technologies (IEEE-DEST'07). IEEE Computer Society, 2007.

[3]   C. J. Date , An introduction to Database Systems (5th edition), Addison Wesley, USA, 1996.

[4]   S. Greenberg and D. Marwood. Real time groupware as a distributed system: concurrency control and its effect on the interface. In Proc. ACM Conference on Computer Supported Cooperative Work, pages 207– 217. ACM Press, Nov. 1994.

[5]   L. McGuffin and G. Olson. ShrEdit: A Shared Electronic Workshpace. CSMIL Technical Report, 13, 1992.

[6]   C. Sun and C. Ellis. Operational transformation in real-time group editors: Issues, algorithms, and achievements. In Proceedings of ACM Conference on Computer Supported Cooperative Work, pages 59–68. ACM Press, Nov. 1998.

[7]   P. Bernstein, N. Goodman, and V. Hadzilacos. Concurrency Control and Recorvery in Database Systems. Addision-Welsley, 1987.

[8]   J. Gray, A. Reuter. Transaction processing: Concepts and Techniques, Morgan Kaufmann Publishers, USA, 1993.

[9]   C. Sun, X. Jia, Y. Zhang, Y. Yang, and D. Chen. Achieving convergence, causality-preservation, and intention-preservation in real-time cooperative editing systems. ACM Transactions on Computer-Human Interaction, 5(1):63 – 108, Mar. 1998.

[10] A. Elmagarmid , Database Transaction Model for Advanced applications, Morgan – Kaufmann, 1994.

[11] J. E. B. Moss , Nested transaction an approach to Reliable Distributed Computing, MIT Press, USA, 1985.

[12] T. Kakeshita, Xu Haiyan , "Transaction sequencing problems for maximal parallelism", Second International Workshop on Transaction and Query Processing (IEEE), 2-3 Feb. 1992, pp: 215 – 216, 1992.

[13] M.S. Haghjoo, M.P. Papazoglou, "TrActorS: a transactional actor system for distributed query processing", Proceedings of the 12th International Conference on Distributed Computing Systems (IEEE CNF), 9-12 June 1992, pp: 682 – 689, 1992.

[14] L.F. Cabrera, G. Copeland, W. Cox et al. Web Services Business Activity Framework (WS-BusinessActivity). August 2005. Available http://www128.ibm.com/developerworks/webservices [19 Sep 2006]

[15] A. Razavi, P.J. Krause and S.K. Moschoyiannis. DBE Report D24.28, Universtiy of Surrey, 2006.

[16] J. Yang, M. Papazoglou and W-J. van de Heuvel. Tackling the Chal-lenges of Service Composition in E-Marketplaces. In Proc. 12th RIDE-2EC, pp. 125-133, IEEE Computer Society, 2002.

[17] M.P. Papazoglou. Service-Oriented Computing: Concepts, Charac-teristics and Directions. In Proc. WISE'03, IEEE, pp. 3-12, 2003.

[18] A. Razavi, P. Malone, S.Moschoyiannis, B.Jennings, P.Krause. A Distributed Transaction and Accounting Model for Digital Ecosystem Composed Services. In Proc. IEEE Int'l Conf on Digital Ecosystems and Technologies (IEEE-DEST'07). IEEE Computer Society, 2007.

[19] P. Furnis, S. Dala, T. Fletcher et al. Business Transaction Protocol, version 1.1.0, November 2004. Available at http://www.oasisopen. org/committes/downaload.php [19 September 2006]

[20] L.F. Cabrera, G. Copeland, J. Johnson and D. Langworthy. Coordinating Web Services Activities with WS-Coordination, WSAtomicTransaction, and WS-BusinessActivity. January 2004. Available: http://msdn.microsoft.com/webservices/default.aspx [19 September 2006]

[21] L.F. Cabrera, G. Copeland, M. Feingold et al. Web Services Coordination (WS-Coordination). August 2005. Available http://www-128.ibm.com/developerworks/webservices/library/specification/ws-tx [19 September 2006]

[22] P. Furnis and A. Green. Choreology Ltd. Contribution to the OASIS WS-TX Technical Committee relating to WS-Coordination, WSAtomicTransaction and WS-BusinessActivity. November 2005.

[23] F.H. Vogt, S. Zambrovski, B. Grushko et al. Implementing Web Ser-vice Protocols in SOA: WS-Coordination and WS-BusinessActivity. In Proc.7th IEEE Conf on E-Commerce Technology Workshops, pp. 21-26, IEEE Computer Society, 2005.

[24] van der Meer, D. Datta, A. Dutta, K. Ramamritham, K. Navathe, S.B. (2003), "Mobile user recovery in the context of Internet transactions", IEEE Transactions on Mobile Computing, Volume: 2, Issue: 2, April-June 2003, pp: 132 – 146.

[25] Wikipedia, 'Infectious disease', http://en.wikipedia.org/wiki/Infectious_disease (last access: 08/03/2007).

[26] US Department of Health and Human Services, 'Fact Sheet: Isolation and Quarantine', Department of health and Human Services; Centers for Disease Control and Prevention, <http://www.cdc.gov/ncidod/dq/isolationquarantine.htm>, last access: 08/03/2007.

Communicating Process Architectures 2007
Alistair A. McEwan, Steve Schneider, Wilson Ifill, and Peter Welch
IOS Press, 2007

trancell - an Experimental ETC to Cell BE Translator

Ulrik SCHOU JØRGENSEN and Espen SUENSON

Department of Computer Science, University of Copenhagen,
Universitetsparken 1, 2100 Kbh. Ø, Denmark.

`ulriksj@diku.dk`, `expen@diku.dk`

Abstract. This paper describes `trancell`, a translator and associated runtime environment that allows programs written in the occam programming language to be run on the Cell BE microarchitecture. `trancell` cannot stand alone, but requires the front end from the KRoC/Linux compiler for generating Extended Transputer Code (ETC), which is then translated into native Cell SPU assembly code and linked with the `trancell` runtime. The paper describes the difficulties in implementing occam on the Cell, notably the runtime support required for implementing channel communications and true parallelism. Various benchmarks are examined to investigate the success of the approach.

Keywords. occam, Cell BE, parallelism, concurrency, translation

Introduction and Motivation

Parallel techniques are an important element in the ongoing quest for increasing performance of computing machinery. Since the seventies microprocessor manufacturers have devoted a great deal of effort to developing techniques that exploit instruction-level parallelism. These techniques include, for example, pipelining and superscalar execution as well as various techniques for out-of-order execution. It seems like this approach might be on the decline due to both memory latency and the difficulties associated with the high energy consumption and operating frequencies that mainstream microprocessors exhibit [1]. One possible way to remedy this situation is to exploit language-level parallelism instead of instruction-level parallelism. This has the distinct advantage that it is possible to use several processing cores running in parallel to increase performance instead of making each individual core faster and thus more complex. However, in order to exploit language-level parallelism programs must usually be written or rewritten specifically with concurrency in mind.

There are basically two ways of writing concurrent programs: Either by making them parallel from the start by writing in a language or framework supporting concurrency, or by writing a sequential program and then rewriting it automatically or manually. Manual rewrite of sequential programs is quite costly. Automatic rewriting methods have their uses but they also have their limitations [2]. The greatest gain in performance is possible when writing programs that are concurrent from the start. A lot of coordination languages, frameworks and libraries have been developed for this purpose, however, we feel that for concurrent programming to become a truly integrated habit for the programmer, the programming language itself needs to have concurrent constructs. A number of concurrent general purpose programming languages exists, notably Ada, Cilk, Erlang and occam (excluding proof-of-concept and experimental languages). None of these has seen widespread use in the public at large as general purpose languages. Java has some language support for concurrency, but it is based on an unsafe thread model that does not aid the programmer much.

The dominant model for concurrent programming at the time of writing is that of threads. This is possibly due to the fact that the concept of threads is close to the underlying hardware of many microprocessors. Thread-based programming is inherently unsafe and complex, and this has given concurrent software, and by derivation parallel hardware, an image of being difficult to program and exploit efficiently [2]. By using a mathematical foundation, e.g. a process calculus, as the basis for designing concurrent programming languages it is possible to design a safe, easy to use and efficient language. Several of the aforementioned languages have such a foundation. occam is based on the process calculus CSP[1], which allows the compiler to make certain guarantees about the generated code such as that it is free of race conditions [6]. This paper will focus exclusively on the occam programming language and concurrency model.

The occam language is extraordinary in that it was co-developed with the transputer microarchitecture, ensuring an excellent performance of the combination. This means that occam already has an associated architectural model for implementation. The Cell BE microarchitecture is a powerful novel and highly parallel microprocessor. It has nine cores, of which eight are identical specialized SIMD processing units, intended to deliver the computational power of the processor. The Cell BE promises great performance, but commentators are concerned about whether programmers will be able to exploit the Cell's many cores. The purpose of the trancell is double-sided. Firstly to provide the basis for a programming environment on the Cell that can ensure greater programmer productivity than C/C++ environments due to a safer concurrency model and language support for concurrency. Secondly to promote the use of concurrent languages in general, and specifically occam, by providing an easy-to-use and high performance development platform for a microarchitecture that commentators predict to become very popular.

1. Implementing occam on the Cell BE

To enable occam code to run on the Cell, it is necessary to translate occam code to Cell assembly and to provide a run-time environment that emulates the transputer hardware support for occam that the Cell lacks. To save some work on development, the occ21 compiler from the Linux/x86 version of KRoC is used to generate transputer assembly, which is then translated by trancell into Cell SPU assembly. This assembly is then linked with the trancell runtime, which is written in a combination of GNU C and SPU assembly.

No attempt is made to make the system portable as is the case with KRoC. This is because the Cell is quite unique so it is unlikely that the system should be ported to other architectures. Another reason is that it allows trancell to be very architecture-specific for performance reasons. However, it is attempted to make the system as modular as possible.

It has not been a priority to be able to utilize the processing power of the PPE, as it would complicate the project to write a translator for two different target instruction sets. The implementation has focused on speed of the generated programs, since trancell can be viewed as a proof-of-concept implementation regarding performance. This means that the current version of trancell implements a quite limited amount of the ETC instructions due to development time constraints. For example, no floating point instructions are implemented since their implementation would not add new concepts to trancell but mainly be useful in practice.

---

[1]Communicating Sequential Processes [3]. The derivative language occam-π also incorporates ideas from the π-calculus [4][5].

1.1. Overview of the Transputer Architecture

The assembly of the original transputer microprocessors has been modified extensively over the years by the KRoC team [4][5][7]. The source of `trancell` is the textual ETC (Extended Transputer Code) produced by the combination of `occ21` and `tranx86` of the current KRoC distribution[4][10]. The virtual transputer that executes ETC is a stack machine with a three-level integer stack and a three-level floating point stack as well as some additional registers for handling concurrency and memory allocation. It features limited addressing modes as well as some complex instructions, notably for parallelism, channel communication and loops. The original `occam` did not support transparent usage of several processor units, but the transputer did have hardware for transparent channel communication with other processors.

1.2. Overview of the Cell BE Architecture

The Cell Broadband Engine microarchitecture is a joint development by Sony, Toshiba and IBM. It has been developed for use in the PlayStation 3 video game console, but it is intended to be used also in embedded systems as for example digital television sets. The term "broadband engine" has been coined because the Cell is intended to be used in settings requiring high network performance.

The Cell delivers computational power by having nine cores on a single chip. This helps spreading out the power consumption over a larger amount of the chip surface, alleviating power consumption and heat generating complications. One of the cores is intended as a central controller, running an operating system and distributing work to the other cores. This is the Power processor element (PPE). The other eight cores are intended to be the workhorses of the processor, doing most of the actual computations. These are called the Synergistic processor elements (SPEs) or Synergistic processor units (SPUs)[2]. The cores all operate at 3.2 GHz.

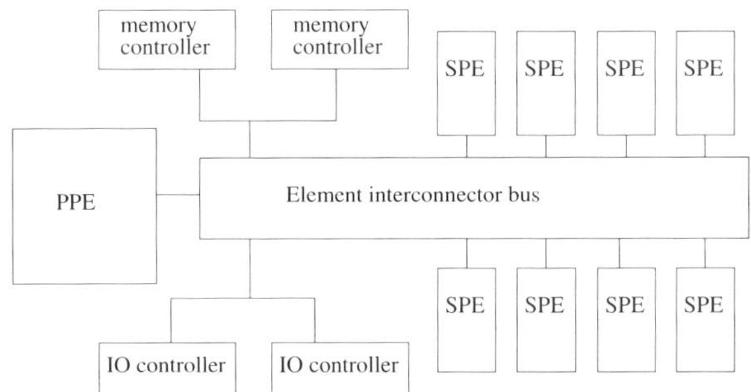

Figure 1. Diagram of the main components of the Cell BE architecture.

The PPE is based on a standard 64-bit Power PC design, but has a significantly simpler implementation compared to other Power PC processors. It has 32 KB first-level and 512 KB second-level cache. It provides two simultaneous threads of execution by interleaving instruction sequences and duplicating most architectural resources. The PPE is not used by `trancell` for running `occam` processes so the architectural details of the PPE is not of great importance to the discussion.

The eight SPEs are identical. They feature a somewhat conventional load-store RISC instruction set. The SPEs have no cache, but instead each have 256 KB of local memory

---

[2]The distinction being that the SPU is the processing unit itself, while the term SPE denotes the SPU with its associated memory and interfaces.

(which could alternatively be regarded as a fully programmer-managed cache). Crucial to the SPEs' performance is their SIMD capacity. The SPE has 128 general purpose registers, each of 128 bit. Load-store instructions move 128 bit at a time - the data must be 128-bit-aligned. Arithmetic and logical instructions operate on quantities of 16, 32 or 64 bit at a time. This means that each instruction manipulate from eight to two quantities simultaneously (apart from load/stores). Addresses are 32 bit.

To keep the SPEs simple, they have shallow pipelines. Instead of speculative execution of branches, compilers can aid the hardware by using a special hint-for-branch instruction. The instructions are split in two groups[3] so up to two instructions can be issued each cycle.

The SPEs each have a DMA controller for main memory access. Main memory is coherent and all cores share address space. The cores and the main memory interface communicate by way of a high speed ring bus, the Element interconnector bus (EIB). The SPEs can communicate in several ways. They can themselves issue DMA commands to transfer between local and main memory. The PPE can also issue these commands on behalf of the SPEs, freeing the SPE for other tasks. The SPEs can also issue DMA commands that transfer between the local memory of two SPEs. Lastly, the SPEs can be interrupted by the PPE or other SPEs and have messages in special mailbox registers.

2. Translator

The translator part of `trancell` is written in Standard ML, which is a language well suited to the implementation of translators.

A program written in **occam** is translated to binary ETC format by `occ21`. `tranx86` is then used to translate the ETC binary into textual ETC, after this `trancell` translates from textual ETC to SPU assembly. The GCC assembler is then used to generate binary SPU objects which are linked with the runtime.

The translation from ETC to SPU assembly is multipass. First the textual ETC is lexed, then parsed into a rather direct abstract representation. Instruction by instruction the textual ETC is then translated into SPU assembly. Figure 2 shows an overview of the translation process.

2.1. Register Handling

The register stack of the transputer is simulated by assigning a virtual register to each occurrence of a stack access in the code. The virtual registers are then assigned to actual SPU registers by a register-colouring phase. This is the same approach as is used in `tranx86` [10].

2.2. The Register Stack and ETC-procedure Calls

The specification of the CALL and RET instructions states that the stack is the same before and after the instruction. Since the procedures are register-coloured independently, it is unknown what virtual registers are assigned to what SPU registers after a call and after a return. To solve this, a prologue and an epilogue is introduced before and after CALL and RET that saves and restores the register stack in designated registers.

2.3. SIMD Instructions

One of the main strengths of the Cell SPE is the SIMD instructions. This could be exploited in loops, especially over arrays. This is not implemented in the current version of `trancell`.

---

[3]Arithmetic and floating point instructions, and memory and branch instructions, roughly.

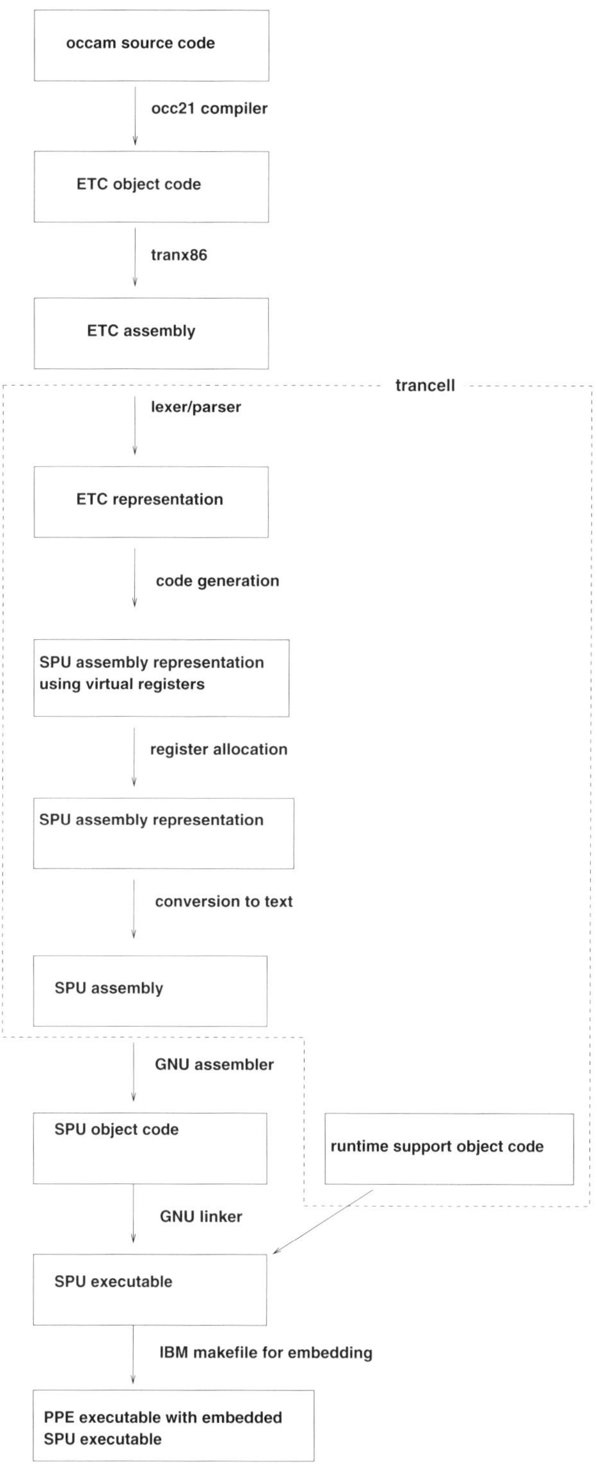

Figure 2. Overview of the translation process.

2.4. Memory Alignment

The SPE requires that all memory accesses are 128-bit aligned. Both for ease of implementation and for speed, `trancell` aligns all variables at 128 bit, leaving some portion of every memory location unused. This is wasteful of space but efficient in terms of execution speed since we avoid generating extra instructions to access variables that are not 128-bit aligned.

2.5. Memory layout

The implementation of `trancell` does not presently make use of the Cell's main memory. All text and data reside in the SPEs' local memories. Amongst other things, this means that programs are limited in size to less than 256 KB. Clearly, to be practically useful the runtime system has to be augmented with facilities for doing swapping to main memory. This task is fortunately aided by the **occam** compiler's guarantee that the program is free of race conditions.

2.6. Omissions

Out of the more than 100 ETC instructions, only 44 have been implemented. The instructions that have been left out consist mainly of floating point instructions, various arithmetic and logical operations, and some instructions concerning the transputer state e.g. the error flags. There is no support for floating point numbers.

Constants are passed directly as arguments to the SPU assembler instructions; no instructions are generated to construct large constants. This means that there are limits to the size of constants depending on the specific instructions they are used with.

There is no array support, as the **occam** notion of *vector space* (a heap) is not supported. In addition, the translator cannot handle data sections in the ETC.

The translator does not generate hint-for-branch instructions to provide branch prediction for the SPE. This means that jumps and function calls incur inordinately large penalties. However, the techniques to use for generating these instructions lie outside the main scope of this paper.

3. Runtime Environment

The principal responsibilities of the runtime are to provide parallelism through scheduling of processes and channel communication between processes.

At the moment, the runtime environment is not able to migrate processes at run time (see section 5). This means that a distinct **occam** program must be written for each SPE that the programmer wishes to use. Channel communication is transparent to the programmer, although the programmer does have to decide on which SPE the channel words should reside. In addition, the **occam** compiler will not check for correct usage of the channels since the programs for the SPEs are compiled separately.

We have aimed for a symmetrical implementation on the SPEs. To be independent of the exact number of available SPEs, we have also aimed for an implementation of autonomous, self-contained SPEs. The PPE has quite little to do - it starts and shuts down the SPEs and relay communication as described in section 3.2.

3.1. Scheduling

Scheduling takes place on each SPE according to the transputer model of scheduling. The scheduler will not cause processes to be moved to another SPE.

```
ENBC    -- sequence of channel enabling instructions
ENBC
ENBC
...

ALTWT   -- wait for a channel to become ready
...

DISC    -- sequence of channel disabling instructions
DISC
DISC
...

ALTEND  -- direct control flow depending on
        -- which channels are ready
```

Figure 3. Pseudocode for a typical ALT sequence in ETC.

3.2. Communication

One of CSP's and thereby occam's main mechanisms is channel communication. Two processes on the same chip simply communicate through a channel word located in local memory, as described in [8]. The Cell lacks the transputer hardware for external (inter-core) communication, so this must be emulated in the runtime environment. As in intra-core communication, the two processes communicate via a channel word in memory, located in the local memory of the SPE that the programmer has designated.

The best way of doing external communication would be to use the SPEs' signalling mailboxes to communicate directly between them using either interrupts or polling (or perhaps a combination). However, for simplicity all SPE-to-SPE communication is relayed via the PPE. The PPE will constantly poll the SPEs in a round-robin fashion, buffering outgoing messages and delivering them when the addressee wishes it. The SPEs send outgoing messages instanly (stalling if need be) and poll for incoming messages each time the scheduler runs.

3.3. The ALT Construction

The ALT construction is one of the most powerful features of occam. The pseudocode of a typical ALT sequence in ETC can be seen in figure 3.

The channels are enabled asynchronously and the whole construct synchronizes on the ALTWT instruction. The semantics of the construct is such that it might be important to disable the channels in sequence (in the case of a PRI ALT). Thus, each DISC instruction is synchronized, only executing after the channel has been disabled. This is inefficient if more than one SPE is involved, since the process might be descheduled a number of times in succession. A better solution would be to disable the channels asynchronously, only synchronizing on the final ALTEND instruction, but due to the added complexity of keeping track of the order of DISC instructions this has not been implemented in trancell.

4. Benchmarks

4.1. Commstime

The "commstime" benchmark is described by R. Peel [9] and measures communication overhead. The benchmark was run both with all processes on a single SPE and with the four processes on separate SPEs and the channel words located on a fifth SPE (the worst performing situation possible). For comparison, the same benchmark was conducted on an Intel Pentium III 1 GHz processor running KRoC. The results can be seen in table 1.

The reason the single-SPE version of `trancell` performs so badly in comparison with KRoC is that the SPEs poll the PPE for incoming messages during each context switch. Profiling information from the IBM Full System Simulator [13] shows that over 85 percent of the SPE cycles are spent waiting for communication with the PPE. This overhead becomes less significant with higher workloads on the SPEs. If local channel communication latency is important for an application the overhead could be reduced by increasing the latency of global channel communication.

Table 1. Benchmarking of commstime

	iterations	time	overhead
KRoC	160,000,000	63.5 s	99.1 ns
`trancell` single SPE	320,000	11.8 s	9.25 μs
`trancell` 5 SPEs	320,000	497 s	388 μs

4.2. ALT Overhead

To measure the overhead of the ALT construct, a simple benchmark has been constructed. It consists of three interconnected processes of which each repeatedly ALTs on input from the two other. The results are given in table 2.

Table 2. Benchmarking of the ALT construct

	iterations	time	overhead
KRoC	250,000,000	26.1 s	348 ns
`trancell` single SPE	4,000,000	218 s	18.2 μs
`trancell` 3 SPEs	32,000	59.9 s	624 μs

4.3. Parallel Speedup

To investigate to what extent the desired ability of `trancell` to achieve parallel speedup has been met, we have constructed a simple embarrassingly parallel benchmark, that represents the best possible case of parallelisation. Five processes perform a dummy computation and the program only communicates to synchronize upon completion of the processes. A real-life example of this type of program could be Monte Carlo simulation.

To have something to compare against, a C program performing roughly the same computation has been constructed. The most notable difference from the occam program is that the C program controls and synchronizes the computing processes from the PPE instead of one of the SPEs.

The calculated efficiency of the program can be seen in table 3. The reason that efficiency is low in the benchmark with 3 SPEs is the uneven distribution of 5 processes on the 3 SPEs. The running times of the benchmark can be seen in figure 4.

The benchmark is very satisfying, but it should be noted that it represents the optimal case. Programs that are not so obviously parallelisable might very well exhibit much lower gains or even degrading performance due to communication overhead.

Table 3. Efficiency of parallel speedup in percent

	single SPE	3 SPEs	5 SPEs
trancell	100	83.3	99.9
C	100	95.4	100

5. Extending `trancell` to Support Process Migration

Currently, the programmer has to assign processes to SPEs and they will not move during execution. This requires the programmer to do a careful analysis of the program to obtain reasonable performance. In this section we describe how `trancell` could be enhanced to support process migration between SPEs at runtime. This would enable us to do dynamic load balancing easing the programmer's task as well as possibly leading to better performance (depending on the balancing scheme used).

5.1. Migrating Processes

The main difficulty in moving processes between SPEs is that the transputer memory model dictates a continuous stack of workspaces, where each process can refer back to variables located in an ancestor's workspace. Thus, it is not possible simply to move a process and its associated workspace.

Instead of a stack of workspaces, `trancell` should allocate memory for each process separately. Ancestral variables that are referred should also be given space, and upon termination of a process the variables should be written back to the parent workspace.

Since a process can in principle refer to all of the workspace stack, an analysis must be carried out on the ETC code to determine which variables should be copied for read and which should be copied for write upon process creation and termination. There will be no problems due to shared memory since the occam compiler ensures that there are no race conditions.

The analysis of the ETC code should also determine which memory locations contain channel words, since these are not to be treated as workspace variables but as global addresses.

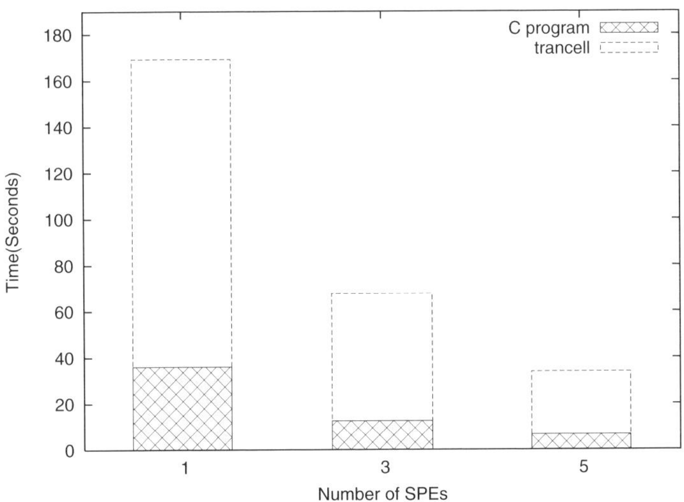

Figure 4. Benchmarking of an embarrassingly parallel program. The benchmark was run for 10^9 iterations.

This scheme would allow us to move any process that is on a ready queue simply by copying the associated memory of the process and enqueuing the process on another SPE. The drawback, compared to transputers, is that an additional amount of copying takes place when processes are created and terminated.

5.2. Load Balancing

For the load balancing itself we envisage a so-called envious policy. If a SPE has no more processes in its ready queue it asks its neighbours (according to the network topology) for a process to take over.

The network topology is to be implemented in the runtime environment as to be easily changeable. This allows experimentation with different topologies. All processes would initially be started on one core, but as the program is run processes that have been migrated could start new processes on other cores.

5.3. Channel Words

The processes need to know the channels they refer to. Since the processes can move about arbitrarily, the easiest solution would be to keep the channel words at the SPE of the process that initialised them. This has a potential performance drawback if processes on two SPUs are communicating via a third SPU, but the solution avoids the overhead of having a protocol for exchanging information about channel whereabouts.

Conclusion and Further Work

In this paper the `trancell` translator and runtime is presented, a program that in conjunction with the **occam** compiler of KRoC can run **occam** on the multi-core Cell platform using ETC as intermediate code.

`trancell` supports programmer transparent channel communication between SPEs, but for multi-core applications separate programs have to be compiled for each core. The paper describes how `trancell` could be enhanced to support process migrating, which in turn would allow programmer transparent multi-core applications and dynamic load balancing. In the authors' opinion, this approach should be investigated and is the most important further work on `trancell` to be undertaken.

To investigate the success of the approach, benchmarks have been examined showing a 388 μs communication overhead and 624 μs ALT overhead for multi-core applications. Experiments with a benchmark timing the best possible parallel case show a satisfying parallel speedup.

To make a truly efficient implementation of `trancell` some further work is required:

- Support for swapping to main memory.
- Support for SIMD instructions.
- Support for the full ETC instruction set - possibly also the **occam-**π extensions.
- Support for running processes on the PPE as it is the most powerful core in the Cell.
- Generating hint-for-branch instructions for the SPEs.
- Array support.

`occ21` and ETC was chosen to be the front end and source of the translator for pragmatic reasons. However, it is not an ideal choice for implementing **occam** on the Cell. To make a well performing implementation some information is needed that is present at the **occam** level but not the ETC level. Moreover, ETC introduces a lot of quirks of the transputer and KRoC that we have had to work around. All in all, an implementation of a full new **occam** compiler would be a major benefit to the project of running **occam** on the Cell BE.

There are other approaches to porting occam to the Cell that might prove interesting. Dimmich et al. have investigated the prospect of interpreting the ETC [11]. Yet another approach is to compile occam to another high-level language such as C and then use existing compilers for the Cell to make executables. The SPOC compiler [12] could be used for this purpose, though it requires some investigation to find out how easily the runtime environment can be ported to the Cell.

References

[1] J.A. Kahle et al., Introduction to the Cell multiprocessor, *IBM Journal of Research and Development* vol. 49 (2005), 589–604.

[2] Edward A. Lee, The Problem with Threads, *Computer* vol. 39 no. 5, (2006), 33-42.

[3] C.A.R. Hoare, Communicating Sequential Processes, Prentice Hall, 1985.

[4] P.H. Welch and D.C. Wood, The Kent Retargetable occam Compiler, Parallel Processing Developments, *Proceedings of WoTUG 19*, IOS Press, 1996.

[5] F.R.M. Barnes and P.H. Welch, Communicating Mobile Processes, *Communicating Process Architectures*, IOS Press, 2004.

[6] SGS-THOMSON Microelectronics Limited, occam 2.1 Reference Manual, 1995.
     Available at:
     `www.wotug.org/occam/documentation/oc21refman.pdf`.

[7] M.D. Poole, Extended Transputer code - a Target-Independent Representation of Parallel Programs, Architectures, Languages and Patterns for Parallel and Distributed Applications, *Proceedings of WoTUG 21*, IOS Press, 1998.

[8] D.A.P. Mitchell et al., Inside the Transputer, Blackwell Scientific Publications, 1990.

[9] R.M.A. Peel, Parallel Programming for Hardware/Software Co-Design, 2001.
     Available at:
     `http://www.computing.surrey.ac.uk/personal/st/R.Peel/research/bcs-220201-4.pdf`.

[10] F.R.M. Barnes, tranx86 – an Optimising ETC to IA32 Translator, *Communicating Process Architectures 2001*, IOS Press, 2001.

[11] D.J. Dimmich, C. Jacobson and M.C. Jadud, A Cell Transterpreter, *Communicating Process Architectures 2006*, IOS Press, 2006.

[12] M. Debbage et al., Southampton's portable occam compiler (SPOC), *WOTUG-17*, 1994.

[13] IBM Corporation, Performance Analysis with the IBM Full-System Simulator, 2006.

Communicating Process Architectures 2007
Alistair A. McEwan, Steve Schneider, Wilson Ifill, and Peter Welch
IOS Press, 2007

A Versatile Hardware-Software Platform for In-Situ Monitoring Systems

Bernhard H. C. SPUTH, Oliver FAUST, and Alastair R. ALLEN

Department of Engineering, University of Aberdeen, Aberdeen AB24 3UE, UK
{b.sputh, o.faust, a.allen}@abdn.ac.uk

Abstract. In-Situ Monitoring systems measure and relay environmental parameters. From a system design perspective such devices represent one node in a network. This paper aims to extend the networking idea from the system level towards the design level. We describe In-Situ Monitoring systems as network of components. In the proposed design these components can be implemented in either hardware or software. Therefore, we need a versatile hardware-software platform to accommodate the particular requirements of a wide range of In-Situ Monitoring systems. The ideal testing ground for such a versatile hardware-software platform are FPGAs (Field Programmable Gate Arrays) with embedded CPUs. The CPUs execute software processes which represent software components. The FPGA part can be used to implement hardware components in the form of hardware processes and it can be used to interface to other hardware components external to the processor. In effect this setup constitutes a network of communicating sequential processes within a chip. This paper presents a design flow based on the theory of CSP. The idea behind this design flow is to have a CSP model which is turned into a network of hardware and software components. With the proposed design flow we have extended the networking aspect of sensor networks towards the system design level. This allows us to treat In-Situ Measurement systems as sub-networks within a sensor network. Furthermore, the CSP based approach provides abstract models of the functionality which can be tested. This yields more reliable system designs.

Keywords. Embedded systems, System on Chip, Network on Chip, Hardware Software co-Design, Multi-core, Water Monitoring, in-situ sensors, libCSP2

Introduction

Clean drinking water is one of the most important, if not the most important food for humans and animals alike [1]. Furthermore, it is under constant danger of being polluted by environmental threats [2]. This is the reason why 9 institutions from 6 European countries formed a consortium to carry out the WARMER (WAter Risk Management in EuRope) project. The WARMER project is funded by the Sixth Framework Programme for European Research and Development (FP6). FP6 emphasises the problems of food quality, and pollution of the environment. WARMER is a follow up of SEWING (System for European Water monitorING) [3]. The focus of the SEWING consortium was the development of sensors to measure water quality. WARMER aims to enhance the work done by the SEWING consortium by creating a flexible in-situ monitoring system (IMS) and integrating remote sensing measurements obtained from satellites.

Figure 1 shows a brief overview of the system proposed by the WARMER project. The system consists of the following components:

- In-situ measurement systems (IMS) — The IMS measures the properties of the environment through its sensor probes. The obtained measurement data is interpreted

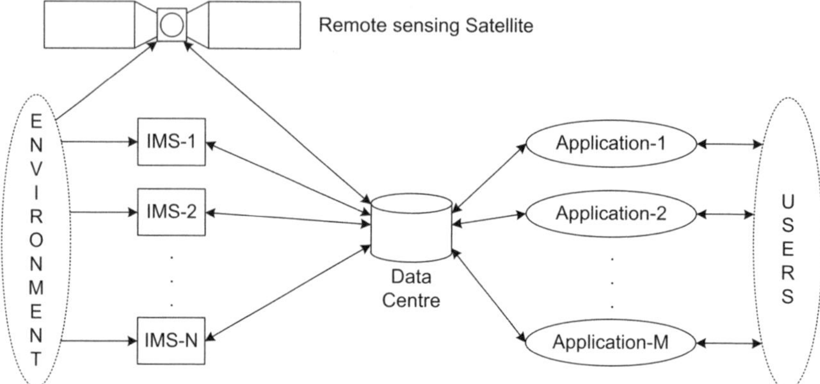

Figure 1. Overall system proposed by the WARMER project

by a built-in processing platform. The processing platform decides whether or not a critical level of pollution is reached. Under normal conditions, i.e. no critical level of pollution is detected, the measurement system sends its sensor data at predefined intervals, for instance every 30 minutes, to the Data Centre. However, if a critical level of pollution is detected, the measurement system sends a direct alarm to the Data Centre. This is only a simple example, naturally one should be able to define much more complicated monitoring schemes, for instance the measurement result of one sensor may influence the measurement strategy of another one.

- Remote sensing Satellite — It periodically acquires images of the wider area of interest. Unfortunately, these images have a low resolution, one pixel represents a square of $100m \times 100m$, and the update frequency varies between 1 and 3 days depending on the location of interest. This limits the use of remote sensing satellites. However, the WARMER consortium wants to combine satellite and IMS data to offer a more complete overview. In the following text the satellite is of no concern.
- Data Centre — A Data Centre aggregates satellite images and measurements from in-situ measurement systems over a long period of time. Furthermore, it analyses the data and exposes interfaces for user Applications. The interfaces provide access to the analysis results and the raw data. Long term data integrity is very important in order to detect slow degradations in the environment.
- Applications — These are the user interface of the system. Both IMS and Data Centre operate without human participation. The Applications interact with the Data Centre to obtain the measurement data of interest and present it to the users.

The project is interdisciplinary involving specialists in chemistry [4], environmental engineering [1,2], remote sensing [5], computer science, electronics [6], and semiconductor technology [7]. These requirements justify the large group of international collaborators.

With this paper we propose the creation of a new processing platform for the IMS. The IMSs will be deployed at remote locations, such as river beds and drinking water reservoirs. This implies that batteries power the system. The total cost of ownership for such a monitoring solution depends largely on the length of the IMS service interval. Needless to say, a longer interval is more cost effective. It is therefore of great importance that the IMS consumes as little power as possible. Before one can think of optimising the power consumption of a system, it is necessary to understand the role of the processing platform within the IMS.

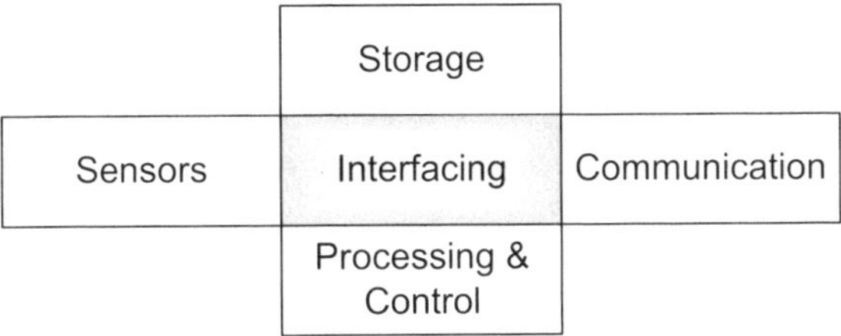

Figure 2. The building blocks of the in-situ measurement station

Figure 2 gives an overview of the IMS structure, it consists of four major building blocks: Sensors, Storage, Processing & Control, and Communication. Currently these building blocks are implemented as individual hardware components. These hardware components have incompatible and sometimes conflicting interfaces. Therefore, elaborate interfacing or glue logic is necessary to enable data exchange between these components. In general the system design has a low level of integration. This low level of integration directly translates into high energy requirements. Furthermore, there is no clear strategy which outlines the interaction between the individual components. This makes the whole design very inflexible and error prone.

Our goal is to outline a flexible design strategy which leads to energy efficient systems. In effect we propose a design strategy for NoC (Network on Chip) [8,9]. In a hardware and software co-design environment NoCs consist of networked hardware and software processes. We abstract both process types and model the network in terms of Communicating Sequential Processes (CSP). This model describes the functionality in a very compact manner. Furthermore, it is possible to make profound statements about stability and security of the system.

The first step in our proposed design strategy is to create and test a functional CSP model. In a second step a designer has the flexibility to implement a CSP process in either hardware or software. On the system level this strategy leads to simpler and higher integrated designs, because it is possible to balance the workload between software and hardware without additional considerations about functionality or stability. Higher integration removes many of the stand alone components present in the current system design. So, the proposed system design has a lower component count, and hence a lower power consumption. This is true if we assume that: (a) each component requires some energy for housekeeping tasks like memory refreshes and (b) data exchange between individual components is less energy efficient than data exchange within an integrated system. Furthermore, most of the literature about higher integrated systems supports this claim [10].

We present in Section 1 the current, the desired and our proposed hardware for the IMS processing platform. Section 2 discusses the processing platform design. Software implementation aspects for the proposed processing platform are discussed in Section 3. In Section 4 we demonstrate how to implement a NoC using the design flow introduced in this paper. The paper closes with conclusions and further work in Section 5.

1. Proposed In-Situ Processing Platform

Before we propose a new processing platform for the IMS it is necessary to evaluate the existing systems from our partners in terms of storage and communications. This is good practice in order to prevent incompatibilities with existing structures.

1.1. Current In-Situ Monitoring System

A wide variety of electrical interfaces are used to connect sensor probes with the processing platform, ranging from analogue interfaces over RS-232 to USB. Similarly, the connection between IMS and Data Centre can be implemented using various different communication standards, but presently there is a bias towards using the mobile phone infrastructures, such as GSM (Global System for Mobile communications) [11]. As processing platform, the partners currently use a wide variety of processors ranging from off the shelf PC processors, over microcontrollers such as Intel 8051 or TI MSP430, to CPLDs (Complex Programmable Logic Devices) and FPGAs.

1.2. Desired In-Situ Monitoring System

The desired features of the next generation IMS include service intervals of 6 months and longer, smaller physical size of the system, and also more compatibility between sensor probes and processing platform is desired. Furthermore, our partners would like to have up to 20 sensor probes per in-situ measurement system. Additionally, one proposes to use IEEE 1451.x [12,13] to interface sensors with the processing platform. However, this standard is still in development and hence modifications on its interface are still possible. To prevent any loss of measurement data in case of a communication breakdown, the partners desire local data storage of up to 10 MB. To link IMS and Data Centre mobile phone infrastructure is still desired. However, a little flexibility on this side is still appreciated, especially with the background of GSM being phased out during the next 10 years and UMTS (Universal Mobile Telecommunications System) [14,15] taking over. Another interesting communication standard which can be used for this link is WiMax (Worldwide Interoperability for Microwave ACCess (IEEE 802.16)) [16,17], of which deployment has started in urban areas already. Taking all these points into consideration resulted in the processing platform for the in-situ measurement system we propose.

1.3. System Requirements

The biggest design constraint for the next generation in-situ measurement station is the service interval of 6 months and longer. This requires a very energy efficient measurement system design. One way to reduce the energy consumption is higher system integration. In the best case this results in a SoC (System on Chip) where sensors, processing platform, communication system, and battery are integrated in a single piece of silicon. Because all the components are specially designed for this SoC, unnecessary and power consuming abstraction layers can be avoided. A SoC can be seen as the ultimate long term goal. However, in the short term this solution has a number of drawbacks for the WARMER project:

- *High initial investment* – producing a SoC requires custom made components, i.e. ASICs (Application Specific Integrated Circuits) and the setup of specialised manufacturing lines. This does make sense for mass market products with high production volumes, but not for products like a measurement system of which only a few thousand units will be deployed across Europe.
- *Inflexible* – once production of systems has started, it is very hard to change anything. This makes the system not future secure: imagine what happens if GSM is finally replaced by UMTS and the SoC is designed to use GSM. The result is partial redevelopment. Similarly, we would be unable to utilise newly available sensor technology.

All these points make a SoC an unsuitable design approach for the desired in-situ measurement systems. However, one main point still holds:

A system on chip consumes less power because it avoids unnecessary abstraction layers.

For the proposed processing platform this means: we must find a way to avoid unnecessary abstraction layers and at the same time we need the ability to interface to different building blocks of the IMS.

1.4. Proposed Design

The flexibility of FPGAs permits us to accommodate all electrical interfaces used by our partners, ignoring any necessary voltage conversions for the moment. Furthermore, FPGAs can also perform signal processing tasks. Due to their truly parallel nature FPGAs can be clocked much slower, compared to machine-architectures, while performing the same processing. In general, higher clock speeds result in higher energy consumption. On the other hand there are some parts of the system, for instance the measurement scheduler, which are best implemented using a machine-architecture. Furthermore, having a built-in machine-logic lowers the barriers for our partners not used to FPGAs to use the system. To solve this problem we propose to utilise an FPGA with a built-in processor, such FPGAs are readily available from vendors such as Xilinx, Inc. Presently, we are evaluating a Xilinx Virtex 4 (XC4VFX12) [18] which has an embedded PowerPC 405. Figure 3 shows an overview of the hardware of the proposed practical solution.

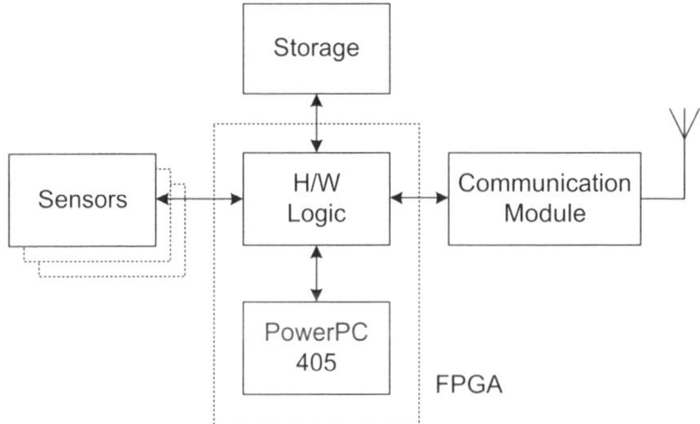

Figure 3. Hardware setup of the in-situ measurement station

The IMS hardware is a mix of hardware-logic and machine-logic, which must be combined in order to achieve our goal of a low power processing platform for the in-situ measurement system. Furthermore, this system is a true parallel system where new samples are aggregated, while older ones are analysed, and the results of previous analysis are sent to a Data Centre, at the same time. To avoid any race, deadlock, or livelock conditions we decided to follow the principles of CSP (Communicating Sequential Processes) and treat the different entities as CSP processes. This means, these entities only communicate over CSP style channels. What still remains to do is to design the interface between the different components, especially between the software executed by the machine-logic and the hardware-logic.

2. System Design Inside the FPGA

The previous section detailed the proposed hardware setup for the IMS. This is only the outside view of the system, what happens inside the FPGA, is much more interesting. Figure 4 shows one possible configuration / use of the FPGA. In the centre of the FPGA is a native PowerPC 405 core, which acts as central controller of the IMS, communicating over FDLs (Fast Duplex Links)[1] with the Sensor Controllers 1–N, Storage Controller, Comms

---

[1] A fast duplex link consists of two FSLs (Fast Simplex Links) [19], one for each direction.

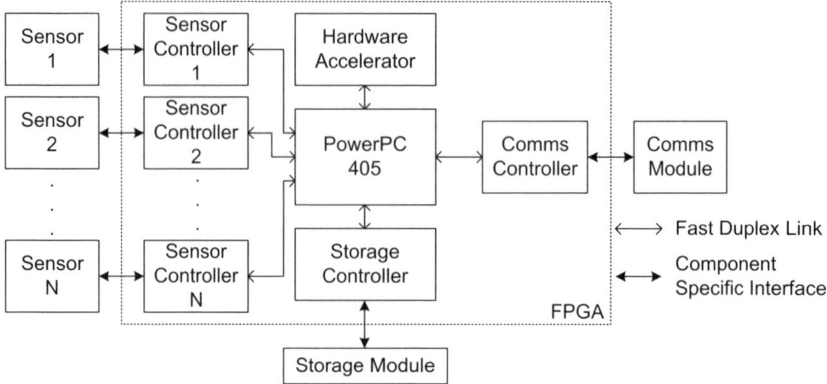

Figure 4. Inside view of the processing platform

Controller, and a Hardware Accelerator. In the following we detail the role of each of these components:

- Sensor Controllers – These interface the PowerPC core to the individual sensors. This interfacing involves not only protocol translation but also includes the electrical interface of the sensor. The PowerPC core communicates with the Sensor Controllers using a unified interface. This unified interface is a dedicated protocol between the Sensor Controllers and the PowerPC core. Furthermore, the Sensor Controllers can perform sensor specific signal processing. This avoids doing this processing in the sensor and hence allows a higher integration of the complete system.
- Comms Controller – The task of the Comms Controller is to interface the PowerPC core to a Comms Module. Similar to the Sensor Controller the Comms Controller performs not only a protocol translation but also provides the necessary electrical interface. The Comms Controller and the PowerPC core communicate over a FDL using a standardised protocol. This allows us to exchange the Comms Module, for instance to move from GSM to UMTS.
- Storage Controller – The Storage Controller abstracts the interface of the Storage Module and provides it in the form of a predefined protocol over a FDL.
- Hardware Accelerator – The Hardware Accelerator performs signal processing tasks which are too complex for the PowerPC core. The PowerPC core communicates with the Hardware Accelerator using a FDL. The system, shown in Figure 4, contains only a single Hardware Accelerator. However, there is no reason to limit the number of hardware accelerators in the system.

All these components are implemented as hardware-logic cores using the normal development tools. In the following we detail how the PowerPC core will be integrated in the design.

2.1. Integration of the PowerPC Core

The hardware-logic setup is fairly simple, each component represents one process, connected via FDLs to the PowerPC core. The controllers interface to hardware entities outside the processing platform, using hardware specific interfaces. Figure 5 shows the software process structure executed by the PowerPC. This structure is very similar to the hardware-process structure, detailed in Figure 4. In the center of the design is the *IMS Control Process*, which controls the IMS.

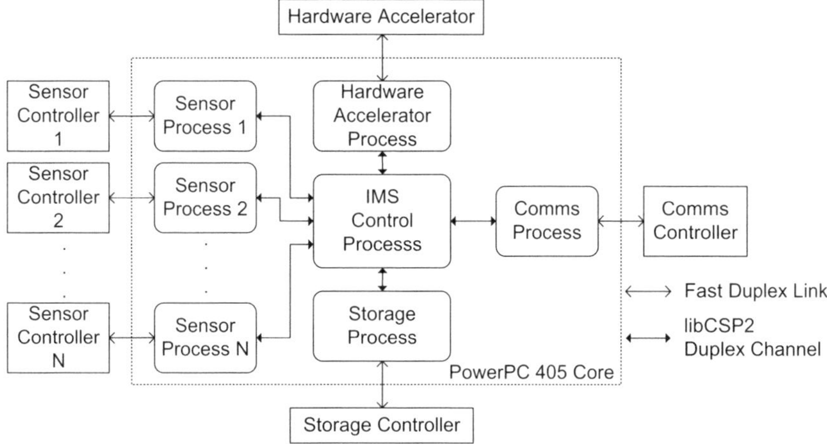

Figure 5. Process network within the PowerPC

Each present Sensor Controller is represented by a Sensor Process in software. This Sensor Process translates requests from the IMS Control Process into requests for the Sensor Controller, furthermore it may do additional signal processing. Possible applications for this sensor specific signal processing are, for instance: sensor calibration, detection of faulty sensors, or data type conversions.

The IMS Control Process communicates with the Comms Controller via the Comms Process. This process performs the translation, and contains all necessary configuration / authentication information for the chosen communication network. For instance when using GSM to connect to the Data Centre, the authentication to the GSM network provider is handled by this process. Furthermore, this process handles the identification of the IMS with the Data Centre, and ensures that no messages are lost between IMS and Data Centre.

To communicate with the Storage Controller the IMS Control Process communicates with the Storage Process. Initially, this process will only perform a simple request translation. However, in future we can add a file system like functionality, for instance by appending the current time and date to each entry. Furthermore, it could support data encryption to prevent others reading the stored information.

Finally, the Hardware Accelerator Process translates requests between the IMS Control Process and the Hardware Accelerator. This process will expose a call-channel interface. This allows users to utilise the functions the Hardware Accelerator offers just like a normal function call.

3. Implementation of the Software

The previous sections detailed the proposed structure of the IMS processing platform. A vital aspect of the proposed system is the duality between hardware and software. In the following we discuss how we plan to implement the software system run by the PowerPC core. There are a number of constraints which we need to take into consideration when implementing the software:

- Amount of memory available to the PowerPC; The currently proposed FPGA (XC4VFX12 [18]) offers 81KB of BRAM (Block RAM) memory, i.e. if any hardware-logic requires memory, this memory is deducted from these 81 KB. It is of course possible to use external RAM, however this requires more energy, more space on the PCB (Printed Circuit Board), and finally it costs more money. Another possibility is to choose an FPGA with more internal memory, but even then the memory footprint of the software remains an issue.

- Utilisation of non-standard interfaces of the PowerPC; The proposed processing platform relies on the use of FDLs to communicate with the hardware-logic. Hence, the chosen operating system must allow us to access the FDL interface provided by the PowerPC core.
- The choice of programming languages is limited; We are not the only ones developing software for the proposed processing platform. In fact most of the extensions to it will be developed by our partners in the consortium. The questionnaire revealed that most people are familiar with C and C++. It is safe to assume that they have already legacy code which they would like to reuse.

3.1. Operating System

In order to comply with these constraints, we decided to use XMK (Xilinx Microkernel) [20] as OS (Operating System). XMK is a small OS for the Xilinx MicroBlaze SoftCPU and hard wired PowerPC 405 cores. This OS abstracts the access to the FDL interfaces for both MicroBlaze and PowerPC 405.

After choosing an applicable OS it is time to choose a CSP environment to implement our processes on the machine-architecture. One possible choice is to port and extend the Transterpreter [21,22] to our chosen hardware platform. Choosing the Transterpreter implies that the software will be developed in Occam. While this is no problem for us, our partners don't have that background and furthermore they have legacy code in other languages such as C. Therefore, we need a C based solution. We therefore propose to use libCSP2 [23] as CSP environment for the processing platform. There are two reasons for this: firstly, libCSP2 has already built-in support for FDLs, secondly it allows one to develop CSP style software in C. This ensures a flat learning curve for our partners, when they want to develop their own extensions to our system.

3.2. Recent Developments of libCSP2

There have been a number of small enhancements to libCSP2 recently. First, libCSP2 now abstracts FSLs as normal channels, i.e. a process can not determine whether it uses a software channel or an FSL based channel. This allows developers greater freedom when doing multi-core designs with libCSP2.

Furthermore, we changed the build system from autotools to CMake [24]. This step allows users now to build the library outside of their current software project. This means the libCSP2 source code is not present in the users software project.

Presently, we are working on a formal verification of the implementation of libCSP2 on XMK. However, this work will still take some time to complete.

4. Example: Sensor Integration

This section demonstrates the proposed CSP based design flow for NoCs works. Our design goal is the integration of a new sensor into the processing platform. The desired functionality is straightforward: the IMS Control Process acquires a measurement value form a sensor and triggers an alarm is the acquired measurement value is above a certain threshold. We implement this simple example with libCSP2 technology on a Virtex-4 FPGA.

The first step in the design is to create and test a functional CSP model of the system. Subsequently, the CSP processes are mapped onto the available processors, in this case FPGA and PowerPC. After the mapping step follows the implementation. The individual processes are implemented using processor specific tools and networked using domain specific channels. The following text concentrates on the implementation of the processes on the PowerPC using libCSP2 and how to link them with processes located within the FPGA.

4.1. CSP Model

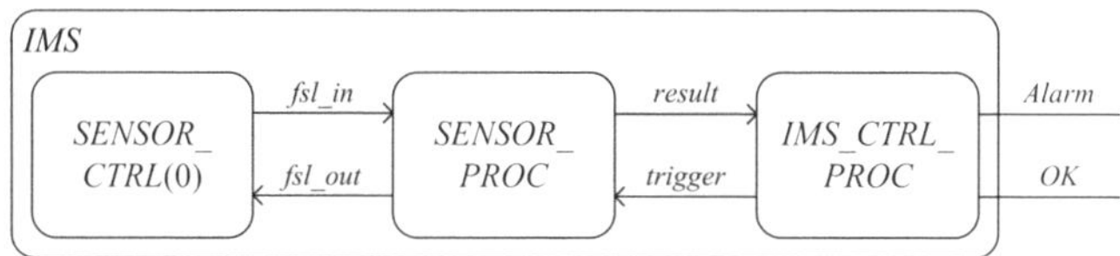

Figure 6. IMS process network

Figure 6 shows the *IMS* process network which represents the system functionality. The main task of the process network is to raise an *Alarm* when a sensor detects a harmful pollution. We model this functionality with three processes: *SENSOR_CTRL*, *SENSOR_PROC*, and *IMS_CTRL_PROC*. These communicate over the channels: *fsl_in*, *fsl_out*, *trigger*, and *result*. We did not do extensive tests on the CSP model, because it is very simple. The following paragraphs explain the CSP model for each of these processes.

The process *SENSOR_CTRL*(i) (Equation 1) represents the Sensor Controller. This process expects to receive the command value 48^2 from the channel *fsl_out* and then returns a measurement value in the range $[0..49]$. Any other value on channel *fsl_out* will be ignored.

$$SENSOR\_CTRL(i) =$$
$$fsl\_out?x : \{48\} \rightarrow fsl\_in!(i \bmod 50) \rightarrow SENSOR\_CTRL(i + 1) \tag{1}$$

Equation 2 specifies the process *SENSOR_PROC* which represents the Sensor Process. The process waits for any message on channel *trigger* and then requests a measurement value from the Sensor Controller, by sending the value 48 over channel *fsl_out*. Then it waits for a message from the Sensor Controller on channel *fsl_in*. This message represents the measurement value. The process then relays this value to the IMS Control Process over the channel *result*. The process is now ready to process the next request.

$$SENSOR\_PROC =$$
$$trigger?x \rightarrow fsl\_out!48 \rightarrow fsl\_in?x \rightarrow result!x \rightarrow SENSOR\_PROC \tag{2}$$

Process *IMS_CTRL_PROC* (Equation 3) represents the IMS Control Process. The IMS Control process requests a measurement value from the Sensor Process, by sending a message over the channel *trigger*. After that it waits for the measurement value to arrive on the channel *result* and then compares the received value with 42. If the measurement value is smaller or equal 42 everything is OK and the process issues an *OK* event. Otherwise, the process issues an *Alarm* event. In both cases the process recurses to start a new round of measurement.

---

^{2}The command value, as well as the range of measurement values ($[0..49]$) and the threshold value (42), are arbitrary chosen values.

$$IMS\_CTRL\_PROC =$$

$$trigger!1 \rightarrow result?x \rightarrow \begin{cases} Alarm \rightarrow IMS\_CTRL\_PROC & \text{if } x > 42 \\ OK \rightarrow IMS\_CTRL\_PROC & \text{otherwise} \end{cases} \quad (3)$$

Process *IMS* (Equation 4) represents the complete IMS, which consists of the processes: *SENSOR_CTRL*(0), *SENSOR_PROC*, and *IMS_CTRL_PROC*. To avoid any outside interference all transactions on channels: *fsl_in*, *fsl_out*, *trigger*, and *result* are hidden. Only the events *Alarm* and *OK* are visible to the outside world.

$$IMS = SENSOR\_CTRL(0) \parallel SENSOR\_PROC \parallel IMS\_CTRL\_PROC$$
$$\setminus \{| \, fsl\_in, fsl\_out, trigger, result \, |\} \quad (4)$$

4.2. Mapping the Processes onto the Available Processors

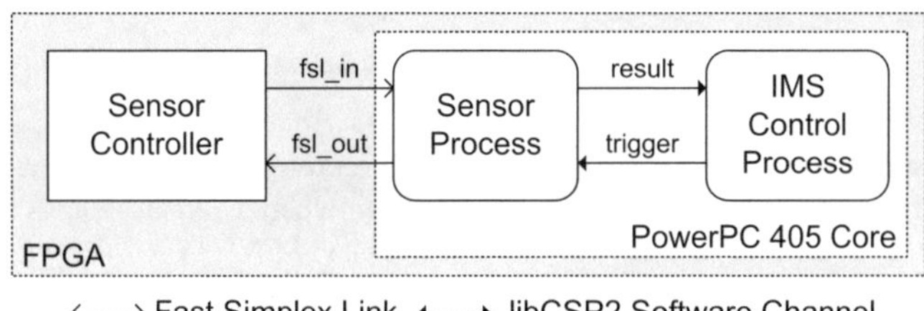

Figure 7. Process mapping onto the processors of the processing platform

Figure 7 illustrates the mapping of the CSP processes onto the processors of the processing platform. The Sensor Controller gets implemented outside the PowerPC, in order to interface directly with the Sensor. The Sensor Process is located within the PowerPC core together with the IMS Control Process. The Sensor Controller and the Sensor Process communicate using FSLs (Fast Simplex Links), while Sensor Process and IMS Control Process use libCSP2 software channels.

4.3. Implementing the Processes Located within the PowerPC

Listing 1 is the libCSP2 implementation of process *SENSOR_PROC* (Equation 2).

```
void SensorProcess(pChannel trigger, pChannel result, \
                   pChannel fsl_in, pChannel fsl_out){
  int msg = 0;
  while(1){
    ChanInInt(trigger, &msg);
    ChanOutInt(fsl_out, 48);
    ChanInInt(fsl_in, &msg);
    ChanOutInt(result, msg );
} }
```

Listing 1: Sensor Process Implementation using libCSP2

Listing 2 gives the libCSP2 implementation of process *IMS_CTRL_PROC* defined in Equation 3. From a functional point of view, the main difference between the functional model and the implementation lies in the handling of the alarm. The implementation does not issue an alarm event over a channel, instead it outputs the corresponding strings onto the console, (lines 7 and 9).

```
   void ControlProcess(pChannel trigger, pChannel result){
2    int value = 0;
     while(1){
4      ChanOutInt(trigger, 1); // trigger a new measurement value
       ChanInInt(result, &value); // receive the new value
6      if(42 < value){
         xil_printf("Alarm\r\n");
8      }else{
         xil_printf("OK\r\n");
10 } } }
```

Listing 2: IMS Control Process Implementation

The processes located in the PowerPC core have now to be instantiated, linked with the Sensor Controller and then executed. Listing 3 demonstrates how this is done using libCSP2. The listing consists of three sections: one declaring necessary variables (lines 2 – 7), a definition section (lines 10 and 11). The last section is the function **void*** shell_main(**void**) which uses these declarations and definitions (lines 13 – 21).

```
   //Channel and process declarations;
2  pProcess sensor  = NULL;
   pProcess control = NULL;
4  pChannel trigger = NULL;
   pChannel result  = NULL;
6  pChannel fsl_in  = NULL;
   pChannel fsl_out = NULL;
8
   // Defining necessary intermediate functions;
10 void procSensor(void) {SensorProcess(trigger, result, fsl_in, fsl_out);}
   void procControl(void) {ControlProcess(trigger, result);}
12
   void* shell_main(void*   dummmy){
14   CSP_ChanAllocInit( &trigger, CSP_ONE2ONE_CHANNEL);
     CSP_ChanAllocInit( &result, CSP_ONE2ONE_CHANNEL);
16   FSL_AllocInitChannelInput( &fsl_in, CSP_ONE2ONE_CHANNEL, 0);
     FSL_AllocInitChannelOutput( &fsl_out, CSP_ONE2ONE_CHANNEL, 0);
18   ProcAllocInit(&sensor, procSensor);
     ProcAllocInit(&control, procControl);
20   ProcPar(control, sensor, NULL);
   }
```

Listing 3: IMS Processing Platform Setup

Lines 10 and 11 define intermediate functions, which represent the Sensor Process and the IMS Control Process. The reason for these intermediate functions is that libCSP2, in its current state, only allows un-parametrised functions to act as processes. Unfortunately, the functions which represent the Sensor Process and the IMS Control Process have parameters, making these intermediate functions necessary.

The function **void*** shell_main(**void**) (line 13) represents the program entry point. Once started it allocates and initialises the channels: trigger and result as normal software channels (lines 14 and 15). To connect the Sensor Process with the Sensor Controller the function allocates and initialises two FSL channel-ends: fsl_in and fsl_out. The channel-end fsl_in gets allocated as FSL channel input for FSL-ID 0 (line 16). This means that the process using this channel end may only input data from the FSL, but not output data to it. The function then allocates and initialises the channel-end fsl_out as FSL channel output for FSL-ID 0 (line 16). The statements that follow (lines 18 – 20), allocate and initialise the two processes and then execute them in parallel. This completes the implementation of the process network located in the PowerPC.

This example demonstrated how to implement NoCs using libCSP2 and Xilinx FSL. Furthermore, it supports our claim of unification of channels within libCSP2. To move the Sensor Process outside the PowerPC no change of the IMS Control Process is necessary, only the software channels `trigger` and `result` have to be replaced with FSL channel-ends.

5. Conclusions and Further Work

This paper proposed a processing platform design for the WARMER in-situ monitoring system. One of the requirements of this in-situ monitoring system is: Long service intervals in the range of 6 to 12 months. Another aspect is the flexibility to work with or replace various existing systems used by other WARMER collaborators. One last requirement is to design the system such that our partners can reuse their code and extend the system without our help. In the practical part of the paper we demonstrated how FPGA technology can be used to achieve higher system integration with more flexibility. These goals were achieved with a network of software and hardware processes.

A big advantage of the proposed system is the sheer ease with which it allows the designer to create hybrids of hardware- and machine-logic, when using CSP style communications between these processes. Another benefit of Communicating Sequential Processes is the duality of hardware- and machine-logic. Each hardware-logic core has a process representation within the machine-logic. This allows the designer to choose the processing platform which executes specific data or control centric algorithms. This freedom leads to optimised systems, because of an optimal use of processing resources. Furthermore, it is easy to follow the data flow within the system. This makes the system easy to understand and extend. The use of libCSP2 as CSP environment for the software part of the system allows the partners to reuse previously developed algorithms without too much difficulty.

The approach we present in this paper is not restricted to processing platforms embedded in in-situ monitoring systems but is generally applicable to hardware-software co-design.

5.1. Further Work

This project is still in the drafting stage, and there is still a lot of work to be done. Nevertheless, we already see a number of areas to be explored. In its current state the processing platform needs to be designed / compiled specifically for the used sensors. While this is fine for prototyping and small scale use it becomes a nuisance once the system is deployed out in the field, because it is not possible to plug and play the sensors. To solve this, two areas need to be investigated: partial reconfiguration of the FPGA, and partial process network reconfiguration of the libCSP2 process network. Here we see again the duality of the system design, where hardware- and machine-logic are closely coupled. For libCSP2 these requests mean to implement stateful poisoning. Furthermore, libCSP2 needs an extension that provides call-channels. However, the previously mentioned further work items are long term work items, in the short term we need to start developing the protocols used between the PowerPC 405 core and the hardware-logic cores. Not to forget convincing our partners of the advantages of this design approach.

Acknowlededements

This work was supported by the European FP6 project "WARMER" (contract no.: FP6-034472).

References

[1] Amara Gunatilaka. Groundwater woes of Asia. *Asian Water*, pages 19–23, January 2005.

[2] Amara Gunatilaka. Can EU directives show Asia the Way. *Asian Water*, pages 14–17, December 2006.

[3] Results of the IST-2000-28084 Project SEWING: System for European Water monitorING. Available (23.04.2007) at: http://www.sewing.mixdes.org/downloads/final_results.pdf, December 2004.

[4] Renata Mamińska and Wojciech Wróblewski. Solid-state microelectrodes for flow-cell analysis based on planar back-side contact transducers. *Electroanalysis*, 18(13–14):1347–1353, July 2006.

[5] V. V Malinovsky and S. Sandven. SAR monitoring of oil spills and natural slicks in the Black Sea. Submitted to *Remote Sensing of Environment.*, 2007.

[6] A Legin, A Rudnitskaya, B Seleznev, and D Kirsanov. Chemical sensors and arrays for simultaneous activity detection of several heavy metal ions at low ppb level. In *Proceeding of Pittcon 2004*. Pittsburgh Conference, March 2004.

[7] M. T. Castañeda, B. Pérez, M. Pumera, A. Merkoçi, and S. Alegret. Sensitive stripping voltammetry of heavy metals by using a composite sensor based on a built-in bismuth precursor. *Analyst*, 130(6):971–976, 2005.

[8] Luca Benini and Giovanni De Micheli. Networks on Chips: A New SoC Paradigm. *Computer*, 35(1):70–78, 2002.

[9] Grant Martin. Book Reviews: NoC, NoC ... Who's there? *IEEE Design and Test of Computers*, 23(6):500–501, 2006.

[10] Hugo De Man. System-on-Chip Design: Impact on Education and Research. *IEEE Des. Test*, 16(3):11–19, 1999.

[11] M. Rahnema. Overview of the GSM System and Protocol Architecture. *IEEE Communications Magazine*, 31(4), April 1993.

[12] Richard D. Schneeman. Implementing a standards-based distribution measurement and control application on the internet. Technical report, U.S. Department of Commerce, Gaithersburg, Maryland 20899 USA, June 1999.

[13] James D. Gilsinn and Kang Lee. Wireless interfaces for IEEE 1451 sensor networks. In *Proceedings of the First ISA/IEEE Conference*, pages 45–50. IEEE, November 2001.

[14] Antonios Alexio, Dimitrios Antonellis, and Christos Bouras. Evaluating Different One to Many Packet Delivery Schemes for UMTS. In *WOWMOM '06: Proceedings of the 2006 International Symposium on on World of Wireless, Mobile and Multimedia Networks*, pages 66–72, Washington, DC, USA, 2006. IEEE Computer Society.

[15] Xiao Xu, Yi-Chiun Chen, Hua Xu, Eren Gonen, and Peijuan Liu. Parallel and distributed systems: simulation analysis of RLC timers in UMTS systems. In *WSC '02: Proceedings of the 34th conference on Winter simulation*, pages 506–512. Winter Simulation Conference, 2002.

[16] S.J. Vaughan-Nichols. Achieving wireless broadband with WiMax. *Computer*, 37(6):10–13, June 2004.

[17] Teri Robinson. WiMax to the world? *netWorker*, 9(4):28–34, 2005.

[18] Xilinx, Inc., 2100 Logic Drive San Jose, CA 95124-3400, United States of America. *Virtex-4 Family Overview*, DS12 (v2.0) edition, January 2007.

[19] Xilinx, Inc. *Fast Simplex Link (FSL) Bus (v2.00a)*, 1 December 2005.

[20] Xilinx, Inc, 2100 Logic Drive San Jose, California 95124 United States of America. *OS and Libraries Document Collection*, 24 October 2005.

[21] Christian Jacobson and Matthew C. Jadud. The Transterpreter: A Transputer Interpreter. In Ian R. East, David Duce, Mark Green, Jeremy M. R. Martin, and Peter H. Welch, editors, *Communicating Process Architectures 2004*, pages 99–106, September 2004.

[22] Damian J. Dimmich, Christian Jacobson, and Matthew C. Jadud. Native Code Generation using the Transterpreter. In Frederick R. M. Barnes, Jon M. Kerridge, and Peter H. Welch, editors, *Communicating Process Architectures 2006*, pages 269–280, September 2006.

[23] Bernhard Sputh, Oliver Faust, and Alastair R. Allen. Portable CSP Based Design for Embedded Multi-Core Systems. In Frederick R. M. Barnes, Jon M. Kerridge, and Peter H. Welch, editors, *Communicating Process Architectures 2006*, pages 123–134, September 2006.

[24] Ken Martin and Bill Hoffman. *Mastering CMake 2.2 Edition*. Kitware, Inc., Clifton Park NY, USA, 24 February 2006.

Communicating Process Architectures 2007
Alistair McEwan, Steve Schneider, Wilson Ifill and Peter Welch (Eds.)
IOS Press, 2007

High Cohesion and Low Coupling: the Office Mapping Factor

Øyvind TEIG
Autronica Fire and Security (A UTC Fire and Security company)
Trondheim, Norway
`http://home.no.net/oyvteig`

Abstract. This case *observation* describes how an embedded industrial software architecture was "mapped" onto an office layout. It describes a particular type of program architecture that does this mapping rather well. The more a programmer knows what to do, and so may withdraw to his office and do it, the higher the *cohesion* or completeness. The less s/he has to know about what is going on in other offices, the lower the *coupling* or disturbance. The project, which made us aware of this, was an embedded system built on the well-known process data-flow architecture. All interprocess communication that carried data was on synchronous, blocking channels. In this programming paradigm, it is possible for a process to refuse to "listen" on a channel while it is busy doing other things. We think that this in a way corresponds to closing the door to an office. When another process needs to communicate with such a process, it will simply be blocked (and descheduled). No queuing is done. The process, or the programmer, need not worry about holding up others. The net result seems to be good isolation of work and easier implementation. The isolation also enables faster pinpointing of where an error may be and, hence, in fixing the error in one place only. Even before the product was shipped, it was possible to keep the system with close to zero known errors. The paradigm described here has become a valuable tool in our toolbox. However, when this paradigm is used, one must also pay attention should complexity start to grow beyond expectations, as it may be a sign of too *high* cohesion or too *little* coupling.

Keywords. Case study, embedded, channel, Office Mapping Factor, cohesion, coupling, black-box encapsulation

Introduction

The system we are describing here has been discussed in two published papers [1-2]. It was a rather small project with up to four programmers, running for some time. The result was several hundred KB of fully optimized code in an 8-bit microcontroller. The product we discuss has high commercial value for our company. It is *part of* a new fire detection panel, with one such unit per cabled loop. A loop contains addressable units, for fire detection and other inputs or outputs. (Autronica pioneered "addressable fire detectors" in the late seventies.) Together with fire detectors and fire panels it completes Autronica Fire and Security's range of products. The product described here is called *AutroLooper* and is not available as a separate product.

Several AutroLoopers communicate (over a text based protocol) with a "main" processor on the board. The team, which developed that system partly, used object orientation, UML and automatic code generation. The degree of "office mapping factor" in that system is not discussed here.

Working in industry on rather small projects, we seem to get on with the next project as fast as possible when one project has been properly tested, documented and then "closed". This paper is meant as a reflection over an observation, squeezed into a glitch between the project proper and its follow-up, a linear expansion of the product we describe here.

This case study tries to list and explain our experience. We have not done any comparative study or discussed what we could have done differently. There are no metrics involved in this industrial "case observation".

Observe that in this document we often refer to what "we" did and how "our" implementation is. Please see [1-2] for all these cases.

In this document, a *process* is what some embedded real-time systems mostly consist of, closely related to the *thread* and *task* terms.

1. The Office Mapping Factor

By *office mapping factor* we mean the degree to which a program architecture may be mapped to separate offices for individual programmers to implement, and the degree to which that implementation work is complete (high cohesion) and undisturbed (low coupling). We assume that module properties like high cohesion and low coupling cause high office mapping factor, and that high office factor is wanted.

Prior to the mapping to offices, the program architecture would have gone through a partitioning phase mostly based on functional decomposition. We believe that high office mapping factor gives high satisfaction in the office, as the programmer knows what he should do, and does not have to think about how the others do their parts.

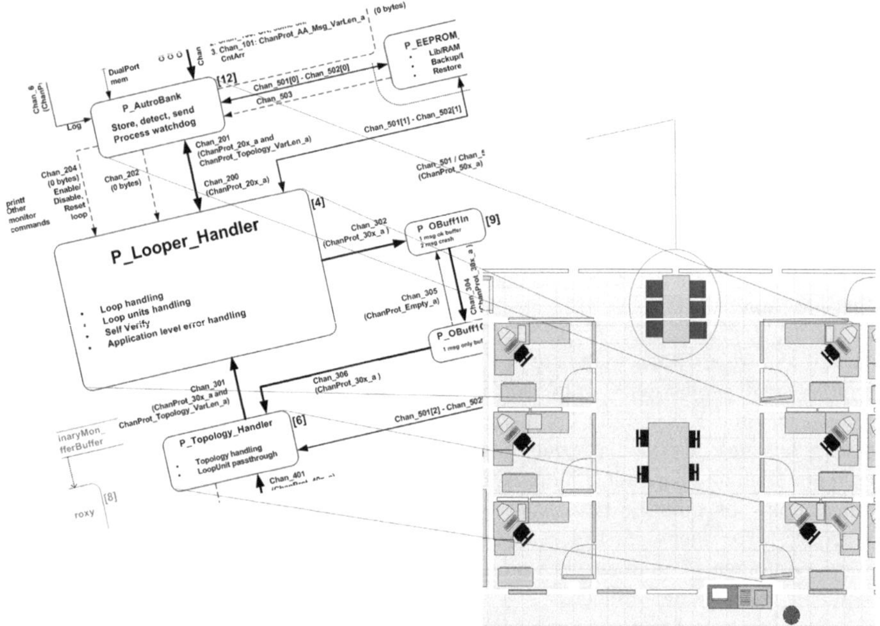

Figure 1. Software architecture and office floor plan

The architecture we built (glimpsed in Figure 1) was, to a large extent, a consequence of informal meetings around a whiteboard, and an understanding of the semantics of our processes and channels (pictures of those whiteboards are our minutes). This way the architecture itself reflected each team member's speciality. Management had picked people with some, but not much overlap in knowledge. We believe that this contributed to a higher office mapping factor. Not only beneficial for development, we also think that as time passes and maintenance needs to be done, getting different people on the project will be easier with this high office mapping factor.

Office mapping could also allow that one programmer does more than one process. It would mean that he would mostly need to relate to the communication pattern between his own processes. Role wise he would first do his one job, exit the office and enter it again for another. And mostly forget about the internals of his finished work only to concentrate on the present.

2. High Cohesion and Low Coupling and the Office

In the context of a confined office, having high cohesion means that the programmer knows what to do and is able to fulfil the task having little communication or coupling with the others in the team. He would not need to know how the others solve their processes.

Cohesion and coupling in this case seem to be inversely related. The less complex the protocols between the processes are, the more complete is a process' work.

However, the programmer must understand that the protocol or contract must be adhered to 100%, and he must know that he cannot "cheat" by sharing state with the other processes – other than by concrete communication. Going by the agreed-upon cross-office rules (the protocol message contents and sequence semantics) also gives a concerted feeling: one is part of a real team.

But, is this not is how programming has been done since the fifties? For procedural programming languages a function has always taken parameters and returned values. A function has had high cohesion, and coupling has been through the parameters. However, concurrent constructs (or even Object Oriented constructs) may in some cases be at stake with the cohesion and coupling matters. Processes may be implemented more or less as black boxes and may have subtle communication patterns. The lesson learned with **occam** [3] in the eighties and nineties was that the clean process and communication model was worth building on. This is what we did in this product. **occam** (without pragmas) ensured complete "black-box" encapsulation.

3. Process Encapsulation Check List and the Office Mapping Factor

A check list for process encapsulation might be like this (below). One could say that "wear" of the office mapping factor may be caused by:

1. For a process, not being able to control when it is going to be used by the other processes. Serving "in between" or putting calls in a local queue makes it much more complicated to have "quality cohesion". *[Java classes, for example, cannot prevent their methods being called (except, through 'synchronized', where other synchronized methods can't run if one is already running). But that does not prevent the method from happening. It just delays it. It cannot be stopped, even if the object is not in a position to service it (like a "get" on an empty buffer). Not being able to control when it may be used by other processes means that things of*

importance to a process may change without it being aware of it. This trait is further discussed here, since the other points (below) only to a small degree are valid in our project.]

2. Incorrect or forgotten use of protection for inter-process communication. *[So, use a safe pattern for it – as we have used in this project.]*

3. Communication buffer overflows would most often cause system crashes. *[We use synchronous channels, which cannot cause buffer overflow during inter-process communication. However, buffer overflows on external I/O are handled particularly by link level protocols.]*

4. Mixing objects and processes in most languages and operating systems, since most languages have been designed to allow several types of patterns to be implemented. *[We use the process definition from a proprietary run-time system which gives us occam-like processes and intercommunication in ANSI C.]*

5. Too much inheritance in OO. There is a well documented tension between inheritance and encapsulation, since a subclass is exposed to the details of its parent's implementation. This has been called "white-box" encapsulation [4]. This is especially interesting here, if a process is an instance of an object.

6. Aliasing of variables or objects. Aliasing is to have more than one name on the same memory cell simultaneously. This type of behaviour is required in doubly linked lists, but would cause subtle errors found well into the product's life cycle. *[We don't think we have these.]*

7. Tuning with many priorities. Priority inversion may soon happen. How to get out of a potential priority inversion state may be handled by the operating system. However, many smaller systems do not have this facility. Therefore design with many priorities is difficult to prove not to have errors. *[We have medium priority for all Processes (scheduling of them is paused when the run queue is empty), low for Drivers (called when ready queue is empty), and high priority for all interrupts (which never cause any rescheduling directly). This is a scheme, which holds for the rather low load that our non pre-emptive system needs to handle.]*

8. Not daring to do assert programming and instead leave unforeseen states unhandled or incorrectly handled. System crashes by assert programming puts the pain up front, hopefully before any real damage is done. However, it also removes subtle errors at the end. *[We have used a globally removable macro for most asserts, and we have hundreds of them. Overall, they seem to cause so little overhead and such high comfort that we have not felt it correct to remove them. This author thinks of them as self-repairing genes of our "software cell": on each iteration with the programmer, the cell enters longer and longer life cycles.]*

4. Mapping

The mapping of the processes was easily done, since the team members in our case had their specialities. After all, that is why management had co-located us in the new facilities with subunits of 6-and-6 offices around small squares (Fig.1).

The application proper was handled by two persons, the intricacies of the loop protocol by a third and the internal data store and high level text protocol by a fourth. And importantly, the fifth – a proper working project leader. One of the five also had responsibility for the architecture (this author). His experience (of which much was with occam and the SPoC implementation [5]) projected, of course, that way of thinking on the architecture. This was much discussed in the team prior to the decision, but with the some

20 years of average embedded real-time experience for the rest of the team, the concepts were soon understood and accepted. Even, with some enthusiasm.

With the mapping of processes to offices (in most respects here, "office" really means "person"), we had a parallel architecture that also enabled parallel development. We think this shortened development time – the higher the office factor, the greater the shortening.

5. The Closed Door Metaphor

The first point in the numbered list above mentions "not being able to control when it is going to be used by the other processes" as a potential problem for the office mapping factor.

With the mapping scheme, terms from computer science become metaphors for office layout or human behaviour. Below, when a *door* is mentioned, it both means the physical door to the office, a person's possibility to be able to work undisturbed for some time, and that an embedded process is able to complete its work before it goes on to handle other requests. It is, as one can see in this paper, difficult not to anthropomorphise the behaviour of the computer.

When we talk about being undisturbed, we mean both undisturbed programmers and processes. Low cohesion means a good protocol between processes and a corresponding small need to discuss further with the other team members, because it is all in the protocol description.

Below, we outline three important facets, which in our opinion, may influence the office mapping factor: *sloping*, *non-sloping* and *sender-side sloping* message sequence diagrams. These terms are probably invented here, as eidetic tools.

5.1 Sloping Message Sequence Diagram

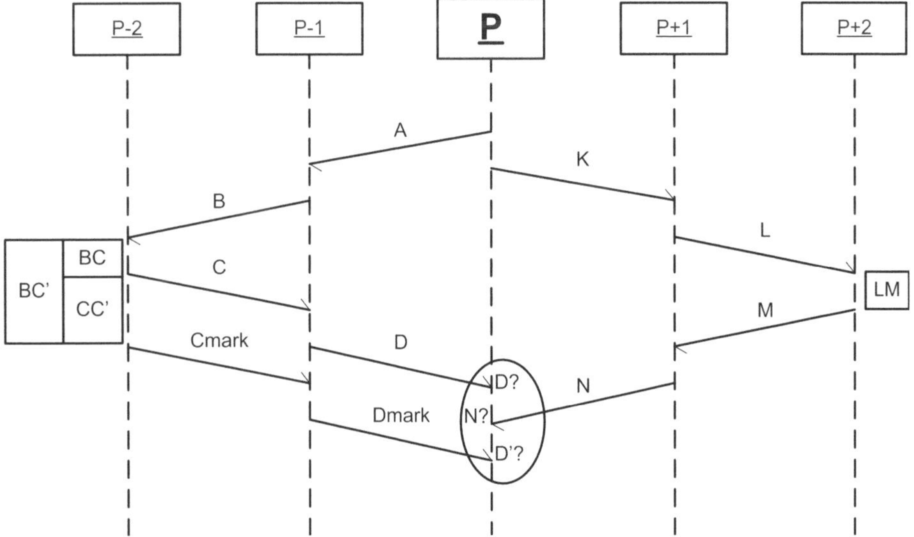

Figure 2. Asynchronous communication is "wide open"

A sloping message diagram describes a communication scheme where a process (the top boxes in Figure 2 show 5 of them) sends with no blocking or synchronization. This is called asynchronous communication. Here sending takes time, meaning that the time line (the vertical lines, where time flows downwards) is present both for sender and receiver. Sender sends at a time, the run-time system buffers the message, and the receiver receives it some time later. The important thing here is that neither party blocks or is descheduled for the sake of the communication. *Time flows and exists* for them – to do other things as required.

This communication scheme is much used. However, for a concurrent design it is only one important tool in the toolbox. If the asynchronous behaviour is wanted, it is the right tool. Otherwise, there may be certain advantages for *not* using this scheme.

In Figure 2, P sends two orders: A (first) and K for which it needs to have confirmation messages D and N. The middle, left box shows that the time for P-2 to respond is either time B to C ("BC") causing reply C, or BC' causing the same data to be sent later as reply C$_{mark}$. Depending on whether P-2 has to have the extra CC' time, the confirmation ordering back to P (of its original A then K messages) will be switched.

Not knowing which response comes first is illustrated by the question marks ("D?", "N?" or "D'?") in the centre bottom ellipse – to indicate that the acknowledges are indeterminate with respect to when they arrive.

Sometimes, the order and the confirmation must be in phase. Either it *must*, or it is *simpler* this way – with less internal complexity. Relying on *any* order of the replies could be equally problematic. With synchronised channel communication, we can be in charge on this point: we could decide to listen (and hold) any way we want.

With the scheme above, it would be better to make the design able to handle order swapping. Easy: just get on with the next job in P when the number of pending replies have reached zero.

But what if, instead of merely a swapped order, completely unrelated messages arrive from other senders? Then, it is not so easy: the process soon becomes a scheduler for itself. This adds complexity, because in the deepest sense every program has to know something about how the other parties with which it communicates behave internally. *"Can you wait a little in this case?" / "I will send you an extra confirmation when you may go on."* This kind of out-of-office conversation could be a warning sign that the next time the programmers enter their offices, it will take longer. And then, longer again. We do not have *WYSIWYG* semantics.

5.2 Non-sloping Message Sequence Diagram

Here we describe another tool in the toolbox: the *synchronous blocking communication* scheme. (Note that *blocking* means descheduled or set aside until the communication has happened. It does not mean making the processor slower or unable to do any meaningful work. The throughput of a synchronous system might even be higher than an asynchronous system, provided adequate buffering and asynchronicity is applied at the terminals.)

In Figure 3, we see messages drawn by an offline log client that we had made for us. Here, each time-stamp has a complete message; it is the time of the communication. The "rendezvous" happens there. It is *not* the time when the first process on the synchronous one-way channel gets descheduled. (In our case, we have a non-preemptive run-to-completion scheduler beneath.) At this point, *time stops* for this process. Time, of course, flows; but not for the descheduled process. It may only be scheduled again when the second process has been at the other end of the channel, and the run-time system has **memcpy**'d data across, directly from inside the state space of the sender to inside the state space of the receiver.

Although deadlocks may happen with synchronous systems unless safe patterns to avoid them are used [1-2], the synchronous communication scheme has some advantages.

Firstly, there is no possibility of buffer overflow.

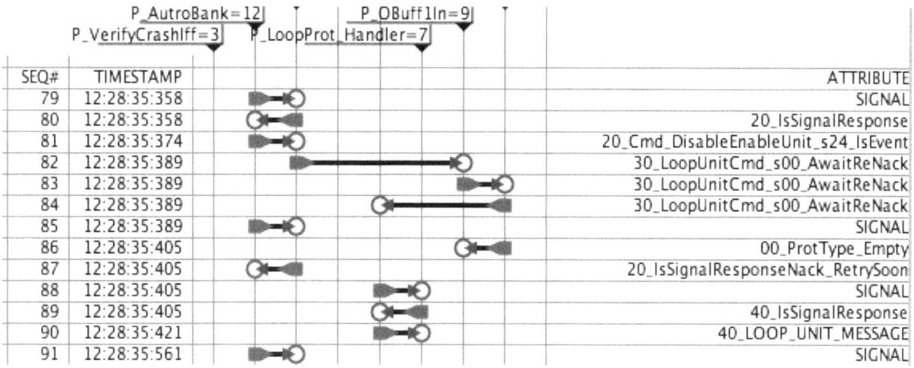

SEQ#	TIMESTAMP					ATTRIBUTE
79	12:28:35:358					SIGNAL
80	12:28:35:358					20_IsSignalResponse
81	12:28:35:374					20_Cmd_DisableEnableUnit_s24_IsEvent
82	12:28:35:389					30_LoopUnitCmd_s00_AwaitReNack
83	12:28:35:389					30_LoopUnitCmd_s00_AwaitReNack
84	12:28:35:389					30_LoopUnitCmd_s00_AwaitReNack
85	12:28:35:389					SIGNAL
86	12:28:35:405					00_ProtType_Empty
87	12:28:35:405					20_IsSignalResponseNack_RetrySoon
88	12:28:35:405					SIGNAL
89	12:28:35:405					40_IsSignalResponse
90	12:28:35:421					40_LOOP_UNIT_MESSAGE
91	12:28:35:561					SIGNAL

Figure 3. Synchronous communication has "door"

Secondly, and this is stressed here, the receiving process need not "listen on the channel" when it is active doing other work. It may communicate with other processes and not worry about whether there is any other process waiting on the ignored channel. Observe that it does not care, the process and data-flow architecture has been designed so that the waiting of one process does not have any consequences which the busy process need worry about. If, far above these processes there is a conversation going on with another processor, which needs a reply even if several processes are blocked further below, the design must have considered this[1].

Not listening on the channel is equal to having the office door shut. Building our system with this paradigm, we believe, has given lower coupling and higher quality cohesion. All communication in the system we describe here is based on this. We believe that this is one of the reasons why we seem to have a high office mapping factor[2].

Observe that the ALT construct makes it possible to listen to a set of channels or an array of channels, with or without timeout. This listening is blocking and – according to the door metaphor – individually closable. So, there is no busy-polling of channels (if this is not what we really want – at some asynchronous I/O terminal).

5.3 Sender-side-sloping Message Sequence Diagram: Pipes

It is possible to have asynchronous *sending* and blocking *reception* if we use *pipes*. With pipes there is one queue per pipe. A listener may then choose not to listen on a pipe. Most often a pipe may have at least one "buffer". Some times they block when they have received N elements, some times they just return a "full" state. Often a pipe cannot have zero buffers – which would have allowed for true synchronous messaging.

It is possible to build this kind of system also with a composite buffer process and synchronous blocking channels. We have one in our system, and it contains two small processes (it may be spotted in Figure 1 as *P_OBuff1In* and the process below it).

---

[1] In our case, it is handled with active process "role thinking". An in-between process is a *slave* both "up" and "down" and a mix of synchronous blocking data-rich and asynchronous data-less signals is used.

[2] The run-time layer we used to facilitate this we built on top of an asynchronous system. This was considered (correct or not, at the time) to be the only viable way to introduce this paradigm into the then present psycho-technological environment.

A pipe construction is a versatile tool. However, using it *may* give a little lower office mapping factor. We may have to know more about the sender: *"Does it block now? Should I treat it now? When does he send?"*. And the receiver: *"Is it listening now? May I be too eager a producer? How do I handle it if I have too much to send? Should I change state and go over to polled sending then?"*.

The fact that time has not stopped for the sender, after a sending, may therefore be a complicating factor.

6. Scope

The system we have described contains medium to large grained processes, which contain long program sequences. Whether the office mapping factor has any significance for a system with small state machines realised as communicating processes, we have not investigated.

Also, as mentioned, we have not done any comparative studies of other paradigms, like OO/UML. For the scope of this article, whether more traditional programming or OO/UML is used *inside* each office or process, is not discussed. It is the mapping of the full process data-flow architectural diagram onto offices that is discussed.

Taking a 100% OO/UML architecture, with only the necessary minimum of processes, and investigate the office mapping factor would be interesting.

7. Warnings

7.1 High Cohesion Could Cause Too High Internal Complexity

With high cohesion, there is of course a possibility that a person may sit so protected in the office that the system would organically grow more than wanted. Also, inside a process one has to watch out for the *cuckoo* entering the nest. It is hard to see every situation beforehand, but still it is also a good idea to analyse and design to some depth. Within a real-time process, any methodology that the programmer is comfortable with should be encouraged. This, of course, could include OO and UML.

7.2 Low Coupling Could Also Cause Too High Internal Complexity

We saw during our development phase that, if we modified the architecture, we were able to serve the internal process "applications" to a better extent. The first architecture is described in [1] and the second in [2]. However, not even [2] needs be the final architecture. With low coupling, we then have tools to insert new processes or new channels, or to remove some. This could be necessary if we discover that we do too much in a process. To split (and kick out the cuckoo) may be to rule – but it does not have to be. These considerations should be done any time an unforeseen complexity arises, if one has a feeling that it is an architectural issue. On the second architecture we introduced an *asynchronism* with the introduction of a two element (composite and *synchronous*) data buffer process. This led to *more* coupling (communication) and *less* cohesion (state) in the connected processes – but ultimately to lower complexity.

8. Testing

Inside each office individual testing was done, in the more traditional way, on smaller functions, with debugger and printouts.

However, interesting to see was that testing of the processes was almost always done *in vivo*, with all the other processes present – on each office's *build*. The reason that this was possible was that with the parallel implementations, the protocols were incrementally made more and more advanced, on a need to have basis. It seemed like the tasks were well balanced, because there was not much waiting for each other. Programming and testing was – almost, synchronous.

We kept track of each error and functional point. Before release of version 1.0 (yet to come) we have zero to a few known bugs to fix. It seems like it is easy to determine which office should do an error fix. There have been little errors in interprocess communication. It has been easy to determine where an error might be located.

9. Other Teams

We released incremental new beta versions for the other team to use, mostly on set dates. The date was the steering parameter, not a certain amount of functionality. We felt it was easier to keep the enthusiasm this way, and that it helped the office mapping factor. This has briefly been described in Norwegian in [6].

10. Conclusion

It seems that a successful mapping from a process data-flow architecture to offices is possible. Simultaneous programming with high cohesion (in process and office) and low coupling (between processes and offices) is defined as high "Office Mapping Factor", a term coined here. It seems like the product we have developed, described here and in two other publications ([1-2]), has benefited from the architecture chosen. We have not studied whether other methodologies would be better or worse off, since this paper is an industrial *case observation*.

References

[1] Ø. Teig, "From message queue to ready queue (Case study of a small, dependable synchronous blocking channels API – Ship & forget rather than send & forget)". In *ERCIM Workshop on Dependable Software Intensive Embedded Systems*, in cooperation with *31ˢᵗ. EUROMICRO Conference on Software Engineering and Advanced Applications (SEAA)*, Porto, Portugal, August/September 2005. Proceedings: ISBN 2-912335-15-9, IEEE Computer Press. [Read at `http://home.no.net/oyvteig/pub/pub_details.html#Ercim05`]

[2] Ø. Teig, "No Blocking on Yesterday's Embedded CSP Implementation (The Rubber Band of Getting it Right and Simple)". In *Communicating Process Architectures 2006*, P. Welch, J. Kerridge, and F.R.M. Barnes (Eds.), pp. 331-338, IOS Press, 2006. [Read at `http://home.no.net/oyvteig/pub/pub_details.html#NoBlocking`]

[3] Inmos Ltd.: "Occam-2 Reference Manual", Prentice Hall, 1998.

[4] E. Gamma, R. Helm, R. Johnson and J. Vlissides, "Design Patterns: Elements of Reusable Object-Oriented Software", Addison-Wesley, ISBN 0-201-63361-2, 1995.

[5] M. Debbage, M. Hill, S.M. Wykes and D.A. Nicole, "Southampton's Portable Occam Compiler (SPOC). In *Proceedings of WoTUG-17: Progress in Transputer and occam Research*, R. Miles and A.G. Chalmers (Eds.) , pp. 40-55, IOS Press, ISBN 90-5199-163-0, March 1994.

[6] Ø. Teig, "Så mye hadde vi. Så mye rakk vi. Men får de?" (In Norwegian) ["We had so much. We made it to this much. But how about them?" In *Teknisk Ukeblad* internett February 2006 and #15, May 2006, page 71. Read at `http://home.no.net/oyvteig/pub/pub_details.html#TU_feb_06`

Øyvind Teig is Senior Development Engineer at *Autronica Fire and Security, a UTC Fire and Security company*. He has worked with embedded systems for some 30 years, and is especially interested in real-time language issues. See `http://home.no.net/oyvteig/` for publications.

Communicating Process Architectures 2007
Alistair A. McEwan, Steve Schneider, Wilson Ifill, and Peter Welch
IOS Press, 2007

323

A Process Oriented Approach
to USB Driver Development

Carl G. RITSON and Frederick R.M. BARNES

Computing Laboratory, University of Kent,
Canterbury, Kent, CT2 7NF, England.
{cgr,frmb}@kent.ac.uk

Abstract. Operating-systems are the core software component of many modern computer systems, ranging from small specialised embedded systems through to large distributed operating-systems. The demands placed upon these systems are increasingly
complex, in particular the need to handle concurrency: to exploit increasingly parallel
(multi-core) hardware; support increasing numbers of user and system processes; and
to take advantage of increasingly distributed and decentralised systems. The languages
and designs that existing operating-systems employ provide little support for concurrency, leading to unmanageable programming complexities and ultimately errors in
the resulting systems; hard to detect, hard to remove, and almost impossible to prove
correct.Implemented in occam-π, a CSP derived language that provides guarantees
of freedom from race-hazards and aliasing error, the RMoX operating-system represents a novel approach to operating-systems, utilising concurrency at all levels to simplify design and implementation. This paper presents the USB (universal serial bus)
device-driver infrastructure used in the RMoX system, demonstrating that a highly
concurrent process-orientated approach to device-driver design and implementation is
feasible, efficient and results in systems that are reliable, secure and scalable.

Keywords. occam-pi, operating-systems, RMoX, concurrency, CSP, USB, embedded-
systems, PC104

Introduction

The RMoX operating-system, previously presented at this conference [1], represents an interesting and well-founded approach to operating-systems development. Concurrency is utilised
at the lowest level, with the operating-system as a whole comprised of many interacting parallel processes. Compared with existing systems, that are typically sequential, RMoX offers
an opportunity to easily take advantage of the increasingly multi-core hardware available —
it is scalable. Development in occam-π [2,3], based on CSP [4] and incorporating ideas of
mobility from the π-calculus [5], gives guarantees about freedom from race-hazard and aliasing error — problems that quickly become unmanageable in existing systems programmed
using sequential languages (which have little or no regard for concurrency), and especially
when concurrency is added as an afterthought.

Section 1 provides an overview of the RMoX system, its motivation, structure and operation. Section 2 provides a brief overview of the USB hardware standard, followed by details
of our driver implementation in section 3. An example showing the usage of the USB driver
is given in section 4, followed by initial conclusions and consideration for future and related
work in section 5.

1. The RMoX Operating System

The RMoX operating-system is a highly concurrent and dynamic software system that provides an operating-system functionality. Its primary goals are:

- *reliability*: that we should have some guarantees about the operation of the system components, possibly involving formal methods.
- *scalability*: that the system as a whole should scale to meet the availability of hardware and demands of users; from embedded devices, through workstations and servers, to massively parallel supercomputers.
- *efficiency*: that the system operates using a minimum of resources.

The majority of existing operating-systems fail to meet these goals, due largely to the nature of the programming languages used to build them — typically C. Reliability within a system utilising concurrency requires that we have a solid understanding of that concurrency, including techniques for formal reasoning. This is simply not the case for systems built with a *threads-and-locks* approach to concurrency, as most operating-systems currently use. The problem is exasperated by the use of 3rd-party code, such as device-drivers provided by specific hardware vendors — the OS cannot guarantee that code being "plugged in" interacts in a way that the OS expects. Getting this right is up to the hardware vendor's device-driver authors, who are unlikely to have access to every possible configuration of hardware and other device-drivers which the OS uses, in order to test their own drivers.

Scalability is a highly desirable characteristic for an OS. Most existing operating-systems are designed with specific hardware in mind, and as such, there is a wealth of OSs for a range of hardware. From operating-systems specific to embedded devices, through general-purpose operating-systems found on workstations and servers, up to highly concurrent and job-based systems for massively-parallel supercomputers. Unfortunately, most operating-systems fail to scale beyond or below the hardware for which they were originally intended. Part of the scalability problems can be attributed to concurrency — the mechanisms that existing systems use to manage concurrency are themselves inherently unscalable.

A further issue which RMoX addresses is one of efficiency, or as seen by the user, performance. *Context-switching* in the majority of operating-systems is a notoriously heavyweight process, measured in thousands of machine cycles. Rapid context-switching is typically required to give 'smooth' system performance, but at some point, the overheads associated with it become performance damaging. As such, existing systems go to great lengths to optimise these code paths through the OS kernel, avoiding the overheads of concurrency (specifically context-switches) wherever possible. The resulting code may be efficient, but it is hard to get right, and almost impossible to prove correct given the nature of the languages and concurrency mechanisms used. Furthermore, the OS cannot generally guarantee that loaded code is well behaved — either user processes or 3rd-party drivers. This results in a need for complex hardware-assisted memory protection techniques.

In contrast, the RMoX OS *can* make guarantees about the behaviour of foreign code — we insist that such code conforms. Fortunately, the **occam-π** compiler does this for us — it is one-time effort for the compiler writer. Clearly there are issues relating to *trust*, but those are orthogonal to the issues here, and are well addressed in other literature (related to security and cryptography). Having assurances that code running within the OS is well-behaved allows us to do away with many overheads. Most notably, the context-switch (including communication) can be measured in tens of machine cycles, orders of magnitude smaller than what currently exists. With such small overheads, we can use concurrency as a powerful tool to simplify system design. Furthermore, the resulting systems are scalable — we can run with as few or as many processes are required.

1.1. Structure

The structure of the RMoX operating-system is shown in figure 1, with detail for the "driver.core" shown. The network is essentially a client-server style architecture, giving a guarantee of deadlock freedom [6].

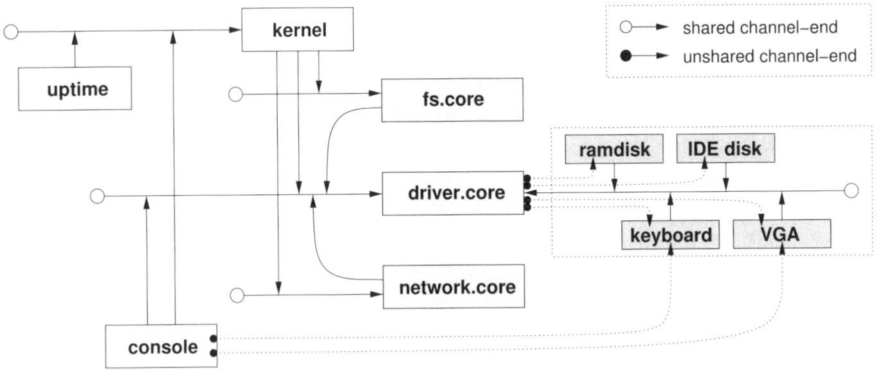

Figure 1. RMoX operating-system process network.

There are three core services provided by the RMoX system: device-drivers, file-systems and networking. These simply provide management for the sub-processes (or sub-process networks) that they are responsible for. When a request for a resource is made, typically via the 'kernel' process, the relevant 'core' process routes that request to the correct underlying device. Using mobile channels, this allows *direct* links to be established between low-level components providing a particular functionality, with high-level components using them. Protocols for the various types of resource (e.g. file, network socket, block device-driver) are largely standardised — e.g. a file-system driver (inside "fs.core") can interact with any device driver that provides a block-device interface. Since such protocols are well defined, in terms of interactions between processes, building pipelines of processes which layer functionality is no issue. Some consideration must be given to shutting these down correctly (i.e. without inducing deadlock); fortunately that process is well understood [7].

As the system evolves, links established between different parts of the system can result in a fairly complex process network. However, if we can guarantee that individual components interact with their environments in a 'safe' way (with a per-process analysis performed automatically by the compiler), then we can guarantee the overall 'safe' behaviour of the system — a feature of the compositional semantics of CSP as engineered into the occam-π language. This type of formalism is already exploited in the overall system design — specifically that a client-server network is deadlock free; all we have to do is ensure that *individual* processes conform to this.

The remainder of this paper focuses on the USB device-driver architecture in RMoX. Supporting this hardware presents some significant design challenges in existing operating-systems, as it requires a dynamic approach that layers easily — USB devices may be plugged-in and unplugged arbitrarily, and this should not break system operation. The lack of support for concurrency in existing systems can make USB development hard, particularly when it comes to guaranteeing that different 3rd-party drivers interact correctly (almost impossible in existing systems). RMoX's USB architecture shows how concurrency can be used to our benefit: breaking down the software architecture into simple, understandable, concurrent components; producing a design that is scalable, and an implementation that is reliable and efficient.

2. The Universal Serial Bus

The *Universal Serial Bus* (USB) [8,9] first appeared in 1996 and has undergone many revisions since. In recent years it has become the interface of choice for low, medium and high speed peripherals, replacing many legacy interfaces, e.g. RS232, PS/2 and IEEE1284. The range of USB devices available is vast, from keyboards and mice, through flash and other storage devices, to sound cards and video capture systems. Many classes of device are standardised in documents associated with the USB, these include human-interface devices, mass-storage devices, audio input/output devices, and printers. For these reasons adding USB support to the RMoX operating system increases its potential for device support significantly. It also provides an opportunity to explore modelling of dynamic hardware configurations within RMoX.

2.1. USB Hardware

The USB itself is a 4-wire (2 signal, 2 power) half-duplex interface, supporting devices at three speeds: 1.5 Mbps (low), 12 Mbps (full) and 480 Mbps (high). There is a single bus master, the host controller (HC), which controls all bus communication. Communication is strictly controlled — a device cannot initiate a data transfer until it has been offered the appropriate bandwidth by the HC. The topology of a USB bus is a tree, with the HC at the root. The HC provides a root hub with one or more ports to which devices can be connected. Additional ports can be added to the bus by connecting a hub device to one of the existing bus ports. Connected hubs are managed by the USB driver infrastructure, which maintains a consistent view of the topology at all times. Figure 2 shows a typical arrangement of USB hardware.

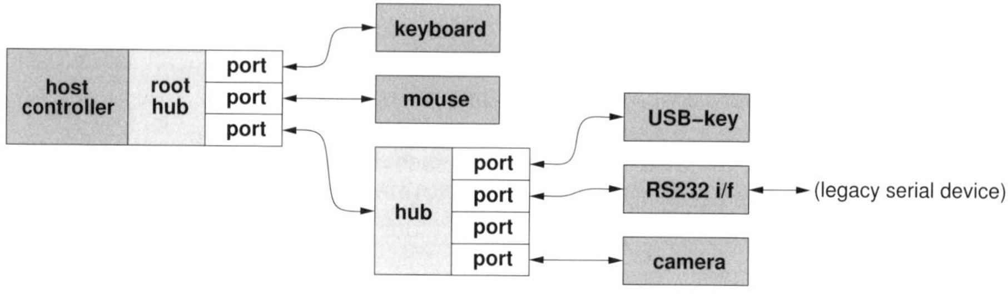

Figure 2. Example USB hardware tree.

Unlike more traditional system busses, such as PCI [10], the topology the USB is expected to change at run-time. For this and the reasons above, access to bus devices is via communication primitives provided by the USB driver infrastructure, rather than CPU I/O commands or registers mapped into system memory. Although it should be noted that this difference does not preclude the use of DMA (direct memory access) data transfers to and from bus devices.

2.2. USB Interfaces

Each device attached to the bus is divided into interfaces, which have zero or more endpoints, used to transfer data to and from the device. Interfaces model device functions, for example a keyboard with built-in track-pad would typically have one interface for the keyboard, and one for the track-pad. Interfaces are grouped into configurations, of which only one may be active at a time. Configurations exist to allow the fundamental functionality of the device to change. For example, an ISDN adapter with two channels may provide two configurations: one configuration with two interfaces, allowing the ISDN channels to be used independently; and another with a single interface controlling both channels bound together (*channel bonding*).

Individual *interfaces* may also be independently configured with different functionality by use of an "alternate" setting. This is typically used to change the transfer characteristics of the interface's endpoints. For example, a packet-based device interface, such as a USB audio device, may have alternate settings with different packet sizes. Depending on the bus load or other conditions, the driver can select the most appropriate packet size using an "alternate" setting.

Figure 3 illustrates the hierarchy of configurations, interfaces and endpoints, with an active configuration, interface and endpoint, shown down the left-hand side of the diagram.

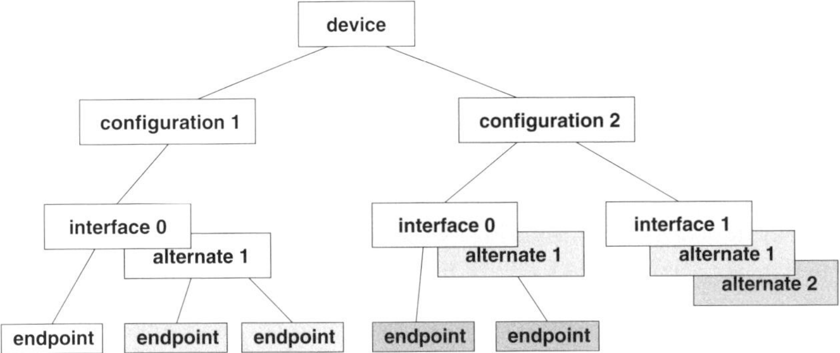

Figure 3. USB configuration, interface and endpoint hierarchy.

2.3. USB Interface Endpoints

Endpoints are the sinks and sources for communications on the bus. Bus transactions are addressed first to the device, then to an endpoint within it. A software structure known as a *pipe* is used to model the connection between the host and an endpoint, maintaining the state information (not entirely dissimilar to the structure and state maintained by a TCP connection). With a few exceptions (detailed later), communication on these pipes is logically the same as that on **occam** channels: unidirectional, synchronous and point-to-point. At the lower bus protocol level, acknowledgements, sequence numbers and CRC checks exist which reinforce these characteristics.

There are four different types of endpoint defined by the USB standards, each of which specifies how the communication 'pipe' should be used:

- *Control*, uses a structured message protocol and can exchange data in either direction. A setup packet containing the request is transferred from the host to the device, followed by zero or more bytes of data in a direction defined by the request type. These are used to enumerate and configure devices, and are also used by many USB device classes to pass information, such as setting the state of keyboard LEDs.
- *Bulk*, exchanges data unidirectionally on demand, no structure is imposed on the data. These are the most similar to a traditional Unix 'pipe'. They are used by storage devices, printers and scanners.
- *Interrupt*, these act similarly to *bulk* except data is exchanged on a schedule. At a set interval, the host offers bus time to the device and if it has data to transfer, or is ready, then it accepts the bandwidth offered. Alternatively, the device delays using a negative acknowledgement, and the transfer is tried again at the next specified interval. This

process continues for as long as the host desires. For example, the typical keyboard is offered a transfer every 10ms, which it uses to notify key-state changes.

- *Isochronous*, like *interrupt* these also use a schedule. The difference is that isochronous transfers are not retried if the device is not ready or a bus error occurs. Since isochronous transfers are not retried, they are permitted to use larger packets than any of the other types. Isochronous transfer are used where data has a constant (or known maximum) rate and can tolerate temporary loss; audio and video are the typical uses.

2.4. Implementation Challenges

There are a variety of considerations when building a USB device-driver 'stack'. Firstly, the dynamic nature of the hardware topology must be reflected in software. Traditional operating systems use a series of linked data-structures to achieve this, with embedded or global locks to control concurrent access. The implementation must also be fault-tolerant to some degree — if a user unplugs a device when in use, the software using that device should fail gracefully, not deadlock or livelock.

As USB is being increasingly used to support legacy devices (e.g. PS/2 keyboard adaptors, serial and parallel-port adapters), the device-driver infrastructure needs to be able to present suitable interfaces for higher-level operating system components. These interfaces will typically lie *underneath* existing high-level device-drivers. For instance, the 'keyboard' driver (primarily responsible for mapping scan-codes into characters and control-codes, and maintaining the shift-state), will provide access to any keyboard device on the system, be it connected via the onboard PS/2 port or attached to a USB bus. Such low-level connectivity details are generally uninteresting to applications — which expect to get keystrokes from a 'keyboard' device, regardless of how it is connected (on-board, USB or on-screen virtual keyboards). Ultimately this results in a large quantity of internal connections within the RMoX "driver.core", requiring careful design to avoid deadlock.

In addition to handling devices and their connectivity, the USB driver is responsible for managing power on the bus. This essentially involves disallowing the configuration of devices which would cause too much current to be drawn from the bus. Devices are expected to draw up to 100 mA by default (in an unconfigured state), but not more than 500 mA may be drawn from any single port.

3. Software Architecture

All device-driver functionality in RMoX is accessed through the central "driver.core" process (figure 1), which directs incoming requests (internal and external) to the appropriate driver within. To support the dynamic arrival and removal of devices, a new "dnotify" device-driver has been added. This is essentially separate from the USB infrastructure, and is responsible for notifying registered listeners when new devices become available or old ones are removed.

The USB driver infrastructure is built from several parts. At the lowest level is a *host controller driver* (HCD), that provides access to the USB controller hardware (via I/O ports and/or memory-mapping). The implementation of one particular HCD is covered in section 3.3. At the next level is the "usb.driver" (USBD) itself. This process maintains a view of the hardware topology using networks of sub-processes representing the different USB busses, acting as a client to HCD drivers and as a server to higher-level drivers. Figure 4 shows a typical example, using USB to provide the 'console' with access to the keyboard.

The "usb.keyboard" process uses the USBD to access the particular keyboard device, and provides an interface for upstream "keyboard" processes. Such a "keyboard" process might actively listen for newly arriving keyboard devices from "dnotify", managing them all together — as many existing systems do (e.g. pressing 'num-lock' on *one* of the keyboards causes *all* num-lock LEDs to toggle).

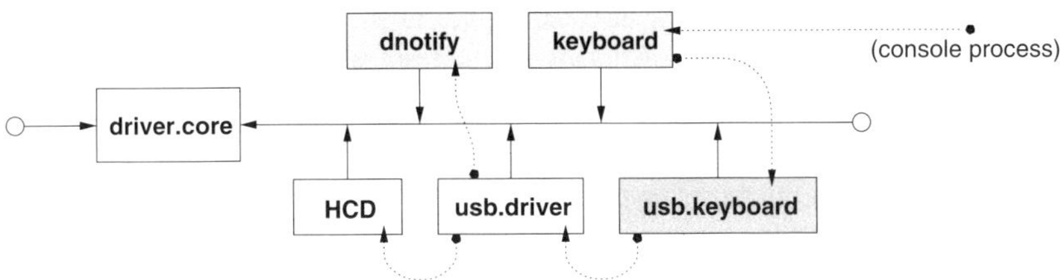

Figure 4. USB device-driver top-level components.

3.1. USB Driver Structure

Processes outside the USB driver can gain access to the USB at three levels: bus-level, device-level and interface-level. The "usb.driver" contains within it separate process networks for each individual bus — typically identified by a single host controller (HC). These process networks are highly dynamic, reflecting the current hardware topology. When a host controller driver instance starts, it connects to the USB driver and requests that a new bus be created. Mobile channel bundles are returned from this request, on which the host controller implements the low-level bus access protocol and the root hub. Through this mechanism the bus access hardware is abstracted. Figure 5 shows the process network for a newly created bus, with three connected USB devices, one of which is a hub. For clarity, some of the internal connections have been omitted.

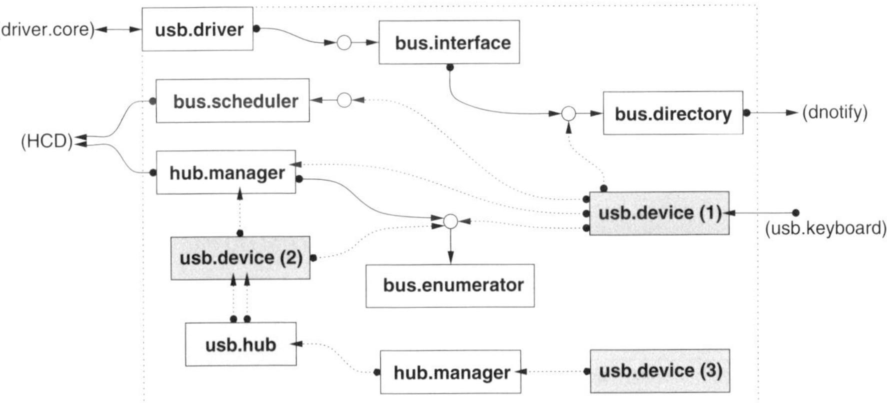

Figure 5. USB device-driver bus-level components.

Within each bus sub-network are the following processes:

- "bus.interface" provides mediated access to the bus, specifically the bus directory. It services a channel bundle shared at the client-end, which is held by the USB driver and other processes which request bus-level access.
- "bus.directory" maintains a list of all devices active on the bus and client channel-ends to them. Attempts to open devices and interfaces pass through the directory which resolves them to channel requests on specific devices. When devices and interfaces are added or removed from the directory, their information is propagated to the 'dnotify' driver which acts as a system wide directory of *all* devices (not just USB).
- "bus.enumerator" is responsible for assigning device addresses (1-127), and acts as a mutex lock for bus enumeration. The lock functionality is necessary as only one de-

vice maybe enumerated on the bus at any given time. When a device is first connected it does not listen to the bus. After its port is reset it begins listening to the bus and responding to queries on the default address (0). The USB driver then sends a "set address" request to the default address.

- "bus.scheduler" is responsible for managing bus bandwidth and checking the basic validity of bus transactions. The USB standard dictates that certain types of traffic may only occupy a limited percentage of the bus time (specific values depend on the bus revision). If there is sufficient bandwidth and the request is deemed valid then it is passed to the HCD for execution.

- "hub.manager", of which there may be many instances, one for each hub and one for the root hub, are responsible for detecting device connection, disconnection, and initiating associated actions such as enumeration or device shutdown.

From Figure 5, it is possible to see that a hierarchy exists between the "hub.manager", "usb.hub" and "usb.device" processes. The "usb.hub" process converts the abstract hub protocol used by the "hub.manager" process into accesses to the hub's device endpoints. The root hub, not being an actual USB device, is implemented directly by the HCD in the abstract protocol of the "hub.manager" and hence no "usb.hub" process is necessary.

During the enumeration of a port, the "hub.manager" process *forks* a "usb.device" process, passing it the client-end of a channel bundle. The channel bundle used is client/server plus notify, and contains three channels: one from client to server, and two from server to client. The client is either listening on the 'notify' channel or making a request using the client/server channels. The server process normally requests on the client/server channel pair; if it wishes to 'notify' the client then it must do so in parallel, in order to maintain deadlock freedom.

Client/server plus notify channel bundles, already mentioned, are used between hubs and devices. When the "hub.manager" detects that a port has been disconnected, it notifies the devices attached to it. This is done by passing the server-end of the channel bundle to a newly forked process, in order to prevent the hub blocking whilst it waits for the device to accept the disconnect notification. The forked process performs the aforementioned parallel service of client/server and notify channels. A similar pattern is also used between the underlying hub driver ("usb.hub" or "HCD") and the "hub.manager" to notify of changes in the hub state (port change or hub disconnect).

3.2. USB Device Structure

Figure 6 shows the internal structure of the "usb.device" processes, and within these 'interface' and 'endpoint' processes. With the exception of the default control endpoint, these form the structure described in 2.2 (figure 3), and model the hierarchy defined in the USB specification directly as processes. When a device is configured (non-zero configuration selected), it forks off interface processes to match those defined in the configuration (read from the device). The interfaces in turn fork endpoints to match their current alternate setting. Changing an interface's alternate setting causes the endpoints to be torn down, and changing the configuration of the device tears down all interfaces and endpoints.

Devices, interfaces and endpoints maintain a channel bundle, the client-end of which is given out when they are "opened". This channel-end is *not* shared, so that the process can track the active client. If the device is disconnected, or the interface or endpoint is torn down, then it continues to respond to requests (with errors) until the client-end of this "public" channel bundle is returned, after which it may shutdown (and release its resources). As the USB topology is expected to change during normal system operation (adding and removing devices), so the process network must not only grow, but safely shrink. Maintaining these public channel-ends as exclusive (unshared) allows us to guarantee this safety.

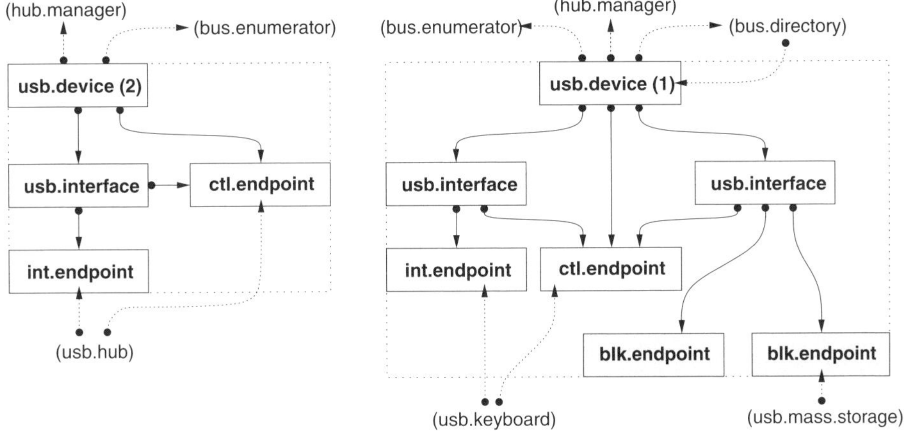

Figure 6. USB device-driver device-level components.

It is however, still possible to safely share resources if the need arises by issuing a separate channel bundle to each client that opens it. When all channel-ends have been returned, the resource may safely terminate. This pattern is used for control endpoints, which due to their structured data transfers can be safely used by many clients at once. Additionally, the default control endpoint must be accessible to all interfaces and their clients. Shared access to devices, interfaces and other endpoints does not typically make sense (given the nature of devices), and hence is not implemented. If we do later decide to introduce sharing, it can be added at a 'higher-level' within the process network.

Requests to open a device come in over the device-directory interface channel. If the device is not already open then it returns that client channel-end via the directory. Requests to open interfaces are passed first to the associated device, which in turn queries the interface over its internal channel. Interfaces may also be opened through the device's channel-end. Using the first approach it is possible open an interface without first opening its associated device (which may already be open). This allows interfaces to function independently and separates functions from devices — i.e. the keyboard driver only uses the keyboard interface, without communicating with the associated device. Endpoints are only accessible through their associated interface — this makes sense as a driver for a function will typically be the only process using the interface and its endpoints.

Care must be taken when implementing the main-loop of the endpoint processes, such that the channel from the interface is serviced at a reasonable interval. This is mainly a concern for interrupt endpoints, where requests to the bus could wait for a very long period of time before completing. For all other endpoint types, bus transactions are guaranteed to finish within a short period of time, hence synchronous requests are guaranteed to complete promptly. The consequence of ignoring this detail would be that the system could appear to livelock until some external event (e.g. key press, or device removal) occurs, causing a pending interrupt request to complete.

3.3. USB UHCI

A number of host controller standards exist, of which UHCI (*Universal Host Controller Interface*) is one. These allow a single USB host controller driver to be written such that it supports a range of host controller hardware. RMoX has drivers for the UHCI, OHCI and EHCI standards. The UHCI [11] standard, released by Intel in 1996, is the simplest and shall be used as an example to explore how data is transferred efficiently from endpoints to the bus. Figure 7, expands the HCD part of Figure 4, as implemented by the UHCI driver.

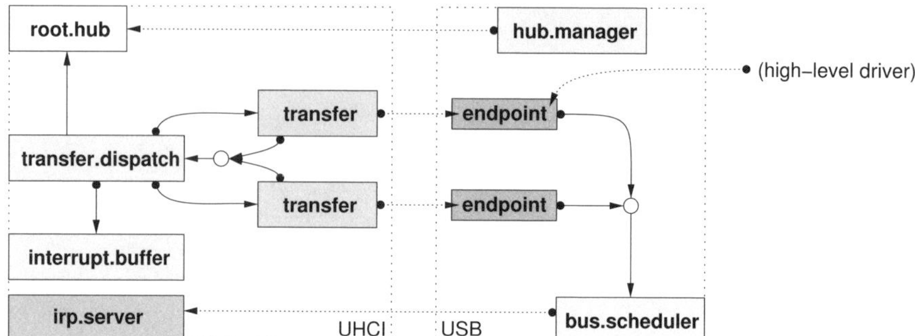

Figure 7. Overview of the 'uhci.driver' host controller driver.

The "uhci.driver" is broken down into four main processes (ignoring transfers, which are forked in response to demand as explained below):

- "root.hub" provides access to the hardware registers which implement the ports of the root hub, and receives relevant interrupt information from "transfer.despatch".
- "interrupt.buffer" receives interrupts from the underlying interrupt routing subsystem (part of the PCI driver). When an interrupt is received, the hardware status register is read, then cleared before the interrupt subsystem is told it can unblock the interrupt line the UHCI hardware is using. Status register flags are buffered and passed to the "transfer.despatch" process on request. The "interrupt.buffer" is similar in function to an interrupt handler subroutine in a traditional OS kernel, such as Linux.
- "transfer.despatch" manages all other registers of the UHCI hardware not handled by other processes. It also manages a set of linked data structures in system memory which are accessed by the UHCI hardware and is used to initiate and control bus transfers.
- "irp.server" (I/O request packet server) implements the HC protocols which the "bus.scheduler" process uses to schedule traffic. On receiving a transfer request from the "bus.scheduler" it forks off a transfer to handle that request.

From the descriptions above it is clear that the UHCI hardware registers are partitioned between the cooperating processes. This ensures that there are no shared resource race-hazards between processes within the driver. To further reinforce this, there are no shared memory buffers; all memory used is *mobile* and is *moved* between processes as appropriate.

As previously mentioned, the "irp.server" forks off a transfer process to handle each bus transfer request. As part of each request received from the "bus.scheduler" is a client channel-end. This is also passed to the transfer process during the fork. The endpoint that initiated the transfer holds the server-end of the channel bundle, and so provides a direct path between the endpoint and the driver.

The transfer process builds a set of linked data structures to describe the packets which will be exchanged on the bus. These data structures are then registered with the despatch process which links them into the hardware accessible data structures it maintains. In the same request, the transfer process also passes a client channel-end on which the despatch process can communicate with it. When the despatch process detects a hardware condition, and associated data structure changes that suggest the state of a transfer has changed, then it contacts the associated transfer process passing back any associated memory buffers. The transfer process then examines the data structures. Not all of the data structures which must be examined are accessible to the despatch process, hence the transfer process implements this check.

Based on the state of the transfer data structures, the transfer process, when queried, tells the despatch process to continue, suspend or remove its transfer. If the transfer is complete or has failed then the transfer process notifies the endpoint, which in turn can decide to issue a new transfer or terminate the transfer process. This allows the network between the endpoint and despatch process, and any allocated data structures, to persist across multiple transfers, reducing communication and memory management overheads. This is legal in bandwidth scheduling terms as only interrupt and isochronous transfers are allocated bus bandwidth, based on their schedule, which cannot be changed once a request has begun. When the transfer is finally terminated the endpoint will notify the "bus.scheduler" that the bandwidth is once again free. However, it should be noted that for hardware reasons, control and bulk transfers do not use this *persistence* feature with the UHCI driver.

Memory buffers from the client are passed directly from endpoint to transfer process, and are used for DMA with the underlying hardware. This creates an efficient *zero-copy* architecture, and has driven investigation into extending the occam-pi runtime allocator to be aware of memory alignment in DMA memory positioning requirements.

4. Using the USB Driver

As an example of using the USB driver, we consider a version of the "usb.keyboard" process. Instead of connecting directly to "usb.driver", the USB keyboard driver registers the client-end of a "CT.DNOTIFY.CALLBACK" channel-bundle with the "dnotify" driver, requesting that it be notified about USB keyboard connections. This involves setting up a data-structure with details of the request and passing it along with the notification channel-end to the "dnotify" driver, using the following code:

```
-- USB device classes (HID or boot-interface) and protocol (keyboard)
VAL INT INTERFACE.CLASS.CODE IS ((INT USB.CLASS.CODE.HID) << 8) \/ #01:
VAL INT INTERFACE.PROTOCOL IS 1:

CT.DNOTIFY.CALLBACK? cb.svr:
SHARED CT.DNOTIFY.CALLBACK! cb.cli:
MOBILE []DEVICE.DESC intf.desc:
INT notification.id:
SEQ
  cb.cli, cb.svr := MOBILE CT.DNOTIFY.CALLBACK   -- allocate callback bundle
  intf.desc := MOBILE [1]DEVICE.DESC             -- allocate descriptor array

  intf.desc[0][flags]    := DEVICE.MATCH.TYPE \/
                            (DEVICE.MATCH.CLASS \/ DEVICE.MATCH.PROTOCOL)
  intf.desc[0][type]     := DEVICE.TYPE.USB.INTERFACE
  intf.desc[0][class]    := INTERFACE.CLASS.CODE
  intf.desc[0][protocol] := INTERFACE.PROTOCOL

  CLAIM dnotify!
    SEQ
      dnotify[in] ! add.notification; DNOTIFY.INSERTION; cb.cli; intf.desc
      dnotify[out] ? CASE result; notification.id
```

The resulting network setup is shown in the left-hand side of figure 8. The "usb.keyboard" driver then enters its main-loop, waiting for requests from either the driver-core, or "dnotify". When a USB keyboard is subsequently connected (or if one was already present), the notification is sent and "usb.keyboard" responds by forking off a driver process ("keyboard.drv").

This initially connects to the USB *interface* specified in the notification (which will be for the connected keyboard), as shown in the right-hand side of figure 8. The code for this is as follows:

```
PROC keyboard.drv (VAL DEVICE.DESC device, SHARED CT.OUTPUT! keyboard,
                   SHARED CT.BLOCK! usb)
  CT.USB.INTERFACE! intf:
  INT result:
  SEQ
    -- connect to interface
    CLAIM usb!
      SEQ
        usb[in] ! ioctl; IOCTL.USB.OPEN.INTERFACE; device[address]
        usb[out] ? CASE result; result
        IF
          result = ERR.SUCCESS
            usb[device.io] ? CASE intf
          TRUE
            SKIP
    ... get endpoints and start main loop
  :
```

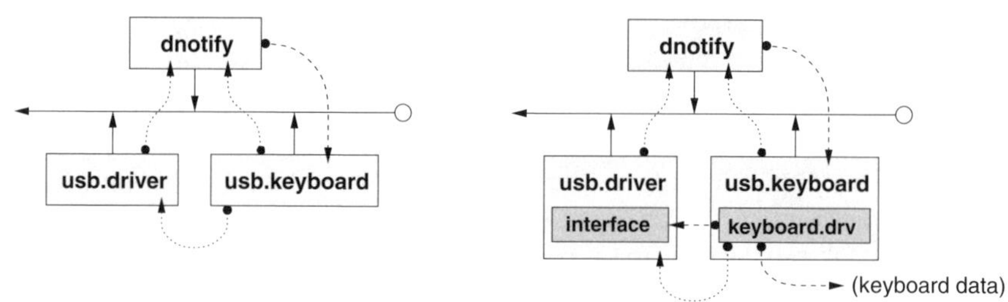

Figure 8. Setup of the 'usb.keyboard' device-driver

4.1. Using USB Interfaces

With a connection to the USB interface (in the variable 'intf'), the keyboard driver requests connections to the *control* and *interrupt* endpoints of the USB interface. Discovering the identifier of the interrupt endpoint first involves querying the interface, simply:

```
MOBILE []BYTE endpoints:
SEQ
  intf[in] ! list.endpoints
  intf[out] ? CASE endpoints; endpoints
```

The returned mobile array is expected to be of length 1, containing the interrupt endpoint identifier. The control endpoint is identified separately, as there is at most one per interface. Connections to the endpoints are then established, resulting in connectivity similar to that shown in figure 6. The following code is used for this, omitting error-handling for brevity:

```
CT.USB.EP.CTL! ep0:
CT.USB.EP.INT! int.ep:
SEQ
  intf[in] ! open.endpoint; 0               -- request control endpoint
  intf[out] ? CASE ctl.ep; ep0
  intf[in] ! open.endpoint; endpoints[0]    -- request interrupt endpoint
  intf[out] ? CASE int.ep; int.ep
```

In addition to listing and connecting to specific endpoints, the interface-level connection is used for listing and switching between alternative interfaces, retrieving information about the device, and other USB specific control.

4.2. Using Interrupt and Control Endpoints

From this point, the USB keyboard driver uses the two endpoint connections to receive keyboard data and control the keyboard. The receiver loop (using the interrupt endpoint) is structured in the following way:

```
packet := MOBILE [8]BYTE
INITIAL BOOL done IS FALSE:
WHILE NOT done
  SEQ
    int.ep[in] ! dev.to.host; packet            -- request 8 byte input
    int.ep[out] ? CASE complete; result; packet -- response
    IF
      result > 0                                -- received data
        process.packet (packet, keyboard!)      -- send keys to terminal
      result = 0                                -- no data
        SKIP
      TRUE
        done := TRUE                            -- interrupt pipe error (exit)
```

The control endpoint is used to set the keyboard LEDs and keyboard rate, in addition to other USB control. The following code example is used to set the keyboard LEDs:

```
VAL BYTE type IS USB.REQ.TYPE.HOST.TO.DEV \/
                 (USB.REQ.TYPE.CLASS \/ USB.REQ.TYPE.INTERFACE):
MOBILE []BYTE data:
INT result:
SEQ
  data := MOBILE [1]BYTE
  data[0] := leds                              -- each bit represents an LED

  ep0[in] ! type; HID.REQ.SET.REPORT; (INT16 HID.REPORT.OUTPUT) << 8;
           INT16 (device[address] /\ #FF); data
  ep0[out] ? result; data                      -- get response

  IF
    result >= 0
      SKIP                                      -- success
    TRUE
      ...  report error
```

As can be seen, using control endpoints is moderately cumbersome, but this is to be expected given the vast range of USB devices available. However, general device I/O through the interrupt endpoint is largely straightforward.

Concurrency is a significant advantage in this environment, allowing a single device-driver to maintain communication with multiple endpoints simultaneously, without significant coding complexity. This particularly applies to situations where a single driver uses multiple USB devices, which may operate and fail independently. One example would be a software RAID (redundant storage) driver, operating over many USB mass storage devices, and presenting a single block-level interface in the RMoX device layer. Expressing such behaviours in non-concurrent languages in existing operating systems is complex and error-prone, primarily due to the lack of an explicit lightweight concurrency mechanism.

5. Conclusions and Future Work

In conclusion, we have designed and developed a robust and efficient process-orientated USB driver. Significantly, the process networks we have developed bare an almost *picture perfect* resemblance to the hierarchy presented in the USB standards and the network which exists between physical devices. Furthermore, as a feature of the development language and process-orientated approach, our driver components are scheduled independently. This allows us, as developers, freedom from almost all scheduling concerns. For example "hub.manager" processes can make synchronous device calls, without causing the entire system to cease functioning.

RMoX itself still has far to go. The hardware platform for which we are developing is a PC104+ *embedded PC* — a standardised way of building embedded PC systems, with stackable PCI and ISA bus interconnects [12]. This makes a good initial target for several reasons. Firstly, the requirements placed on embedded systems are substantially less than what might be expected for a more general-purpose (desktop) operating-system — typically acting as hardware management platforms for a specific application (e.g. industrial control systems, ATM cash machines, information kiosk). There is, however, a strong requirement for reliability in such systems. Secondly, the nature of the PC104+ target makes the RMoX components developed immediately reusable when targeting desktop PCs in the future. Additionally, USB is being increasingly used for device connectivity within embedded PC104 systems, due to its versatility. Assuming a future RMoX driven ATM cash machine, adding a survellience camera would simply involve plugging in the USB camera, installing the appropriate video device-driver and setting up the application-level software (for real-time network transmission and/or storage on local devices) — this could be done without altering the existing system code at all, it simply runs in parallel with it. The builds are routinely tested on desktop PCs and in emulators as standard, exercising the *scalability* of RMoX. We also have a functional PCI network interface driver, and hope to experiment with distributed RMoX systems (across several nodes in a cluster) in the not too distant future.

In addition to the RMoX operating-system components is development work on the tool-chain and infrastructure. Developing RMoX has highlighted a need for some specific language and run-time features, such as the aforementioned allocation of aligned DMA-capable memory. A new occam-π compiler is currently being developed [13] which will allow the easy incorporation of such language features. There is also a need to stabilise existing occam-π language features, such as nested and recursive mobile data types, and port-level I/O.

5.1. Related Work

The most significant piece of related research is Microsoft Research's Singularity operating system [14], which takes a similarly concurrent approach to OS design. Their system is programmed in a variant of the *object-orientated* C# language, which has extensions for efficient communication between processes — very similar in principle and practice to occam-π's mobilespace [15]. The times reported for context-switching and communication in Singularity are some 20 times slower than what we have in RMoX, though their justification for it is incorrect in places (e.g. assuming occam processes can only wait on a single channel — not considering the 'ALT' construct). Some of the difference is correctly attributed to RMoX's current lack of support for multi-core/multi-processor machines. Fortunately, we know how to build these CSP-style schedulers for multi-processor machines, with comparatively low overheads, using techniques such as *batch-scheduling* [16], and are currently investigating this.

More generally, there is a wide range of related research on novel approaches to operating-system design. Most of these, even if indirectly, give some focus to the language and programming paradigm used for implementation — something other than the *threads-and-locks* procedural approach of C. For example, the Haskell operating-system [17] uses a functional paradigm; and the Plan9 operating-system [18] uses a concurrent variant of C ("Alef"). However, we maintain the view that the *concurrent process-orientated* approach of occam-π is more suitable — as demonstrated by the general scalability and efficiency of RMoX, and the ease of conceptual understanding in the USB driver hierarchy — software organisation reflects hardware organisation.

A lot of ongoing research is aimed at making current languages and paradigms more efficient and concrete in their handling of concurrency. With RMoX, we are starting with something that is already highly concurrent with extremely low overheads for managing that concurrency — due in part to years of experience and maturity from CSP, occam and the Transputer [19].

Acknowledgements

We would like to thank the anonymous reviewers who provided valuable feedback and suggestions for improvement. This work was funded by EPSRC grant EP/D061822/1.

References

[1] F.R.M. Barnes, C.L. Jacobsen, and B. Vinter. RMoX: a Raw Metal occam Experiment. In J.F. Broenink and G.H. Hilderink, editors, *Communicating Process Architectures 2003*, WoTUG-26, Concurrent Systems Engineering, ISSN 1383-7575, pages 269–288, Amsterdam, The Netherlands, September 2003. IOS Press. ISBN: 1-58603-381-6.

[2] Frederick R.M. Barnes. *Dynamics and Pragmatics for High Performance Concurrency.* PhD thesis, University of Kent, June 2003.

[3] P.H. Welch and F.R.M. Barnes. Communicating mobile processes: introducing occam-pi. In A.E. Abdallah, C.B. Jones, and J.W. Sanders, editors, *25 Years of CSP*, volume 3525 of *Lecture Notes in Computer Science*, pages 175–210. Springer Verlag, April 2005.

[4] C.A.R. Hoare. *Communicating Sequential Processes.* Prentice-Hall, London, 1985. ISBN: 0-13-153271-5.

[5] R. Milner. *Communicating and Mobile Systems: the Pi-Calculus.* Cambridge University Press, 1999. ISBN-10: 0521658691, ISBN-13: 9780521658690.

[6] P.H. Welch, G.R.R. Justo, and C.J. Willcock. Higher-Level Paradigms for Deadlock-Free High-Performance Systems. In R. Grebe, J. Hektor, S.C. Hilton, M.R. Jane, and P.H. Welch, editors, *Transputer Applications and Systems '93, Proceedings of the 1993 World Transputer Congress*, volume 2, pages 981–1004, Aachen, Germany, September 1993. IOS Press, Netherlands. ISBN 90-5199-140-1. See also: http://www.cs.kent.ac.uk/pubs/1993/279.

[7] P.H. Welch. Graceful Termination – Graceful Resetting. In *Applying Transputer-Based Parallel Machines, Proceedings of OUG 10*, pages 310–317, Enschede, Netherlands, April 1989. Occam User Group, IOS Press, Netherlands. ISBN 90 5199 007 3.

[8] Compaq, Intel, Microsoft, and NEC. Universal Serial Bus Specification - Revision 1.1, September 1998.

[9] Compaq, Hewlett-Packard, Intel, Lucent, Microsoft, NEC, and Philips. Universal Serial Bus Specification - Revision 2.0, April 2000. URL: http://www.usb.org/developers/docs/usb_20_05122006.zip.

[10] PCI Special Interests Group. PCI Local Bus Specification - Revision 2.2, December 1998.

[11] Intel. Universal Host Controller Interface (UHCI) Design Guide, March 1996. URL: http://download.intel.com/technology/usb/UHCI11D.pdf.

[12] PC/104 Embedded Consortium. PC/104-Plus Specification, 2001. URL: http://pc104.org/.

[13] F.R.M. Barnes. Compiling CSP. In P.H. Welch, J. Kerridge, and F.R.M. Barnes, editors, *Communicating Process Architectures 2006*, volume 64 of *Concurrent Systems Engineering Series*, pages 377–388, Amsterdam, The Netherlands, September 2006. IOS Press. ISBN: 1-58603-671-8.

[14] M. Fahndrich, M. Aiken, C. Hawblitzel, O. Hodson, G. Hunt, J.R. Larus, and S. Levi. Language support for Fast and Reliable Message-based Communication in Singularity OS. In *Proceedings of EuroSys 2006*, Leuven, Belgium, April 2006. URL: http://www.cs.kuleuven.ac.be/conference/EuroSys2006/papers/p177-fahndri%ch.pdf.

[15] F.R.M. Barnes and P.H. Welch. Mobile Data, Dynamic Allocation and Zero Aliasing: an occam Experiment. In Alan Chalmers, Majid Mirmehdi, and Henk Muller, editors, *Communicating Process Architectures 2001*, volume 59 of *Concurrent Systems Engineering*, pages 243–264, Amsterdam, The Netherlands, September 2001. WoTUG, IOS Press. ISBN: 1-58603-202-X.

[16] K. Debattista, K. Vella, and J. Cordina. Cache-Affinity Scheduling for Fine Grain Multithreading. In James Pascoe, Peter Welch, Roger Loader, and Vaidy Sunderam, editors, *Communicating Process Architectures 2002*, WoTUG-25, Concurrent Systems Engineering, pages 135–146, IOS Press, Amsterdam, The Netherlands, September 2002. ISBN: 1-58603-268-2.

[17] Thomas Hallgren, Mark P. Jones, Rebekah Leslie, and Andrew Tolmach. A principled approach to operating system construction in haskell. In *ICFP '05: Proceedings of the tenth ACM SIGPLAN international conference on Functional programming*, pages 116–128, New York, NY, USA, September 2005. ACM Press.

[18] Rob Pike, Dave Presotto, Sean Dorward, Bob Flandrena, Ken Thompson, Howard Trickey, and Phil Winterbottom. Plan 9 from bell labs, 1995. Available from `http://www.cs.bell-labs.com/plan9dist/`.

[19] M.D. May, P.W. Thompson, and P.H. Welch. *Networks, Routers and Transputers*, volume 32 of *Transputer and occam Engineering Series*. IOS Press, 1993.

Communicating Process Architectures 2007
Alistair A. McEwan, Steve Schneider, Wilson Ifill, and Peter Welch
IOS Press, 2007

A Native Transterpreter for the LEGO Mindstorms RCX

Jonathan SIMPSON, Christian L. JACOBSEN and Matthew C. JADUD

Computing Laboratory, University of Kent,
Canterbury, Kent, CT2 7NZ, England.

{jon, christian, matt} @transterpreter.org

Abstract. The LEGO Mindstorms RCX is a widely deployed educational robotics platform. This paper presents a concurrent operating environment for the Mindstorms RCX, implemented natively using occam-pi running on the Transterpreter virtual machine. A concurrent hardware abstraction layer aids both the developer of the operating system and facilitates the provision of process-oriented interfaces to the underlying hardware for students and hobbyists interested in small robotics platforms.

Introduction

At the University of Kent, we have access to over forty LEGO Mindstorms RCX robotics kits for use in teaching. Additionally, it is our experience through outreach to local secondary schools and competitions like the FIRST LEGO League[1] that the RCX is a widely available educational robotics platform. For this reason, we are interested in a fully-featured occam-π interface to the LEGO Mindstorms.

The Transterpreter, a portable runtime for occam-π programs, was originally developed to support teaching concurrent software design in occam2.1 on the Mindstorms[2]. In its original implementation, the Transterpreter ran on top of BrickOS, a POSIX-compliant operating system for the RCX[3]. However, as the Transterpreter has grown to support all of occam-π, it has used up all of the available space on the RCX. Given that the Transterpreter will no longer fit onto the RCX along with BrickOS, a new approach is required for a renewed port to the system.

To resolve the memory space issues, we can create a direct hardware interface for the Transterpreter that removes the underlying BrickOS operating system, freeing space to accommodate the now larger virtual machine. To achieve this, we can interact both with routines stored in the RCX's ROM as well as directly with memory-mapped hardware. While it was originally imagined that a C 'wrapper' would need to bind the virtual machine to a given hardware platform, we have discovered that much of this work can instead be done directly from occam-π, thus providing a concurrency-safe hardware abstraction layer.

1. Background

The LEGO Mindstorms Robotics Command eXplorer (RCX) is a widely available educational robotics platform. It takes the form of a large LEGO 'brick' containing a Renesas H8/300 processor running at 16MHz, 16KB of ROM, and 32KB of RAM shared by the firmware image and user programs. There are three input ports for connecting a variety of sensors, three output ports for motors, and an infra-red port used for uploading of firmware and programs. This infra-red port can also be used for communicating with other robots.

1.1. The Transterpreter

The Transterpreter is a virtual machine for occam-π written in ANSI C. At its inception, the virtual machine was designed to bring the occam2.1 programming language to the RCX as an engaging environment for teaching concurrency. The Transterpreter has platform-specific wrappers which link the portable core of the interpreter to the world around it[4]. In the case of the LEGO Mindstorms, a wrapper was originally written to interface with BrickOS[3]. However, there is limited memory space on the RCX, as shown in Figure 1. The choice of building on top of BrickOS was made because it was the quickest and easiest way to get the Transterpreter running on the LEGO Mindstorms; however, it proved impractical for running all but the smallest and simplest of programs.

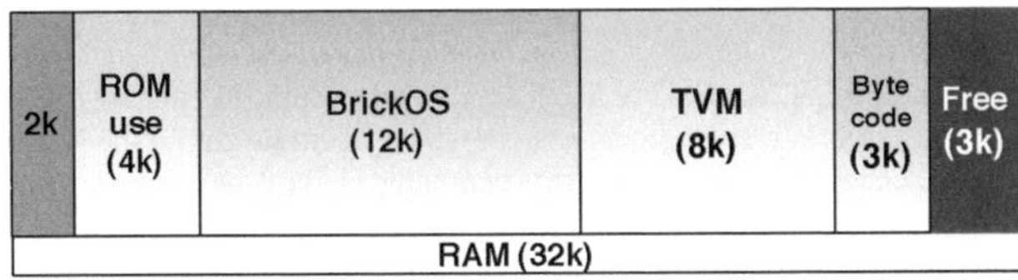

Figure 1. The memory distribution of the original Transterpreter RCX wrapper, using BrickOS.

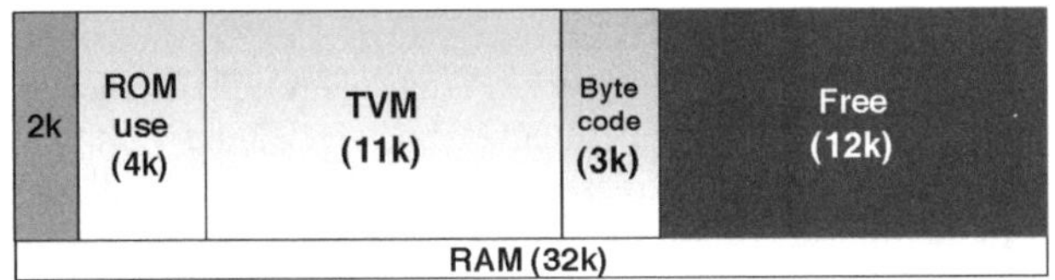

Figure 2. The memory distribution of the native Transterpreter RCX wrapper.

It should be noted that the remaining 3KB of memory space shown in Figure 1, left available after uploading the firmware and user program, was shared to meet the runtime needs of BrickOS, the Transterpreter, and the user's occam2.1 program. As a user's programs grew, this 3KB would be used both by the increased bytecode size of their program as well as a likely increase in memory usage for the larger program.

The Transterpreter virtual machine has grown to support the occam-π programming language[5], an extension of occam2.1 [6]. The extended occam-π feature set is extremely useful for concurrent robotics programming[7]. Unfortunately, supporting these extensions grew the compiled Transterpreter binary by 3KB, and as a result, running the Transterpreter on top of BrickOS is no longer a possibility. By running the Transterpreter natively on the RCX, as shown in Figure 2, we leave 12KB of free memory for the execution of user programs on the virtual machine.

2. Design Considerations

Our design goal is to implement a runtime and process interface for the LEGO Mindstorms RCX, and as such we must provide hardware interfaces to the occam-π programmer. When writing code to interface with the RCX hardware there are three main approaches which can be taken: running with an existing operating system providing the hardware interface, using a C library to interface with ROM functions, or interfacing directly with ROM functions and memory from occam-π.

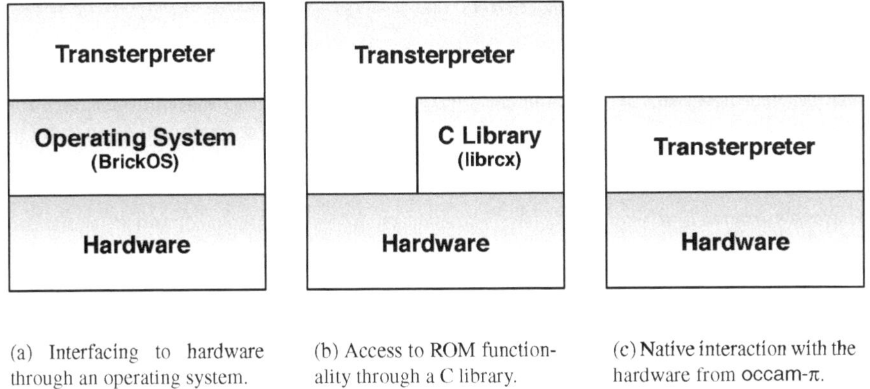

(a) Interfacing to hardware through an operating system.

(b) Access to ROM functionality through a C library.

(c) Native interaction with the hardware from occam-π.

Figure 3. Potential design choices for a new RCX port of the Transterpreter.

2.1. On Top of an Existing Operating System

Running on top of an existing operating system for the RCX was previously explored by running the Transterpreter on top of BrickOS (Figure 3(a)). This saved a great deal of work, as BrickOS exposed a high-level hardware API to the Transterpreter. However, this approach introduces additional storage and run-time memory space penalties, and is not practical given the current size of the virtual machine. In Figure 1 on the facing page, BrickOS is reported as occupying 12KB of space on the LEGO Mindstorms; this is *after* 7KB of unnecessary code had been removed from the operating system; reducing this further would be extremely challenging. To support occam-π on the Mindstorms, another approach must be taken, and a new hardware abstraction layer developed.

2.2. Through a C Library

The ROM image supplied with the LEGO Mindstorms contains the 'RCX executive' (Figure 4), which loads at power on and contains routines for interfacing with the hardware. These ROM routines are used by the standard LEGO firmware supplied with the Mindstorms robotics kit.

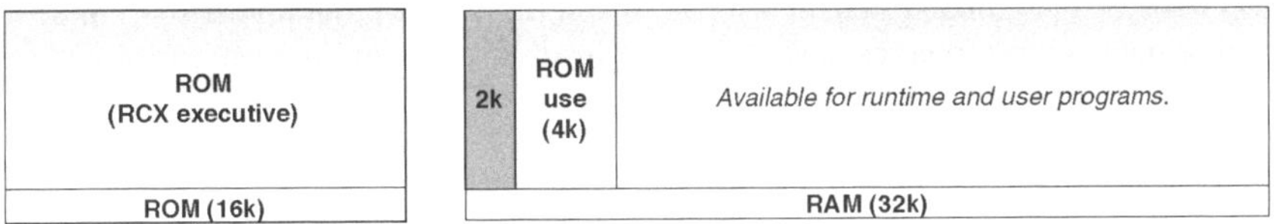

Figure 4. The RCX Executive with both ROM and RAM components, loaded at power-on.

These ROM routines can be exploited to give low-level control over the device without additional memory space penalties, as they are always present in the RCX ROM. However, these routines are not suitable for end-users to program against; they are far too low-level for everyday use.

librcx is a C library that wraps all of the available ROM routines, and provides C programmers with a slightly more usable interface to the hardware[8]. One possible approach to porting the Transterpreter to the LEGO Mindstorms would be to do all of the hardware abstraction in C, as shown in Figure 3(b) on the preceding page. The problem with this approach is that librcx was designed for use from C and not from a concurrent programming language. Any hardware abstraction layer written in C would not interact correctly with the occam-π scheduler, which could lead to race hazards (or worse), the likes of which are avoided if the abstraction layer is written in occam-π.

2.3. Native Interaction

At its core, librcx has five assembly code blocks, each of which calls a ROM routine accepting a specific number of parameters. By exposing these five function calls to occam-π, we can write virtually all of the operating system replacement code in without resorting to C, and leverage the concurrency primitives provided by occam-π (Figure 3(c) on the page before). This also allows a process interface to the hardware to be exposed naturally, and the 'operating system' components to benefit from a safe, concurrent runtime.

By layering processes, some which provide low-level access to hardware and others that form a higher level API for programmers, we can offer different interfaces to different types of programmer. Novice users might work with the higher level processes, unaware that these processes hide details of the underlying functionality. More advanced users or the system programmer may wish to use the low-level processes to perform specific tasks or interact with the hardware more directly. We discuss this further in Section 4 on page 344.

3. A Concurrent Hardware Abstraction Layer

There is one simple reason for wanting to write as much of our code in occam-π as possible: safety. BrickOS[3] and LeJOS[9], two particularly prominent examples of third-party runtimes for the RCX, both use a time-slicing model of concurrency, where multiple 'tasks' are run on the system at the same time. This time-slicing model is then mapped to a threaded programming model for the user. This is a fundamentally unsafe paradigm to program in, regardless of how careful one is[10].

This would not be a problem, except that robotics programming naturally tends to involve multiple tasks running concurrently. For this reason, threading finds its way into all but the most trivial programs written for BrickOS or LeJOS. In "Viva la BrickOS," Hundersmarck et al. noted that that the default scheduling mechanisms in BrickOS are prone to priority inversion under heavy load[11]. By developing from the hardware up in occam-π, we protect both the operating system developer as well as the end-programmer from these kinds of basic concurrency problems, and strive to provide a safer environment for programming robots like the RCX.

3.1. Implementation Considerations

There are a number of implementation challenges that arise given that we have chosen to natively interface the Transterpreter with the RCX. occam-π provides two ways to access underlying hardware functionality: the Foreign Function Interface (FFI) and placement of variables at memory locations. When used correctly, both allow us to safely interact with the underlying hardware from occam-π.

3.1.1. The Foreign Function Interface

The RCX's ROM routines are made available through five core C functions, which we can access through occam-π's Foreign Function Interface mechanism[12]. Unfortunately, the RCX hardware is big-endian, while the Transterpreter runs as a little-endian virtual machine. This means that considerable byte-swapping is required on values and addresses being passed back and forth between the occam-π and C, as can be seen in Listing 1.

```
void rcall_1 (int w*)
{
  __rcall1 (SwapTwoBytes(w[0]), SwapTwoBytes(w[1]));
}
```

Listing 1. rcall_1, a FFI call that passes its parameters to the RCX's ROM.

The five core calls to LEGO ROM routines, once provided to occam-π via the FFI, allow the majority of the ROM's functionality to be accessed. In cases where return values are required, such as when reading from a sensor, individual FFI calls must be written that marshal the values correctly to and from C (eg. swapping from big-endian to little-endian on their way back into occam-π). For example, the C function rcall_1() shown in Listing 1 can be accessed via the FFI from occam-π as shown in Listing 2.

```
-- ROM addresses for sensor access.
VAL [3]INT sensor.addr IS [#1000, #1001, #1002]:
-- Constants for system developer & user programming.
DATA TYPE SENSOR.NUM IS INT:
VAL SENSOR.NUM SENSOR.1 IS 0:
VAL SENSOR.NUM SENSOR.2 IS 1:
VAL SENSOR.NUM SENSOR.3 IS 2:

#PRAGMA EXTERNAL "PROC C.tvmspecial.1.rcall.1 (VAL INT addr, param) = 0"
INLINE PROC rcall.1 (VAL INT addr, param)
  C.tvmspecial.1.rcall.1 (add, param)
:

PROC sensor.active (VAL SENSOR.NUM sensor)
  rcall.1(#1946, sensor.addr[(INT sensor)])
:
```

Listing 2. sensor.active sets a sensor on the RCX 'active' through the occam-π FFI.

3.1.2. Variable Placement in Memory

occam-π supports the placement of variables at specific addresses in memory. As inputs and outputs on the RCX are memory-mapped, occam-π processes can be written that interface directly with the hardware by reading and writing to specific locations. Use of variable placement speeds up the system significantly, as the interpreter can read values directly rather than making calls into the RCX's ROM routines through C.

Endianness continues to be an issue when using variable placement with multi-byte variables, as values must again be byte-swapped due to the difference in endianness between hardware and virtual machine. Additionally, as functions of the RCX's ROM are being called, the firmware works with the same memory addresses and care must be taken not to disturb memory values that are in use by the ROM.

```
PROC run.pressed (CHAN BOOL pressed!)
  INITIAL INT port4.dr.addr IS #FFB7:
  [1]BYTE port4.dr:
  PLACE port4.dr AT port4.dr.addr:
  #PRAGMA DEFINED port4.dr
  WHILE TRUE
    IF
      -- Masking bit 2 of the byte value.
      (port4.dr[0] /\ #02) = 0
        out ! TRUE
      TRUE
        SKIP
:
```

Listing 3. `run.pressed` uses a variable placed in memory to read the 'Run' button state.

The use of variable placement to read the button values from the RCX, as shown in Listing 3, is an example of how hardware interactions can be simplified and the number of calls through C to the ROM can be reduced. Additionally the memory read operation can happen much more quickly than an equivalent FFI call and the necessary byte-swapping between occam-π and C that ensues. Endianness issues are avoided in this particular case, as the value of button presses are stored as individual bit flags in a BYTE value.

3.2. Advantages of Concurrency

By working with a concurrent language all the way from the hardware up there are advantages gained in both safety and simplicity. The LEGO Mindstorms RCX contains a segmented LCD display, including two segments used to draw a walking person on the screen (Figure 5). When debugging occam-π code running on the RCX it can be hard to tell if the runtime environment has crashed or deadlocked, as printing is frequently not possible once an error has occurred.

Figure 5. The 'walking figure' on the LCD display of the RCX

By running the `debug.man` process in parallel with other code being tested (like the process `foo()`, shown in Listing 4 on the facing page), it is possible to see that the VM is running, executing instructions and scheduling correctly. Using threading to get the same effect from a C program would have introduced additional complexity, whereas in occam-π it is natural to use concurrency for developing and debugging programs on what is otherwise a "black box" system.

4. Toward a Process Interface

Our goal is to have a complete, process-oriented interface to the LEGO Mindstorms RCX. This involves developing a hierarchy of processes, starting with an API for programmers

```occam
#INCLUDE "LCD.occ"

PROC debug.man ()
  WHILE TRUE
    SEQ
      -- Sleeping causes us to deschedule
      sleep (500 * MILLIS)
      lcd.set.segment (LCD.STANDING)

      sleep (500 * MILLIS)
      lcd.set.segment (LCD.WALKING)
:

PROC main (CHAN BYTE kyb?, scr!, err!)
  PAR
    debug.man()
    foo()
:
```

Listing 4. The debug.man process helps detect VM faults

to use down through direct access to the hardware. Looking just at input, and particularly the LEGO light sensor, we can see the stacking of one process on top of another to provide a concurrent interface to the underlying, sequential hardware. The occam-π code for light.sensor is shown in listing 5. This process provides a simple and logical end user interface for reading values from a light sensor, connected to one of the input ports on the RCX.

The light.sensor process abstracts across a more generic sensor process. Each type of sensor for the LEGO Mindstorms has its own read mode, and may be active or passive. Hiding these details from the end user lets them develop programs in terms of the robotics hardware sitting in front of them, rather than generic interfaces. Layering the processes in this way also means that more advanced programmers can use the sensor process directly, as they may have created their own 'homebrew' sensors for the RCX and want to have explicit control over the combination of parameters used to set up the sensor.

```occam
PROC light.sensor (VAL SENSOR.NUM num,
                   VAL INT delay,
                   CHAN SENSOR.VAL out!)
  CHAN SENSOR.VAL values:
  PAR
    sensor(num, delay, SENSOR.LIGHT, SENSOR.MODE.PERCENT, values!)
    SENSOR.VAL value:
    WHILE TRUE
      SEQ
        values ? value
        out ! value
:
```

Listing 5. The light.sensor process abstracts over a generic sensor process.

5. Leveraging **occam-π**: A small example

The most challenging part of robotic control—the scheduling and interleaving of sensor readings, computation over that data, and the control of one or more actuators—is handled transparently when developing programs for the RCX in **occam-π** running on the Transterpreter. While the example show here is simple, it provides a taste for the kinds of things that are possible when we target a concurrent programming language at a small robotics platform like the RCX.

Figure 6 illustrates a process network where each sensor on the LEGO Mindstorms communicates to a `work` process, which performs a calculation over the sensor data and then sends commands on to the associated `motor` process. Specifically, if the light sensor is reading a particularly light reading (a value greater than 512), the motor is set to spin forwards; otherwise, it is set to spin backwards.

Listing 6 provides the code for this network, and demonstrates the use of a replicated `PAR` for initializing these nine concurrent processes. Furthermore, it illustrates a few aspects of the concurrent API provided for interfacing the LEGO Mindstorms. Types have been defined for all sensor and motor channels: sensors communicate `SENSOR.VAL`s, while motors expect to receive `MOTOR.CMD`s, a tagged protocol that encourages the programmer to be clear about whether a motor is running in a forward or backwards direction. This helps keep our programs semantically clear, and let the type checker help make sure programs are correct. Additionally, the `light.sensor` process allows the programmer to determine how often the sensor will be sampled; in this example, we are sampling the three sensors once every one, two, and three seconds (respectively).

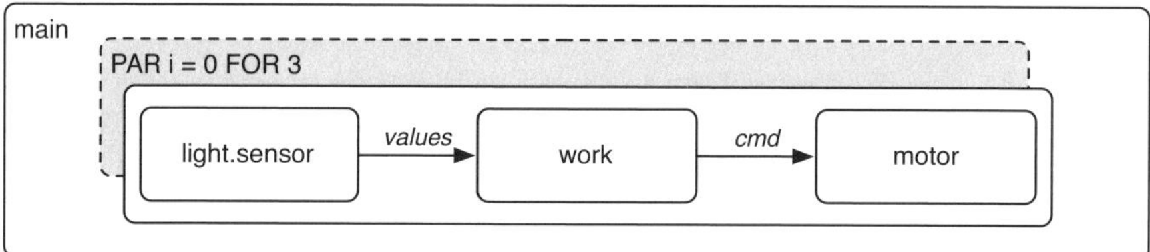

Figure 6. A process network connecting sensors to motors.

This small example does not illustrate any of the more advanced features of **occam-π**: `SHARED` channels, `MOBILE` data, `BARRIER`s, and so on. It does demonstrate, however, that we can quickly and easily set up many concurrent tasks and execute them directly on the LEGO Mindstorms. As our code grows more complex (as described in [7]), the benefits of a concurrent language and runtime for robotics becomes more apparent.

6. Conclusions and Future Work

Our initial goal was to resuscitate the LEGO Mindstorms RCX as a full-featured platform for **occam-π** robotics. To achieve this, we had to explore and overcome a number of challenges in developing a new wrapper for the Transterpreter and creating a concurrent, process-oriented interface for the RCX's functionality. However, a great deal more work is required before we have a platform that is casually usable by a robotics hobbyist or novice programmer.

The porting of the virtual machine and development of a concurrent hardware abstraction layer is only the first step towards providing a generally usable **occam-π** robotics environment. On top of the hardware abstraction layer, we need to write a small operating system

```
#INCLUDE "Sensors.occ"
#INCLUDE "Motors.occ"
#INCLUDE "common.occ"

PROC work (CHAN SENSOR.VAL in?, CHAN MOTOR.CMD out!)
  SENSOR.VAL x:
  WHILE TRUE
    SEQ
      in ? x
      IF
        x > 512
          out ! forward; 5
        TRUE
          out ! backward; 5
:

PROC main ()
  [3]CHAN SENSOR.VAL values:
  [3]CHAN MOTOR.CMD cmd:
  PAR i = 0 FOR 3
    PAR
      light.sensor(i, ((i + 1) * SECONDS), values[i]!)
      work(values[i]?, cmd[i]!)
      motor(i, cmd[i]?)
:
```

Listing 6. A sample program that maps sensor values to motor speeds in parallel.

or monitor that will run along side user programs and provide a basic user interface for the RCX. For example, there are four buttons on the RCX: On-Off, View, Prgm, and Run. At the least, we need to allow users to turn the brick on and off as well as start and stop their programs. The monitor would also need to handle the upload of new programs; the RCX maintains its memory state while powered down, and therefore it is possible to keep the runtime and monitor on the RCX, while the user might upload new bytecode to be executed. This saves the user from the slow and tedious process of uploading a complete virtual machine every time they change their program.

Even with a simple operating system running along side user programs, there is still more work to be done to provide a usable robotics programming environment. Currently, we provide a simplified IDE for programming in occam-π on Mac OSX, Windows, and Linux platforms. This IDE, based on JEdit[1], is extensible through plugins. Our old plugin must be updated to support the uploading of our new Transterpreter-based firmware to the RCX, as well as the compilation of programs for running in this 16-bit environment. This is not hard, but handling the inevitable errors that will occur (failed uploads over IR, and so on) and reporting them to the user in a clear and meaningful manner is subtle, but critical work. We say "critical" because the success of a language is determined as much by the quality of its end-user tools as well as the quality and expressive power of the language itself.

With a usable programming environment in place, we would then like to develop a set of introductory programming exercises using our process-oriented interface to the LEGO Mindstorms. We believe the RCX is an excellent vehicle for teaching and learning about concurrency. While the existing API is already clearly documented, additional materials are absolutely necessary to support novice learners encountering concurrent robotics programming in occam-π for the first time.

---

[1] http://www.jedit.org/

In this vein, we are ultimately interested in the combination or creation of a visual process layout tool like gCSP[13], POPExplorer[14], or LOVE[15] that supports our process-oriented interface to the RCX. The semantics of occam-π nicely lend themselves to visualization, and a toolbox of pre-written occam-π processes to enable graphical, concurrent robotics programming feel like a natural combination. This could potentially offer an environment where novices could begin exploring concurrency without having to (initially) write any occam-π code at all. In the long run, our goal is to reduce the cost of entry for new programmers to explore occam-π in problem spaces that naturally lend themselves to process-oriented solutions.

Acknowledgements

Many people continue to contribute to the Transterpreter project in many ways. David C. Wood was kind enough to supervise this work as a final year project at the University of Kent. The University of Kent Computing Laboratory and Peter Welch have provided support for hardware and travel for presenting our work. Damian Dimmich continues to develop a native big-endian Transterpreter, and Adam Sampson contributed excellent code that has considerably reduced the size of (and increased the speed of) the core interpreter. Matthew Jadud was supported during this time by the EPSRC-funded DIAS project.

References

[1] FIRST LEGO League. http://www.firstlegoleague.org/.

[2] Christian L. Jacobsen and Matthew C. Jadud. Towards Concrete Concurrency: occam-pi on the LEGO Mindstorms. In *SIGCSE '05: Proceedings of the 36th SIGCSE technical symposium on Computer science education*, pages 431–435, New York, NY, USA, 2005. ACM Press.

[3] brickOS™Homepage. http://brickos.sourceforge.net/.

[4] Christian L. Jacobsen and Matthew C. Jadud. The Transterpreter: A Transputer Interpreter. In *Communicating Process Architectures 2004*, pages 99–107, 2004.

[5] P.H. Welch and F.R.M. Barnes. Communicating Mobile Processes: introducing occam-pi. In A.E. Abdallah, C.B. Jones, and J.W. Sanders, editors, *25 Years of CSP*, volume 3525 of *Lecture Notes in Computer Science*, pages 175–210. Springer Verlag, April 2005.

[6] INMOS Limited. *occam2 Reference Manual*. Prentice Hall, 1984. ISBN: 0-13-629312-3.

[7] Jonathan Simpson, Christian L. Jacobsen, and Matthew C. Jadud. Mobile Robot Control: The Subsumption Architecture and occam-pi. In Frederick R. M. Barnes, Jon M. Kerridge, and Peter H. Welch, editors, *Communicating Process Architectures 2006*, pages 225–236. IOS Press, September 2006.

[8] Kekoa Proudfoot. librcx. http://graphics.stanford.edu/ kekoa/rcx/tools.html, 1998.

[9] LeJOS: Java for LEGO Mindstorms. http://lejos.sourceforge.net.

[10] Hans-J. Boehm. Threads cannot be implemented as a library. In *PLDI '05: Proceedings of the 2005 ACM SIGPLAN conference on Programming language design and implementation*, pages 261–268, New York, NY, USA, 2005. ACM Press.

[11] Christopher Hundersmarck, Charles Mancinelli, and Michael Martelli. Viva la brickOS. *Journal of Computing Sciences in Colleges*, 19(5):305–307, 2004.

[12] David C. Wood. KRoC – Calling C Functions from occam. Technical report, Computing Laboratory, University of Kent at Canterbury, August 1998.

[13] Jan F. Broenink, Marcel A. Groothuis, and Geert K. Liet. gCSP occam Code Generation for RMoX. In *Communicating Process Architectures 2005*, pages –, sep 2005.

[14] Christan L. Jacobsen. *A Portable Runtime for Concurrency Research and Application*. Doctoral thesis, University of Kent, 2007.

[15] Adam Sampson. LOVE. https://www.cs.kent.ac.uk/research/groups/sys/wiki/LOVE, 2006.

Communicating Process Architectures 2007
Alistair A. McEwan, Steve Schneider, Wilson Ifill, and Peter Welch
IOS Press, 2007

Integrating and Extending JCSP

Peter WELCH [a], Neil BROWN [a], James MOORES [b],
Kevin CHALMERS [c] and Bernhard SPUTH [d]

[a] *Computing Laboratory, University of Kent, Canterbury, Kent, CT2 7NF, UK.*
[b] *23 Tunnel Avenue, London, SE10 0SF, UK.*
[c] *School of Computing, Napier University, Edinburgh, EH10 5DT, UK.*
[d] *Department of Engineering, University of Aberdeen, Scotland, AB24 3UE, UK.*

Abstract. This paper presents the extended and re-integrated JCSP library of CSP packages for Java. It integrates the differing advances made by Quickstone's JCSP Network Edition and the *"core"* library maintained at Kent. A more secure API for connecting networks and manipulating channels is provided, requiring significant internal re-structuring. This mirrors developments in the occam-pi language for mandated direction specifiers on channel-ends. For JCSP, promoting the concept of channel-ends to first-class entities has both semantic benefit (the same as for occam-pi) and increased safety. Major extensions include *alting barriers* (classes supporting external choice over multiple multi-way synchronisations), channel *output guards* (straightforward once we have the alting barriers), channel *poisoning* (for the safe and simple termination of networks or sub-networks) and *extended rendezvous* on channel communications (that simplify the capture of several useful synchronisation design patterns). Almost all CSP systems can now be directly captured with the new JCSP. The new library is available under the LGPL open source license.

Keywords. JCSP, Alting Barriers, Output Guards, Extended Rendezvous, Poison

Introduction

JCSP (*Communicating Sequential Processes for Java*[1]) [1,2,3,4] is a library of Java packages providing a concurrency model that is a judicious combination of ideas from Hoare's CSP [5] and Milner's π-calculus [6]. It follows many of the principles of occam-π [7,8,9,10], exchanging compiler enforced security for programmer checked rules, losing some ultra-low process management overheads but winning the model for a mainstream programming language. Along with CTJ [11], JCSP is the forerunner of similar libraries for other environments – such as C++CSP [12], CTC++ [13] and the .NET CSP implementations [14,15].

JCSP enables the dynamic and hierarchic construction of process networks, connected by and synchronising upon a small set of primitives – such as message-passing channels and multiway events. Each process manages its own state and engages in patterns of communication with its environment (represented by channels, barriers etc.) that can be formally contracted (in CSP). Each process is independently constructed and tested without concern for multiprocessing side-effects – there is no need for locking mechanisms. In this way, our long developed skills for *sequential* design and programming transfer directly into *concurrent* design and programming. Whole system (multiprocessing) behaviour yields no surprises and can be analysed for bad behaviour (e.g. deadlock) formally, with the option of assistance from automated model checkers (such as FDR [16]). The model works unchanged whether the concurrency is *internal* to a single machine (including multicore architectures) or *distributed* across many machines (including workstation clusters and the Internet).

---

[1] Java is a trademark of Sun Microsystems

JCSP is an *alternative* concurrency model to the threads and monitor mechanisms built into Java. It is also *compatible* with it – indeed, it is currently implemented on top of it! With care, the two models can profitably be mixed[2]. Java 1.5 includes a whole new set of concurrency primitives – some at a very low level (e.g. the *atomic* swaps and counts). These also provide an alternative to threads and monitors. Depending on the relative overheads between the 1.5 and classical methods, it may be worthwhile re-implementing JCSP on the lowest level 1.5 primitives. Meanwhile, we are confident in the current implementation, which has been formalised and model checked [17].

JCSP was developed following WoTUG's *Java Threads Workshop* [18] in 1996. Using ideas kicked around at that workshop [19], the first library (JCSP 0.5, [20]) was designed and put together by Paul Austin, a Masters student at Kent, some time in 1997. It has been under continuous development ever since by a succession of undergraduate/Masters/PhD students (Neil Fuller, Joe Aldous, John Foster, Jim Moores, David Taylor, Andrew Griffin) together with the present authors. A major undertaking was the spin-off of *Quickstone Technologies Limited* (QTL), that crafted the JCSP Network Edition. This enables the dynamic distribution of JCSP networks across any network fabric, with no change in semantics (compared with a single JVM version) – only a change in performance and the size of the system that can be run. Sadly, QTL is no more – but its work survives and is being re-integrated with the core version (which had made several independent advances, some reported here) to form the LGPL open-source new JCSP 1.1 release.

JCSP was designed for use with anything above and including Java 1.1. This compatibility with Java 1.1 has been maintained up to the current *core* release: JCSP 1.0-rc7. Given that most modern mobile devices support at least Java 1.3, we may relax this self-imposed constraint (and start, for example, using collection classes in the revised implementation). Other new mechanisms available in Java 1.5 (e.g. *generics*) and their binding into the future of JCSP are discussed in section 6.

In section 1 of this paper, we describe and motivate small changes in API and the re-factoring of the channel classes and interfaces resulting from the merger of the JCSP Network Edition and JCSP 1.0-rc7. Section 2 presents the *alting barriers* that are completely new for JCSP, together with some implementation details. Section 3 shows how these facilitate channels that allow *output guards* in external choice (*alting*). The addition of *extended rendezvous* to JCSP is given in section 4, including how this works with buffered channels of various kinds. Section 5 presents the addition of channel poisoning for the safe and simple termination of networks (or sub-networks). Finally, Section 6 considers opportunities for the future of JCSP.

1. Class Restructure

1.1. JCSP 1.0-rc7

In JCSP 1.0-rc7, there are two interfaces for channel-ends: `ChannelInput` and `ChannelOutput`. There is also the abstract class `AltingChannelInput`, which extends the abstract class `Guard`[3] and the interface `ChannelInput` and enables channels to be used as *input guards* in external choice (*alting*). All this remains in JCSP 1.1.

---

[2]For *straightforward* management of a shared resource, we have sometimes employed direct visibility with `synchronized` blocks to serialise access – rather than accept the overheads of a very simple server process. For more sophisticated management, we would always use a process. Using and reasoning about an object's `wait`, `notify` and `notifyAll` methods should be avoided at all costs!

[3]This defines a public *type* with a set of method headers visible and used only within `org.jcsp.lang` – sadly, Java does not permit such things in an `interface`.

JCSP 1.0-rc7 channel classes, such as `One2OneChannel`, implement the `AltingChannelInput` and `ChannelOutput` classes/interfaces and all the corresponding methods. Processes take channel-end types, such as `ChannelOutput` or `AltingChannelInput`, as arguments to their constructor. Actual channel instances are passed directly to these constructors – with Java implicitly casting them down to the expected interface types.

This structure allows misuse: a process, having been given a `ChannelInput`, can cast it to a `ChannelOutput` – and vice-versa! Such tricks do enable a channel to be used in both directions, but would probably lead to tears. They are prevented in JCSP 1.1.

Classical zero-buffered fully synchronising channels are provided along with a variety of buffered versions (blocking, overwriting, overflowing). Zero-buffered channels are implemented with a different (and faster) logic than the buffered ones. A memory inefficient feature of the JCSP 1.0-rc7 implementation is that the buffered channels *sub-class* the zero-buffered classes, although that is not relevant (or visible) to the API. So, buffered classes retain fields relevant only to the unused superclass logic. This does not happen in JCSP 1.1.

1.2. JCSP Network Edition

In the JCSP Network Edition, the channel-end interfaces and abstract classes are the same as above. There are also extended interfaces, `SharedChannelInput` and `SharedChannelOutput`, that do not reveal any extra functionality but indicate that the given channel-end can be safely shared (internally) between multiple concurrent sub-processes. Channels with unshared ends, such as `One2OneChannel`, cannot be plugged into them.

A significant change is that channels, such as `One2OneChannel` and `Any2OneChannel`, are now *interfaces* (not *classes*) with two methods: `in()` for extracting the reading-end and `out()` for the writing-end. Implementations of these channel-end interfaces are package-only known classes returned by *static* methods of the `Channel` class (or actual instances of *class factories*, such as `StandardChannelFactory`).

In fact, those package-only known channel-end implementing classes are the same as the package-only known classes implementing channels – so, processes can still cast channel inputs to outputs and vice-versa!

1.3. JCSP 1.1

JCSP 1.1 merges the two libraries. Channel-end interfaces and abstract classes remain the same. Channels themselves are interfaces, as in the JCSP Network Edition. This time, however, channel-end implementations are package-only known classes that *delegate* their methods to *different* package-only known classes implementing the channels. Further, the input-end implementing classes are different from the output-end classes. So, input-ends and output-ends can no longer be cast into each other. Apart from this improvement in security, the change is not apparent and the API remains the same as that for JCSP Network Edition.

Users of the library are only exposed to interfaces (or abstract classes) representing the functionality of channels and channel-ends. Implementation classes are completely hidden. This also allows for easier future changes without affecting the visible API.

1.4. Using Channels from within a Process

The JCSP *process* view of its external channels is unchanged. Here is a simple, but *fair*, multiplexor:

```java
public final class FairPlex implements CSProcess {

  private final AltingChannelInput[] in;
  private final ChannelOutput out;
```

```
public FairPlex (AltingChannelInput[] in, ChannelOutput out) {
  this.in = in;
  this.out = out;
}

public void run () {
  final Alternative alt = new Alternative (in);
  while (true) {
    final int i = alt.fairSelect ();
    out.write (in[i].read ());
  }
}

}
```

1.5. Building Networks of Processes

To build a network, channels must be constructed and used to wire together (concurrently running) process instances. In JCSP 1.0-rc7, channels were directly plugged into processes. Now, as in **occam-π** and the JCSP Network Edition, we must specify which *ends* of each channel to use.

All channels are now constructed using `static` methods of the `Channel` class (*or* an instance of one the specialist channel factories):

```
final One2OneChannel[] a = Channel.one2oneArray (N);    // an array of N channels
final One2OneChannel b = Channel.one2one ();            // a single channel
```

Here is a network consisting of an array of `Generator` processes, whose outputs are multiplexed through `Fairplex` to a `Consumer` process[4]. They are connected using the above channels:

```
final Generator[] generators = new Generator[N];
for (int i = 0; i < N; i++) {
  generators[i] = new Generator (i, a[i].out ());
}

final FairPlex plex = new FairPlex (Channel.getInputArray (a), b.out ());

final Consumer consumer = new Consumer (b.in ());

new Parallel (new CSProcess[] {new Parallel (generators), plex, consumer}).run ();
```

In JCSP 1.0-rc7, the actual channels (`a` and `b`) are passed to the process constructors. Now, we must pass the correct *ends*. The *input-end* of a channel is extracted using the `in()` method; the *output-end* using `out()`[5]. `FairPlex` needs an array of channel *input-ends*, which we could have constructed ourselves, applying `in()` to the individual channel elements. However, this is simplified through the `static` helper methods, `getInputArray()` and `getOutputArray()`, provided by the `Channel` factory.

---

[4]This example is to illustrate the use of channels, including channel arrays, in network construction. If we really only need fair and straightforward multiplexing of individual messages, it would be much simpler and more efficient to connect the generators directly to the consumer using a single `Any2OneChannel`.

[5]These correspond to the *direction specifiers* (? and !) mandated by **occam-π**. The method names `in()` and `out()` must be interpreted from the point of view of the *process* – not the *channel*. The *input-end* is the end of the channel from which a process inputs messages – *not* the end of the channel into which message are put. JCSP is a *process-oriented* model and our terms are chosen accordingly.

2. Alting Barriers

JCSP has long provided a `Barrier` class, on which multiple processes can be enrolled. When *one* process attempts to *synchronise* on a barrier, it blocks until *all* enrolled processes do the same thing. When the last arrives at the barrier, *all* processes are released. They allow dynamic enrollment and resignation, following mechanisms introduced into occam-π [8,21].

This corresponds to fundamental multiway *event synchronisation* in CSP. However, although CSP allows processes to offer multiway events as part of an *external choice*, JCSP does not permit this for `Barrier` synchronisation. Once a process engages with a `Barrier`, it cannot back off (e.g. as a result of a timeout, an arriving channel communication or another barrier). The reason is the same as why *channel output guards* are not allowed. Only *one* party to any synchronisation is allowed to withdraw (i.e. to use that synchronisation as a guard in external choice – *alting*). This enables event choice to be implemented with a simple (and fast) *handshake* from the party making the choice to its chosen partner (who is committed to waiting). Relaxing this constraint implies resolving a choice on which all parties must agree and from which anyone can change their mind (after initially indicating approval). In general, this requires a *two-phase commit* protocol, which is costly and difficult to get right [22].

This constraint has been universally applied in all practical CSP implementations to date. It means that CSP systems involving external choice over multiway events cannot, generally, be directly executed. Instead, those systems must be transformed (preserving their semantics) into those meeting the constraints – which means adding many processes and channels to manage the necessary two-phase commit.

JCSP 1.0-rc7 and 1.1 introduce the `AltingBarrier` class that overcomes that constraint, allowing multiple barriers to be included in the guards of an `Alternative` – along with skips, timeouts, channel communications and *call channel accepts*. Currently, this is supported only for a single JVM (which can be running on a multicore processor). It uses a *fast* implementation that is not a two-phase commit. It has overheads that are linear with respect to the number of barrier offers being made. It is based on the *Oracle* mechanism described at [23,24,25] and summarised in section 2.5.

2.1. User View of Alting Barriers

An *alting* barrier is represented by a family of `AltingBarrier` *front-ends*. Each process using the barrier must do so via its own front-end – in the same way that a process uses a channel via its channel-end. A new alting barrier is created by the `static create` method, which returns an array of front-ends – one for each enrolled process. If additional processes need later to be enrolled, further front-ends may be made from an existing one (through `expand` and `contract` methods). As with the earlier `Barrier` class, processes may temporarily `resign` from a barrier and, later, re-`enrol`.

To use this barrier, a process simply includes its given `AltingBarrier` front-end in a `Guard` array associated with an `Alternative`. Its index will be selected if and only if all parties (processes) to the barrier similarly select it (using their own front-ends).

If a process wishes to commit to this barrier (i.e. not offer it as a choice in an `Alternative`), it may `sync()` on it. However, if all parties only do this, a *non-alting* `Barrier` would be more efficient. A further shortcut (over using an `Alternative`) is provided to *poll* (with timeout) this barrier for completion.

An `AltingBarrier` front-end may only be used by one process at a time (and this is checked at run-time). A process may communicate a *non-resigned* front-end to another process; but the receiving process must *mark* it before using it and, of course, the sending process must not continue to use it. If a process terminates holding a front-end, it may be recycled for use by another process via a *reset*.

Full details of expanding/contracting the set of front-ends, temporary resignation and re-enrolment, communication, marking and resetting of front-ends, committed synchronisation and time-limited polling are given in the JCSP documentation (on-line at [26]).

2.2. Priorities

These do not – *and cannot* – apply to selection between barriers. The `priSelect()` method works locally for the process making the offer. If this were allowed, one process might offer barrier x with higher priority than barrier y ... and another process might offer them with its priorities the other way around. In which case, it would be impossible to resolve a choice in favour of x or y in any way that satisfied the conflicting priorities of both processes.

However, the `priSelect()` method is allowed for choices including barrier guards. It honours the respective priorities defined between non-barrier guards ... and those between a barrier guard and non-barrier guards (which guarantees, for example, immediate response to a timeout from ever-active barriers). Relative priorities between barrier guards are *inoperative*.

2.3. Misuse

The implementation defends against misuse, throwing an `AltingBarrierError` when riled. Currently, the following bad things are prevented:

o different threads trying to operate on the same front-end;
o attempt to enrol whilst enrolled;
o attempt to use as a guard whilst resigned;
o attempt to sync, resign, expand, contract or mark whilst resigned;
o attempt to contract with an array of front-ends not supplied by expand.

Again, we refer to the documentation, [26], for further details and explanation.

2.4. Example

Here is a simple gadget with two modes of operation, switched by a *click* event (operated externally by a *button* in the application described below). Initially, it is in *individual* mode – represented here by incrementing a number and outputting it (as a string to change the label on its controlling button) as often as it can. Its other mode is *group*, in which it can only work if all associated gadgets are also in this mode. Group work consists of a single decrement and output of the number (to its button's label). It performs group work as often as the group will allow (i.e. until it, or one of its partner gadgets, is clicked back to *individual* mode).

```
import org.jcsp.lang.*;

public class Gadget implements CSProcess {

  private final AltingChannelInput click;
  private final AltingBarrier group;
  private final ChannelOutput configure;

  public Gadget (
    AltingChannelInput click, AltingBarrier group, ChannelOutput configure
  ) {
    this.click = click;
    this.group = group;
    this.configure = configure;
  }
```

```
public void run () {

  final Alternative clickGroup =
    new Alternative (new Guard[] {click, group});

  final int CLICK = 0, GROUP = 1;                    // indices to the Guard array

  int n = 0;
  configure.write (String.valueOf (n));

  while (true) {

    configure.write (Color.green)                    // indicate mode change

    while (!click.pending ()) {                       // individual work mode
      n++;                                            // work on our own
      configure.write (String.valueOf (n));          // work on our own
    }
    click.read ();                                    // must consume the click

    configure.write (Color.red);                      // indicate mode change

    boolean group = true;                             // group work mode
    while (group) {
      switch (clickGroup.priSelect ()) {              // offer to work with the group
        case CLICK:
          click.read ();                              // must consume the click
          group = false;                              // back to individual work mode
        break;
        case GROUP:
          n--;                                        // work with the group
          configure.write (String.valueOf (n));       // work with the group
        break;
      }
    }

  }
 }
}
```

The front-end to the alting barrier shared by other gadgets in our group is given by the `group` parameter of the constructor, along with `click` and `configure` channels from and to our button process.

Note that in the above – and for most uses of these alting barriers – no methods are explicitly invoked. Just having the barrier in the guard set of the `Alternative` is sufficient.

This gadget's offer to work with the group is made by the `priSelect()` call on `clickGroup`. If all other gadgets in our group make that offer before a mouse click on our button, this gadget (together with *all* those other gadgets) proceed together on their joint work – represented here by decrementing the count on its button's label. All gadgets then make another offer to work together.

This sequence gets interrupted if any button on any gadget gets clicked. The relevant gadget process receives the click signal and will accept it in preference to further group synchronisation. The clicked gadget reverts to its *individual* mode of work (incrementing the count on its button's label), until that button gets clicked again – when it will attempt to rejoin the group. While any gadget is working on its own, no group work can proceed.

Here is complete code for a system of buttons and gadgets, synchronised by an *alting barrier*. Note that this *single* event needs an *array* of `AltingBarrier` front-ends to operate – one for each gadget:

```java
import org.jcsp.lang.*;

public class GadgetDemo {

  public static void main (String[] argv) {

    final int nUnits = 8;

    // make the buttons

    final One2OneChannel[] event = Channel.one2oneArray (nUnits);

    final One2OneChannel[] configure = Channel.one2oneArray (nUnits);

    final boolean horizontal = true;

    final FramedButtonArray buttons =
      new FramedButtonArray (
        "AltingBarrier: GadgetDemo", nUnits, 120, nUnits*100,
        horizontal, configure, event
      );

    // construct an array of front-ends to a single alting barrier

    final AltingBarrier[] group = AltingBarrier.create (nUnits);

    // make the gadgets

    final Gadget[] gadgets = new Gadget[nUnits];
    for (int i = 0; i < gadgets.length; i++) {
      gadgets[i] = new Gadget (event[i], group[i], configure[i]);
    }

    // run everything

    new Parallel (
      new CSProcess[] {
        buttons, new Parallel (gadgets)
      }
    ).run ();

  }

}
```

This example only contains a single alting barrier. The JCSP documentation [26] provides many more examples – including systems with intersecting sets of processes offering multiple multiway barrier synchronisations (one for each set to which they belong), together with timeouts and ordinary channel communications. There are also some *games*!

2.5. Implementation Oracle

A fast resolution mechanism of choice between multiple multiway synchronisations depends on an *Oracle* server process, [23,24,25]. This maintains information for each barrier and each process enrolled. A process offers *atomically* a set of barriers with which it is prepared to engage and blocks until the *Oracle* tells it which one has been breached. The *Oracle* simply keeps counts of, and records, all the offer sets as they arrive. If a count for a particular barrier becomes complete (i.e. all enrolled processes have made an offer), it informs the lucky waiting processes and *atomically* withdraws all their other offers – *before* considering any new offers.

2.5.1. *Adapting the Oracle for JCSP (and* occam-π*)*

For JCSP, these mechanics need adapting to allow processes to make offers to synchronise that include *all* varieties of `Guard` – not just `AltingBarriers`. The logic of the *Oracle* process is also unravelled to work with the usual *enable/disable* sequences implementing the `select` methods invoked on `Alternative`. Note: the techniques used here for JCSP carry over to a similar notion of alting barriers for an extended occam-π [27].

The `AltingBarrier.create(n)` method first constructs a hidden *base* object – the actual alting barrier – before constructing and returning an array of `AltingBarrier` front-ends. These front-ends reference the base and are chained together. The base object is not shown to JCSP users and holds the first link to the chain of front-ends. It maintains the number of front-ends issued (which it assumes equals the number of processes currently enrolled) and a count-down of how many offers have *not* yet been made to synchronise. It has methods to expand and contract the number of front-ends and manage temporary resignation and re-enrolment of processes. Crucially, it implements the methods for *enabling* (i.e. receiving an offer to synchronise) and *disabling* (i.e. answering an enquiry as to whether the synchronisation has completed and, if not, withdrawing the offer). These responsibilities are delegated to it from the front-end objects.

Each `AltingBarrier` front-end maintains knowledge of the process using it (*thread id* and resigned status) and checks that it is being operated correctly. If all is well, it claims the monitor lock on the base object and delegates the methods. Whilst holding the lock, it maintains a reference to the `Alternative` object of its operating process (which might otherwise be used by another process, via the base object, upon a successful completion of the barrier).

The *Oracle* logic works because each full offer set from a process is handled atomically. The *select* methods of `Alternative` make individual offers (*enables*) from its guard array in sequence. A global lock, therefore, must be obtained and held throughout any enable sequence involving an `AltingBarrier` – to ensure that the processing of its set of offers (on `AltingBarriers`) are not interleaved with those from any other set. If the *enables* all fail, the lock must be released before the *alting* process blocks. If an offer (*enable*) succeeds in completing one of the barriers in the guard set, the lock must continue to be held held throughout the subsequent *disable* (i.e. withdraw) sequence *and* the disable sequences of all the other partners in the successful barrier (which will be scheduled by the successful *enable*)[6]. Other disable sequences (i.e. those triggered by a successful non-barrier synchronisation) do not need to acquire this lock – even if an alting barrier is one of the guards to be disabled.

2.5.2. *Distributing the Oracle*

The current JCSP release supports `AltingBarriers` only *within* a single JVM. Extending this to support them across a distributed system has some issues.

A simple solution would be to install an actual *Oracle* process at a network location known to all. At the start of any *enable* sequence, a network-wide lock on the *Oracle* is obtained (simply by communicating with it on a shared claim channel). Each *enable/disable* then becomes a communication to and from the *Oracle*. The network lock is released follow-

---

[6]This means that multiple processes will need to hold the lock in parallel, so that a counting semaphore (rather than monitor) has to be employed.

ing the same rules outlined for the single JVM (two paragraphs back). However, the network overheads for this (per *enable/disable*) and the length of time required to hold the network-wide lock look bad.

A better solution may be to operate the fast *Oracle* logic locally within each JVM – except that, when a local barrier is potentially overcome (because all local processes have offered to engage with it), the local JCSP kernel negotiates with its partner nodes through a suitable two-phase commit protocol. This allows the local kernel to cancel safely any network offer, should local circumstances change. Only if the network negotiation succeeds are the local processes informed.

2.5.3. Take Care

The logic required for correct implementation of external choice (i.e. the `Alternative` class) is not simple. The version just for channel input synchronisation required formalising and model checking before we got it right [17]. Our implementation has not (yet) been observed to break under stress testing, but we shall not feel comfortable until this has been repeated for these multiway events. Full LGPL source codes are available by request.

3. Output Guards

It has long been an accepted constraint of occam-π and its derivative frameworks (e.g. JCSP, C++CSP, the CSP implementations for .NET) that channels only support input guards for use in alternatives, and not output guards. The decision allows a much faster and simpler implementation for the languages/frameworks [23].

Now, however, alting barriers provide a mechanism on which channels with both input and output guards can easily be built, as described in [22]. Because there are still extra run-time costs, JCSP 1.1 offers a *different* channel for this – for the moment christened `One2OneChannelSymmetric`.

This *symmetric* channel is composed of two internal synchronisation objects: one standard non-buffered one-to-one channel and one alting barrier. Supporting this, a new channel-end interface (actually abstract class), `AltingChannelOutput`, has been added and derives simply from `Guard` and `ChannelOutput`. We are only providing zero-buffered one-to-one symmetrically alting channels for the moment.

The reading and writing processes are the only two enrolled on the channel's internal barrier – on which, of course, they can *alt*.

For any *committed* communication, a process first commits to synchronise on the internal barrier. When/if that synchronisation completes, the real communication proceeds on the internal one-to-one channel as normal.

If either process wants to use the channel as a guard in an alternative, it *offers* to synchronise on the internal barrier – an offer that can be withdrawn if one of the other guards fires first. If its offer succeeds, the real communication proceeds on the internal channel as before.

Of course, all these actions are invisible to the using processes. They use the standard API for obtaining channel-ends and reading and writing. Either channel-end can be included in a set of guards for an `Alternative`.

Here is a pathological example of its use. There are two processes, A and B, connected by two opposite direction channels, c and d. From time to time, each process offers to communicate on both its channels (i.e. an offer to read and an offer to write). They do no other communication on those channels. What must happen is that the processes resolve their choices in compatible ways – one must do the writing and the other the reading. This is, indeed, what happens. Here is the A process:

```
class A implements CSProcess {

  private final AltingChannelInput in;
  private final AltingChannelOutput out;

  ...  standard constructor


  public void run () {
    final Alternative alt = new Alternative (new Guard[] {in , out});
    final int IN = 0, OUT = 1;
    ...  other local declarations and initialisation
    while (running) {
      ...  set up outData
      switch (alt.fairSelect ()) {
        case IN:
          inData = (InDataType) in.read ();
          ...  reaction to this input
        break;
        case OUT:
          out.write (outData);
          ...  reaction to this output
        break;
      }
    }
  }

}
```

The B process is the same, but with different initialisation and reaction codes and types. The
system must be connected with *symmetric* channels:

```
public class PathologicalDemo {

  public static void main (String[] argv) {

    final One2OneChannelSymmetric c = Channel.one2oneSymmetric ();
    final One2OneChannelSymmetric d = Channel.one2oneSymmetric ();

    new Parallel (
      new CSProcess[] {
        new A (c.in (), d.out ()),
        new B (d.in (), c.out ())
      }
    ).run ();

  }

}
```


4. Extended Rendezvous

Extended rendezvous was an idea originally introduced in occam-π [28]. After reading from
a channel, a process can perform some actions *without* scheduling the writing process –
extending the rendezvous between writer and reader. When it has finished those actions (and
it can take its own time over this), it must then schedule the writer. Only the reader may
perform this extension, and the writer is oblivious as to whether it happens.

Extended rendezvous is made available in JCSP through the `ChannelInput.startRead()` and `ChannelInput.endRead()` methods. The `startRead()` method starts the extended rendezvous, returning with a message when the writer sends it. The writer now remains blocked (engaged in the extended rendezvous) until, eventually, the reader invokes the `endRead()` method. They can be used in conjunction with *alternation* – following the (input) channel's selection, simply invoke `startRead()` and `endRead()` instead of the usual `read()`.

4.1. Examples – a Message Logger and Debugging GUI

Consider the (unlikely) task of tracking down an error in a JCSP system. We want to delay and/or observe values sent down a channel. We could insert a special process into the channel to manage this, but that would normally introduce buffering into the system. In turn, that changes the synchronisation behaviour of the system which could easily mask the error – especially if that error was a deadlock.

However, if the inserted process were to use extended rendezvous, we can arrange for there to be no change in the synchronisation. For example, the following *channel tapping* process might be used for this task:

```java
class Tap implements CSProcess {

  private ChannelInput in;                    // from the original writer
  private ChannelOutput out;                  // to the original reader
  private ChannelOutput tapOut;               // to a message logger

  ... standard constructor

  public void run () {
    while (true) {
      Cloneable message = in.startRead ();  // start of extended rendezvous
      {
        tapOut.write (message.clone ());
        out.write (message);
      }
      in.endRead ();                          // finish of extended rendezvous
    }
  }
}
```

This process begins an extended rendezvous, copies the message to its *tapping* channel before writing it to the process for which it was originally intended. Only when this communication is complete does the extended rendezvous end. So long as the report to the message logger is guaranteed to succeed, this preserves the synchronisation between the original two processes: the original writer is released *if-and-only-if* the reader reads.

The extra code block and indentation in the above (and below) example is suggested to remind us to invoke the `endRead()` method, matching the earlier `startRead()`.

Instead of a message logger, we could install a process that generates a GUI window to display passing messages. As these message are only held during the extended rendezvous of `Tap`, that process no longer needs to *clone* its messages. For example:

```java
class MessageDisplay implements CSProcess {

  private ChannelInput in;                     // from the tap process

  ... standard constructor
```

```
public void run () {

   while (true) {
     Object message = in.startRead ();      // start of extended rendezvous
     {
        ... display message in a pop-up message box
        ... only return when the user clicks OK
     }
     in.endRead ();                          // finish of extended rendezvous
   }
}


}
```

Instead of performing communication in its extended rendezvous, the above process interacts with the user through a GUI. The rendezvous is not completed until the user has seen the data value and clicked OK. This in turn delays the tap process until the user clicks OK, which in turn prevents the original communication between the original two processes until the user has clicked OK.

The addition of these two processes has not altered the semantics of the original system – apart from giving the GUI user visibility of, and delaying ability over, communications on the tapped channel.

With trivial extra programming (e.g. writing a `null` to the tapping channel at the end of the extended rendezvous in `Tap`), the `MessageDisplay` could also clear its message box when the reader process takes the message. If this were done for all channels, a deadlocked system would show precisely where messages were stuck.

Such advanced debugging capabilities can be built entirely with the public API of JCSP. There is no need to delve into the JCSP implementation.

4.2. Rules

The `endRead()` function must be called exactly once after each call to `startRead()`. If the reader *poisons* the channel (section 5) between a `startRead()` and `endRead()`, the channel *will* be poisoned; but the current communication is deemed to have happened (which, indeed, it has) and no exception is thrown. In fact, `endRead()` will never throw a poison exception. Poison is explained in section 5.

4.3. Extended Rendezvous on Buffered Channels

Extended rendezvous and buffered channels have not previously been combined. occam-π, which introduced the extended rendezvous concept, does not support buffered channels. C++CSP originally disallowed extended rendezvous on buffered channels using a badly-designed exception[7]. To distinguish between channel-ends that did, and did not, support extended rendezvous, a more complicated type system would have been necessary. In addition to `AltingChannelInput` and `ChannelInput`, we would need `AltingExtChannelInput` and `ExtChannelInput`. Similarly, there would need to be two more classes for the shared versions.

Instead, we took the decision to allow extended rendezvous on buffered channels, thereby eliminating any divide. The semantics of extended rendezvous on a buffered channel are dependent on the semantics of the underlying buffer. The semantics for (some of) the standard buffers provided with JCSP are explained in the following sub-sections.

---

[7]In the new C++CSP2 [29], the classes have been restructured and the implementation is identical to the new JCSP implementation described here

4.3.1. Blocking FIFO Buffers

The reasoning behind the implemented behaviour of extended rendezvous on FIFO buffered channels with capacity N comes from the semantically equivalent pipeline of N 'id' processes (i.e. *one-place* blocking buffers) connected by non-buffered channels. When an extended rendezvous is begun by the process reading from the buffered channel, the first available (that is, the oldest) item of data is read from the channel, *but not removed from its internal buffer*. If no item of data is available, the process must block. Data is only removed from the channel buffer when the extended rendezvous is completed. This mirrors the semantics of an extended rendezvous on the (unbuffered) output channel of the one-place buffer pipeline.

4.3.2. Overwriting (Oldest) Buffers

When full, writing to these channels does not block – instead, the new data overwrites the *oldest* data in the channel. Thus, the channel always holds the freshest available data – which is important for real-time (and other) systems.

There is no simple equivalent of such an overwriting buffer made from unbuffered channels, so we have no simple guidance for its semantics. Instead we choose to follow the principle of least surprise. As with the FIFO buffers, when an extended rendezvous begins, the least recent data item is read from the buffer but not removed. At any time, the writer writes to the buffer as normal, overwriting data when full – the first such one overwritten being the data just read. When the extended rendezvous completes, the data item is removed – *unless* that data 'slot' has indeed been overwritten. This requires the channel buffer to keep track of whether the data being read in an extended rendezvous has been overwritten or not.

An overwriting buffered channel breaks most of the synchronisation between reader and writer. The writer can always write. The reader blocks when nothing is in the channel, but otherwise obtains the latest data and must accept that some may have been missed. Extended rendezvous is meant to block the writer for a period after a reader has read its message – but the writer must never block!

The above implementation yields what should happen if the writer had come along after the extended rendezvous had completed. Since the writer's behaviour is independent from the reader in this case, we take the view that an earlier write (during the rendezvous) is a scheduling accident that should have no semantic impact – i.e. that it is proper to ignore it.

4.3.3. Zero Buffers

Extended rendezvous on a channel using a `ZeroBuffer` is, of course, identical to extended rendezvous on a normal unbuffered channel.

5. Poison and Graceful Termination

In [30], a general algorithm for the deadlock-free termination (and resetting) of CSP/**occam** networks (or sub-networks) was presented. This worked through the distribution of *poison* messages, resulting in poisoned *processes* having to take a defined set of termination actions (in addition to anything needed for process specific tidyness). This logic, though simple, was tedious to implement (e.g. in extending the channel protocol to introduce poison messages). Furthermore, the poison could not distribute against the flow of its carrying channels, so special changes had to be introduced to reach processes *upstream*.

The poison presented here applies to *channels* rather than processes – and it can spread upstream. When a channel is poisoned, any processes waiting on the channel are woken up and a poison exception thrown to each of them. All future reads/writes on the channel result in a poison exception being thrown – there is no antidote! Further attempts to poison the channel are accepted but ignored. This idea was orignally posted by Gerald Hilderink [31].

Poison is used to shutdown a process network – simply and *gracefully*, with no danger of deadlock. For example, processes can set a single poison exception catch block for the whole of their normal operation. The handler responds just by poisoning all its external channels. It doesn't matter whether any of them have already been poisoned.

Poison spreads around a process network viewed as an undirected graph, rather than trying to feed poison messages around a directed graph. These ideas have already been implemented in C++CSP, and by Sputh and Allen for JCSP itself [32]. This revised JCSP 1.1 poison builds on these experiences.

5.1. API Rationale

One option for adding poison to JCSP would have been to add poisonable channel-ends as separate additional interfaces. This would cause a doubling in the number of channel-end interfaces for JCSP. The reasoning presented in [33] still holds; a separation of poisonable and non-poisonable channel-ends in the type system would lead to complex common processes, that would need to be re-coded for each permutation of poisonable and non-poisonable channel-ends. Therefore, *all* channel-ends have `poison(strength)` methods.

Although all channel-ends have the poison methods, they do not have to be functional. Some channels do not permit poisoning – for example, the default ones: attempts to poison them are ignored.

5.2. Poison Strength

In [32], Sputh and Allen proposed the idea of two levels of poison – *local* and *global*. Channels could be constructed immune to local poison. Thus, networks could be built with sub-networks connected only by *local-immune* channels. Individual sub-networks could then be individually terminated (and replaced) by one of their components injecting *local* poison. Alternatively, the whole system could be shut down by *global* poison.

These ideas have been generalised to allow arbitrary (positive integer) levels of poison in JCSP 1.1. This allows many levels of nested sub-network to be terminated/reset at any of its levels. Poisonable channels are created with a specific level of *immunity*: they will only be poisoned with a poison whose *strength* is greater than their immunity. Poison exceptions carry the strength with which the channel has been poisoned: their handlers propagate poison with that same strength.

Channels carry the current strength of poison inside them: zero (poison-free) or greater than their immunity (poisoned). That strength can increase with subsequent poisoning, but is not allowed to decrease (with a weaker poison).

Note that using different strengths of poison can have non-deterministic results. For example, if different waves of poison, with different strengths, are propagating in parallel over part of a network whose channels are not immune, the strength of the poison exception a process receives will be scheduling dependent – which wave struck first! If a lower strength were received, it may fail to propagate that poison to some of its (more immune) channels before it terminates: without, of course, dealing with the stronger poison arriving later. Care is needed here.

5.3. Trusted and Untrusted Poisoners

Channel-ends of poisonable channels can be created specifically *without* the ability to poison (as in C++CSP [34]): attempts will be ignored (as if their underlying channel were not poisonable). Disabling poisoning at certain channel-ends of otherwise poisonable channels allows networks to be set up with trusted and untrusted poisoners. The former (e.g. a server process) has the ability to shut down the network. The latter (e.g. remote clients) receive the network poisoning but cannot initiate it.

5.4. Examples

Here is a standard *running-sum integrator* process, modified to support network shutdown after poisoning:

```java
public class IntegrateInt implements CSProcess {

  private final ChannelInput in;
  private final ChannelOutput out;

  public IntegrateInt (ChannelInput in, ChannelOutput out) {
    this.in = in;
    this.out = out;
  }

  public void run () {
    try {
      int sum = 0;
      while (true) {
        sum += in.read ();
        out.write (sum);
      }
    } catch (PoisonException e) {              // poison everything
      int strength = e.getStrength ();
      out.poison (strength);
      in.poison (strength);
    }
  }
}
```

A guard for a channel is considered *ready* if the channel is poisoned. This poison will only be detected, however, if the channel is selected and the channel communication attempted. Here is a modification of the `FairPlex` process (from section 1.4) to respond suitably to poisoning. The only change is the addition of the `try/catch` block in the `run()` method:

```java
public final class FairPlex implements CSProcess {

  private final AltingChannelInput[] in;
  private final ChannelOutput out;

  ...  standard constructor

  public void run () {
    try {
      final Alternative alt = new Alternative (in);
      while (true) {
        final int i = alt.fairSelect ();
        out.write (in[i].read ());
      }
    } catch (PoisonException e) {              // poison everything
      int strength = e.getStrength ();
      out.poison (strength);
      for (int i = 0; i < in.length; i++) {
        in[i].poison (strength);
      }
    }
  }
}
```

If the `out` channel is poisoned, the poison exception will be thrown on the next cycle of `FairPlex`. If any of the `in` channels is poisoned, its guard becomes *ready* straight away. This *may* be ignored if there is traffic from unpoisoned channels available and `FairPlex` will continue to operate normally.. However, the *fair* selection guarantees that no other input channel will be serviced twice before that poisoned (and ready) one. In the worst case, this will be after (`in.length` - 1) cycles. When the poisoned channel is selected, the exception is thrown.

5.5. Implementation

The central idea behind adding poison to all the existing channel algorithms is simple. Every time a channel wakes up from a `wait`, it checks to see whether the channel is poisoned. If it is, the current operation is abandoned and a `PoisonException` (carrying the poison strength) is thrown.

However, with just the above approach, it would be possible for a writing process (that was late in being rescheduled) to observe poison added by a reader *after* the write had completed successfully. This was discovered (by one of the authors [35]) from formalising and (FDR [16]) model checking this (Java) implementation against a more direct CSP model, using techniques developed from [17].

Therefore, an extra field is added so that a successfully completed communication is always recorded in the channel, regardless of any poison that may be injected afterwards. Now, the writer can complete normally and without exception – the poison remaining in the channel for next time. This correction has been model checked [35]. It has also been incorporated in the revised C++CSP [36].

6. Conclusions and Future Work

The latest developments of JCSP have integrated the JCSP Network Edition and JCSP 1.0-rc7, keeping the advances each had made separately from their common ancestor. New concepts have been added: choice between multiple multiway synchronisations (alting barriers), output guards (symmetric channels), extended rendezvous and poison. The revised library is LGPL open sourced. We are working on further re-factorings to allow third parties to add new *altable* synchronisation primitives, without needing to modify existing sources. We list here a few extensions that are have been requested by various users and are likely for future releases. Of course, with open source, we would be very pleased for others to complete these with us.

6.1. Broadcast Channels

Primitive events in CSP may synchronise *many* processes. Channel communications are just events and CSP permits any number of readers and writers. Many readers implies that all readers receive the *same* message: either *all* receive or *none* receive – this is multiway synchronisation. Many writers is a little odd: *all* must write the *same* message or *no* write can occur – still multiway synchronisation.

All channels currently in JCSP restrict communications to point-to-point message transfers between *one* writer and *one* reader. The `Any` channels allow any number of writers and/or readers, but only one of each can engage in any individual communication.

Allowing CSP *many-reader* (broadcasting) channels turns out to be trivial – so we may as well introduce them. The only interesting part is making them as efficient as possible.

One way is to use a process similar to `DynamicDelta` from `org.jcsp.plugNplay`. This cycles by waiting for an input and, then, outputting *in parallel* on all output channels. That introduces detectable buffering which is easily eliminated by combining the input and outputs in an extended rendezvous (Section 4). We still do not have multiway synchronisation, since the readers do not have to wait for each other to take the broadcast. This can be achieved by the *delta* process outputting twice and the readers reading twice. The first message can be `null` and is just to assemble the readers. Only when everyone has taking that is the real message sent. Getting the second message tells each reader that every reader is committed to receive. The *delta* process can even send each message *in sequence* to its output channels, reducing overheads (for unicore processors).

The above method has problems if we want to allow *alting* on the broadcast. Here is a simpler and faster algorithm that shows the power of *barrier synchronisation* – an obvious mechanism, in retrospect, for broadcasting!

```
public class One2ManyChannelInt

  private int hold;
  private final Barrier bar;

  public One2ManyChannelInt (final int nReaders) {
    bar = new Barrier (nReaders + 1);
  }

  public void write (int n) {          -- no synchronized necessary
    hold = n;
    bar.sync ();                       -- wait for readers to assemble
    bar.sync ();                       -- wait for readers to read
  }

  public int read () {                 -- no synchronized necessary
    bar.sync ();                       -- wait for the writer and other readers
    int tmp = hold;
    bar.sync ();                       -- we've read it!
    return tmp;
  }

}
```

The above *broadcasting channel* supports only a fixed number of readers and no alting. This is easy to overcome using the dynamics of an `AltingBarrier`, rather than `Barrier` – but is left for another time. For simplicity, the above code is also not *dressed* in the full JCSP mechanisms for separate channel-ends, poisoning etc.. It also carries integers. Object broadcasting channels had better be carefully used! Probably, only *immutable* objects (or clones) should be broadcast. Otherwise, the readers should only ever read (never change) the objects they receive (and anything that they reference).

The above code uses the technique of *phased barrier synchronisation* [8,21,37]. Reader and writer processes share access to the `hold` field inside the channel. That access is controlled through phases divided by the barriers. In the first phase, only the writer process may write to `hold`. In the second, only the readers may read. Then, it's back to phase one. No locks are needed.

Most of the work is done by the first barrier, which cannot complete until all the readers and writer assemble. If this barrier were replaced by an *alting* one, that could be used to enable external choice for all readers and the writer.

Everyone is always committed to the second barrier, which cannot therefore stick. It's only purpose is to prevent the writer exiting, coming back and overwriting `hold` before all the readers have taken the broadcast. If the first barrier were replaced by an `AltingBarrier`, the second could remain as this (faster) `Barrier`.

However, other optimisations are possible – for example, by the readers decrementing a *reader-done* count (either atomically, using the new Java 1.5 concurrency utilities, or with a standard monitor lock) and with the last reader resetting the count and releasing the writer (waiting, perhaps, on a 2-way `Barrier`).

6.2. Java 1.5 Generics

Java 1.5 (also known as Java 5) was a major release that introduced many new features. The three main additions pertinent to JCSP are generics, autoboxing, and the new `java.util.concurrent` package (and its subpackages).

Generics in Java are a weak form of *generic* typing. Their primary use is to enhance semantic clarity and eliminate some explicit type casting (whilst maintaining type safety). They have been particularly successful in the revised collection classes.

Generics can be used to type more strongly JCSP channels (and avoid the cast usually needed on the return `Object` from a `read/startRead()` method). It would make the type of the channel explicit and enforced by the compiler. Generics require a Java compiler of version 1.5 or later, but they can be compiled into earlier bytecode versions executable by Java 1.3.

6.3. Java 1.5 Autoboxing

Autoboxing is the term for the automatic conversion from primitive types (such as `int` or `double`) into their class equivalents (`Integer` and `Double` respectively). Particularly when combined with generics, this allows primitive types directly to be used for communicating with generic processes through object-carrying channels. For example, if both autoboxing and generics are used in future versions of JCSP, the following codes would be legal. First, we need a generic channel:

```
One2OneChannel<Double> c = Channel.<Double>one2one (new Buffer<Double> (10));
```

Then, a writing process could execute:

```
out.write (6.7);
```

where `out` is the output-end of the above channel (i.e. `c.out()`). A reading process could execute:

```
double d = in.read ();
```

where `in` is the input-end of the above channel (i.e. `c.in()`). Note the lack of any casts in the above codes.

Like generics, autoboxing requires a 1.5 compiler but can be compiled to be executable by earlier versions, such as 1.3. This makes generics and autoboxing a potential candidate for inclusion in JCSP that would still allow Java 1.3 compatibility to be maintained – although it would mean that JCSP developers would need a Java 1.5 compiler.

6.4. Java 1.5 New Concurrency Utilities

The `java.util.concurrent` package contains new concurrency classes. Some classes complement JCSP well: the `CopyOnWriteArrayList` and `CopyOnWriteArraySet` classes can be safely shared between processes to increase efficiency.

Some classes have close similarity to certain JCSP primitives. `CyclicBarrier` is one such class, implementing a barrier (but with a useful twist in its tail). However, it does not support dynamic enrolment and resignation, nor any form of use in anything resembling *external choice*. Its support for the thread *interruption* features of Java makes it, arguably, more complex to use.

`BlockingQueue` looks similar to a FIFO-buffered channel, with `Exchanger` similar to an unbuffered channel. However, they are not direct replacements since neither class supports *external choice*.

The *atomic* classes (in `java.util.concurrent.atomic`) are tools on which JCSP primitives might profitably be built. This is an avenue for future work.

6.5. Networking

Consideration must also be taken as to how the new features in the core can be implemented into JCSP Network Edition. One of the strengths provided in JCSP is the transparency (to the process) of whether a channel is networked or local. If (generic) typed channels are to be implemented, then a method of typing *network* channels must also be available. This brings with it certain difficulties. Guarantees between two nodes must be made to ensure that the networked channel sends and receives the expected object type. However, of more importance at the moment is the implementation of networked barriers, and also networked alting barriers, to allow the same level of functionality at the network level as there is at the local level. Extended rendezvous and guarded outputs on network channels are also considerations.

If the move to exploit Java 1.5 is made in JCSP, then certain features of Java can be taken advantage of in the network stack to improve resource usage, and possibly performance. Java 1.4 introduced a form of *'channel'*, in its `java.nio.channels` package, that can be used to have the native system do some of the work for us. These channels can be used for multiplexing. Since they can represent network connections, we may be able to prune the current networking infrastructure of JCSP to reduce the number of processes needed to route things around – saving memory and run-time overheads.

Attribution

The original development of JCSP was done by Paul Austin and Peter Welch. Further contributions came from Neil Fuller, John Foster and David Taylor. The development of JCSP Network Edition was done by Jim Moores, Jo Aldous, Andrew Griffin, Daniel Evans and Peter Welch. The implementation of poison (and proof thereof) was done by Bernhard Sputh and Alastair Allen. Alting barriers were designed and implemented by Peter Welch. The addition of extended rendezvous, and the merging of all these strands was done by Neil Brown, Peter Welch and Kevin Chalmers.

The authors remain in debt to the CPA/WoTUG community for continual encouragement, feedback and criticism throughout this period. We apologise unreservedly to any individuals not named above, who have nevetheless made direct technical inputs to JCSP.

References

[1] P.H. Welch and P.D. Austin. The JCSP (CSP for Java) Home Page, 1999. Available at: `http://www.cs.kent.ac.uk/projects/ofa/jcsp/`.

[2] P.H.Welch. Process Oriented Design for Java: Concurrency for All. In H.R.Arabnia, editor, *Proceedings of the International Conference on Parallel and Distributed Processing Techniques and Applications (PDPTA'2000)*, volume 1, pages 51–57. CSREA, CSREA Press, June 2000.

[3] P.H. Welch, J.R. Aldous, and J. Foster. CSP networking for java (JCSP.net). In P.M.A. Sloot, C.J.K. Tan, J.J. Dongarra, and A.G. Hoekstra, editors, *Computational Science - ICCS 2002*, volume 2330 of *Lecture Notes in Computer Science*, pages 695–708. Springer-Verlag, April 2002. ISBN: 3-540-43593-X. See also: `http://www.cs.kent.ac.uk/pubs/2002/1382`.

[4] P.H. Welch and B. Vinter. Cluster Computing and JCSP Networking. In James Pascoe, Peter Welch, Roger Loader, and Vaidy Sunderam, editors, *Communicating Process Architectures 2002*, WoTUG-25, Concurrent Systems Engineering, pages 213–232, IOS Press, Amsterdam, The Netherlands, September 2002. ISBN: 1-58603-268-2.

[5] C.A.R. Hoare. *Communicating Sequential Processes*. Prentice-Hall, 1985.

[6] R. Milner. *Communicating and Mobile Systems: the Pi-Calculus*. Cambridge University Press, 1999. ISBN-10: 0521658691, ISBN-13: 9780521658690.

[7] P.H. Welch and F.R.M. Barnes. Communicating mobile processes: introducing occam-pi. In A.E. Abdallah, C.B. Jones, and J.W. Sanders, editors, *25 Years of CSP*, volume 3525 of *Lecture Notes in Computer Science*, pages 175–210. Springer Verlag, April 2005.

[8] F.R.M. Barnes, P.H. Welch, and A.T. Sampson. Barrier synchronisation for occam-pi. In Hamid R. Arabnia, editor, *Proceedings of the 2005 International Conference on Parallel and Distributed Processing Techniques and Applications (PDPTA'05)*, pages 173–179, Las Vegas, Nevada, USA, June 2005. CSREA Press.

[9] F.R.M. Barnes. occam-pi: blending the best of CSP and the pi-calculus. http://www.occam-pi.org/, 10 February 2007.

[10] The occam-pi programming language, June 2006. Available at: http://www.occam-pi.org/.

[11] J.F. Broenink, A.W.P. Bakkers, and G.H. Hilderink. Communicating Threads for Java. In Barry M. Cook, editor, *Proceedings of WoTUG-22: Architectures, Languages and Techniques for Concurrent Systems*, pages 243–262, 1999.

[12] N.C.C. Brown and P.H. Welch. An Introduction to the Kent C++CSP Library. In Jan F. Broenink and Gerald H. Hilderink, editors, *Communicating Process Architectures 2003*, pages 139–156, 2003.

[13] B. Orlic. and J.F. Broenink. Redesign of the C++ Communicating Threads Library for Embedded Control Systems. In F. Karelse STW, editor, *5th Progress Symposium on Embedded Systems*, pages 141–156, 2004.

[14] A. Lehmberg and M.N. Olsen. An Introduction to CSP.NET. In Frederick R. M. Barnes, Jon M. Kerridge, and Peter H. Welch, editors, *Communicating Process Architectures 2006*, pages 13–30, 2006.

[15] K. Chalmers and S. Clayton. CSP for .NET Based on JCSP. In Frederick R. M. Barnes, Jon M. Kerridge, and Peter H. Welch, editors, *Communicating Process Architectures 2006*, pages 59–76, 2006.

[16] Formal Systems (Europe) Ltd., 3, Alfred Street, Oxford. OX1 4EH, UK. *FDR2 User Manual*, May 2000.

[17] P.H. Welch and J.M.R. Martin. Formal Analysis of Concurrent Java Systems. In Peter H. Welch and Andrè W. P. Bakkers, editors, *Communicating Process Architectures 2000*, pages 275–301, 2000.

[18] WoTUG: Java Threads Workshop, 1996. Available at: http://wotug.ukc.ac.uk/parallel/groups/wotug/java/.

[19] P.H.Welch. Java Threads in the Light of occam/CSP. In P.H.Welch and A.W.P.Bakkers, editors, *Architectures, Languages and Patterns for Parallel and Distributed Applications*, volume 52 of *Concurrent Systems Engineering Series*, pages 259–284, Amsterdam, April 1998. WoTUG, IOS Press.

[20] P. Austin. JCSP: Early Access, 1997. Available at: http://www.cs.kent.ac.uk/projects/ofa/jcsp0-5/.

[21] P.H. Welch and F.R.M. Barnes. Mobile Barriers for occam-pi: Semantics, Implementation and Application. In J.F. Broenink, H.W. Roebbers, J.P.E. Sunter, P.H. Welch, and D.C. Wood, editors, *Communicating Process Architectures 2005*, volume 63 of *Concurrent Systems Engineering Series*, pages 289–316, Amsterdam, The Netherlands, September 2005. IOS Press. ISBN: 1-58603-561-4.

[22] A.A. McEwan. *Concurrent Program Development*. DPhil thesis, The University of Oxford, 2006.

[23] P.H. Welch. A Fast Resolution of Choice between Multiway Synchrnoisations (Invited Talk). In Frederick R. M. Barnes, Jon M. Kerridge, and Peter H. Welch, editors, *Communicating Process Architectures 2006*, pages 389–390, 2006.

[24] P.H. Welch, F.R.M. Barnes, and F.A.C. Polack. Communicating complex systems. In Michael G Hinchey, editor, *Proceedings of the 11th IEEE International Conference on Engineering of Complex Computer Systems (ICECCS-2006)*, pages 107–117, Stanford, California, August 2006. IEEE. ISBN: 0-7695-2530-X.

[25] P.H. Welch. TUNA: Multiway Synchronisation Outputs, 2006. Available at: http://www.cs.york.ac.uk/nature/tuna/outputs/mm-sync/.

[26] P.H. Welch. JCSP: AltingBarrier Documentation, 2006. Available at: http://www.cs.kent.ac.uk/projects/ofa/jcsp/jcsp1-0-rc7/jcsp-docs/jcsp/l%ang/AltingBarrier.html.

[27] F.R.M. Barnes. Compiling CSP. In P.H. Welch, J. Kerridge, and F.R.M. Barnes, editors, *Proceedings of Communicating Process Architectures 2006 (CPA-2006)*, volume 64 of *Concurrent Systems Engineering Series*, pages 377–388. IOS Press, September 2006.

[28] F.R.M. Barnes and P.H. Welch. Prioritised Dynamic Communicating Processes - Part I. In James Pascoe, Roger Loader, and Vaidy Sunderam, editors, *Communicating Process Architectures 2002*, pages 321–352, 2002.

[29] N.C.C. Brown. C++CSP2. http://www.cppcsp.net/, 10 February 2007.

[30] P.H. Welch. Graceful Termination – Graceful Resetting. In *Applying Transputer-Based Parallel Machines,*

Proceedings of OUG 10, pages 310–317, Enschede, Netherlands, April 1989. Occam User Group, IOS Press, Netherlands. ISBN 90 5199 007 3.

[31] G.H. Hilderink. Poison, 2001. Available at: `http://occam-pi.org/list-archives/java-threads/msg00528.html`.

[32] B.H.C. Sputh and A.R. Allen. JCSP-Poison: Safe Termination of CSP Process Networks. In *Communicating Process Architectures 2005*, 2005.

[33] N.C.C. Brown. Rain: A New Concurrent Process-Oriented Programming Language. In Frederick R. M. Barnes, Jon M. Kerridge, and Peter H. Welch, editors, *Communicating Process Architectures 2006*, pages 237–252, 2006.

[34] N.C.C. Brown. C++CSP Networked. In Ian R. East, David Duce, Mark Green, Jeremy M. R. Martin, and Peter H. Welch, editors, *Communicating Process Architectures 2004*, pages 185–200, 2004.

[35] B.H.C. Sputh. *Software Defined Process Networks*. PhD thesis, University of Aberdeen, August 2006. Initial submission.

[36] N.C.C. Brown. C++CSP2: A Many-to-Many Threading Model for Multicore Architectures. In Alistair A. McEwan, Steve Schneider, Wilson Ifill, and Peter Welch, editors, *Communicating Process Architectures 2007*, volume 65 of *Concurrent Systems Engineering Series*, 2007.

[37] C. Ritson and P.H. Welch. A Process-Oriented Architecture for Complex System Modelling. In Alistair A. McEwan, Steve Schneider, Wilson Ifill, and Peter Welch, editors, *Communicating Process Architectures 2007*, volume 65 of *Concurrent Systems Engineering Series*, 2007.

Communicating Process Architectures 2007
Alistair McEwan, Steve Schneider, Wilson Ifill and Peter Welch (Eds.)
IOS Press, 2007

371

Hardware/Software Synthesis and Verification Using Esterel

Satnam SINGH
Microsoft Research, Cambridge, CB3 0FB, United Kingdom

Abstract. The principal contribution of this paper is the demonstration of a promising technique for the synthesis of hardware and software from a single specification which is also amenable to formal analysis. We also demonstrate how the notion of synchronous observers may provide a way for engineers to express formal assertions about circuits which may be more accessible then the emerging grammar based approaches. We also report that the semantic basis for the system we evaluate pays dividends when formal static analysis is performed using model checking.

Keywords. Hardware/software co-design, synthesis, verification, synchronous languages.

Introduction

Conventional approaches for the design and implementation of systems that comprise hardware and software typically involve totally separate flows for the design and verification of the hardware and software components. The verification of the combined hardware/ software system still remains a significant challenge. The software design flow is based around imperative languages with semantics that model Von Neuman style architectures. The most popular software languages have semantics that can only be expressed in an operational way which is rather unsatisfactory from a formal verification and static analysis viewpoint. Conventional hardware description languages have semantics based on an event queue model which also lends itself to an operational style semantic description which is different from the semantic model typically used for software. This immediately poses a problem for any verification techniques based on static analysis (e.g. formal verification) that need to analyse a system that comprises software and hardware. Is it possible to come up with a way of describing hardware and software based on the same semantic model or semantic models that can be easily related to each other?

This paper explores a possible answer to this question by investigating a formalism that has already proved itself as a mature technology for the static analysis of software. The formalism is captured in the Esterel V7 programming language [2] and we report the result of experiments which evaluate this methodology for the synthesis of both hardware and software. We also report our experience of performing static analysis of hardware systems with properties expressed as synchronous observers which are checked using an embedded model checker.

1. Hardware Software Trade-offs

Given that Esterel has a semantic basis that is appropriate for synthesizing either hardware or software we attempted an experiment which uses Esterel to describe a system which is first implemented entirely in software and then entirely in hardware. If we can get such a flow to work then we would have achieved several desirable properties including the ability to produce corresponding hardware and software from the same specification and the ability to formally analyse the hardware or software. Furthermore, we would have a lot of flexibility to partition the system description so that some of it is realized in hardware and the rest is mapped to software.

There are many useful applications for the ability to obtain either a hardware or software implementation from a single specification (or a hybrid of both). We are particularly interested in the case where dynamic reconfiguration [3] of programmable logic devices is used to swap in and out hardware blocks to perform space/time trade-offs. A hardware block may be swapped out to make way for a more important calculation in hardware or because it implements a calculation which can now be performed in software. In this case the function of the hardware block is replaced by a software thread on a processor (also on the reconfigurable logic device). This novel application requires the hardware and software components to have identical semantics. Conventional approaches involve designing both the hardware and software independently followed by an informal verification process to establish their equivalence. The problem of trying to produce matching software and hardware blocks is one of the major impediments to the research efforts in the field of task-based dynamic reconfiguration. We avoid the duplication of implementation effort, ensure that both resulting implantations have the same behaviour and we can also prove formal properties about our generated circuits.

As an example of a quite simple hardware/software trade-off experiment we present the case of peripheral controller which can be implemented either in hardware or software. We chose a reconfigurable fabric realized by Xilinx's Virtex™-II FPGA and it is on this device that we perform the hardware/software trade-offs. We use a specific development board manufactured by Xilinx called the MicroBlaze Multimedia Development Board, as shown in Figure 1 which contains a Virtex-II XC2V2000 FPGA.

Figure 1. A Xilinx Virtex-II development board

Software threads execute on a 32-bit soft processor called MicroBlaze which is realized as a regular circuit on the Virtex-II FPGA. For the purpose of this experiment we need to choose an interface that runs at a speed which can be processed by a software thread running on a soft processor. We selected the RS232 interface on this board which has all its wires (RX, TX, CTS, RTS) connected directly to the FPGA (there is no dedicated UART chip on the board). Now we have the choice to read and write over the RS232 serial port either by creating a UART circuit on the FPGA fabric or by driving and reading the RX and TX wires from software.

The send and receive portions of an RS232 interface were described graphically using Esterel's safe state machine notation. The receive portion is illustrated in Figure 2. This version does not use hardware flow control.

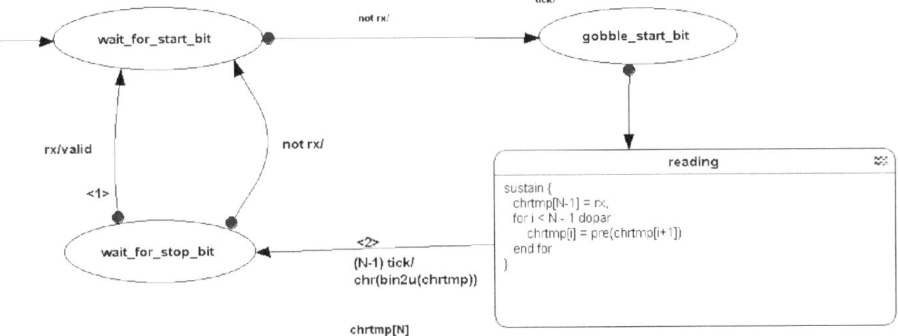

Figure 2. The receive component of the RS232 interface

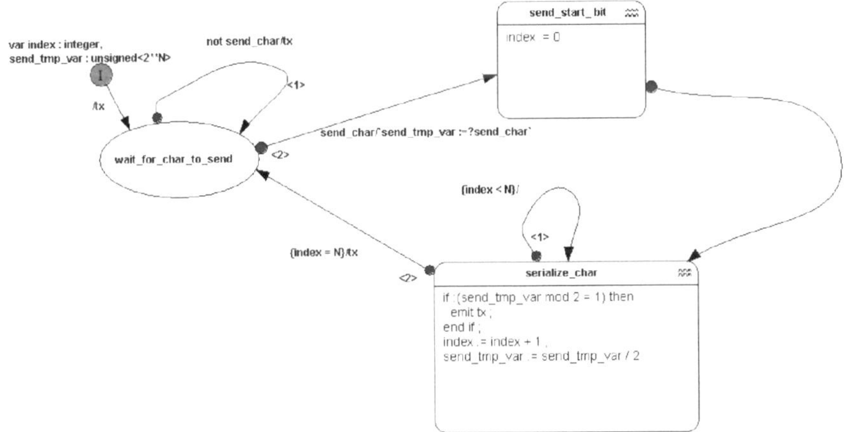

Figure 3. The send component of the RS232 interface.

This state machine waits for a start bit and then assembles the parallel character representation from the serial data from the RX wire and if a valid parity is produced it emits the consumed character. Not all the operations required to recognize a valid character on the RX serial line are convenient to describe using a graphical notation.

For example, here we describe the notion if shifting in a new character bit into an internal buffer using a text (in a textual macrostate). The send component is shown in Figure 3

It is possible to hierarchically build layers on top of these descriptions to add additional features e.g. a FIFO for the read and send channels. Graphical and textual descriptions can be freely mixed with the graphical descriptions being automatically converted into their equivalent Esterel textual equivalents.

Esterel has been used for some time for the synthesis of C software from either Esterel textual descriptions of the graphical state machine representations. Recently the ability to generate hardware from Esterel has become available with various hardware descriptions languages supported. We experimented with Esterel Technology's implementation (called Esterel Studio) by generating VHDL netlists and pushing them through Xilinx's flow which performs synthesis, mapping, placement and then produces a final implementation bitstream.

The generated VHDL simulated without any problems using the commercial Modelsim simulator and produced the same waveforms as the built-in simulator provided by Esterel Technolgoies (which can dump and display VCD files). The generated VHDL was also processed without complain by Xilinx's implementation tools and required no adjustment for an implementation bitstream to be produced.

The receive circuit is implemented using 21look-up tables which represents less than 1% of the capacity available on a XC2V2000 FPGA and is competitive with hand crafted implementations. We have experimented with the synthesis of several types and sizes of designs from Esterel. We noted that in all cases the generated VHDL results in implementations which are comparable to hand crafted behavioural descriptions or readily available IP-blocks. This is an important requirement since the inability to produce reasonable quality circuits would rule out this approach for making hardware/software trade-offs.

We give below the interface for the VHDL generated in the case when only the receive component is synthesized.

```
01  library IEEE;
02  use IEEE.STD_LOGIC_1164.all;
03  use IEEE.NUMERIC_STD.all;
04  use work.receive_data_type_pkg.all;
05  use work.receive_data_pkg.all;
06  entity receive is
07  port (clk: in std_logic;
08        rst: in std_logic;
09        rx: in std_logic;
10        valid: out std_logic;
11        chr: out std_logic;
12        chr_data: out unsigned (7 downto 0)
13  );
14  end receive;
```

One point to note is that the signal which is emitted by a character is recognized is represented in Esterel as a *value* signal. A value signal can be either present or absent (as represented by the value of the `chr` signal) and it also has a value which is represented by the bit-vector `chr_data`. In this case the `chr` signal can be used as an interrupt or flag to some other circuit (e.g. a FIFO) or software process (e.g. an interrupt handler) which can capture the character that has just been read. The `clk` signal corresponds to the baud rate of the RX input.

If this circuit were to be used as a peripheral on the OPB bus then an extra component is required to relate the clock frequency of the OPB bus to the baud rate of the peripheral. A clock divider circuit for performing this function is easily described in Esterel.

We have tested the resulting circuit by executing it on the MicroBlaze Multimedia board. Using an integrated logic analyser and by connecting an external terminal to the RS232 port we were able to observe the correct characters being recognized by the receive circuit.

Next we configured the Esterel software to generate C rather than VHDL. This generated code which generates code to implement the UART functionality. Our target embedded system was still chosen to be the MicroBlaze Multimedia board and we instantiated a soft MicroBlaze processor on the FPGA. We also instantiated a timer circuit which generated interrupts at the same frequency as the baud rate. The interrupt handler code sampled the input of the RX input wire and used this value as an input to the Esterel generated state-machine. For each signal that can be emitted we define a call-back handler routine. In our case we defined a routine that simply wrote out the character that was read by the state-machine. We performed the same experiment as before and observed the embedded software correctly reading characters from a remote terminal. This demonstrated that for in this case the Esterel flow successfully yielded both a hardware and software implementation from the same specification. In the software case the complete functionality of the UART was realised in code: the only input was the RX bit which was sampled at the baud rate. Now the developer can choose between a hard or soft implementation of a UART depending on constraints like area and speed. We successfully tested the UART at up to 19200 baud.

2. Assertions Using Synchronous Observers

Given that we have the basic ability to synthesize either hardware of software from the same specification based on a clean semantic we next decided to see if we could stay within the Esterel methodology to try and prove properties about our circuits.

Emerging techniques for specifying assertions typically involve using an extra language which has suitable operators for taking about time (past, present and future) and logic relationships between signals. These languages are often concrete representations of formal logics and assertion languages are really temporal logics which can be statically analysed. Can the graphical safe state machine notation provide an alternative way of specifying properties about circuits which has the advantage of being cast in the same language as the specification notation? And can these circuit properties be statically analysed to formally prove properties about circuits?

The investigate these questions we performed another experiment where we design a peripheral for IBM's OPB bus which forms part of IBM's CoreConnect™ IP bus [1]. We chose the OPB bus because it is used by the MicroBlaze soft processor. This makes it easy for us to test and configure an implementation of this peripheral from the soft processor implemented on an FPGA.

An example of a common transaction on the OPB-bus is shown in Figure 4. The key feature of the protocol that we will verify with an example is that a read or write transaction should be acknowledged within 16 clock ticks. Unless a control signal is asserted to allow for more time if a peripheral does not respond within 16 ticks then an error occurs on the bus and this can cause the system to crash. Not shown is the OPB_RNW signal which determines whether a transaction performs a read or a write.

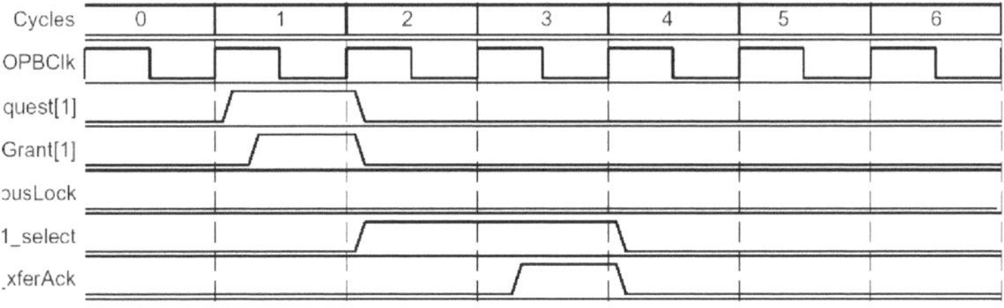

Figure 4. A sample OPB transaction

We considered the case of a memory mapped OPB slave peripheral which has two device registers that a master can write into and a third device register that a master can read from. The function performed by the peripheral is to simply add the contents of the two 'write' registers and make sure that the sum is communicated by the 'read' register. A safe state machine for such a peripheral is shown in Figure 5.

This generated VHDL for this peripheral was incorporated into Xilinx's Embedded Developer Kit and it was then used as a building block of a system which also included a soft processor, an OPB system bus and various memory resources and interfaces. The successfully incorporation of the generated peripheral into the vendor tool flow is illustrated in Figure 6. We wrote test programs to check the operation of the peripheral with a 50MHz OPB system bus. The peripheral always produced the correct answer.

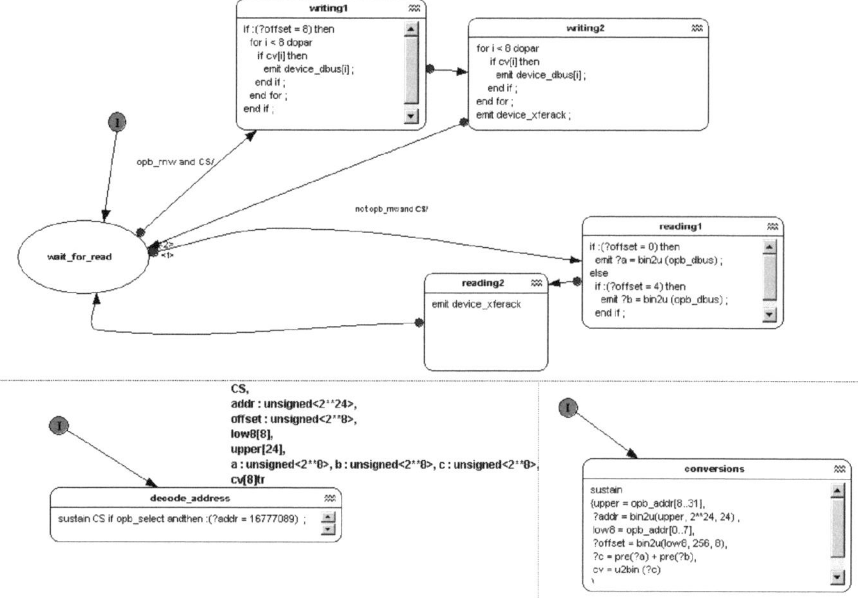

Figure 5. An OPB-slave peripheral

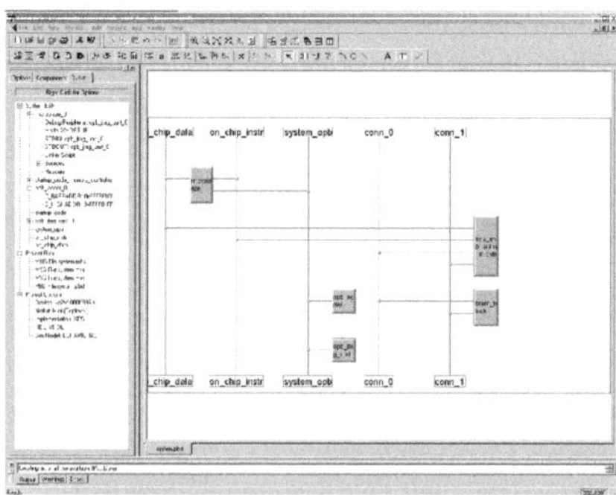

Figure 6. OPB slave incorporated into vendor design flow

Having successfully implemented an OPB peripheral from the Esterel specification we then attempted to prove an interesting property about this circuit. We choose to try and verify the property that this circuit will always emit an OPB transfer acknowledge signal two clock ticks after it gets either a read or a write request. If we can statically prove this property we know that this peripheral can never be the cause of a transfer acknowledge timeout event.

We expressed this property as a regular Esterel safe state machine as shown in Figure 7. This synchronous observer tracks the signal emission behaviour in the implementation description and emits a signal if the system enters into a bad state i.e. a read or write request is not acknowledged in exactly two clock ticks.

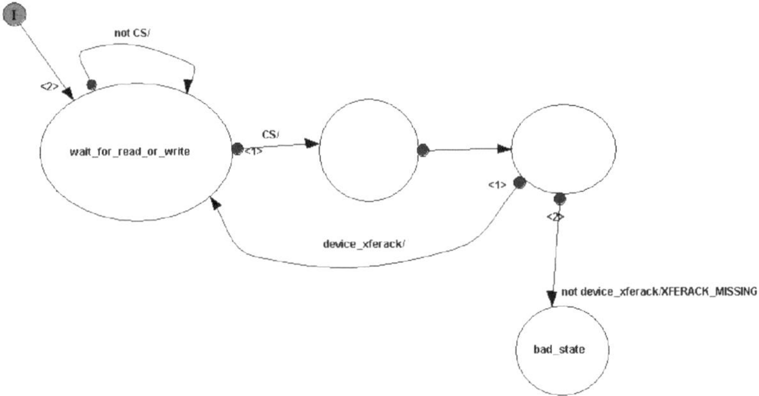

Figure 7. An assertion expressed as a synchronous observer

One way to try and check this property is to try and use it in simulations to see if an error case can be found. Esterel Studio supports this by either simulation directly within the Esterel framework or by the automatic generation of VHDL implementation files and test benches which can check properties specified as synchronous observers.

However, the Esterel Studio also incorporates a built-in model checker (Prover-SL from Prover Technology) which can be used to try to *prove* such properties. We use the latest version 7 of Esterel, which allows reasoning about data *and* control - an improvement on previous versions of the language. We configured the model check to see if the error signal corresponding to a bad state being entered is ever emitted i.e. might the circuit take longer than two clock ticks to acknowledge a transfer? It took Esterel Studio under two seconds on a Sun Sparc Ultra-60 workstation to prove this signal is never emitted.

```
15   esverify -v OPB.eid -checkis0 XFERACK_MISSING
16   --- esverify: Reading model from file "OPB.eid".
17   --- esverify: Checking if output "XFERACK_MISSING" is 0
18   --- esverify: Start model-checking properties
19   --- esverify: Verification complete for signal XFERACK_MISSING: --- esverify:
     --
20   --- esverify: Model-Checking results summary
21   --- esverify: --- esverify: Status of output "XFERACK_MISSING": Never
     emitted.
```

We then produced a deliberately broken version of the peripheral which did not acknowledge read requests. Within two seconds the software was able to prove that there is a case when the acknowledge signal is not asserted after a transaction and provided a counter-model and VCD file.

A conventional approach to catching such approach bugs involves either simulation (which has poor coverage) or the use of bus monitors which snoop the bus at execution time looking for protocol violations. A failure to acknowledge a transaction is one of the types of bugs that such systems can be configured to catch. However, it is far more desirable to catch such problems with a static analysis. We are currently trying to convert a list of around 20 such bug checks used in a commercial OPB bus monitor into a collection of Esterel synchronous observers to allow us to check peripheral protocol conformance with static analyses.

3. Conclusions

The approach of using Esterel to produce hardware and software seems to show some promise. Initial experiments show that serviceable hardware and software can be produced and implemented on real hardware and embedded processors. The possibility to enter system specifications graphically makes this method much more accessible to regular engineers than competing formalisms which uses languages which are quite different to what engineers are used to. For any realistic system the developer still has to write some portions textually and become aware of the basic underlying principles of Esterel. It remains to be seen if the cost of learning this formalism is repaid by increased productivity, better static analysis and the ability to trade off hardware and software implementations.

However, there are many refinements that need to be made to the Esterel language to properly support hardware description. Most of these requirements are easily met without upsetting the core design of the language. Examples include a much more flexible way of converting between integers and bit-vectors and to allow arbitrary precision bit-vectors. Currently performing an integer-based address decode for a 64-bit bus is possible in Esterel but one has to process the bus in chunks not larger than 31 bits.

Another appealing aspect of this flow is the ability to write assertions in the same language as the system specification. This means that engineers do not need to learn yet another language and logic. Furthermore, the formal nature of Esterel's semantics may help to make static analysis easier. Our initial experiments with using the integrated model checker are certainly encouraging. However, we need to design and verify more complex systems before we can come to a definitive conclusion about this promising technology for the design and verification of hardware and software from a single specification.

A very useful application of this technology would be to task-based dynamic reconfiguration. This method would avoid the need to duplicate implementation effort and it would also allow important properties of dynamic reconfiguration to be statically analysed to ensure that reconfiguration does not break working circuits.

"Virtex-II" is a trademark of Xilinx Inc. "CoreConnect" is a trademark of IBM.

References

[1] IBM, "The CoreConnect™ Bus Architecture", http:// www.chips.ibm.com/product/coreconnect/ docscrcon_wp.pdf, 1999.

[2] Multiclock Esterel. Gérard Berry and Ellen Sentovich. *Correct Hardware Design and Verification Methods*. CHARME 2001.

[3] Markus Weinhardt and Wayne Luk. *Task-Parallel Programming of Reconfigurable Systems*. Field-Programmable Logic and Applications. Belfast, UK. Springer-Verlag. 2000.

Communicating Process Architectures 2007
Alistair A. McEwan, Steve Schneider, Wilson Ifill, and Peter Welch
IOS Press, 2007

Modeling and Analysis of the AMBA Bus Using CSP and B

Alistair A. McEWAN [1] and Steve SCHNEIDER,

University of Surrey, U.K.

Abstract. In this paper, we present a formal model and analysis of the AMBA Advanced High-performance Bus (AHB) on-chip bus. The model is given in CSP∥B—an integration of the process algebra CSP and the state-based formalism B. We describe the theory behind the integration of CSP and B. We demonstrate how the model is developed from the informal ARM specification of the bus. Analysis is performed using the model-checker ProB. The contribution of this paper may be summarised as follows: presentation of work in progress towards a formal model of the AMBA AHB protocol such that it may be used for inclusion in, and analysis of, co-design systems incorporating the bus, an evaluation of the integration of CSP and B in the production of such a model, and a demonstration and evaluation of the ProB tool in performing this analysis. The work in this paper was carried out under the Future Technologies for Systems Design Project at the University of Surrey, sponsored by AWE.

Keywords. CSP∥B, AMBA, formal modeling, ProB, co-design

Introduction

In this paper we present a model of the AMBA Advanced High-performace Bus (AHB) in the formalism CSP∥B, and investigate analysis of the model using the model-checker and animator ProB. The AMBA bus, produced by ARM, is a freely available standard for on-chip busses in embedded systems. Implementations are available, and tools are available for the testing of components. Our aim is to show that CSP∥B can be used to model the bus, and that models such as this can be used in the design, development, and formal analysis of hardware/software co-design systems. It is our belief that the combination of the state based formalism B-Method, and the process algebra CSP permits accurate descriptions of the implementation of such systems that can be refined both to hardware and software; and the necessary potential for more abstract models for development and analysis purposes. This work has been carried out within the AWE funded project 'Future Technologies for System Design' at the University of Surrey, which is concerned with formal approaches to co-design.

The paper begins in section 1 by presenting some background information on CSP∥B, ProB, and the AMBA bus. including notes on the main AMBA protocols. This is followed in section 2 by a description of the protocol about which this paper is concerned. Section 3 presents the CSP∥B/ProB model. A discussion about the types of analysis that can be done on this model is presented in section 4, and some conclusions are drawn in section 5.

The contribution of this paper can be summarised as follows: a demonstration of modeling components used in a typical co-design environment using CSP∥B, an evaluation of ProB in the development and analysis of CSP∥B modeling, and the presentation of an AMBA AHB model that can be used for the formal analysis and development of components to be attached to an implementation of the bus.

---

[1]Corresponding Author: *Alistair A. McEwan, Department of Computing, University of Surrey, Guildford, U.K. GU2 7XH*. E-mail: `a.mcewan@surrey.ac.uk`

1. Background

1.1. Combining CSP and B

CSP$\|$B [1,2] is a combination of the process algebra CSP [3,4,5] and the language of abstract machines supported by the B-Method [6,7]. A *controlled component* consists of a B machine in parallel with a CSP process which is considered as the controller. Their interaction consists of synchronisations of B operations with corresponding events in the CSP controller. Consistency of the combination requires that operations are called only within their preconditions. Other properties of the combination may also be considered, such as deadlock-freedom, or various safety or liveness properties. Previous work has developed theory to verify controllers[8], and to combine them into larger systems[9]. The approach taken in this paper differs in that it applies a model-checker to the CSP$\|$B in order to achieve verification.

1.2. B Machines

The B-Method develops systems in terms of *machines*, which are components containing state and supporting operations on that state. They are described in a language called *Abstract Machine Notation*. The most important aspect of B to understand for this paper is that B operations are associated with preconditions, and if called outside their preconditions then they diverge. A full description of the B-Method can be found in [6,7], and tool support is provided by [10,11].

A machine is defined using a number of clauses which each describe a different aspect of the machine. The MACHINE clause declares the abstract machine and gives its name. The VARIABLES clause declares the state variables used to carry the state information within the machine. The INVARIANT clause gives the type of the state variables, and more generally it also contains any other constraints on the allowable machine states. The INITIALISATION clause determines the initial state of the machine. The OPERATIONS clause contains the operations that the machine provides: these include query and update operations on the state.

Example 1 *The format of a B operation*

$$oo \longleftarrow op(ii) = \text{PRE } P \text{ THEN } S \text{ END}$$

$\square$

The format of a B operation is given in example 1. The declaration $oo \longleftarrow op(ii)$ introduces the operation: it has name op, a (possibly empty) output list of variables oo, and a (possibly empty) input list of variables ii. The precondition of the operation is predicate P. This must give the type of any input variables, and can also give conditions on when the operation can be called. If it is called outside its precondition then divergence results. Finally, the body of the operation is S. This is a *generalised substitution*, which can consist of one or more assignment statements (in parallel) to update the state or assign to the output variables. Conditional statements and nondeterministic choice statements are also permitted in the body of the operation. Other clauses are also allowed, for instance regarding machine parameters, sets and constants. For an example B machine, see section 3 where the B machine that is the subject of this paper is introduced.

1.3. CSP

CSP processes are defined in terms of the *events* that they can and cannot do. Processes interact by synchronising on events, and the occurrence of events is atomic. The set of all events is denoted by Σ. Events may be compound in structure, consisting of a *channel name* and some (possibly none) *data values*. Thus, events have the form $c.v_1...v_n$, where c is the channel name associated with the event, and the v_i are data values. The *type* of the channel c is the set of values that can be associated with c to produce events. For instance, if *trans* is a channel name, and $\mathbb{N} \times \mathbb{Z}$ is its type, then events associated with *trans* will be of the form $trans.n.z$, where $n \in \mathbb{N}$ and $z \in \mathbb{Z}$. Therefore $trans.3.8$ would be one such event.

CSP has a number of semantic models associated with it. The most commonly accepted are the *Traces* model, and the *Failures/Divergences* model. Full details can be found in [4,5]. A *trace* is a finite sequence of events. A sequence tr is a trace of a process P if there is some execution of P in which exactly that sequence of events is performed. The set $traces(P)$ is the set of all possible traces of process P. The traces model for CSP associates a set of traces with every CSP process. If $traces(P) = traces(Q)$ then P and Q are equivalent in the *traces model*, and we write $P =_T Q$. A *divergence* is a finite sequence of events tr. Such a sequence is a *divergence* of a process P if it is possible for P to perform an infinite sequence of internal events (such as a livelock loop) on some prefix of tr. The set of divergences of a process P is written $div(P)$. A *failure* is a pair (tr, X) consisting of a trace tr and a set of events X. It is a failure of a process P if either tr is a divergence of P (in which case X can be any set), or (tr, X) is a stable failure of P: a trace tr leading to a stable state in which no events of X are possible. The set of all possible failures of a process P is written $failures(P)$. If $div(P) = div(Q)$ and $failures(P) = failures(Q)$ then P and Q are equivalent in the *failures-divergences model*, written $P =_{FD} Q$.

Verification of CSP processes typically takes the form of refinement checking: where the behaviour of one process is entirely contained within the behaviour of another within a given semantic model. Tool support for this is offered by the model-checker FDR[12].

1.4. CSP Semantics for B Machines

Morgan's CSP-style semantics [13] for event systems enables the definition of such semantics for B machines. A machine M has a set of traces $traces(M)$, a set of failures $failures(M)$, and a set of divergences $div(M)$. A sequence of operations $\langle e_1, e_2 \ldots e_n \rangle$ is a *trace* of M if it can possibly occur. This is true precisely when it is not guaranteed to be blocked, in other words it is not guaranteed to achieve *false*. In the wp notation of [13] this is $\neg wp(e_1;\ e_2;\ \ldots;\ e_n, false)$, or in Abstract Machine Notation $\neg([e_1;\ e_2;\ \ldots;\ e_n]false)$. (The empty trace is treated as *skip*). A sequence does not diverge if it is guaranteed to terminate (i.e. establish *true*). Thus, a sequence is a divergence if it is *not* guaranteed to establish *true*, i.e. $\neg([e_1;\ e_2;\ \ldots;\ e_n]true)$. Finally, given a set of events X, each event $e \in X$ is associated with a guard g_e. A sequence with a set of events is a *failure* of M if the sequence is not guaranteed to establish the disjunction of the guards. Thus, $(e_1;\ e_2;\ \ldots;\ e_n, X)$ is a failure of M if $\neg[e_1;\ e_2;\ \ldots;\ e_n](\bigvee_{e \in X} g_e)$. More details of the semantics of B machines appear in [1]. The CSP semantics for B machines enables the parallel combination of a B machine and a CSP process to be formally defined in terms of the CSP semantics.

The term *CSP controller* P means a process which has a given set of control channels (events) C. The controlled B machine will have exactly $\{|\ C\ |\}$ [2] as its alphabet: it can communicate only on channels in C where a channel name corresponds to an operation in the machine. To interact with the B machine, a CSP controller makes use of control channels which have both input and output, and provide the means for controllers to synchronise with B machines. For each operation $w \longleftarrow e(v)$ of a controlled machine with v of type T_1 and w of type T_2 there will be a channel e of type $T_1 \times T_2$, so communications on e are of the form $e.v.w$. The operation call $e!v?x \to P$ is an interaction with an underlying B machine:

---

[2]The notation $\{|\,|\}$ is used to fully qualify channel sets in CSP. For instance, assuming channel $X\ :\ Bool$, $\{|\ X\ |\}$ is the set $\{X.true, X.false\}$.

the value v is passed from the process as input to the B operation, and the value x is accepted as output from the B operation.

In previous work, controllers were generated from a sequential subset of CSP syntax[2], including prefixing, input, output, choice, and recursion. The motivation for this restriction was verification. Various consistency results were possible for combinations of B machines with such controllers by identifying control loop invariants which held at recursive calls. In this paper there is no need for such restrictions on the syntax of CSP controllers as we do not applying those techniques. Instead we use the ProB model-checker to establish results. This means that the full range of CSP syntax supported by ProB is available for expressing the CSP controllers. This includes parallel and interleaving operators, as well as prefixing, sequential composition, recursion, and the various forms of choice.

1.5. ProB Tool Support

ProB [14] is an animator and model-checker for the B-Method. A B machine can be model-checked against its invariants, with counter-examples given when an invariant is violated. The latest version of ProB also includes support for a model incorporating a B machine and a CSP controller. The B machine captures state, and the CSP characterises interactions with the environment, normally restricting the states in which a related B operation may be invoked. The result is a combination of the two formalisms that is very similar in approach to CSP‖B. Although there are some differences to the way CSP‖B combines CSP and B, it is still a useful tool for developing, investigating, and animating CSP‖B models. In this paper we regard the combination of CSP and B as supported by ProB as the same as CSP‖B, although we remark where differences are significant. [3]

The version of CSP that is implemented in ProB bears a resemblance to, and draws some inspiration from, the CSP_M of FDR. Despite this, there are several differences to CSP_M. For instance, there are no replicated operators, channel type declarations are not supported, and there is no support for the functional language included in FDR. A reader familiar with CSP_M will easily comprehend the CSP supported by ProB, although will notice some of these differences. In this paper, we remark on the differences between the ProB CSP and CSP_M where differences are significant.

1.6. The AMBA bus

The *Advanced Microcontroller Bus Architecture)* (AMBA) is an on-chip communication standard for embedded micro controllers[15]. The standard is presented in an informal manner; and is intended to assist engineers connecting components to, or designing components for, the bus; and to support the modular development of complex *systems on a chip*. Freely available implementations of the bus are available. The three protocols described in [15] are:

- *Advanced High Performance Bus* (AHB) is a system backbone bus, intended for the connection of devices such as processors and on chip memory caches.
- *Advanced System Bus* (ASB) is similar to AHB, but is not specifically targeted at high performance systems.
- *Advanced Peripheral Bus* (APB) is designed for low power peripherals; and has a correspondingly simpler functionality.

A fourth protocol, *AXB*, is also used in high performance systems but is not considered in this paper. Table 1 presents comparisons of the three protocols described above.

---

[3]The differences are in the theoretical basis of the combination, and a discussion is not within the scope of this paper. The interested reader is referred to [14] and [8,9].

Table 1. High level description of properties of an AMBA bus

	AHB	ASB	APB
High performance	✓[15],1.1,1.3	✓[15],1.3	
High clock rate	✓[15],1.1		
System backbone	✓[15],1.1	✓[15],1.1	
On-chip memories	✓[15],1.1	✓[15],1.1	
Off-chip memories	✓[15],1.1	✓[15],1.1	
External memory interfaces	✓[15],1.8	✓[15],1.8	
Low power optimised			✓[15],1.1,1.3
Used in conjunction with AHB			✓[15],1.1
Used in conjunction with ASB			✓[15],1.1
Pipelined operation	✓[15],1.3	✓[15],1.3	
Multiple bus masters	✓[15],1.3	✓[15],1.3	
Burst transfers	✓[15],1.3		
Split transactions	✓[15],1.3		
Latched address and control			✓[15],1.3
Simple interface			✓[15],1.3
Suitable for many peripherals			✓[15],1.3

2. Components in the AMBA AHB Protocol

In this paper we model AHB. This is because, unlike APB, it is intended for on-chip components as a system backbone, and is therefore more fundamental to co-design systems; and it is a newer, more advanced protocol than ASB.

An AHB bus is essentially a central multiplexor and controller. Components connected to the bus request transfers and the bus arbitrates to whom, when, and under what conditions the bus is granted. It is also responsible for multiplexing data, address, and control signals to the correct destinations. A typical AHB system contains the following components:

- *AHB master*: A master initiates read and write operations by providing address and control information. Only one master may actively use the bus at a time.
- *AHB slave*: A slave responds to a read or write operation within a given address-space. The slave signals back to the master the success, failure, or waiting of the transfer.
- *AHB arbiter*: The arbiter ensures only one master at a time initiates data transfers. Even though the arbitration protocol is fixed, any arbitration algorithm, such as *highest priority* or *fair access* can be implemented depending on application requirements.
- *AHB decoder*: The decoder is used to decode the address of each transfer and provide a select signal for the slave that is involved in the transfer. It may be thought of as multiplexing shared lines of communication.

An AHB system consists of a collection of masters, slaves, a single arbiter, and a decoder managing accesses to the communication interconnect lines. A component which has a master interface may also have a slave interface.

A transaction starts with a master requesting the bus. When appropriate, the arbiter grants a master control of the bus. The master then drives control and address information and handshakes this with the destination slave, before driving the actual transaction data—which may be from the master to the slave (a write transaction) or from a slave to a master (a read transaction). The transaction completes either when the slave has transfered all of the data that the master required, or when the arbiter has called it to a halt for some overriding reason.

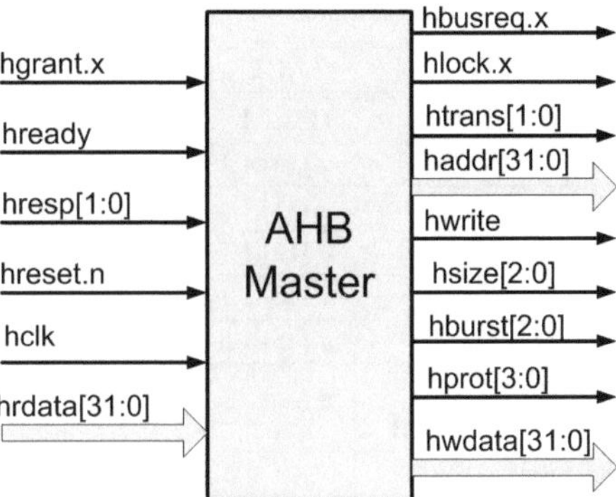

Figure 1. The AHB master interface diagram

In the following sections, we construct the interfaces of each component, allowing for a prototype construction that is readily checkable against the ARM specification.

2.1. The Interface of an AHB Master

In figure 1 an AHB master is shown in terms of inputs and outputs. An input is an arrow leading into the master, and an output is an arrow leading out of a master. The width in terms of bit indices is given. Where this is a single bit—therefore either high or low—no width is given.

A master requests the bus by setting its *hbusreq.x* signal high (where x is a unique identifier); and may indicate that it does not wish its allocation to be interleaved with other transactions by also setting its *hlock.x* signal. The transfer type is denoted by a range of signals on *htrans*, and the direction of the transfer by setting *hwrite* either high or low. The size is given on *hsize*, the number of beats on *hburst*, and *hprot* is used if there is further user level protection required. A master is told it is the highest priority waiting when *hgrant.x* is high, and the bus is ready for use when *hready* is high. Responses from the active slave are on *hresp*, and data can be read from a slave on *hrdata*. Each master has a clock pulse and reset line.

This is described in terms of sets of CSP channels in definition 1 for a given master x. The set of channels leading to all masters would be achieved by disregarding the identifier x for an individual master. This distinction between channels global to the masters, and channels individual to each master is important as it dictates synchronization sets and interleaving when processes are composed in the CSP model. [4]

Definition 1 *AHB Master x actuates and senses*

```
OUTPUTS(x) = {| hbusreq.x, hlock.x, htrans, haddr, hwrite, hsize,
               hburst, hprot, hwdata |}
INPUTS(x)  = {| hgrant.x, hready, hresp, hreset.x, hclk, hrdata  |}
```

□

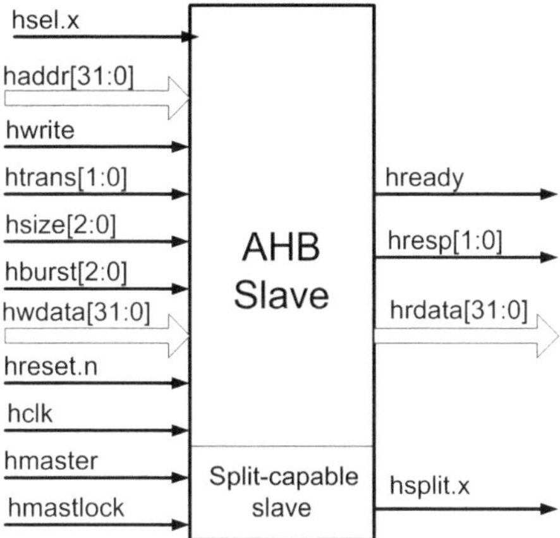

Figure 2. The AHB slave interface diagram

2.2. The Interface of an AHB Slave

When a slave has finished a current transaction, it sets *hready* high. Other responses, such as error conditions, can be relayed back to the master on *hresp*. If the transaction is a read transaction, data is placed on the *hrdata* line The *hsel.x* is unique to a given slave *x*, and when high indicates the current transfer is intended for that slave. The signals *hwrite*, *htrans*, *hsize* and *hburst* are slave ends of master outputs mentioned previously. Each slave has a reset and clock line. This is described in terms of sets of CSP channels in definition 2. The signals *hmaster*, *hmastlock*, and *hsplit.x* are concerned with split transactions and are not considered in our model, although we include them in the definition for completeness.

Definition 2 *AHB Slave x actuates and senses*

```
OUTPUTS(x) = {| hready, hresp, hrdata, hsplit.x |}
INPUTS(x)  = {| hset.x, haddr, hwrite, htrans, hsize, hburst,
               hwdata, hreset.x, hclk, hmaster, hmastlock |}
```

□

2.3. The Interface of an AHB Arbiter

The arbiter ensures that only one master believes it has access to the bus at any one given time (and this may be a default master if necessary). It achieves this by monitoring request lines from masters wishing access, and selecting a master to grant the bus to from those requests. The description in [15] does not prescribe a resolution strategy; in this model we abstract using non-determinism. Figure 3 shows an AHB arbiter in terms of inputs and outputs. This is described in terms of sets of CSP channels in definition 3.

Definition 3 *AHB arbiter actuates and senses*

```
OUTPUTS = {| hgrant, hmaster, hmastlock |}
INPUTS  = {| hbusreq, hlock, haddr, hsplit, htrans,
            hburst, hresp, hready, hreset.x, hclk |}
```

□

---

[4]In considering channels, sets, and types, the first difference between ProB CSP and CSP_M appears. CSP_M requires channels to be typed. For instance, the single-bit channel *hwrite* could be declared *chan hwrite* : 0 | 1; however ProB does not support typing, and instead infers types from values being passed.

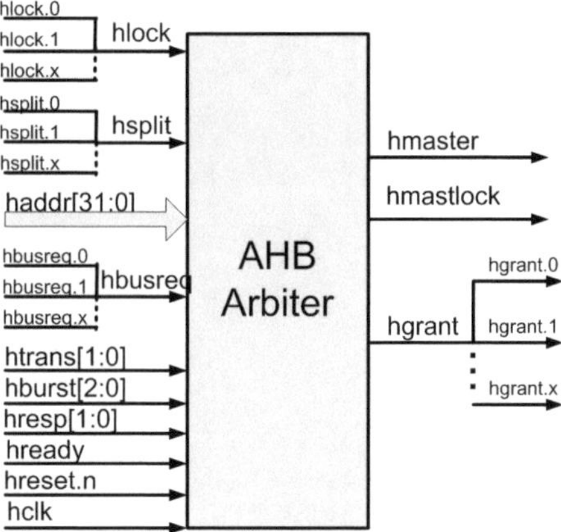

Figure 3. The AHB arbiter interface diagram

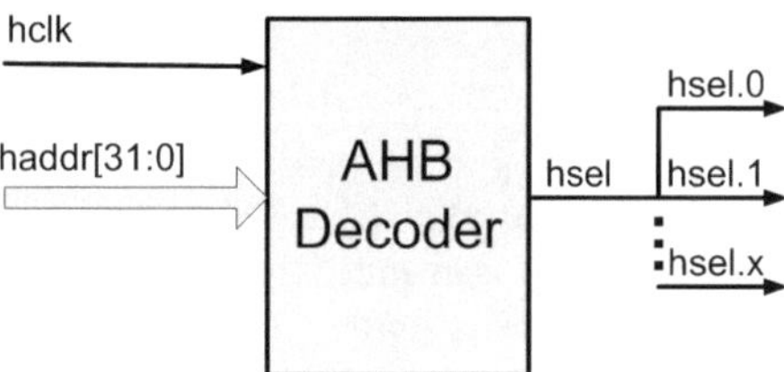

Figure 4. The AHB decoder interface diagram

2.4. The Interface of an AHB Decoder

The decoder acts as a multiplexor, and decodes data on the address bus identifying the slaves that transactions are intended for; and setting the relevant slave select line high. Figure 4 shows a decoder in terms of inputs and outputs, with the CSP channels in definition 4.

Definition 4 *AHB Decoder actuates and senses*

```
OUTPUTS = {| hsel |}
INPUTS   = {| haddr, hclk |}
```

□

2.5. An Example AHB Network

Figure 5 shows an example AHB network, comprising a master, slave, arbiter, and decoder. The master and slave are identified by their individual x tags—a more complex system would have more tagged lines unique to given masters and slaves. The diagram shows the various signals communicating between components. Where a line connects exactly two components (in this case because only one master and slave have been included) a simple arrow is used; where a signal is common to more than two components the lines fan out with a solid dot. Dashed lines are used in the diagram where lines cross solely to avoid confusion. For further clarity in the diagram, the signals *hclk* and *hreset*, which are common to all components are listed in the box for each component. Arrows connecting components in this diagram are implemented as synchronizations in the CSP. Care must be taken with arrows parameterized

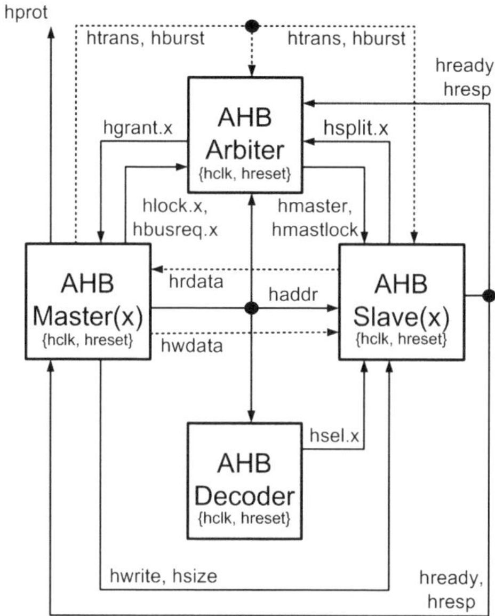

Figure 5. An example AHB system with one master and one slave

with master and slave numbers though, as these are implemented as interleavings unique to each master as per the previous comments. The model of the bus can be seen to emerge from this diagram as a CSP process with an alphabet corresponding to the interface of the arbiter and decoder, controlling a B machine which captures the internal state.

3. A Model of the AHB Components

In this section, we develop the model of the bus. The B machine is given in section 3.1, and the CSP controller in section 3.2. For each, the syntax used is as accepted by ProB. For the B, this is valid input to the B-Toolkit.

3.1. A B Machine Describing Internal State

Local state is modeled in terms of clocked, synchronous registers. That is, each register (or variable) has a value on a given clock cycle which does not alter on that clock cycle. If written to on a clock cycle, it takes on the new value *only when the clock ticks*. If it is not written to, the value does not change on the next cycle. Every register updates simultaneously. The invariant given in definition 5 contains the type declarations for each local piece of state; and a further conjunct that is used (and described later in section 4.2) for verification purposes. [5]

Definition 5 *Local variables (registers) and types*

```
SETS BurstType = { SINGLE, INCR, WRAP4,INCR4, WRAP8,INCR8, WRAP16,INCR16 }
VARIABLES XX, YY, ZZ, YYlatched, ZZlatched,
          Burst, Burstlatched, BurstCount, BurstCountlatched

INVARIANT
    XX <: 0..15          & YY <: 0..15          & ZZ <: 0..15 &
```

---

[5] For a full reference of the syntax of a B machine and notation, the reader is referred to [7].

```
YYlatched <: 0..15  & ZZlatched <: 0..15  &

Burst        : BurstType &
Burstlatched : BurstType &
BurstCount : 0..17      &
BurstCountlatched : 0..17 &
((BurstCountlatched > 0) => (Burst = Burstlatched))
```

$\square$

A master lodges a request by setting its request line high, and the arbiter chooses from all masters requesting the bus on a given cycle. If a master does not have its request line high it is assumed not to want the bus—i.e. a request is current *only when the request line is high.* Requests may be for either a locked, or an unlocked transaction. YY records all the masters that have set their request line on the current cycle. In this clocked synchronous model, this request is not stored in the arbiter until the clock ticks. *YYlatched* contains all of the masters that lodged a request on the previous cycle: it is the value of YY on the previous cycle— the successful master will be drawn from the set *YYlatched*. The same clock synchronous behaviour is true of ZZ and *ZZlatched*, used to record requests for locked transactions. XX records which masters have not lodged a request on the current cycle (ensuring that each master may only lodge one request per cycle); while *Burst* and *BurstCount* relate to control for the current transaction.

Initially, no masters have lodged a request on the current cycle, and no masters could have lodged a request on the previous cycle. Curiously though, *YYlatched* is non-empty: this corresponds to a default master (0) always being assumed to have requested the bus.[6]

Definition 6 *Initialisation*

```
INITIALISATION XX                  := 0..15   ||
               YY                  := {}       ||
               ZZ                  := {}       ||
               YYlatched           := {0}      ||
               ZZlatched           := {}       ||
               Burst               := INCR     ||
               Burstlatched        := INCR     ||
               BurstCount          := 0        ||
               BurstCountlatched   := 0
```

$\square$

When a master requests the bus, it is recorded by removing its index from XX and placing it in YY. No assumption is made about whether or not this was a locked request or not. The variables recording requests on the previous cycle remain unchanged.

Definition 7 *Recording a master's request for the bus*

```
Request(xx)  =
PRE xx : 0..15  THEN
   XX          := XX - {xx}   ||
   YY          := YY \/ {xx}
END;
```

$\square$

A master may request that a transaction is locked, and this is recorded by placing its index in ZZ. No assumption is made about whether or not this master has actually requested a transaction, and requests recorded from the previous clock cycle remain unchanged.

---

[6]In B, || denotes that the assignments all happen concurrently.

Definition 8 *Recording a master's request to lock the transaction*

```
LockedRequest(xx) =
PRE xx : 0..15 THEN
   ZZ         := ZZ \/ {xx}
END;
```

□

YYlatched records all masters on the previous clock cycle who requested an *unlocked* transaction, and *ZZlatched* records all of those requesting a *locked* transaction. When the arbiter chooses which master is to be granted the bus on the next cycle, it non-deterministically selects an element from the union of these two sets.

Definition 9 *Choosing a master to which to grant the bus*

```
xx <-- Choose =
BEGIN
   xx :: YYlatched \/ ZZlatched
END;
```

□

One artefact of ProB is an inability to directly access state in the B machine from CSP—an operation must be invoked with a return value. *TestLock* returns a boolean value indicating if master identifier passed to it is currently granted rights to a locked transaction. *GetBurstType* performs a similar function to test the burst type of a transaction, and *GetBurstCount* indicates whether or not we are on the last element of a given burst. [7]

Definition 10 *Testing for locked transactions, bust types, and burst sizes*

```
xx <-- TestLock(yy) =
PRE yy : 0..15 THEN
   IF yy : ZZlatched THEN xx := TRUE ELSE xx := FALSE END
END;

xx <-- GetBurstType =
BEGIN
   IF Burst=SINGLE THEN xx:= TRUE ELSE xx:= FALSE END
END;

xx <-- GetBurstCount =
BEGIN
   IF BurstCount > 0 THEN xx := TRUE ELSE xx := FALSE END
END;
```

□

When the type of burst is specified, a fixed length is assumed, and recorded by the operation *SetBurst*. These lengths are one larger than might be expected due to the synchronous nature of the clocked assignments: they only really take on a meaningful value on the next clock cycle, by which time they will have been decremented by 1. A variable length transaction— given by the type *SINGLE*—is assumed to be a fixed length of one burst, and the controlling master is responsible for retaining the bus by continually re-asserting the request.

---

[7] These values are boolean rather than $\mathbb{N}$ as ProB has not fully implemented CSP guards using tests on natural numbers—so this functionality must be moved into the B machine.

Definition 11 *Setting the burst type*

```
SetBurst(xx) =
PRE xx : BurstType THEN
  Burst := xx ||
  IF xx = INCR16 or xx = WRAP16 THEN BurstCount := 17 ELSE
  IF xx = INCR8  or xx = WRAP8  THEN BurstCount := 9   ELSE
  IF xx = INCR4  or xx = WRAP4  THEN BurstCount := 5   ELSE
  IF xx = SINGLE or xx = INCR   THEN BurstCount := 2   END END END END
END;
```

☐

The operation *tock* is carried out exactly when the clock ticks, and implements the clocked synchronous behaviour. When the clock ticks, a new cycle begins. No masters may have requested the bus yet on this new cycle, so XX is maximal, and YY and ZZ are emptied. *YYlatched* takes on the value that YY held, ignoring all those who had also set the lock line high. It therefore holds all of those masters who requested an unlocked transaction on the clock cycle just ending. *ZZlatched* takes on all those masters who set the lock line high *and* requested the bus: the effect being that if a master erroneously set the lock line high but did not request the bus, it will be ignored. In case there were no requests lodged, it is assumed that the default master (0) must have lodged a request for an unlocked transaction. Finally, the type of the bus on the current cycle is stored, along with a note about any new burst type that may have been input. This information is used for verification purposes in section 4.2.

Definition 12 *Synchronous clocked updates*

```
tock =
BEGIN
  XX             := 0..15                                            ||
  YY             := {}                                               ||
  ZZ             := {}                                               ||
  IF YY={} THEN YYlatched:= {0} ELSE YYlatched := YY - ZZ END        ||
  ZZlatched := YY /\ ZZ                                              ||
  IF BurstCount > 0 THEN  BurstCount := BurstCount - 1 END           ||
  BurstCountlatched := BurstCount                                    ||
  Burstlatched := Burst
END
```

☐

3.2. The CSP Controller

In CSP‖B and ProB, CSP controls when, and under what conditions, B operations can be invoked. An invocation of a B operation corresponds to the CSP controller engaging in an event of the same name, and parameters to the operation, and results of the operation, are passed in the types of the event.

The process $COLLECT\_REQUESTS$ listens on the request lines.[8] When one goes high (indicated by the process engaging in a *hbusreq* event) it calls an operation in the B machine that records this. A lock line (*hlock*) may also go high, and when it does, another B operation is called. The *hready* signal may also go high, indicating that the *current* transaction is ending. The *hgrant* signal is used by the arbiter to indicate the highest priority request on the *previous* clock cycle, and this is achieved by calling the B operation *Choose* that returns the value to the CSP. Finally, the clock may tick—indicated by the event *tock*.[9]

---

[8]Where behaviour is the same for multiple masters, we only include the CSP for master 0 to conserve space. This is because the CSP in ProB does not implement replicated operators, as a user of FDR may expect.

[9]We use the event *tock* to denote a clock tick as "tick" '($\checkmark$) is commonly used in CSP to denote termination.

Definition 13 *Collecting requests*

```
COLLECT_REQUESTS =
  hbusreq.0 -> Request!0 -> COLLECT_REQUESTS
  [] hlock.0 -> LockedRequest!0 -> COLLECT_REQUESTS
  [] hready -> COLLECT_REQUESTS
  [] Choose.HighPri -> hgrant.HighPri -> COLLECT_REQUESTS
  [] tock -> COLLECT_REQUESTS
;;
```

□

Definition 13 does not constrain how many times on each clock cycle an event may occur, but the B machine assumes a master may only record one request per cycle.[10] This constraint is captured in the CSP by placing definition 13 in parallel with processes describing this constraint. This process insists that when a request is lodged, the clock must tick before it may be lodged again; however the clock may tick an indeterminate number of times without a request being lodged. Other constraints are that *hready* may go high at most once per cycle, and that the arbiter must choose and grant the highest priority master on each cycle.

　　Writing the behavioural constraints in separate parallel processes in this way is a stylistic choice: they could have been added in a more implicit manner. However, in adopting this style the behavioural constraints are up-front: readily identifiable and easily changed should the model require adaptation or further development.

Definition 14 *Constraining requests*

```
REG_HREADY = hready -> tock -> REG_HREADY [] tock -> REG_HREADY;;
REG_CHOOSE = Choose.HighPri -> hgrant.HighPri -> tock -> REG_CHOOSE;;
REG_REQ0 = hbusreq.0 -> tock -> REG_REQ0 [] tock -> REG_REQ0;;
REG_LOCK0 = hlock.0 -> tock -> REG_LOCK0 [] tock -> REG_LOCK0 ;;

REGULATE =
  ( REG_HREADY [|{tock}|] REG_CHOOSE )
  [|{tock}|]
  ( REG_REQS [|{tock}|] REG_LOCKS )
;;
```

□

Definition 15 presents the CSP process controlling locked transactions. A new transaction begins when the previous transaction ends with an *hready* signal. There are two possibilities here, corresponding to the first external choice in this process: that the arbiter may receive the *hready* signal before issuing an *hgrant* signal on a given clock cycle, or vica-versa. Subsequent behaviour is dependent upon whether or not the B machine indicates it is a locked transaction. At this point the clock ticks and the controller evolves into the transaction phase.

Definition 15 *Locked transactions*

```
LOCKED_TRANS =
  hready -> (
    hgrant.0 ->
      TestLock!0?RR ->
        tock -> (if RR then LOCKED_CTRL_INFO(0) else LOCKED_TRANS) )
  []
  hgrant.0 -> (
    hready ->
```

---

[10]The piece of syntax ; ; indicates the end of a process definition.

```
    TestLock!0?RR ->
      tock-> (if RR then LOCKED_CTRL_INFO(0) else LOCKED_TRANS)
    []
    tock -> LOCKED_TRANS )
;;
```

$\square$

In the control phase, the arbiter ensures the master locks the bus using *hmastlock*, and asserts control with *hmaster*. The master then dictates the burst type for the transfer—either fixed or variable length. Behaviour branches after the clock has ticked depending upon transfer type.

Definition 16 *Control phase of a locked transaction*

```
LOCKED_CTRL_INFO(PP) =
  hgrant?ANY ->
    hmastlock ->
      hmaster!PP ->
        hburst?TT ->
          SetBurst.TT ->
            GetBurstType?UU ->
              tock -> (if UU then LOCKED_VAR(PP) else LOCKED_FIXED(PP))
;;
```

$\square$

In a locked transaction, the master is required to continually assert the lock lines while the transaction is in progress. The arbiter is required to assert the master that will be granted the bus on the next cycle *if the current transaction completes*. If the burst count is zero after the clock has ticked, then behaviour returns to monitoring for the next transaction, otherwise the current transaction continues to control the bus for another cycle.

Definition 17 *Control phase of a locked transaction*

```
LOCKED_FIXED_DATA(PP) =
  hmastlock ->
    hmaster!PP ->
      hburst?TT ->
        hgrant?ANY ->
          tock -> LOCKED_FIXED(PP)
;;

LOCKED_FIXED(PP) =
  GetBurstCount?XX ->
    ( if XX then LOCKED_FIXED_DATA(PP) else LOCKED_TRANS )
```

$\square$

The main process is the process responsible for collecting requests, in parallel with the constraints placed upon it, in parallel with the process that marshals locked requests, and therefore implements the arbiter of figure 3 as well as implicitly implementing the multiplexing of control lines performed by the decoder. In this paper, we omit unlocked requests to simplify the model. The locked transaction marshaled synchronises with the request collector on the *hgrant* and *hready* signals—which is sufficient (in conjunction with the state stored in the B machine) for it to spot wheen the entire system is in a state corresponding to a locked transaction. All processes synchronise on the global clock event *tock*—which also causes the clocked synchronous behaviour in the B machine.

Definition 18 *The main controller*

```
MAIN =
  ( COLLECT_REQUESTS
    [|{ tock, Choose, hgrant, hready, hbusreq, hlock }|]
    REGULATE )
  [|{ tock, hgrant, hready }|]
  LOCKED_TRANS
;;
```

□

4. ProB Analysis of the Model

In this section, we discuss some analysis that can be done on this model using ProB, and show how ProB can be used to check properties either of the B machine in isolation, or of the combination with CSP. We also demonstrate the usefulness of ProB in developing the model because of the way it can be used to animate models. We document some experiences of the tool—some of which are mentioned above in the CSP model. We also discuss some contrasts with how FDR may be used in the development of a model—for instance, how FDR was used in the development of a CSP∥B based Content Addressable Memory using *Circus* in [16].

4.1. Animating Models Using ProB

The initial use of ProB is in the construction of the CSP∥B model, and particularly in the combination of the CSP controller and the underlying B machine. Animation in ProB allows the user to step through the behaviour of the CSP controller, at each step being offered a set of possible next steps to perform. The B machine is updated in the light of operation calls, and the updated state is exhibited.

This ability to step through the behaviour of the combined system supports exploration of its description, and enables immediate feedback on whether it exhibits the expected behaviour. Thus ProB is effective in supporting the construction of the formal model at the point it is being developed, and in ensuring consistency between the CSP and the B.

Figure 6 presents a snapshot of ProB animating the model. The uppermost window is an editor for the B machine. The bottom left window shows the state of each variable in the B machine, and a check on whether or not the state meets the machine's invariant. The bottom center window shows the CSP events currently on offer by the CSP controller (which includes available B operations). The bottom right window shows (from bottom to top) the trace of the animation so far.

Firstly, the machine is initialised with the *initialise_machine* call. As this begins the first clock cycle the default master 0 has been chosen by engaging in the B operation $Choose \rightarrow (0)$ and granted with the CSP event $hgrant(0)$. The *hready* signal has occurred, indicating that new transaction may begin from this point in the trace onwards. As this is simply a default transaction, it is found to be unlocked by the B operation $TestLock(0) \rightarrow (FALSE)$. At this point, the arbiter starts receiving requests from masters wishing to use the bus on the next cycle. Master 1 lodges a request by synchronising with the controller on the event $req(1)$ and the Arbiter records this fact with the B operation $Request(1)$. The master confirms this is a locked transaction request with the event and operation $hlock(1)$ and $LockedRequest(1)$ respectively. Master 2 also lodges a request using the event and operation $req(2)$ and $Request(2)$ respectively. At this point, the clock ticks, updating all the synchronous registers.

The value of the B machine variables in the left window reflects the state of the B machine at this point. The current state satisfies the machine invariants. No masters have yet

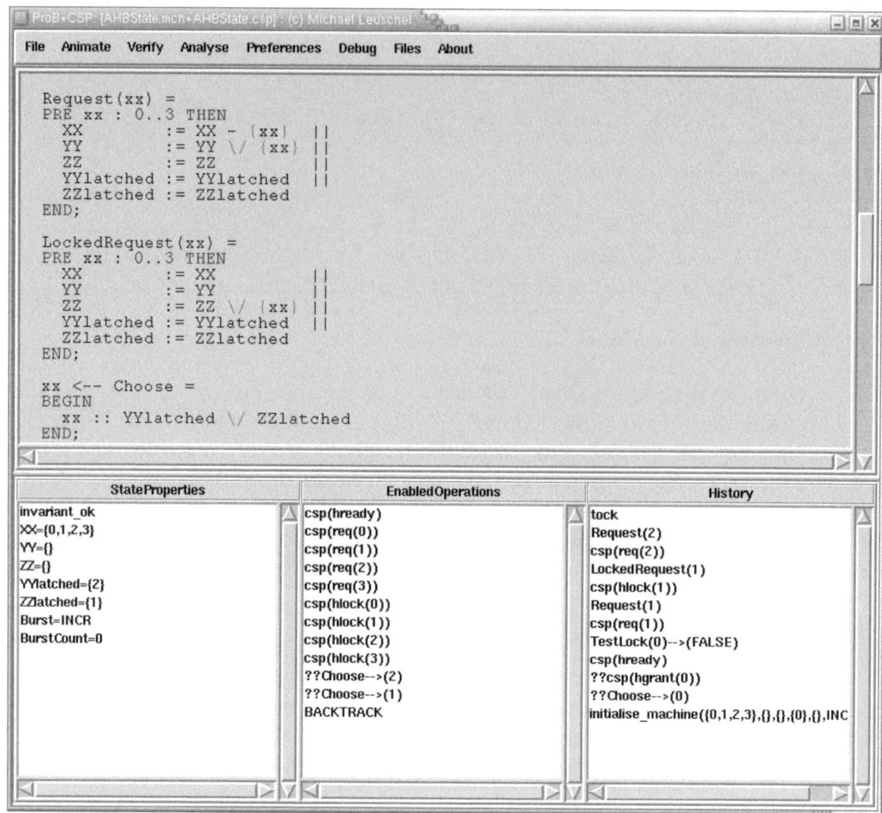

Figure 6. A snapshot of animating the model (with 4 masters) in ProB

requested the bus on this current (i.e, second) cycle, so XX is maximal and YY and ZZ are empty. On the previous (first) cycle, master 2 requested an unlocked transaction and master 1 requested a locked transaction and this is reflected in the values of $YYlatched$ and $ZZlatched$. There is no transaction in progress, so the burst types hold their default value.

Finally we can see the events that can be performed in the current state (including B operations). Each of the masters may request the bus, and request that the transaction is locked by engaging in their respective CSP events *req* and *hlock*. The controller may invoke the operation to choose a master for the next clock cycle, and this operation may return the value 1, or 2. Notification of a current transaction ending may also be received. The $BACKTRACK$ alternative is for stepping backwards through the animation.

4.2. Model-Checking Using ProB

Although ProB supports animation, much of its power derives from its ability to perform model-checking, either on a stand-alone B machine, or else on a CSP and B combination. Various properties can be checked through model-checking. The property that we have focused on in this analysis is invariant checking: that the machine can never reach a state in which its invariant is false. Properties of interest can be expressed as clauses in the invariant, and then investigated through the model-checker.

As an example, we have considered the property that the burst variable should not be reset while a burst is in progress. Recall that a burst value is set when a master obtains a lock on the bus. It will then have control of the bus, and will not release it, until the burst has completed. The value corresponding to the time remaining for the burst is tracked in the variable $BurstCount$ within the B machine: this is set at the same time as $Burst$.

We wish to express this property as a requirement that $Burst$ should not change while an existing burst is underway. To express this, we make use of the variables $Burstlatched$ and $BurstCountlatched$ which track the values of $Burst$ and $BurstCount$ from the previous clock cycle. The property is then captured as the requirement that if the burst had not finished on the previous clock cycle then a new burst should not be set: $Burst$ should be the same as $Burstlatched$. Formally, this is given as the statement $((BurstCountlatched > 0) \Rightarrow (Burst = Burstlatched))$ and incorporated into the invariant of the B machine.

Model-checking the stand-alone B machine with this assertion finds that the invariant is not always true. A trace given by ProB which leads to the violation of the invariant is given in example 2. This trace brings us to a state where $BurstCountlatched = 2$, and yet $Burst = INCR$ and $Burstlatched = SINGLE$ are different, indicating that $Burst$ has just changed. In fact, the same invariant violation can be reached through a shorter sequence of events, given in example 3.

Example 2 *A counter-example produced by ProB*

$$\langle initialise\_machine, LockedRequest(2), LockedRequest(3),$$
$$SetBurst(SINGLE), Request(2), tock, SetBurst(INCR)\rangle$$

□

Example 3 *A shorter counter-example*

$$\langle initialise\_machine, SetBurst(SINGLE), tock, SetBurst(INCR)\rangle$$

□

This violation is not unexpected: the AHB state machine in isolation will not ensure that the desired assertion is met: it is able to accept updates to the burst type at any stage, and this capability is what allows the invariant to be violated.

However, we expect the assertion to be true when the AHBstate machine is controlled by the CSP controller: the aim is that the controller ensures that updates to the burst type cannot occur in the middle of a burst. ProB is also able to model-check the AHBstate when it is under the control of the CSP controller. In this case it turns out that ProB does not find any invariant violations, confirming that the assertion is indeed valid for AHBstate in the appropriate context. This is what we had aimed for in the combined design.

4.3. Experiences of CSP and B in ProB

This case study has exposed a number of experiences with using ProB. In this section we discuss some of these experiences. This discussion is intended to provide the reader with a guide as to practical, and mature, use of ProB in a typical CSP‖B development, and why, and where, it may be of use.

- *Differences with CSP_M*: a number of differences with the CSP_M supported by FDR exist. Some of these are minor, and some more major. For instance, the syntax of the two is subtly different, some constructs in FDR are not supported by ProB, and the functional language in FDR is not supported in ProB. The impact is that a CSP script supported by FDR will not currently be directly supported by ProB and vice-versa. This is unfortunate as there is a wealth of experience and knowledge in using FDR that may not be directly applicable to a ProB script.

- *Structured development*: ProB does not have support for a structured development of a system of B machines, unlike tool support for the B-Method such as the B-Toolkit [10] and Atelier-B [11]. Although the B supported by the two is the same, ProB does not allow, for instance, included B machines in a script—there is only support for one machine per model. Other B machines must be manually included in-line. This is unfortunate as a project in the B-Toolkit requires manual intervention before being loaded into ProB; and this type of intervention should typically be avoided in high assurance systems.
- *Differences with CSP∥B*: a characteristic feature of CSP∥B is that a call to a B operation from a CSP controller can be hidden from external observations, with the result that only observations of the controller are possible. However, ProB handles hiding of controller calls to B operations differently. In hiding a call, the *call itself* becomes non-deterministic—there is no control over the value of parameters. This is in contrast to the CSP∥B approach. This is unfortunate because it is an important semantic difference between ProB and CSP∥B—although in a development/test cycle such as the one in this paper it is of minimal impact.
- *Animation*: the ability to animate models is very useful. This can be done for B machines using the B-Toolkit, or CSP processes using $Probe$[17]; but to be able to animate the combination of the two together means that many errors and inconsistencies can be caught early in the development cycle.
- *Model-checking vs theorem proving*: model-checking invariants in the B is useful. Model-checking is generally considered a more convenient route to verification than theorem proving because of its automatic nature. The B-Toolkit provides a theorem prover; to complement this with a model-checker is extremely valuable for developmental cycles as typically one would like to relieve the proof burden as much as possible.
- *Invariants over CSP processes*: a speculative usage of ProB that we have begun to explore through this case study is the use of invariants over CSP traces (or even failures) rather than just invariants in the B. To a user of FDR, the construction and asserting of a traces refinement in a specification is a useful tool in checking safety requirements[16]. A mechanism for specifying an invariant over traces of a process in ProB would, we expect, be a valuable addition to the tool; although we have not considered the theory about how such an addition could be formulated.

5. Conclusions and Discussion

In this paper we have presented a case study where we modeled an existing on-chip bus protocol using a combination of CSP and B, and performed some analysis of the model using ProB. A driving aim of the paper was to investigate how CSP∥B, and ProB, may be used in a typical co-design development.

An interesting aspect of this case study is that it models an existing implementation, with the aim of providing a platform for formal analysis against components with which it is to be used. Thus in places, the model follows closely the behaviour described in the specification document, rather than some more abstract mathematical model. This has both benefits and drawbacks. Benefits include an easier discussion about the correctness of the model relative to the rather informal specification; while drawbacks include the constraints that this places on the construction of the model.

The AMBA bus is commonly used in co-design systems. Components on the bus may be processors, memory, or bespoke components. In building a model of the bus interacting via a CSP interface with bus components, we have found the combination of CSP and B sufficient to model signals, communications, and registers. The model in this paper is restricted to clocked synchronous hardware; an item of future work is to investigate the combination of CSP and B for asynchronous co-design systems.

We have attempted to remain faithful to the AMBA specification in the construction of our model, but as yet have not cross-checked it with an implementation. In fact, we believe that in doing so, we will discover behaviours that need revision. The model in this paper therefore represents work in progress. An item of future work would be to develop a master (or slave) component using CSP∥B and ProB, verify the correctness with respect to our model, derive an implementation and connect it to an implementation of the AMBA bus. Although subsequent testing of this implementation would not guarantee the correctness of the model, it would provide enough feedback to guide its evolution.

Another aim of this paper was to investigate the usage of ProB in a modeling and development exercise such as this. The conclusions drawn from this are listed in section 4.3; in summary the existence of tool support for proved useful in the development and prototyping phase although there were limitations in what could be achieved and in the compatibility with tools for both CSP and B. A discussion of issues such as these—semantic and syntactic integration of formalisms and impact on associated tool support is held in [16,18].

One of the most interesting results to come out of ProB usage concerns the verification techniques that may be used. ProB produces counter-examples when a machine invariant is violated, as in section 4.2. Using machine invariants to capture safety properties is well understood in (amongst others) the B community; using invariants over traces to capture safety properties proved by refinement checking is well understood in the CSP community. In this paper however, we augmented the B machine with extra information, designed to capture extra interactions with the CSP, such that the machine invariant could capture safe states. An uncontrolled B machine was shown to violate the invariant, whilst the B machine in parallel with the CSP controller was shown to respect the safety invariant. Although this example was simple, the important detail is the technique for lifting information into B. Further understanding and evolution of this technique of capturing traces invariants as properties of the B machine is an important item that we leave for future work.

Acknowledgements

This work was funded by AWE under the 'Future Technologies for System Design' project, and has benefitted from discussions with Wilson Ifill, Neil Evans, and Helen Treharne.

References

[1] H. E. Treharne. *Combining control executives and software specifications*. PhD thesis, Royal Holloway, University of London, 2000.

[2] S.A. Schneider and H.E. Treharne. Communicating B machines. In *ZB2002*, volume LNCS 2272, pages 416–438, 2002.

[3] C. A. R. Hoare. *Communicating Sequential Processes*. Prentice-Hall International Series in Computer Science. Prentice-Hall, 1985.

[4] A. W. Roscoe. *The Theory and Practice of Concurrency*. Prentice-Hall, 1997.

[5] S.A. Schneider. *Concurrent and Real-time Systems: The CSP approach*. Wiley, 1999.

[6] J-R. Abrial. *The B-Book: Assigning Programs to Meanings*. Cambridge University Press, 1996.

[7] S. A. Schneider. *The B-Method: an introduction*. Palgrave, 2001.

[8] Steve Schneider and Helen Treharne. CSP theorems for communicating B machines. *Formal Aspects of Computing*, 17, 2005.

[9] Steve Schneider, Helen Treharne, and Neil Evans. Chunks: Component verification in CSP∥B. In *IFM 2005*, volume LNCS 3771, pages 89–108, 2005.

[10] B-Core. *B-Toolkit*.

[11] Clearsy. *Atelier-B*.

[12] Formal Systems (Europe) Ltd. FDR: User manual and tutorial, version 2.82. Technical report, Formal Systems (Europe) Ltd., 2005.

[13] C. C. Morgan. Of wp and CSP. In W.H.J. Feijen, A. J. M. van Gesteren, D. Gries, and J. Misra, editors, *Beauty is our Business: a birthday salute to Edsger J. Dijkstra*. Springer-Verlag, 1990.

[14] M. Leuschel and M. Butler. ProB: A Model Checker for B. In *FM 2003: The 12th International FME Symposium*, pages 855–874, 2003.

[15] ARM. Introduction to the amba bus. Technical Report 0011A, ARM, 1999.

[16] Alistair A. McEwan. *Concurrent Program Development*. DPhil thesis, The University of Oxford, 2006.

[17] Formal Systems (Europe) Ltd. Probe user manual. Technical report, Formal Systems (Europe) Ltd., 2005.

[18] C. Fischer. How to combine Z with a process algebra. *LNCS*, 1493, 1998.

Communicating Process Architectures 2007
Alistair A. McEwan, Steve Schneider, Wilson Ifill, and Peter Welch
IOS Press, 2007

A Step Towards Refining and Translating B Control Annotations to Handel-C

Wilson IFILL [a,b] and Steve SCHNEIDER [b]

[a] *AWE Aldermaston, Reading, Berks, England;*
[b] *Department of Computing, University of Surrey, Guildford, Surrey, England.*

{ W.Ifill, S.Schneider } @surrey.ac.uk

Abstract. Research augmenting B machines presented at B2007 has demonstrated how fragments of control flow expressed as annotations can be added to associated machine operations, and shown to be consistent. This enables designers' understanding about local relationships between successive operations to be captured at the point the operations are written, and used later when the controller is developed. This paper introduces several new annotations and I/O into the framework to take advantage of hardware's parallelism and to facilitate refinement and translation. To support the new annotations additional CSP control operations are added to the control language that now includes: recursion, prefixing, external choice, if-then-else, and sequencing. We informally sketch out a translation to Handel-C for prototyping.

Keywords. B Method, CSP, Hardware Description Language,

Introduction

Annotating B-Method specifications with control flow directives enables engineers to describe many aspects of design within a single notation. We generate proof obligations (*pobs*) to demonstrate that the set of executions allowable by the annotations of a B [1] [2] machine do not cause operations to diverge. The benefit of this approach is that only the semantics of the machine operations are required in checking the annotations, and these checks are similar in size and difficulty to standard B machine consistency checks. Controllers written in CSP, which describe the flow of control explicitly, can be checked against the annotations. There is no need to check the CSP [3] [4] [5] directly against the full B description. Once the annotations are shown to be correct with respect to the B machine we can evaluate controllers against the annotations without further reference to the machine. Machines can be refined and implemented in the normal way while remaining consistent with the controller. In previous work [6] we presented the NEXT and FROM annotations, which permitted simple annotated B specifications and controllers to be written. Before that [7] we presented a route to VHDL [8], a hardware description language, from B. In this paper we present three more annotations: NEXT_SEQ, NEXT_PAR and NEXT_COND and add input and output to the operations. We also begin to present an informal refinement theory for annotations and a route to implementation via Handel-C. The refinement theory outline in this paper allows the annotations to be independently refined and remain consistent with the Machine.

Previous work obtaining hardware implementations from B approached the problem by using B as a Hardware Description Language (HDL) that translates to VHDL [9] [10]. Our approach achieves the goal of obtaining hardware via Handel-C as an intermediate stepping stone, which means that the B that is translated does not require the same degree of HDL structural conformance as does the B for VHDL translation. Approaches that translate HDLs to B for analysis [11] do not support the development process directly. Event B [12] has been used to support the development of hardware circuits [13] that includes refinement but not the code generation process. Not only are we working towards code generation, but we wish to work with specifications that model both state and control equally strongly. CSP$\|$B [14] [15] has the capability to model state and event behaviour, but the CSP controller must be instantiated with B components to verify the combination. We break the verification of controllers down into manageable stages, and offer an approach to refinement and translation. Integrations of CSP and Z (CSP-Z) by Moto and Sampaio [16] and CSP and Object Z (CSP-OZ) Fischer [17] require a CSP semantics to be given to Z in order for integration to be analysable as a whole. Our approach differs to other formal language integrations in two ways. Firstly, The control flow behaviour is capture during the development of the state operation in the form of annotation. The annotations are control specifications. Only later is a complete controller developed that satisfies the annotations. In this way the developer of the state operations in B can constrain controller behaviour, but full controller development can be postponed and possibly performed by a different engineer. Secondly, there is no notion of executing the models together and analysing this integration for deadlocks. In this approach the different formal notations provide different views of the system, and both views are required to obtain a executable model.

This paper describes extensions to the work presented in B2007 [6]. This papers contribution is the introduction of additional next annotations, incorporation of I/O into the annotations, and an informal treatment of refinement and translation. In Section 1, the general framework is introduced. In Section 2 a B machine is introduced along with the NEXT annotation. The proof obligations associated with the annotations and control language are given in Section 3. The consistency of the annotations are given in Section 4. A refinement and translation outline is given in Section 5. An example illustration of some refinements and translations are given in section 6. A discussion on the benefits and future work is had in Section 7.

We restrict our attention in this paper to correct B machines: those for which all proof obligations have already been discharged. We use I to refer to the invariant of the machine, T to refer to the machine's initialisation, P_i to refer to the precondition of operation Op_i, and B_i to refer to the body of operation Op_i.

Controllers will be written in a simple subset of the CSP process algebraic language [3,5]. The language will be explained as it is introduced. Controllers are considered as *processes* performing *events*, which correspond to operations in the controlled B machine. Thus operation names will appear in the controller descriptions as well as the B machine definitions. The Handel-C translations are shallow and in a few cases performed in accordance with existing translation work [18] [19].

1. The General Framework

The approach proposed in this paper introduces *annotations* on B operations as a mechanism for bridging the gap between B machines and CSP controllers, while maintaining the separation of concerns. The approach consists of the following components:

- **Machine definition**: the controlled component must first be defined.
- **Annotations**: the initialisation and the operations in the machine definition are annotated with fragments of control flow.
- **Annotation proof obligations**: verification conditions that establish consistency of the annotations with the controlled machine. This means that the fragments of control flow captured by the annotations really are appropriate for the machine.
- **Controller**: this is a process that describes the overall flow of control for the B machine.

- **Consistency checking**: establishing that the controller is consistent with the annotations by showing that that every part of the control flow is supported by some annotation.
- **Refine/Translate**: refinement may be needed before a translations can be achieved. The translation is the final step and requires additional annotation directives to set type sizes and I/O ports.

Checking a CSP controller against a machine is thus reduced to checking it against the annotations and verifying that the annotations are appropriate for the machine. The relationship between the different parts of the approach are given in Figure 1.

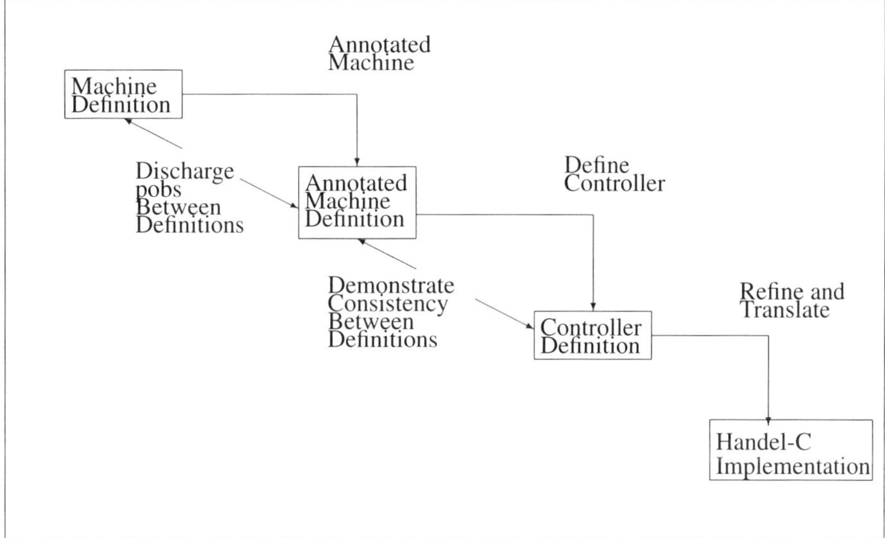

Figure 1. The Process Flow in the Approach.

The framework presented here is quite general, in that it may be applied to both Event-B and classical B. Additional annotations maybe added along with supporting control operations as required. Provided that a consistency argument can be developed. The first step to be taken is therefore to fix on the control language and the associated annotations to be incorporated into the B machine descriptions.

2. The Approach

We will demonstrate the approach with a simple model to illustrate aspects of the approach. The annotation we consider first is the NEXT annotation. An extremely simple controller language consisting only of prefixing, choice, parallel, if-then-else, and recursion is used to develop the example.

2.1. A B Machine

The B-Method [1] has evolved two major approaches: classical B and Event-B. Annotations can be used in either classical B machines, or Event-B systems. Classical B approaches focus on the services that a system might provide, whereas Event-B focuses on the events that occur within the system. B Machines are used in the examples. The generic classical B $MACHINE\ S$, given below, has variables, invariant, initialisation, and a set of operations $OP1$ through to OPn that have inputs and outputs. v describes a set of inputs and y describes a set of outputs to and from an operation, respectively.

$MACHINE\ S$

$VARIABLES\ v$

$INVARIANT\ v$

$INITIALISATION\ v :\in u$

$OPERATIONS$

$\qquad y \longleftarrow OP1(z_1) \mathrel{\widehat{=}} P_1 \mid B_1;$

$\qquad y_2 \longleftarrow OP2(z_2) \mathrel{\widehat{=}} G_2 \implies B_2;$

$\qquad \ldots$

$\qquad y_n \longleftarrow OPn(z_n) \mathrel{\widehat{=}} P_n \mid B_n$

END

The operations are defined in Guarded Substitution Language (GSL). It is asserted that the machine is consistent when each operation can be shown to establish the machine invariant, I, and the machine cannot deadlock. Every operation must be either guarded, G, or have a precondition, P, but all must have a next annotation (not shown). In Event-B, unlike classical B, new operations can be added during refinement. In the examples we anticipate the need for operations in the later stages of refinement by introducing the signature of the operation with a body defined by the *skip* operation. We do not in this paper adapt the proof obligations for Event-B refinement. The refinement process may involve adding detail to the specification in a consistent way to realise an implementation, which is a key notion in B. Refinement involves removing non-determinism and adopting concrete types. We add to the concept of B refinement with the annotations, by adding the notion of annotation control flow refinement.

3. The Annotation with I/O

We annotate operations of B machines with a NEXT annotation that supports operations with I/O. If the conjunction of proof obligations for all the annotations are discharged then we say that the annotations are consistent with the machine. A consistent controller that evolves in accordance with the next annotations steps will not diverge or deadlock. A NEXT annotation on the current operation OP_i (where OP_i represents $y_i \longleftarrow Op_i(z_i)$ and y_i is the output vector, $y_1 \ldots y_n$, and z_i is the input parameter vector, $z_1 \ldots z_m$) introduces another operation OP_j, or set of operations $OP_{j_1}, \ldots, OP_{j_n}$, which will be enabled after OP_i has executed (where an operation in the annotation OP_j represents $Op_j(e_j)$ and e_j is the input expression vector, $e_1 \ldots e_m$). In the NEXT annotation e_j is a list of expressions which serves as inputs on which OP_j can be called next. In this paper we will restrict the expressions to variables v defined in the B machines. The variables become ports in the hardware implementation. The value of these variable is not considered when calculating the proof obligations. Only the type of the variables is checked.

3.1. The Basic NEXT *Annotation*

$OP_i \ \mathrel{\widehat{=}} \ \textbf{PRE} \quad P_i \quad \textbf{THEN} \quad B_i\ \textbf{END} \quad \text{/*}\ \{OP_{j_1}, \ldots, OP_{j_n}\}\ \text{NEXT */} ;$

Definition 3.1 (Proof Obligations of the Basic NEXT on INITIALISATION) *Given the following B initialisation:*

INITIALISATION $\quad$ T /* $\{ Op_j ? v_j \}$ *NEXT* */ ;

the related proof obligations follow:

$$[T]((v_j \in T_j) \Rightarrow P_j)$$

The NEXT annotation following the initialisation indicates the first enabled operation. There can be more than one operation in the annotation. The example illustrates only one next operation. The variables used as input parameters in the annotation ($? v_{j_1} \ldots ? v_{j_m}$) must be of the type required in the operation definition.

Definition 3.2 (Proof Obligations of the Basic NEXT on Operations) *Given the following B operation:*

$$y_i \longleftarrow Op_i(z_i) \quad \widehat{=} \quad \textbf{PRE} \quad P_i \quad \textbf{THEN} \quad B_i \quad \textbf{END}$$
$$\textit{/* } \{ Op_j(v_{j_1}), \ldots, Op_{j_n}(v_{j_n}) \} \textit{ NEXT */} ;$$

the related proof obligations follow:

$$(P_i \wedge I \Rightarrow [B_i]((v_{j_1} \in T_{j_1}) \Rightarrow P_{j_1})) \wedge$$

$$\ldots$$

$$(P_i \wedge I \Rightarrow [B_i]((v_{j_n} \in T_{j_n}) \Rightarrow P_{j_n}))$$

where the elements of v_i and v_j are free in B_i, P_i, and I.

3.2. The NEXT_PAR Annotation

I/O operations can be annotated to indicate parallel execution NEXT_PAR. Two or more sets are introduced (only two illustrated below). Any operation of a respective set can run in parallel with any other operation from any of the other sets.

Definition 3.3 (Proof Obligations of NEXT_PAR) *Given the following B operation:*

$$y_i \longleftarrow Op_i(z_i) \quad \widehat{=} \quad \textbf{PRE} \quad P_i \quad \textbf{THEN} \quad B_i \quad \textbf{END}$$
$$\textit{/* } \{ Op_{j_1}(v_{j_1}), \ldots, Op_{j_n}(v_{j_n}) \}$$
$$\{ Op_{p_1}(v_{p_1}), \ldots, Op_{p_m}(v_{p_n}) \} \textit{ NEXT\_PAR */} ;$$

the related proof obligations follow:

$$(P_i \wedge I \Rightarrow [B_i]((v_{j_1} \in T_{j_1}) \Rightarrow P_{j_1})) \wedge$$

$$\ldots$$

$$(P_i \wedge I \Rightarrow [B_i]((v_{j_n} \in T_{j_n}) \Rightarrow P_{j_n})) \wedge$$

$$(P_i \wedge I \Rightarrow [B_i]((v_{p_1} \in T_{p_1}) \Rightarrow P_{p_1})) \wedge$$

$$\ldots$$

$$(P_i \wedge I \Rightarrow [B_i]((v_{p_n} \in T_{p_n}) \Rightarrow P_{p_n})) \wedge$$

$$variable\_used(\{ OP_{j_1}, \ldots, OP_{j_n} \}) \cap variable\_used(\{ Op_{p_1}, \ldots, OP_{p_n} \}) = \{\}$$

The parallel annotation offers the option to execute two or more operations in parallel after the current operation, provided they do not set or read any variables in common. The proof obligation ensures that all the operations in the annotations are enabled after the current operation. Only one from each set will be executed in parallel.

3.3. The NEXT_SEQ Annotation

Operations can be annotated to indicate a requirement for a particular sequential execution: NEXT_SEQ.

Definition 3.4 (Proof Obligations of NEXT_SEQ) *Given the following B operation:*

$$y_i \longleftarrow Op_i(z_i) \;\; \widehat{=} \;\; \textbf{PRE} \quad P_i \quad \textbf{THEN} \quad B_i \quad \textbf{END}$$
$$\text{/* } \{ \, Op_{j_1}(v_{j_1}), \ldots, Op_{j_n}(v_{j_n}) \, \}$$
$$\{ \, Op_{p_1}(v_{p_1}), \ldots, Op_{p_n}(v_{p_n}) \, \} \; NEXT\_SEQ \text{ */};$$

the related proof obligations follow:

$$(P_i \wedge I \Rightarrow [B_i]((v_{j_1} \in T_{j_1}) \Rightarrow P_{j_1})) \wedge$$

$$\ldots$$

$$(P_i \wedge I \Rightarrow [B_i]((v_{j_n} \in T_{j_n}) \Rightarrow P_{j_n})) \wedge$$

$$(P_{j_1} \wedge I \Rightarrow [B_{j_1}]((v_{p_1} \in T_{p_1}) \Rightarrow P_{p_1})) \wedge$$

$$\ldots$$

$$(P_{j_1} \wedge I \Rightarrow [B_{j_1}]((v_{p_n} \in T_{p_n}) \Rightarrow P_{p_n})) \wedge$$

$$\ldots$$

$$(P_{j_n} \wedge I \Rightarrow [B_{j_n}]((v_{p_1} \in T_{p_1}) \Rightarrow P_{p_1})) \wedge$$

$$\ldots$$

$$(P_{j_n} \wedge I \Rightarrow [B_{j_n}]((v_{p_n} \in T_{p_n}) \Rightarrow P_{p_n}))$$

where the elements of z_i and v_j and v_p are free in B_i, P_i, and I.

The NEXT_SEQ annotation is conceptually different from the NEXT annotation, because it captures specific paths of executions that must exist in a controller. The current operation Op_i must enable each operation in $\{ Op_{j_1}(v_{j_1}), \ldots, Op_{j_n}(v_{j_n}) \}$, and each operation in that set must enable each operation in the set $\{ Op_{p_1}(v_{p_1}), \ldots, Op_{p_n}(v_{p_n}) \}$. Practically, this annotation should be used to depict particular paths: one operation per set.

3.4. The NEXT_COND Annotation

To enable the current operation to conditionally select one set of operations next as opposed to some other set the NEXT_COND annotation is used. The condition NEXT_COND annotation is an extension to the NEXT annotation that supports conditional next path selection.

In definition 3.5 if the output of the current operation is *true* then all the operations OP_{j_1} through to OP_{j_n} are guaranteed to be available to execute. If however the current operation returns false then the operations OP_{p_1} through to OP_{p_n} are guaranteed to be available to execute. The proof of this claim can be verified by discharging the following proof obligation given in definition 3.5:

Definition 3.5 (Proof Obligation of NEXT_COND) *Given the following B operation:*

$$y_i \longleftarrow Op_i(z_i) \;\; \widehat{=} \;\; \textbf{PRE} \quad P_i \quad \textbf{THEN} \quad B_i \quad \textbf{END}$$
$$\text{/* } \{ \, Op_{j_1}(v_{j_1}), \ldots, Op_{j_n}(v_{j_n}) \, \}$$
$$\{ \, Op_{p_1}(v_{p_1}), \ldots, Op_{p_m}(v_{p_m}) \, \} \; NEXT\_COND \text{ */};$$

the related proof obligations follow:

$$(I \wedge P_i \Rightarrow [B_i]((y_i = \mathit{TRUE} \wedge v_{j_1} \in T_{j_1}) \Rightarrow P_{j_1})) \wedge$$

$$\ldots$$

$$(I \wedge P_i \Rightarrow [B_i]((y_i = \mathit{TRUE} \wedge v_{j_n} \in T_{j_n}) \Rightarrow P_{j_n})) \wedge$$

$$(I \wedge P_i \Rightarrow [B_i]((y_i = \mathit{FALSE} \wedge v_{p_1} \in T_{p_1}) \Rightarrow P_{p_1})) \wedge$$

$$\ldots$$

$$(I \wedge P_{1_i} \Rightarrow [B_i]((y_i = \mathit{FALSE} \wedge v_{p_n} \in T_{p_n}) \Rightarrow P_{p_n}))$$

The lists of the NEXT_COND annotation do not have to be the same size. The operation that carries this annotation must have a single boolean output.

3.5. A Simple Controller Language

The next annotation represents a control fragment specification of the whole system. The CSP controller represents a refined view of the annotated B system. The annotated B machine hasn't the fidelity to clearly portray the necessary control detail that the CSP can: the annotations are not clearly laid out as a set of recursive definitions. On translation both the B and the CSP are used to build the implementation, hence the need to develop a controller.

A distinction is drawn between operations that respond to external commands and those that are driven internally. A development will begin with a description of a number of operations: things that the system must do when commanded. During the development refinements will introduce internal operations. We distinguish between external and internal operations by marking the external operations with $/ * ext * /$ annotations, which are discussed in more detail in the refinement and translation section 5.

Definition 3.6 details the CSP subset of control fragments used in this paper: event prefix, choice, interleaving, if-then-else, and recursion control.

Definition 3.6 (Controller Syntax with I/O)

$$R ::= \square_y \, a!y?z \to R \mid$$

$$R1 \, \square \, R2 \mid$$

$$(\square_{y_1} a_1!y_1?z_1 \to skip \mid\mid\mid \ldots \mid\mid\mid \square_{y_n} a_n!y_n?z_n \to skip; \ R) \mid$$

$$\square_y \, e!y \to if \ y \ then \ R1 \ else \ R2 \mid$$

$$S(p)$$

The CSP controller is a different view of the annotated B specification. A more complex arrangement arises if the CSP controller is permitted to carry around local state. The simplified view is represented in figure 2. An annotated B machine output is the same as a CSP controller output. In definition 3.6 the channel a, in the controller fragment $\square_y \, a!y?z \to R$, is an operation name with a choice over all possible outputs y: from the controller's view, if a is called then any output y should be allowed. The outputs are fresh and modelled as a distributed external choice ranging over the type given in the B (the type is not always given in the controller definition). The channel has an input vector z. To accommodate analysis, finite types are used in the CSP. The same restriction does not exist in the B. Hence the CSP representation of the B operation may not be a true representation in terms of input and output, which may be a subset of the B types. $S(p)$ is a parameterised process variable. The external choice operator chooses between two process $R1 \, \square \, R2$ and relates to the $/* \ OP_J \ \text{NEXT}*/$

annotation that has one set. The interleave operator executes the two or more processes concurrently which will not synchronise on any events. The $if - then - else$ operator makes the decision on y; an output of the e operation. Recursive definitions are given as $S \cong R$. In a controller definition, all process variables used are bound by some recursive definition.

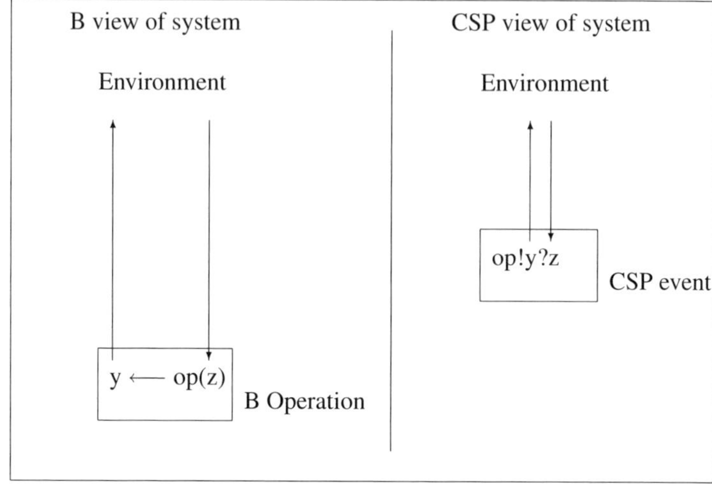

Figure 2. Different views of the same action.

A major constraint is enforced on the way controllers can be written. It facilitates translations, but turns out not to be so troublesome as it first appears. Controllers must start with an initialisation ($R1$), then enter a main loop ($S \cong R2$). This is summarised in definition 3.7. A controller $CTRL$ has a definition, $R1$, given in definition 3.6, in which all the parameterised process variables are the same, S. The definition of S is $R2$ and is also given in definition 3.6. The only recursive calls allowed are to S.

Definition 3.7 (Controller Syntax with I/O)

$$CTRL \cong R1$$

$$S \cong R2$$

where $R1$ and $R2$ are terms from definition 3.6 and

S is the only recursive variable allowed and

$R2$ is guarded as defined in definition 3.9

The results presented in this paper require that all recursive definitions are *guarded*, which means that at least one event must occur before a recursive call. The meaning of consistency between the controller and the annotations is given in terms of the *init* functions. The *init* function returns a set of operations available next and is developed in definition 3.8.

Definition 3.8 (init on CSP controller process with I/O extensions)

$$init(\underset{y}{\square}\, a!y?z \rightarrow R1) = \{a\}$$

$$init(R1 \;\square\; R2) = init(R1)\; \cup\; init(R2)$$

$$init(\underset{y_1}{\square}\, a_1!y_1?z_1 \rightarrow skip \;|||\; \ldots \;|||\; \underset{y_n}{\square}\, a_n!y_n?z_n \rightarrow skip);\; R = \{a_1, a_2, ..., a_n\}$$

$$init(if\ y\ then\ R1\ else\ R2) = init(R1) \cup init(R2)$$

$$init(S(p)) = init(R(p))$$

An action prefix must appear with output on the left. In the first case of the *init* definition the head of the control fragment is extracted. The outputs and inputs of the action are the same as the outputs and inputs of the B operation. The *init* of a prefixed action is the action (event). The *init* of a choice between two processes is the union of the *init* of the individual processes. The *init* of the interleaving is the set of first actions of each interleaving. Annotations clearly show an ordering of operations: an initial operation and a set of next operations. Every operation has a prefix, and is therefore *guarded*. Every control fragment must have a prefix and hence be guarded. The *guard* function is defined in definition 3.9. Prefixed operations are *guarded*. A fragment with an external choice separating the two processes is prefixed if the individual processes are *guarded*. Similarly with the if-then-else. The parameterised process variable is not *guarded*, whereas the recursive definition is guarded if the body is *guarded*.

Definition 3.9 (guarded on CSP controller process with I/O)

$$guarded(\square_y a!y?z \to R1) = true$$

$$guarded(R1 \square R2) = guarded(R1) \wedge guarded(R2)$$

$$guarded((\square_{y_1} a_1!y_1?z_1 \to skip \;|||$$

$$\cdots \;|||$$

$$\square_{y_n} a_n!y_n?z_n \to skip);\; R) = true$$

$$guarded(if\ TRUE\ then\ R1\ else\ R2) = guarded(R1) \wedge guarded(R2)$$

$$guarded(if\ FALSE\ then\ R1\ else\ R2) = guarded(R1) \wedge guarded(R2)$$

$$guarded(S(p)) = false$$

4. I/O NEXT Consistency

Consistency between a *guarded* controller and the annotated B machine is broken down into initial (definition 4.1) and step-consistency (definition 4.2).

Definition 4.1 (Initial-Consistency of M with respect to M_CTRL) *The initial-consistency of the controller fragment R is defined as follows:*

1. $\square_y a!y?z \to R$
   *is initially-consistent with M if $a \in next(INITIALISATION)$ and
   R is step-consistent with M*

2. $R1 \square R2$
   is initially-consistent with M if $R1$ and $R2$ are initially-consistent with M.

3. $S(p)$
   *is initially-consistent with M
   A family of recursive definitions $S \mathrel{\widehat{=}} R$ is initially-consistent with M's annotations
   if each R is initially-consistent with M's annotations.*

   [*We define $next(a)$ as the set of operations in the annotation of a.*]

A controller that starts with an interleaving or a conditional control fragment is not initially-consistent and should be avoided. An initialisation can not have an output which rules out the use of an *if* − *then* − *else* annotation on the initialisation. Ruling out the *interleaving* annotation simplifies initial-consistency checking.

Definition 4.2 (Step-Consistency of M with respect to M_CTRL) *The step-consistency of the controller fragment R is defined as follows:*

1. $\Box_y\, a!y?z \to R$
   is step-consistent with M if $\forall b \bullet b \in init(R) \Rightarrow b \in next(a)$, and R is step-consistent with M.

2. $R1\ \Box\ R2$
   is step-consistent with M if $R1$ and $R2$ are step-consistent with M.

3. $(\Box_y\, a!y_a?z_a \to\ skip\ |||\ \Box_y\, b!y_b?z_b \to\ skip);\ R$
   is step-consistent with M if $\forall e \bullet e \in init(R) \Rightarrow e \in next(a)$ and $e \in next(b)$, and R is step-consistent with M, and $update(a!y_a?z_a)\ \cap\ update(b!y_b?z_b) = \{\}$.

4. $\Box_y\, e \to if\ y\ then\ R1\ else\ R2$
   is step-consistent with M if $y \in BOOL$ and $R1$ and $R2$ are step-consistent with M and
   $\forall\ b\ \in\ init\ (R1)\ \Rightarrow\ b\ \in\ condition\_true(e)$ *and*
   $\forall\ c\ \in\ init\ (R2)\ \Rightarrow\ c\ \in\ condition\_false(e)$
   where condition_true(e) returns the actions that are enabled when y=true and condition_false(e) returns the actions that are enabled when y=false.

5. $S(p)$
   is step-consistent with M
   A family of recursive definitions $S \ \hat{=}\ R$ is step-consistent with M's annotations if each R is step-consistent with M's annotations.

The interleaving operator can only be shown to be consistent in a very limited sense. Two actions are allowed to occur in parallel provided they do not attempt to change the variables used by the other action.

Definition 4.3 (Consistency) *A controller R is* consistent *with the annotations of machine M if it is step-consistent with M's annotations and initially-consistent with M's annotations.*

The main result of this section is that if R is consistent with the annotations of a machine M, and the annotations of M are consistent with machine M, then operations of M called in accordance with the control flow of R will never be called outside their preconditions. We have [6] proven a theorem that shows that this holds for the basic NEXT, and the NEXT_COND annotations. The annotations are lose enough to permit a large set of possible consistent controllers. As such the controller is viewed as a a trace refinement of the annotations. The controllers do not refine the annotations in a failures divergence sense. We believe, but have not yet proven, that the NEXT_PAR and NEXT_SEQ can be rewritten in the basic NEXT form.

The key feature of the proof of this main result is an argument that no trace of R leads to an operation of M called outside its precondition or guard. This is established by building up the traces of R and showing that at each step an operation called outside its precondition cannot be introduced, by appealing to the relevant annotation and applying its proof obligation.

The benefit of this main result is that the details of the operations of M are required only for checking the consistency of the annotations, and are not considered directly in conjunction with the controller. The annotations are then checked against the controller using the definition of consistency above. This enables a separation of concerns, treating the annotations as an abstraction of the B machine.

5. Refinement and Translation to Handel-C

Refining should be considered where an otherwise cumbersome translation would result. Narrowing down the choice of the next operation reduces the size of the implementation, and avoids the translation process making an arbitrary choice to resolve the choice in the annotations. The first set of refinements, given in table 1[1] replace annotated sets with their subsets: non-determinism is reduced. The operation references, like OP_J, quoted in the tables are all sets.

NEXT external choice refinement reduces non-determinism in the choices offered in the next step. The NEXT interleave refinement reduces the non-determinism in one or more branches of the interleave execution. The NEXT sequential refinement reduces the non-determinism in one or more sections of the sequence. The NEXT conditional refinement reduces choice in a similar way.

Table 2 outlines some structural refinements. In case 1 a new set of operations are introduced OP_J. New operations can be introduced into Event-B in subsequent refinements. In classical B *new* operations must be introduced beforehand as operators that implement skip. Case 1 refines a simple NEXT operation into a sequence of detailed operations. The refinement sequence must end in the original next operation, which signifies the end of the refinement chain. In case 2 a next sequence NEXT_SEQ to next interleave refinement NEXT_PAR is depicted. It is possible if the operations that would make up the sequence are independent: they neither read nor write to similar variables.

A translation guide for annotations is given in table 3 and table 4. This is a guide because without the knowledge of the control structure, in particular the points of recursion, a translation can not be automated. However, the annotations do differentiate between internal and external B operations, which has an impact on the final structure of the code. The CSP controller is required to get a full picture for translation and table 6, and to some extent table 5, illustrates how translation of the control can proceed. As mentioned, the translation of a particular annotated operator is dependent on whether the operation is an internal or external operation. Internal operations can execute immediately after invocation. The execution of an external operation must wait for external stimulus: a change in the command input bus. A wait loop is introduced to poll the appropriate input bus until an external operation invocation is detected: $wait\_on\_\ldots$. Some annotated operators have restrictions on their I/O mode. External operators are marked with $/ * ext * /$. The NEXT_PAR can only be associated with internal operations next. The NEXT_SEQ must have an external operator at the head of the sequence and internal operations following. This restriction relates to the way this annotation is used in refinement. The CSP controller does not differentiate between internal and external operations. Hence the tables 3, 4, 7, 6, and 5 are all required to obtain a translation.

In tables 3 and 4 a NEXT annotation with one next operation translates to a sequence of two operations. If the second operation is an internal operation then it is case 1: all its inputs are not ported. If the second operation is an external operation (all inputs are ported) then case 2 is the translation template. The controller will wait until a new command arrives then execute the external operation if it was requested. Case 3, sequential arrangement of external operations, is restricted to external operations only. A translation of a sequence that starts with one operation then has a choice of several external operations will test each input set and execute the first operation for which the input has change since its last execution. (The new input values must be latched in.) Interleave action is only permitted between internal operations (case 4): those that take their input from internal variables. The Handel-C *par* statement ensures that all the branches when complete wait until the longest (in terms of clock cycles) has completed. The conditional operator can be used for internal or external action.

---

[1]All tables for this section are given in the Appendix.

In table 4 case 5 is the translation of the NEXT_SEQ. In the previous section the NEXT_SEQ was introduced to support refinement: a basic NEXT is refined into a sequence of operations NEXT_SEQ. The refine an operation that both inputs and outputs to a sequence of operations must input at the beginning of the sequence and output at the end of the sequence. Case 5 reflects this requirement: the first operation in the sequence is an external operation that inputs and the final operation is an internal operation that outputs.

The translations of Stepney [19], and Phillips and Stilles [18] are given in table 5. Only the translation of parametrisable integer declaration, functions, and recursion are used. This is because our source is not CSP (it is annotated B and CSP) and as such channels are not being used to synchronise events. In the table the CSP language construct and translation are mapped. A tick is inserted if they are supported by Stepney (SS) or Phillips and Stilles (PS). When an operation is invoked it takes its input from a port in the environment. Internal synchronisation of operations within machines is not dealt with in this paper. To guide the B translation, table 7 has been developed. A discussion of the example is given in section 6.

6. Example: Safe Control System

We use the example of a safe locking system to illustrate the ideas introduced in the previous sections. The abstract specification outlines the operations of the environment. The operations that are invoked by the environment are indicated with $/ * ext * /$ annotations. Both the operation output and the operation can be marked with $/ * ext * /$ annotations. All $/ * ext * /$ annotation outputs are ported and become part of the Handel-C interface output. All $/ * ext * /$ operations are associated with a bus port that has a state of the same name as the operation. Variables intended as input are marked with $/ * IN * /$. It is possible to mark the variables as $/ * IN * /$ or $/ * OUT * /$. Along with the mode the width of the type is given in bits. Operations are invoked in two ways. The first way has already been introduced; an $/ * ext * /$ operation will have a input bus associated with it, which when set to the operator name will invoke the operation when it is enabled by the control flow. Operations not labelled with $/ * ext * /$ are internal and are invoked immediately when enabled by the control flow.

6.1. The Example's State and Control Flow

In figure 3 the B Abstract Machine for the safe is given. There are three command states *Locked*, *Unlocked*, and *BrokenOpen* which are represented in two bits. The variable *Door* is drawn from the $COMMAND$ type and initialised to *Unlocked*. The *Lock* operation is enabled after initialisation. It is an external operation with externally ported output. After setting the *Door* state variable to *Locked*, *Unlocked* and *BreakOpen* are enabled. For completeness we introduce two operations that will be used later to develop the detailed functionality of the machine during refinement. These operations are *UnlockR1* and *UnlockR2*. Their bodies are not expanded. The *Unlock* is an external operation and has externally ported output. It non-deterministically decides to set the *Door* variable to *Unlocked* or *Locked*. The next operator to be enabled depends on the outcome of the *Unlock* operation. If *Unlocked* was chosen then the next enabled operation is *Lock*, otherwise *Unlocked* or *BreakOpen* will be offered. The *BreakOpen* operation sets the *Door* state to *BrokenOpen* and offers itself as the next operation available.

The controller $CTRL$, given in figure 4, first performs a *Initialisation* then a *Lock* and then jumps to the S process where it can perform either an *Unlock* or *BreakOpen*. The *Unlock* event has a single output that is used as the conditional test in the if-then-else following the *Unlock* event. If the output of the *Unlock* operation is true then the flow of control is repeated starting again at $CTRL$, if it is false then control is repeated at S.

MACHINE   *Safe*

SETS   *COMMAND* = { *Locked* , *Unlocked* , *BrokenOpen* }/*2*/
VARIABLES   *Door*
INVARIANT   *Door* $\in$ COMMAND /*OUT2*/
INITIALISATION   *Door* := Unlocked   /* { Lock } NEXT */

OPERATIONS

/*ext*/ **Status** $\longleftarrow$ /*ext*/ **Lock** $\widehat{=}$
    PRE   *Door* = *Unlocked* **THEN**   *Door* := *Locked* $\parallel$ *Status* := *Locked* **END**
        /* { Unlock, BreakOpen } NEXT */ ;

UnlockR1 (Comb1a,Comb1b) $\widehat{=}$
    PRE   *Comb1a* $\in$ NAT $\wedge$ *Comb1b* $\in$ NAT $\wedge$ *Door* = *Locked* **THEN**   skip **END**   ;

UnlockR2(Comb2a,Comb2b) $\widehat{=}$
    PRE   *Comb2a* $\in$ NAT $\wedge$ *Comb2b* $\in$ NAT $\wedge$ *Door* = *Locked* **THEN**   skip **END**   ;

/*ext*/ **Status** $\longleftarrow$ /*ext*/ **Unlock** $\widehat{=}$
    PRE   *Door* = **Locked**
    THEN
        ANY   *dd* **WHERE**   *dd* : COMMAND - { *BrokenOpen* }
        THEN
            IF   (*Unlocked* = *dd*) **THEN**   *Status* := *1* **ELSE**   *Status* := 0 **END** $\parallel$
            Door := *dd*
        END
    END   /* { Lock } { UnLock,BreakOpen } NEXT_COND */ ;

/*ext*/ **Alarm** $\longleftarrow$ /*ext*/ **BreakOpen** $\widehat{=}$
    PRE   *Door* $\in$ *COMMAND* **THEN**   *Door* := *BrokenOpen* $\parallel$ *Alarm* := *1* **END**
    /* { BreakOpen } NEXT */ ;

END

Figure 3. Safe Machine

$$CTRL = Initialisation \rightarrow \Box_y Lock!y \rightarrow S$$

$$S = (\Box_y Unlock!y \rightarrow (if\ y\ then\ \Box_y Lock!y \rightarrow CTRL\ else\ S))\Box$$

$$(\Box_y BreakOpen!y \rightarrow B\_CTRL)$$

$$B\_CTRL = \Box_y BreakOpen!y \rightarrow B\_CTRL$$

Figure 4. Safe Machine Controller.

6.2. A Refined Example

A refinement of the *Safe* machine, called SafeR, is given in figure 5 and figure 6 . It is a classical B refinement that mimicking a refinement in Event-B. The operation *UnlockR1* and *UnlockR1* are introduced to refine *Unlock*. The laws of refinement of Event-B are not fully justified. The *Safe* REFINEMENT, *SafeR*, breaks down the Unlocking process into two stages. Firstly, a two new operation are slotted into the control in parallel: *UnlockR1*(*Comb1a*, *Comb1b*) and *UnlockR2*(*Comb2a*, *Comb2b*). Both have a combina-

REFINEMENT   *SafeR*
REFINES   Safe
VARIABLES   *Door, Cx1a, Cx2a, Cx1b, Cx2b,*
                                 Master1, Checked1 Master2, Checked2
INVARIANT

$Cx1a \in$ NAT/*IN16*/ $\land$ $Cx2a \in$ NAT/*IN16*/ $\land$

$Cx1b \in$ NAT/*IN16*/ $\land$ $Cx2b \in$ NAT/*IN16*/ $\land$

$Master1 \in$ NAT/*16*/ $\land$ $Checked1 \in$ NAT/*1*/ $\land$

$Master2 \in$ NAT/*16*/ $\land$ $Checked2 \in$ NAT/*1*/

INITIALISATION

$Door$:=unlocked $\|$ $Cx1a$:=0 $\|$ $Cx2a$:=0 $\|$ $Cx1b$:=0 $\|$ $Cx2b$:=0 $\|$

$Master1$:=67 $\|$ $Checked1$:=0 $\|$ $Master2$:=76 $\|$ $Checked2$:=0   /* { Lock } NEXT */

OPERATIONS

/*ext2*/ **Status** $\longleftarrow$ /*ext1*/ **Lock** $\hat{=}$
   PRE
      $Door = Unlocked$
   THEN
      $Door := Locked \| Status := Locked \| Checked1 := 0 \| Checked2 := 0$
   END
        /* { UnlockR1(Cx1a,Cx1b), UnlockR2(Cx2a,Cx2b) } { Unlock } NEXT_SEQ */
        /* { UnlockR1(Cx1a,Cx1b) } { UnlockR2(Cx2a,Cx2b) } NEXT_PAR */ ;

/*ext1*/**UnlockR1(/*16*/Comb1a,/*16*/Comb1b)** $\hat{=}$
   PRE
      $Comb1a \in$ NAT $\land$ $Comb1b \in$ NAT $\land$ $Door = $ **Locked**
   THEN
      IF
         (Comb1a = Master1)
      THEN
         $Checked1 := 1 \| Master1 := $ Comb1b
      ELSE
         $Checked1 := 0$
      END
   END   /* { Unlock } NEXT */ ;

Figure 5. Safe Refinement Part 1.

tion parameter which is compared against a stored master code and a secondly parameter that is used to create a new master key. The *UnlockR* commands update the master combination if a successful comparison occurs. New input variables are added: $Cx1a$, $Cx2a$, $Cx1b$, and $Cx2b$. These are used to input the combination values and are not used by the B Operations. $Checked1$, $Checked2$, $Master1$ and $Master2$ are new variables used by the operations. The annotations of the *Lock* operation are refined. Two operation are added before the *Unlock*. The extra proof obligations can be discharged. The bodies of the *UnlockR* and $Rekey(Comb2)$ are completed at this level. The body of the *Unlock* operation is refined. The annotations of the *Unlock* are refined: the *BreakOpen* operation is removed as an option. What was one unlock operation has been expanded into three (two in parallel).

/*ext1*/**UnlockR2(/*16*/Comb2a,/*16*/Comb2b)** $\widehat{=}$
    PRE
        Comb2a $\in$ NAT $\wedge$ *Comb2b* $\in$ NAT $\wedge$ *Door* = Locked
    THEN
        IF
            (Comb2a = Master2)
        THEN
            Checked2 := 1 $\|$ *Master2* := Comb2b
        ELSE
            Checked2 := 0
        END
    END   /* { Unlock } NEXT */ ;

/*ext2*/**Status** $\longleftarrow$ **Unlock** $\widehat{=}$
    PRE
        Door = Locked
    THEN
        IF   *(Checked1 = 1)* $\wedge$ *(Checked2 = 1)*
        THEN
            Door := *Unlocked* $\|$ *Status* := 1
        ELSE
            Door := *Locked* $\|$ *Status* := 0
        END
    END
END   /* { Lock } { UnlockR } COND_NEXT */ ;

/*ext*/ **Alarm** $\longleftarrow$ /*ext*/ **BreakOpen** $\widehat{=}$

    PRE   *Door* $\in$ COMMAND **THEN**   *Door* := *BrokenOpen* $\|$ *Alarm* := *1* **END**
        /* { BreakOpen } NEXT */
END

Figure 6. Safe Refinement Part 2

$$CTRL = Initialisation \rightarrow \underset{y}{\square} \, Lock!y \rightarrow S$$

$$S = (\,UnlockR1?\,Cx1a?\,Cx1b \rightarrow \; skip \, ||| \; UnlockR2?\,Cx2a?\,Cx2b \rightarrow skip\,) \rightarrow$$

$$\underset{y}{\square} \, Unlock!y \rightarrow (if \; y \; then \; \underset{y}{\square} \, Lock!y \rightarrow S \; else \; S)$$

Figure 7. Refined Safe Controller.

Before refinement the *Unlock* operation has both input and output. The refined version has the input occurring on the first operations in the refined sequence of operations (*UnlockR1* and *UnlockR2*), and the output occurring on the final operation of the sequence (the original *Unlock* operation).

The controller given in figure 7 starts off like the abstract process with an *Initialisation* and a *Lock* then a jump to S. There is in this refined process no choice to *breakOpen*, only *UnlockR1*and *UnlockR2* are offered with *Cx1a* and *Cx1b* and *Cx2a* and *Cx2b* are offered as an input, respectively. The *UnlockR* process is the first in a sequence of processes that refines the original *UnLock* process. The refined sequence starts with a parallel combination of the *UnlockR1* and the *UnlockR2* events then the original *Unlock* event, at which point the output is given. Both legs of the interleaving must terminate before control is passed to the *Unlock*. As before the outcome of *Unlock* determines what happens next. If the *Unlock* was successful the process will be restarted from the beginning. If the current attempt at locking failed then another go at *Unlock* will occur. It is noted that the *Lock* $\rightarrow$ *S* could have been replaced by *CTRL*. However, the former is easier to translate.

6.3. A Hand Translation into Handel-C

The refined B specification provides the details of the types, variables, and functions. The CSP controller provides the executions details that are use later to construct the Handel-C main section. Summaries of hand translations of the refined B specification and the CSP controller are given in Figures 8, 9, and 10 (in Appendix B).

First we review the B translation. The *SETS* clause is translated into an enumerated type. The *INVARIANT* section is used to create the declarations. Variables annotated with a mode will be created as buses of the appropriate I/O type and size. Other variables will be created. Variables which will be bound to ports are created. Each operation which is external is associated with a command input bus of the same name as the machine. The mechanism for requesting an external operation to execute is to change the data on the command input bus to the same name as the operation required. The last requested operation is latched into variable of the same name as the refined machine with a *._var* post fix. Variables are declared for operation outputs. The names of the output bus variables are a concatenation of the operation output name and the operation name. This avoids clashes with similar operation output names. Buses are defined for each / $* IN *$ / and / $* OUT *$ / annotation, external operation, and operation output. Each operation is translated into a function. If an operation has an output the function will return a value. Functions with outputs will have an assignment in them that assigns to the bus output function variable. The function will also return that output in the final statement of the function. Assigning to the function output variable and writing it to a output port as well allows it to be put out on the output bus, and used internally in the Handel-C program. The bodies are translated in a straightforward manner. Assignments in the operations are put together in a *par* Handel-C statement. Assignment and the *if* − *then* − *else* B constructs have straightforward translations. The refined B example is limited to assignment and *if* − *then* − *else*. The *INITIALISATION* is translated into a function called *Initialisation_fnc*.

The CSP controller is used to construct the main Handel-C body. A summary of the hand translations made on the CSP controller are given in table 6. The controller design was structurally limited to facilitate translation: initialisation and setting up operations are performed before a main loop is entered. The first process definition *CTRL_fnc* is not recursive; it is an open process. It translates to a function call *CTRL_fnc*, which invokes the *Initalisation_fnc* and *lock_fnc* functions. On returning to the main program the next function called is the *S_fnc*, which implements the main loop. *S_fnc* is tail recursive and is implemented with a continuously looping while loop; it is a closed process. The first event in the main loop is the *UnlockR* commands. In the translation the *Unlock_fnc* is preceded by *wait_Unlock_fnc* as it is an external operation. The *UnlockR_fnc* functions inputs from the *Cx1a*, *Cx1b*, *Cx2*,and *Cx2* input buses. The *Unlock_fnc* call follows. *Unlock_fnc* returns a value that is assigned to a variable that is output ported. The value is also used to decide the course of the following if-then-else. Either a *Lock_fnc* or an *UnlockR_fnc* is performed after a wait. Then the process recurses.

7. Discussion

This paper has introduced a way of refining annotations that support Event-B style refinement, and set out a guide for translation to an HDL, within the B annotation framework. We have demonstrated how the framework previously presented can be extended for both classical B and Event-B. Our approach sits naturally with refinement. Refinement and translation are still being considered for CSP∥B. In fact the B annotation approach offerers several approaches to refinement: refinement of control flow only, state only, or control flow and state. The extensions to the annotations are fairly rich and now include annotations to support: next selection, sequencing, conditional, parallel execution, and I/O. The inability to define points of recursion has led to a reliance on a CSP controller. We restricted this paper to the consideration of fixed variables as operation inputs, and permitted no scope for controller state. Work on CSP state and defining recursive points in the annotations is currently ongoing. More work is required to automate the translation and develop the proof of the theorem to cover interleaving.

Acknowledgements

The extensions to the refinement have benefited from conversations with Stefan Hallestede and Helen Treharne. Thank you for the positive comments from the referees and detailed lists of error eta, improvements and additions.

References

[1] J-R. Abrial. *The B-Book: Assigning Programs to Meaning*. Cambridge University Press, 1996.

[2] S. Schneider. *The B-Method: An introduction*. Palgrave, 2002.

[3] C. A. Hoare. *Communicating Sequential Processes*. Prentice-Hall International, Englewood Cliffs, New Jersey, 1985.

[4] A. W. Roscoe. *The Theory and Practice of Concurrency*. Prentice-Hall, 1998.

[5] S. Schneider. *Concurrent and Real-time Systems: The CSP Approach*. John Wiley and Sons, 1999.

[6] W. Ifill, S. Schneider, and H. Treharne. Augmenting B with control annotations. In J. Julliand and O. Kouchnarenko, editors, *B2007:Formal Specification and Development in B*, volume 4355 of *LNCS*. Springer, January 2007.

[7] W. Ifill, I. Sorensen, and S. Schneider. *High Integrity Software*, chapter The Use of B to Specify, Design and Verify Hardware. Kluwer Academic Publishers, 2001.

[8] P. T. Ashenden. *The Designer's Guide to VHDL*. Morgan Kaufmann, 1996.

[9] W. Ifill. Formal development of an example processor (AEP) in AMN, C and VHDL. Computer science, University of London, Computer Science Department, Royal Holloway, University of London, Egham, Surrey TW20 0EX, Sept 1999.

[10] A. Aljer, J. L. Boulanger, P. Devienne, S. Tison, and G. Mariano. BHDL: Circuit design in B. In *Applications of Concurrency to System Design*, pages 241–242. IEEE Computer Society, Elsevier, unknown 2003.

[11] A. Aljer and P. Devienne. Co-design and refinement for safety critical systems. In *19th IEEE International Symposium on Defect and Fault Tolerance in VSLI Systems (DFT'04)*, pages 78–86, 2004.

[12] J-R. Abrial and L. Mussat. *Event B Reference Manual*. ClearSy, 1999.

[13] J-R. Abrial. Event driven circuit construction version 5. MATISSE project, August 2001.

[14] H. Treharne and S. Schneider. Communication B machines. In *ZB2002*, 2002.

[15] H. Treharne. *Combining Control Executives and Software Specifications*. PhD thesis, Royal Holloway, University of London, 2000.

[16] Alexandre Mota and Augusto Sampaio. Model-checking CSP-Z: Strategy, tool support and industrial application. *Science of Computer Programming*, 40(1):59–96, May 2001.

[17] C. Fischer. CSP-OZ: A combination of Object-Z and CSP.

[18] J. D. Phillips and G. S. Stiles. An automatic translation of CSP to Handel-C. In I. East, J. Martin, P. Welch, D. Duce, and M. Green, editors, *Communicating Process Architecures 2004*. IOS Press, 2004, 2004.

[19] S. Stepney. CSP/FDR2 to Handel-C translation. Technical report, University of York, June 2003.

A. Refinement and Translation Tables

Table 1. NEXT Refinements - Reduction of Non-determinism.

	Annotation	Refinement	type
1	$OP_i \mathrel{\hat{=}} ...OP_J\ NEXT$	$OP_i \mathrel{\hat{=}} ...OP'_J\ NEXT$	next external choice refinement
2	$OP_i \mathrel{\hat{=}} ...OP_J\ OP_K\ NEXT\_PAR$	$OP_i \mathrel{\hat{=}} ...OP'_J\ OP'_K\ NEXT\_PAR$	next interleave refinement
	$OP_{j_1} \mathrel{\hat{=}} ...OP_X\ NEXT$ ... $OP_{j_n} \mathrel{\hat{=}} ...OP_X\ NEXT$	$OP_{j_1} \mathrel{\hat{=}} ...OP_X\ NEXT$ ... $OP_{j_n} \mathrel{\hat{=}} ...OP_X\ NEXT$	
	$OP_{k_1} \mathrel{\hat{=}} ...OP_X\ NEXT$ ... $OP_{k_n} \mathrel{\hat{=}} ...OP_X\ NEXT$	$OP_{k_1} \mathrel{\hat{=}} ...OP_X\ NEXT$ ... $OP_{k_n} \mathrel{\hat{=}} ...OP_X\ NEXT$	
3	$OP_i \mathrel{\hat{=}} ...OP_J\ OP_P\ NEXT\_SEQ$	$OP_i \mathrel{\hat{=}} ...OP'_J\ OP'_P\ NEXT\_SEQ$	next sequential refinement
	$OP_{j_1} \mathrel{\hat{=}} ...OP_P\ NEXT$ ... $OP_{j_n} \mathrel{\hat{=}} ...OP_P\ NEXT$	$OP_{j_1} \mathrel{\hat{=}} ...OP_P\ NEXT$ ... $OP_{j_n} \mathrel{\hat{=}} ...OP_P\ NEXT$	
4	$OP_i \mathrel{\hat{=}} ...OP_J\ OP_P\ NEXT\_COND$	$OP_i \mathrel{\hat{=}} ...OP'_J\ OP'_P\ NEX\_COND$	next condition refinement
	$OP_{j_1} \mathrel{\hat{=}} ...OP_P\ NEXT$ ... $OP_{j_n} \mathrel{\hat{=}} ...OP_P\ NEXT$	$OP_{j_1} \mathrel{\hat{=}} ...OP_P\ NEXT$ ... $OP_{j_n} \mathrel{\hat{=}} ...OP_P\ NEXT$	

$$\text{where } OP'_J \subseteq OP_J \text{ and}$$
$$OP'_K \subseteq OP_K$$

Table 2. NEXT Refinements - Structural Refinements.

	Annotation	Refinement	type
1	$OP_i \mathrel{\widehat{=}} ...OP_X\ NEXT$	$OP_i \mathrel{\widehat{=}} ...OP_J\ OP_X\ NEXT\_SEQ$	introduction of
		$OP_{j_1} \mathrel{\widehat{=}} ...OP_X\ NEXT$	new operation
		$OP_{j_n} \mathrel{\widehat{=}} ...OP_X\ NEXT$	
2	$OP_i \mathrel{\widehat{=}} ...OP_J\ OP_P\ NEXT\_SEQ$	$OP_i \mathrel{\widehat{=}} ...OP_J\ OP_P\ NEXT\_PAR$	next sequence
	$OP_{j_1} \mathrel{\widehat{=}} ...OP_P\ NEXT$	$OP_{j_1} \mathrel{\widehat{=}} ...OP_P\ NEXT$	to interleave
	$...$	$...$	refinement
	$OP_{j_n} \mathrel{\widehat{=}} ...OP_P\ NEXT$	$OP_{j_n} \mathrel{\widehat{=}} ...OP_P\ NEXT$	
	$variable\_used(\{OP_j,\ldots,OP_k\})$ $\cap$ $variable\_used(\{Op_p,\ldots,OP_q\})$ $= \{\}$		

Table 3. NEXT Annotation Translation Guide Part 1.

	Annotation	Handel-C Translation Fragment	Comment
1	$OP_i \mathrel{\widehat{=}} ...\{OP_{j_1}\}NEXT$ $op_i!y_i?z_i \rightarrow (op_{j_1}!y_{j_1}?z_{j_1} \rightarrow ...$	$y_i = OP_i(v_i)\,;\ y_{j_1} = OP_{j_1}(v_{j_1})$	internal single next translation
2	$OP_i \mathrel{\widehat{=}} ...\{OP_{j_1}\}NEXT$ $/*ext*/OP_{j_1} \mathrel{\widehat{=}} ...$ $op_i!y_i?z_i \rightarrow (op_{j_1}!y_{j_1}?z_{j_1} \rightarrow ...$	$y_i = OP_i(v_i)\,;$ $wait\_on\_OP_{j_1}\,;$ $if\ in = OP_{j_1}$ $then\ y_{j_1} = OP_{j_1}(v_{j_1})\}$ $else\ delay\,;$	external single next translation
3	$/*ext*/OP_i \mathrel{\widehat{=}} ...$ $\{OP_{j_1}, ..., OP_{j_n}\}NEXT$ $op_i!y_i?z_i \rightarrow (op_{j_1}!y_{j_1}?z_{j_1} \rightarrow ...\ \Box\ ...$ $\Box$ $op_{j_n}!y_{j_n}?z_{j_n} \rightarrow ...)$	$y_i = OP_i(z_i)\,;$ $wait\_on\_OP_{j_1-}...\_OP_{j_n}\,;$ $if\ in = OP_{j_1}$ $then\ y_{j_1} = OP_{j_1}(v_{j_1})$ $else\ ...$ $...$ $if\ in = OP_{j_n}$ $then\ y_{j_n} = OP_{j_n}(v_{j_n})$ $else\ skip$	external multiple next choice translation
4	$OP_i \mathrel{\widehat{=}} ...OP_j\ OP_k\ NEXT\_PAR$ $OP_j \mathrel{\widehat{=}} ...OP_X\ NEXT$ $OP_k \mathrel{\widehat{=}} ...OP_X\ NEXT$ $op_i!y_i?z_i \rightarrow (op_j!y_j?z_j \rightarrow ...)\|$ $(op_k!y_k?z_k \rightarrow ...)$	$seq\{y_i = OP_i(v_i),$ $\quad par\{y_j = OP_j(v_j),$ $\quad\quad y_k = OP_k(v_k)$ $\quad \}$ $\}$	internal next interleave translation

Table 4. NEXT Annotation Translation Guide Part 2.

	Annotation	Handel-C Translation Fragment	Comment
5	$OP_i \mathrel{\widehat{=}} ...OP_J\ OP_K\ NEXT\_SEQ$ $/*ext*/OP_{j_1} \mathrel{\widehat{=}} ...OP_K\ NEXT$ $...$ $/*ext*/OP_{j_n} \mathrel{\widehat{=}} ...OP_K\ NEXT$ $OP_{k1} \mathrel{\widehat{=}} ...$ $OP_{kn} \mathrel{\widehat{=}} ...$ $op_i!y_i?z_i \rightarrow (op_{j_1}!y_{j_1}?z_{j_1} \rightarrow \ldots \square$ $\qquad \ldots \square$ $\qquad op_{j_n}!y_{j_n}?z_{j_n} \rightarrow \ldots);$ $\qquad (op_{k1}!y_{k1}?z_{k1} \rightarrow \ldots \square$ $\qquad \ldots \square$ $\qquad op_{k_n}!y_{k_n}?z_{k_n} \rightarrow \ldots)$	$y_i = OP_i(v_i);\ \ wait\_on\_OP_J$ $if\ in = OP_{j_1}$ $then\ y_{j_1} = OP_{j_1}(v_{j_1})$ $else\ \ldots$ $\quad \ldots$ $if\ in = OP_{j_n}$ $then\ y_{j_n} = OP_{j_n}(v_{j_n})$ $else\ skip$ $\quad ;$ $y_{k1} = OP_{k1}(v_{k1})$	next sequential translation
6	$/*ext*/OP_i \mathrel{\widehat{=}} \ldots$ $\qquad OP_J\ OP_K\ NEXT\_COND$ $OP_{j_1} \mathrel{\widehat{=}} ...OP_K\ NEXT$ $\ldots$ $OP_{j_n} \mathrel{\widehat{=}} ...OP_K\ NEXT$ $OP_{k1} \mathrel{\widehat{=}} ...OP_K\ NEXT$ $\ldots$ $OP_{kn} \mathrel{\widehat{=}} ...OP_K\ NEXT$ $op_i!y_i?z_i \rightarrow (op_{j_1}!y_{j_1}?z_{j_1} \rightarrow \ldots \square$ $\qquad \ldots \square$ $\qquad op_{j_n}!y_{j_n}?z_{j_n} \rightarrow \ldots);$ $\qquad (op_{k1}!y_{k1}?z_{k1} \rightarrow \ldots \square$ $\qquad \ldots \square$ $\qquad op_{k_n}!y_{k_n}?z_{k_n} \rightarrow \ldots)$	$y = OP_i(v_i);$ $if\ y$ $\quad \{wait\_on\_OP_J\ ;$ $\quad if\ in = OP_{j_1}$ $\quad then\ y_{j_1} = OP_{j_1}(v_{j_1})$ $\quad else\ \ldots$ $\quad \ldots$ $\quad if\ in = OP_{j_n}$ $\quad then\ y_{j_n} = OP_{j_n}(v_{j_n})$ $\quad else\ skip$ $\quad \}$ $\ else$ $\{wait,$ $if\ in = OP_{k1}$ $\quad then\ y_{k1} = OP_{k1}(v_{k1})$ $\quad else\ \ldots$ $\quad \ldots$ $if\ in = OP_{kn}$ $\quad then\ y_{kn} = OP_{kn}(v_{kn})$ $\quad else\ skip$ $\}$	external next condition translation

Table 5. Existing CSP to Handel-C Translation Guide.

Feature	CSPM	Handel-C	PS	SS
Channel Declarations (from use)	channel	chan, chanin, chanout	✓	
Channel Declarations	channel c	chan SYNC c;		✓
Typed Structured Channel Declarations	channel d : T.T	chan struct d_DATA d		✓
Input Channel Operations	in?x	in?x;	✓	✓
Output Channel Operations	out!x	out!x;	✓	✓
Integer Declarations		int 8 x;	✓	✓
Parametrisable functions	p(n) = ...	void(n)...	✓	✓
External Choice	[ ]	prialt ...	✓	✓
Synchronous Parallel	[| { | ... | } |]	par ...	✓	✓
Replicated Sharing Parallel	[| $Event$ |] n: { i..j }•P(n)	par (n=i; n$_i$=j; ++n)P(n);		✓
Recursion	P = ... → P	while(1) ...	✓	✓
Conditional Choice	if b then P else Q	if (B) then P(); else Q();		✓
Macros	{- ... -}	...		✓

Table 6. CSP to Handel-C Translation Guide.

Feature	CSP	Handel-C
initialisation processes	P $\widehat{=}$ ... R	P_fnc();Q_fnc(); void P_fnc(void){...;}
main loop processes	R $\widehat{=}$ ... R	R_fnc(); void R_fnc(void){while(1){...;}}
prefix (internal)	$< e → P >$	e_fnc ; $<P>$
prefix (external)	$< e → P >$	wait_on_e; e_fnc ; $<P>$
choice (external)	$< P1 □ P2 >$	$<P1>$
interleaved	$< e1 → skip$ $\|\|\| ... \|\|\|$ $e_n → skip; P >$	PAR$\{< e_1 → skip >;$ $...; < e_n → skip >\}; < P >$
if-then-else	$<$if y then P else Q$>$ where $< P >$ is the translation of P	if y $\{<P>\}$ else $\{<Q>\}$

Table 7. B to Handel-C Translation Guide.

Feature	B	Handel-C
set	SETS SS=	typedef enum { AA =
	AA,...,XX/*n*/	(unsigned n) 0, ..., XX } SS;
declaration		
B variable	INVARIANT	unsigned n Vv;
declaration	Vv ∈ TT /*OUTn*/	interface bus_out()
		Vv1 (unsigned 2 OutPort=Vv);
	INVARIANT	unsigned n Vv;
	Vv ∈ TT /*INn*/	interface bus_in(unsigned n inp) Vv();
	INVARIANT	unsigned n Vv;
	Vv ∈ TT /*n*/	
Function	/*extN*/ **Oo**	unsigned 1 Cc_var;
Declaration	⟵ /*ext*/ **Cc**(/* M */Zz)	interface bus_out ()
		Oo_Cc1 (unsigned N Oo_Cc);
		interface bus_in(unsigned 1 inp) Cc ();
		void wait_on_Cc_fnc()
		{ while (Cc.inp == Cc_var) { delay; }
		Cc_var = Cc.inp;
		}
		unsigned N Cc_fnc(unsigned M Zz){
		par{...;};return exp;
		}
Function	PRE P THEN B END	par{<< B >>}
Body		
	IF b THEN c ELSE d END	if << b >> { << c >> }
		else { << d >> } ;
	$b := c$	<< b >> = << c >> ;
initialisation	INITIALISATION ...	void Initialisation(void){ ...; }
main	OPERATION	void main(void){ Initialisation; ... }

B. Hand Translations

```
// set clock = external "Clock";
#define PAL_TARGET_CLOCK_RATE 25175000
#include "pal_master.hch"
// BreakOPen removed in translation as
// not used and no command default added
typedef enum {Not_Commanded =
  (unsigned 2) 0, Locked, Unlocked} COMMAND;
typedef enum {No_Command =
  (unsigned 2) 0, Lock, UnlockR1, UnlockR2} SafeR;
unsigned 2 Door;                         // B variables
unsigned 1 Checked1;
unsigned 16 Master1;
unsigned 1 Checked2;
unsigned 16 Master2;
SafeR SafeR_Bus_var;                     // latch input bus values to
                                         // request operation execution
unsigned 1 Status_Unlock;                // operation output values
unsigned 2 Status_Lock;
interface bus_in(unsigned 16 inp) Cx1a();   // IN annotations
interface bus_in(unsigned 16 inp) Cx2a();
interface bus_in(unsigned 16 inp) Cx1b();
interface bus_in(unsigned 16 inp) Cx2b();
interface bus_in(SafeR inp) SafeR_Bus();    // ext operations
interface bus_out() Door1 (unsigned 2 OutPort=Door);   // OUT annotations
interface bus_out() Status_Unlock1 (unsigned 1 OutPort=Status_Unlock);
```

Figure 8. SafeR Translation Part 1a.

```
void wait_on_Lock_fnc (){
    while (SafeR_Bus.inp != Lock){delay;}
    SafeR_Bus_var = Lock;
}
unsigned 2 Lock_fnc(void){
    par{
        Door = Locked;
        Status_Lock = Locked;
        Checked1 := 0;
        Checked2 := 0;
        }
    return Status_Lock;
}
void wait_on_UnlockR1_fnc(void){
    while (SafeR_Bus.inp != UnlockR1){delay;}
    SafeR_Bus_var = UnlockR1;
}
void UnlockR1_fnc(unsigned 16 Comb1a, unsigned 16 Comb1b){
    if (Comb1a == Master1) {
        par{Checked1 = 1; Master1 = Comb1b;}
    }
    else
        {Checked1 = 0;}
}
void wait_on_UnlockR2_fnc(void){
    while (SafeR_Bus.inp != UnlockR2){delay;}
    SafeR_Bus_var = UnlockR2;
}
void UnlockR2_fnc(unsigned 16 Comb2a, unsigned 16 Comb2b){
    if (Comb2a == Master2) {
        par{Checked2 = 1; Master2 = Comb2b;}
    }
    else
        {Checked2 = 0;}
}
```

Figure 9. SafeR Translation Part 1b.

```
unsigned 1 Unlock_fnc(void){
    par{
        if ((Checked1 = 1) & (Checked2 = 1)){
            par{Door=Unlocked; Status_Unlock=1;}
        }
        else {par{Door=Locked; Status_Unlock=0;}}
    }
    return Status_Unlock;
}
void Initialisation_fnc(void){
    Checked1 = 0;Master1  = 67;
    Checked2 = 0;Master2  = 76;Door = Unlocked; // INITIALISATION
    Status_Lock = 0;Status_Unlock = 0;         // SET OUTPUT DEFAULT
}
void CTRL_fnc(void){
    Initialisation_fnc(); wait_on_Lock_fnc();
    if (SafeR_Bus_var == Lock){Lock_fnc();}else{delay;}
}
void S_fnc(void){
    while(1){par{
                seq{wait_on_UnlockR1_fnc();
                    if (SafeR_Bus_var==UnlockR1){
                        UnlockR1_fnc(Cx1a.inp,Cx1b.inp);}
                    else
                        {delay;}
                } // seq
                seq{wait_on_UnlockR2_fnc();
                    if (SafeR_Bus_var==UnlockR2){
                        UnlockR_2fnc(Cx2a.inp,Cx2b.inp);}
                    else
                        {delay;}
                } // seq
            } // par
            Status_Unlock = Unlock_fnc();
            if (Status_Unlock){
                wait_on_Lock_fnc();
                if (SafeR_Bus_var==Lock){
                    Lock_fnc();
                }
                else {delay;}
            }
            else {delay;}
    } //while
} // S_fnc

void main(void){CTRL_fnc();S_fnc();}
```

Figure 10. SafeR Translation Part 2.

Communicating Process Architectures 2007
Alistair A. McEwan, Steve Schneider, Wilson Ifill, and Peter Welch
IOS Press, 2007

Towards the Formal Verification of a Java Processor in Event-B

Neil GRANT and Neil EVANS

AWE, Aldermaston, UK.

Abstract. Formal verification is becoming more and more important in the production of high integrity microprocessors. The general purpose formal method called Event-B is the latest incarnation of the B Method: it is a proof-based approach with a formal notation and refinement technique for modelling and verifying systems. Refinement enables implementation-level features to be proven correct with respect to an abstract specification of the system. In this paper we demonstrate an initial attempt to model and verify Sandia National Laboratories' Score processor using Event-B. The processor is an (almost complete) implementation of a Java Virtual Machine in hardware. Thus, refinement-based verification of the Score processor begins with a formal specification of Java bytecode. Traditionally, B has been directed at the formal development of software systems. The use of B in hardware verification could provide a means of developing combined software/hardware systems, i.e. codesign.

Keywords. Java processor, microcoded architecture, Event-B, refinement

Introduction

The Score processor has been designed at Sandia National Laboratories in the United States to be used as an embedded target for use with their components modelling system (called Advanced System Simulation Emulation and Test, or 'ASSET'). Now in its second generation, the processor is a hardware implementation of an almost complete Java Virtual Machine. In fact, the implementation far exceeds Sun's expectation of an embedded target. The SSP (Sandia Secure Processor) project started ten years ago, and the SSP2 (now called the Scalable Core Processor, or 'Score') is the current design. The redesign has allowed the processor architecture to be simplified, and this along with implementation efficiencies has allowed significantly more functionality. The ASSET toolset is written in Java and uses Java to describe the component behaviour; this Java code can be compiled without modification to work on the Score processor.

Currently, Sandia uses the following (informal) validation checks on the Score processor:

- ring fencing (monitoring runtime memory access) in Java to check that opcodes do not do anything outside their remit;
- internal consistency checks (by the class loader) and a tree equivalence check;
- regression testing;
- comparison tests between two independent models - one in Java and the other in the hardware description language VHDL.

The motivation for this paper is to demonstrate initial results from an ongoing collaboration between AWE and Sandia to model and verify the Score processor using an established formal method. We choose the B Method, in particular the Event-B subset, for this purpose because it is a method with exceptional tool support which incorporates a dedicated refine-

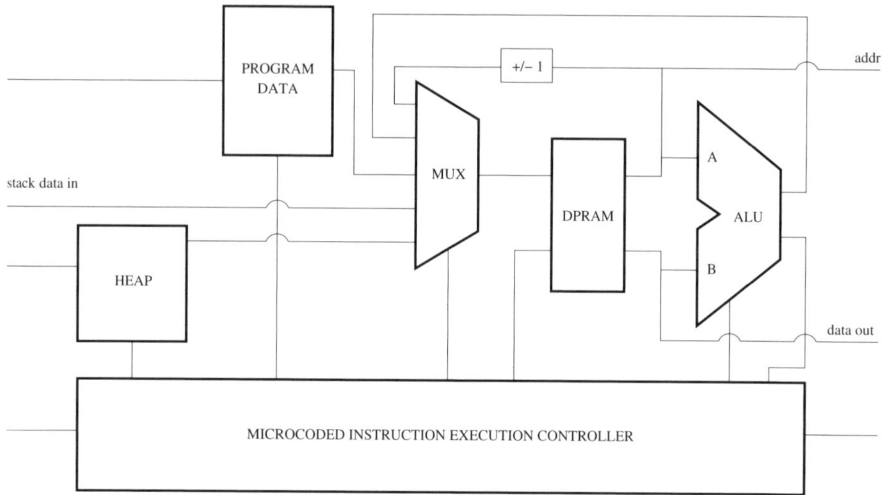

Figure 1. An Abstract Score Architecture

ment technique. We aim to prove that bytecodes are correctly implemented by microcode instructions.

Figure 1 shows a simplified architecture of the Score processor. The specific details of the architecture are not important for the purposes of this paper. When the Score gets a Java bytecode from the program memory interface it is translated into a sequence of microcode instructions from the microcode table (held in the Microcoded Instruction Execution Controller). The power and flexibility of Score comes from the use of a complex microcode table, which can be modified even after the processor has been put onto silicon. In fact, the microcode table can be tailored to contain only the required microcode. The current optimised microcode table (including all the currently supported JVM functionality) is only just over 1600 lines. The original Score processor had a hand-crafted microcode table that was impossible to maintain by anyone other than its creators. Now a systematic methodology takes a structured design for the code and compiles it into a table. Logical names replace numerical values and the microcode is built up from defined fields which are typechecked during compilation.

Sandia's approach allows customisations to be made based on required functionality or runtime requirements. The class loader can determine which bytecodes are used (and hence required) for a particular application, and all the other bytecodes can then be removed from the microcode specification. This allows the table to be reduced to a minimum if necessary. The microcode table flexibility allows the SSP structure to be used more generally than just for the JVM. Non-Java bytecode could also be interpreted on the processor, for example, to emulate another processor.

It is clear from Figure 1 that the microcode is largely responsible for the activities of the processor, although the arithmetic logic unit (ALU) is not transparent: it contains registers that are not under the control of the microcode. The program and heap memories are both 8-bit. However the JVM specification demands a 32-bit stack. The original SSP had an internal 1000 level 32-bit stack, but this was over-specified as typically only 32 levels were ever used. The stack is held in memory that is external to the processor. Within the processor, the state variable memory is a dual port RAM (DPRAM). It stores values and constants including temporary variables and values that represent the stack boundaries.

The next section gives an overview of the Event-B language and its notion of refinement. This is followed by a demonstration of our approach via an example analysis of the JVM instruction *iadd*. Our approach is then put into context with other formal approaches, after which we draw some conclusions. We also discuss how this work could fit in with another AWE-funded project to produce formally verified hardware. This would address the issue of proving correctness with respect to actual (clocked) hardware. It is hoped that the results presented here can be generalised to support the entire lifecycle of hardware development and verification. The longevity of the B Method gives us confidence that well-maintained tool support will be available in the future.

1. Event-B

An abstract Event-B specification [9] comprises a static part called the *context*, and a dynamic part called the *machine*. The machine has access to the context via a **SEES** relationship. This means that all sets, constants, and their properties defined in the context are visible to the machine. To model the dynamic aspects, the machine contains a declaration of all of the state variables. The values of the variables are set up using the **INITIALISATION** clause, and values can be changed via the execution of *events*. Ultimately, we aim to prove properties of the specification, and these properties are made explicit using the **INVARIANT** clause in the machine. The tool support generates the proof obligations which must be discharged to verify that the invariant is maintained. It also has interactive and automated theorem proving capabilities with which to discharge the generated proof obligations.

Events are specialised B operations [1]. In general, an event E is of the form

$$E \; \widehat{=} \; \textbf{WHEN } G(v) \textbf{ THEN } S(v) \textbf{ END}$$

where $G(v)$ is a Boolean guard and $S(v)$ is a generalised substitution (both of which may be dependent on one or more state variables denoted by v)[1]. The guard must hold for the substitution to be performed (otherwise the event is *blocked*). There are three kinds of generalised substitution: *deterministic*, *empty*, and *non-deterministic*. The deterministic substitution of a state variable x is an assignment of the form $x \; := \; E(v)$, for expression E (which may depend on the values of state variables — including x itself), and the empty substitution is *skip*. The non-deterministic substitution of x is defined as

$$\textbf{ANY } t \textbf{ WHERE } P(t, v) \textbf{ THEN } x := F(t, v) \textbf{ END}$$

Here, t is a local variable that is assigned non-deterministically according to the predicate P, and its value is used in the assignment to x via the expression F.

2. Refinement in Event-B

In order to express the desired properties of a system as succinctly as possible, an abstract specification will dispense with many of the implementation details in favour of a more mathematical representation. Refinement is the means by which the artefacts of an implementation can be incorporated into a formal specification whilst maintaining the correct behaviour of the abstract specification. A demonstration of Event-B refinement will be given in the next section.

---

[1]The guard is omitted if it is trivially true.

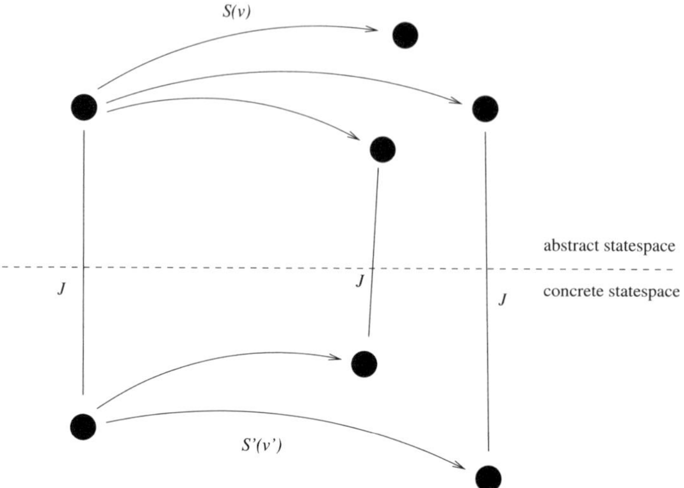

Figure 2. Refinement of an Existing Event

Traditionally, two main kinds of refinement are identified: *data refinement* and *operational refinement*. In data refinement, the aim is to replace abstract state with a more concrete, implementation-like state. Operation refinement aims to replace abstract algorithms (events) comprising abstract constructs with more program-like constructs. Operational refinement addresses the refinement of existing events. Refinement in Event-B also allows the introduction of new events. In many of his talks, Abrial gives a useful analogy for this form of refinement: an abstract specification is comparable to viewing a landscape from a great height. At this level of abstraction we get a good overview of the system without seeing many specific details. Refinement by introducing new events corresponds to moving closer to the ground: fine details that were previously out of sight are now revealed.

The context and machine of an abstract Event-B specification can be refined separately. Refinement of a context consists of adding additional sets, constants or properties (the sets, constants and properties of the abstract context are retained). The link between an abstract machine and its refinement is achieved via a *gluing invariant* defined in the concrete machine. The gluing invariant relates concrete variables to those of the abstract model. Proof obligations are generated to ensure that this invariant is maintained.

The refinement of an existing event is depicted in Figure 2. If, in a state satisfying the gluing invariant J, a concrete event with (refined) generalised substitution S' and variable v' causes a transition to a new state, then the new state is related (via J) to a new state in the abstract world (i.e. a state resulting from the abstract event with generalised substitution S with abstract variable v). Note, the multiple arrows in the diagram indicate that generalised substitutions can be non-deterministic. Also note that it is not necessary for transitions in the abstract world to correspond to transitions in the concrete world (i.e. refinement can reduce the non-determinism).

New events introduced during Event-B refinement are allowed on the proviso that they cannot diverge (i.e. execute forever). This is necessary to ensure that new events cannot take control of the machine, thereby maintaining the visibility of existing events. More formally, divergence freedom is achieved by defining a variant which strictly decreases with the execution of each internal event. Since the variant is a natural number, the execution of internal events must eventually terminate to allow the execution of one or more existing events (after which internal activity may resume).[2] Of course, the desired properties of newly introduced events can be incorporated into the gluing invariant, and a proof that these properties are maintained is required.

3. Example Bytecode: *iadd*

To illustrate our approach using Event-B, we present the arithmetic operation *iadd* which pops two numbers off the stack, adds them together, and then pushes the result on to the stack. This example presents the kind of analysis that would be undertaken for all arithmetic and logical bytecode operations because, in all cases, operands are popped off the stack and the result is pushed back onto the stack. In the interest of simplicity, we only consider the effect of the operation on the data path. For example, we do not model the program counter, nor do we consider how the instruction gets called. In addition, we assume the operands are put on the stack by other instructions that are not considered here.

We begin by specifying the behaviour of *iadd* at a level of abstraction that is independent of the microcode that implements it. This level of abstraction is called the Instruction Set Architecture level (or ISA level). First we define an Event-B context to capture the properties of a stack (of type $\mathbb{N}$). This consists of a static definition of a list and its associated functions (i.e. we define a list as an abstract datatype). This is shown in Figure 3.

In order to specify a stack, we define a deferred set *Stack* and five constants: *null* denotes the empty stack, *cons* produces a new (non-empty) stack by putting an element at the top of an existing stack, *hd* returns the top element of a stack, *tl* returns a stack minus its top element, and *len* returns the length of a stack. The behaviours of these functions are defined as axioms. Note that *hd* and *tl* are partial functions with respect to the set *Stack* because *hd*(*null*) and *tl*(*null*) are undefined. In general, it may also be necessary to define an induction axiom to prove properties of a stack. However, this is not required here.

In addition to *Stack*, we have defined two further sets: *Bytecode* and *Status*. At this stage, only one element of *Bytecode* is declared, namely *iadd*. Further elements can be added via context refinement when necessary. The set *Status* (and its two elements *ACTIVE* and *INACTIVE*) is introduced as a consequence of Event-B refinement. We will see below why this set is necessary.

Next we define the dynamic behaviour as an Event-B machine. This is shown in Figure 4. This machine has access to the static elements via the **SEES** clause. Three variables are defined: *opcode* holds the current JVM instruction, *stack* holds the current state of the stack, and *iadd_status* says whether the execution of *iadd* is in progress (i.e. *ACTIVE*) or not (i.e. *INACTIVE*). The implication statement in the machine's invariant says that there are enough elements on the stack whenever *iadd_status* is *ACTIVE*. The guards of the events guarantee this, but in the real world some other mechanism would be needed to ensure this. (It is the job of the class loader to prevent underflow of the stack.)

The variable *iadd_status* is introduced in anticipation of refinement. This is also the reason for two events: **iAdd_ini** activates the execution (but only when there are enough elements in the stack), and **iAdd** performs the necessary state update. One could imagine an event that would capture the behaviour of *iadd* in one step, i.e.:

---

[2]Since the concrete events only operate on the state variables of the refined model, this form of refinement corresponds to a normal B refinement in which the newly introduced events simply refine the abstract (empty) event *skip*.

CONTEXT *STACK*
SETS
  Stack ;
  Bytecode ;
  Status = { *ACTIVE* , *INACTIVE* }
CONSTANTS
  iadd ,
  null ,
  cons ,
  hd ,
  tl ,
  len
AXIOMS
  $iadd \in Bytecode \wedge$
  $null \in Stack \wedge$
  $cons \in \mathbb{N} \times Stack \to (Stack - \{ null \}) \wedge$
  $len \in Stack \to \mathbb{N} \wedge$
  $hd \in (Stack - \{ null \}) \to \mathbb{N} \wedge$
  $tl \in (Stack - \{ null \}) \to Stack \wedge$
  $\forall n . (n \in \mathbb{N} \Rightarrow \forall s . (s \in Stack \Rightarrow hd (cons (n , s)) = n)) \wedge$
  $\forall n . (n \in \mathbb{N} \Rightarrow \forall s . (s \in Stack \Rightarrow tl (cons (n , s)) = s)) \wedge$
  $len (null) = 0 \wedge$
  $\forall n . (n \in \mathbb{N} \Rightarrow \forall s . (s \in Stack \Rightarrow len (cons (n , s)) = 1 + len (s)))$
END

Figure 3. Abstract stack context

iAdd $\hat{=}$
 WHEN
  $len (stack) > 1 \wedge$
  $opcode = iadd$
 THEN
  $stack := cons (hd (stack) + hd (tl (stack)) , tl (tl (stack)))$
 END

without the need for a status variable. However, events with nontrivial guards and generalised substitutions such as this serve two purposes: the guard says what should hold at the beginning of an execution, and the generalised substitution says what should hold at the end. A refinement that introduces new events would force us to choose between executing the existing event first (to exercise the guard at the appropriate place), or last (to position the generalised substitution appropriately). Since Event-B does not allow events to be split, we are forced to define (at least) two events: one with the nontrivial guard, and another with the generalised substitution. We will say more about this when we consider the refinement itself.

3.1. A Refined Model

The *iadd* operation is broken down into 13 microcoded instructions on the Score processor. An in-depth understanding of the Score processor and how the microcode assembler is struc-

```
MACHINE ISA
SEES STACK
VARIABLES
  opcode ,
  iadd_status ,
  stack
INVARIANT
  opcode ∈ Bytecode ∧
  iadd_status ∈ Status ∧
  stack ∈ Stack ∧
  iadd_status = ACTIVE ⇒ len ( stack ) > 1
INITIALISATION
  opcode :∈ Bytecode  ||
  stack := null  ||
  iadd_status := INACTIVE
EVENTS

    iAdd_ini ≙
     WHEN
       iadd_status = INACTIVE ∧
       len ( stack ) > 1 ∧
       opcode = iadd
     THEN
       iadd_status := ACTIVE
     END  ;

    iAdd ≙
     WHEN
       iadd_status = ACTIVE
     THEN
       stack := cons ( hd ( stack ) + hd ( tl ( stack ) ) , tl ( tl ( stack ) ) )  ||
       iadd_status := INACTIVE
     END

END
```

Figure 4. Abstract machine for *iadd*

tured would be required to fully appreciate the instructions that are used. However, since the aim of this paper is to demonstrate Event-B refinement in this context, we simplify things by breaking the *iadd* operation into 7 *pseudo*-microcoded instructions. This description ignores some features of the processor, but it still incorporates many of the actual implementation details. This compromise allows us to demonstrate the refinement technique involved. By proving this lower-level model is an Event-B refinement of the abstract model, we demonstrate that the low-level behaviour is faithful to the ISA specification of *iadd*. Although we only present one refinement here, the approach is similar for all bytecodes.

The context of the refined model remains the same as the abstract model, so we begin by listing the variables and their associated types in the refined machine. This is shown in Fig-

<table>
<tr><td align="center">VARIABLES</td><td align="center">INVARIANT</td></tr>
<tr><td align="center">opcode1 ,</td><td align="center">opcode1 $\in$ Bytecode $\wedge$</td></tr>
<tr><td align="center">iadd_status1 ,</td><td align="center">iadd_status1 $\in$ Status $\wedge$</td></tr>
<tr><td align="center">SP ,</td><td align="center">SP $\in \mathbb{N} \wedge$</td></tr>
<tr><td align="center">stack1 ,</td><td align="center">stack1 $\in \mathbb{N}_1 \nrightarrow \mathbb{N} \wedge$</td></tr>
<tr><td align="center">stackDataIn ,</td><td align="center">stackDataIn $\in \mathbb{N} \wedge$</td></tr>
<tr><td align="center">ALURegA ,</td><td align="center">ALURegA $\in \mathbb{N} \wedge$</td></tr>
<tr><td align="center">ALURegB ,</td><td align="center">ALURegB $\in \mathbb{N} \wedge$</td></tr>
<tr><td align="center">ALUOutReg ,</td><td align="center">ALUOutReg $\in \mathbb{N} \wedge$</td></tr>
<tr><td align="center">stackDataOut ,</td><td align="center">stackDataOut $\in \mathbb{N} \wedge$</td></tr>
<tr><td align="center">stackDataInSet ,</td><td align="center">stackDataInSet $\in$ BOOL $\wedge$</td></tr>
<tr><td align="center">ALURegASet ,</td><td align="center">ALURegASet $\in$ BOOL $\wedge$</td></tr>
<tr><td align="center">ALURegBSet ,</td><td align="center">ALURegBSet $\in$ BOOL $\wedge$</td></tr>
<tr><td align="center">ALUOutSet ,</td><td align="center">ALUOutSet $\in$ BOOL $\wedge$</td></tr>
<tr><td align="center">stackDataOutSet</td><td align="center">stackDataOutSet $\in$ BOOL</td></tr>
</table>

Figure 5. Refined state variables and their types

ure 5. Three of the variables have counterparts in the abstract model: *opcode*1, *iadd_status*1 and *stack*1. Of these, *stack*1 is most interesting because it is refined by replacing the abstract datatype *Stack* with a partial function mapping (positive) natural numbers to natural numbers. This is closer to the real implementation of a stack because we can think of the domain of the function as memory addresses and the range as the contents of the memory. The variable *SP* is introduced to represent a pointer to the head of the stack.

Other variables introduced here model the registers involved in the computation: *stackDataIn* and *stackDataOut* hold values in transit from/to the stack, and *ALURegA*, *ALURegB* and *ALUOutReg* hold values entering and leaving the ALU. The remaining variables are Boolean flags that are needed to record the state of the registers. They do not correspond to actual components on the Score processor, but they are needed to guard the events so that they are called at the appropriate time.

The invariant shown in Figure 5 only gives the types of the state variables. It says nothing about the correspondence between the concrete variables and the abstract variables. We shall derive the necessary clauses in a systematic way after we have introduced the events. First we consider the refinements of the existing events. These are shown in Figure 6. The event **iAdd_ini** is almost identical to its counterpart in Figure 4, except we use the concrete variables *iadd_status*1 and *opcode*1, and the conjunct $SP > 1$ replaces $len(stack) > 1$. This will impose conditions on the gluing invariant when it is derived. In addition to an *ACTIVE* status, the guard of the refined event **iAdd** now depends on the variable *stackDataOutSet*. This is necessary to block the event until a meaningful value is ready to be pushed onto the stack (which is achieved by the assignment in the generalised substitution). Since the event completes the computation for *iadd*, the flags are reset in preparation for the next arithmetic operation.

The events introduced in this refinement are responsible for updating the state variables so that, when the event **iAdd** executes, *stack*1, *SP* and *stackDataOut* hold the correct values to fulfil the requirements of *iadd*. This happens in a number of stages, which are summarised below:

<table>
<tr><td>

iAdd_ini $\,\widehat{=}$
  WHEN
    iadd_status1 = INACTIVE $\wedge$
    SP > 1 $\wedge$
    opcode1 = iadd
  THEN
    iadd_status1 := ACTIVE
  END

</td><td>

iAdd $\,\widehat{=}$
  WHEN
    iadd_status1 = ACTIVE $\wedge$
    stackDataOutSet = TRUE
  THEN
    stack1 := stack1 $\cup$ { *SP* $\mapsto$ *stackDataOut* } $\|$
    iadd_status1 := INACTIVE $\|$
    stackDataOutSet := FALSE $\|$
    ALURegASet := FALSE $\|$
    ALURegBSet := FALSE $\|$
    ALUOutSet := FALSE
  END

</td></tr>
</table>

Figure 6. Refining the existing events

- **readStackAndDec** pops an element off the stack, decreases the stack pointer and sets *stackDataInSet* to indicate that *stackDataIn* holds a value to be added;
- **writeALURegA** takes the value stored in *stackDataIn* and passes it to the ALU register *ALURegA*, and a flag is set to indicate this;
- **writeALURegB** takes the value stored in *stackDataIn* and passes it to the ALU register *ALURegB*, and a flag is set to indicate this;
- **ALUAdd** adds the values of the two ALU registers and assigns this to *ALUOutReg*;
- **incRegAndLoadStack** assigns *ALUOutReg* to *stackDataOut* in readiness for the stack to be updated.

Note that the order of the events is implicit, and depends on the truth values of the guards. In this case, **readStackAndDec** will occur twice in every execution of *iadd* in order to assign the two input registers of the ALU. The complete set of definitions for the new events is shown in Figure 7.

3.2. Constructing a Gluing Invariant

Ultimately, our gluing invariant should relate the abstract variable *stack* with the concrete variable *stack*1 (i.e. they should be equivalent in some sense). However, before we do this it is necessary to address a number of proof obligations that arise in the refined model. These concern the guards of the events **iAdd_ini** and **iAdd**. The theory underlying the B Method dictates that the guards of refined events must be at least as strong as the guards that they refine. In the case of **iAdd_ini**, we have to prove:

$$(iadd\_status1 = INACTIVE \,\wedge\, SP > 1 \,\wedge\, opcode1 = iadd) \Rightarrow$$
$$(iadd\_status = INACTIVE \,\wedge\, len(stack) > 1 \,\wedge\, opcode = iadd)$$

and in the case of **iAdd**, we have to prove:

$$(iadd\_status1 = ACTIVE \,\wedge\, stackDataOutSet = TRUE) \Rightarrow iadd\_status = ACTIVE$$

One might be tempted to add the following clauses to the invariant:

$$(iadd\_status1 = INACTIVE \Rightarrow add\_status = INACTIVE) \,\wedge$$
$$(iadd\_status1 = ACTIVE \Rightarrow add\_status = ACTIVE) \,\wedge$$
$$SP = len(stack) \,\wedge$$

readStackAndDec $\widehat{=}$
  WHEN
    iadd_status1 = ACTIVE $\wedge$
    SP $\in$ dom (*stack1*) $\wedge$
    stackDataInSet = FALSE $\wedge$
    (*ALURegASet = FALSE* $\vee$
      ALURegBSet = FALSE)
  THEN
    SP := SP − 1 $\parallel$
    stackDataIn := stack1 (SP) $\parallel$
    stack1 := $\{$ *SP* $\}$ $\lhd$ *stack1* $\parallel$
    stackDataInSet := TRUE
  END

writeALURegB $\widehat{=}$
  WHEN
    stackDataInSet = TRUE $\wedge$
    ALURegBSet = FALSE
  THEN
    ALURegB := stackDataIn $\parallel$
    stackDataInSet := FALSE $\parallel$
    ALURegBSet := TRUE
  END

writeALURegA $\widehat{=}$
  WHEN
    stackDataInSet = TRUE $\wedge$
    ALURegASet = FALSE
  THEN
    ALURegA := stackDataIn $\parallel$
    stackDataInSet := FALSE $\parallel$
    ALURegASet := TRUE
  END

ALUAdd $\widehat{=}$
  WHEN
    ALURegASet = TRUE $\wedge$
    ALURegBSet = TRUE $\wedge$
    ALUOutSet = FALSE
  THEN
    ALUOutReg :=
      ALURegA + ALURegB $\parallel$
    ALUOutSet := TRUE
  END

incRegAndLoadStack $\widehat{=}$
  WHEN
    ALUOutSet = TRUE $\wedge$
    stackDataOutSet = FALSE
  THEN
    SP := SP + 1 $\parallel$
    stackDataOut := ALUOutReg $\parallel$
    stackDataOutSet := TRUE
  END

Figure 7. Introducing new events

$$opcode1 = opcode$$

However, the clause $SP = len(stack)$ is not invariant because, in the concrete model, the value of SP changes prior to the execution of **iAdd** whereas the length of *stack* remains the same until (the abstract) **iAdd** event is executed. This illustrates a key feature of the approach: relationships such as this are only relevant in certain states. In this case, we can weaken the clause as follows:

$$iadd\_status1 = INACTIVE \Rightarrow SP = len(stack)$$

The proof of this implication is nontrivial because, even though we are not interested in the active states, we have to analyse them in order to establish the final value of SP. Rather than demonstrating this here, we shall demonstrate a stronger property of the stack.

Our aim is to show that the concrete model captures the behaviour of the *iadd* operation. We do this with a refinement proof in Event-B. The aim, therefore, is to show that the stacks resulting from the computation in the concrete and abstract world are equivalent, on the assumption that the stacks were equivalent prior to the computation. This is depicted in Figure 8. The top half of the diagram represents the abstract world. Two stacks are shown:

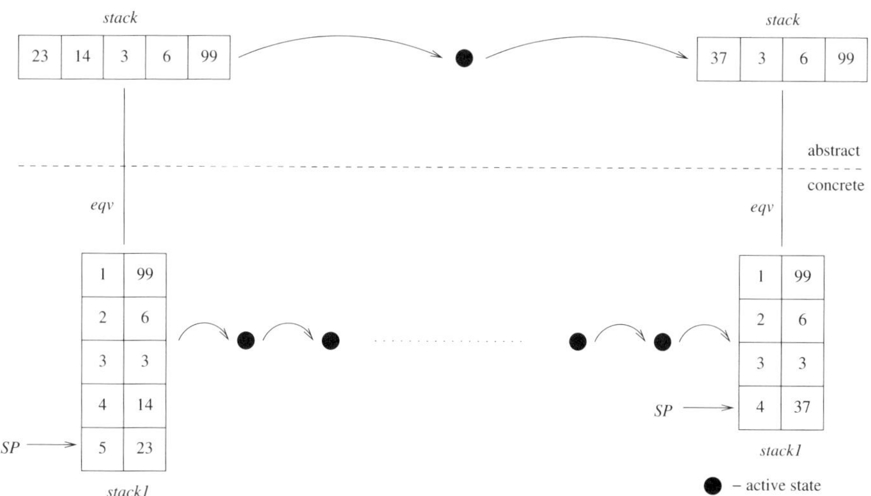

Figure 8. Relating inactive states

one prior to performing *iadd*, and one after. They are connected by two transition arrows and one intermediate (active) state. The leftmost element in both stacks is the head. The bottom half of the diagram gives a concrete representation of the same situation (in this case, the stacks are made up of index/value pairs). Here, the bottom pair is the head, and is labelled by the pointer *SP*. Note that there are more transitions and intermediate states involved in the concrete world.

We define a predicate *eqv* to capture the relationship between these two viewpoints (this is indicated by the vertical lines in the diagram). We begin by declaring the relationship for inactive states. If *eqv* is of type $\mathbb{N} \times (\mathbb{N}_1 \nrightarrow \mathbb{N}) \times Stack \rightarrow BOOL$, then this is written formally (and added to the invariant) as:

$$iadd\_status1 = INACTIVE \;\Rightarrow\; eqv(SP, stack1, stack)$$

where the (well-founded) definition of *eqv* is as follows:

$$eqv(n, s, null) \;=\; n = 0$$
$$eqv(n, s, cons(h, t)) \;=\; n > 0 \wedge n \in \mathsf{dom}\,(\,s\,) \wedge s\,(\,n\,) = h \wedge eqv(n-1, s, t)$$

When applied in the invariant, this definition ensures that the elements of a (non-*null*) abstract stack correspond to the entries in the concrete stack from index *SP* down to index 1, and *SP* is 0 when the abstract stack is empty.

Since the concrete event **iAdd** yields a state in which *iadd_status1* is *INACTIVE*, the following proof obligation is generated by the tool:

$$eqv(SP,\; stack1 \cup \{SP \mapsto stackDataOut\},\; cons(hd(stack) + hd(tl(stack)), tl(tl(stack))))$$

This is due to the generalised substitution arising in the **iAdd** event defined in the refined model:

$$
\begin{aligned}
&\textbf{iAdd} \ \hat{=} \\
&\quad \textbf{WHEN} \\
&\quad\quad \textit{iadd\_status1} = \textit{ACTIVE} \ \wedge \\
&\quad\quad \textit{stackDataOutSet} = \textit{TRUE} \\
&\quad \textbf{THEN} \\
&\quad\quad \textit{stack1} := \textit{stack1} \cup \{\, SP \mapsto \textit{stackDataOut} \,\} \ \parallel \\
&\quad\quad \textit{iadd\_status1} := \textit{INACTIVE} \ \parallel \\
&\quad\quad\quad \vdots \\
&\quad \textbf{END}
\end{aligned}
$$

and the corresponding definition in the abstract model:

$$
\begin{aligned}
&\textbf{iAdd} \ \hat{=} \\
&\quad \textbf{WHEN} \\
&\quad\quad \textit{iadd\_status} = \textit{ACTIVE} \\
&\quad \textbf{THEN} \\
&\quad\quad \textit{stack} := \textit{cons} \,(\, \textit{hd} \,(\, \textit{stack} \,) + \textit{hd} \,(\, \textit{tl} \,(\, \textit{stack} \,)\,) \,,\, \textit{tl} \,(\, \textit{tl} \,(\, \textit{stack} \,)\,)\,) \ \parallel \\
&\quad\quad \textit{iadd\_status} := \textit{INACTIVE} \\
&\quad \textbf{END}
\end{aligned}
$$

It is necessary to prove (or *discharge*) such proof obligations in order to demonstrate that the invariant is maintained. The above proof obligation is true if, prior to the **iAdd** event, the following three subgoals can be proven:

- SP is the 'next' unoccupied position in $stack1$;
- $stackDataOut = hd(stack) + hd(tl(stack))$, i.e. the sum of the first two elements of $stack$;
- $eqv(SP-1,\ stack1,\ tl(tl(stack)))$, i.e. $stack1$ is equivalent to the abstract stack minus its top two elements.

At this stage it is impossible to confirm or refute these subgoals. We have to consider the sequence of events that would have led up to the occurrence **iAdd** in the concrete model. Our approach is to augment the refinement's invariant with any proof obligations that cannot be proven. Then we use the tool to generated additional proof obligations. This process is repeated until no further proof obligations are generated. We begin by adding the above proof obligation to the invariant. However, we do this under the assumption that the guard of **iAdd** (otherwise the event would not have occurred). In particular, we assume $stackDataOutSet = TRUE$:

$$
\begin{aligned}
&stackDataOutSet = TRUE \ \Rightarrow \\
&\quad eqv(SP,\ stack1 \cup \{SP \mapsto stackDataOut\},\ cons(hd(stack) + hd(tl(stack)), tl(tl(stack))))
\end{aligned}
$$

As a consequence, the proof obligation now disappears and a new proof obligation is generated instead:

$$
eqv(SP+1, stack1 \cup \{SP+1 \mapsto ALUOutReg\},\ cons(hd(stack) + hd(tl(stack)), tl(tl(stack))))
$$

This is due to the event **incRegAndLoadStack**, because its generalised substitution is responsible for setting $stackDataOutSet$ to $TRUE$:

incRegAndLoadStack $\widehat{=}$
  WHEN
    $ALUOutSet = TRUE \land$
    $stackDataOutSet = FALSE$
  THEN
    $SP := SP + 1 \;\|$
    $stackDataOut := ALUOutReg \;\|$
    $stackDataOutSet := TRUE$
  END

By performing the substitution on the subgoals, we can derive subgoals that are sufficient to prove this newly generated proof obligation:

- $SP + 1$ is the 'next' unoccupied position in $stack1$;
- $ALUOutReg = hd(stack) + hd(tl(stack))$;
- $eqv((SP + 1) - 1, \; stack1, \; tl(tl(stack)))$.

Of course, the last proof goal simplifies to:

- $eqv(SP, \; stack1, \; tl(tl(stack)))$.

Once again, we cannot confirm or refute these, so we add the generated proof obligation to the invariant. This time, we assume that the guard of **incRegAndLoadStack** holds:

$$ALUOutSet = TRUE \;\land\; stackDataOutSet = FALSE \;\Rightarrow$$
$$eqv(SP+1, stack1 \cup \{SP+1 \mapsto ALUOutReg\}, \; cons(hd(stack)+hd(tl(stack)), tl(tl(stack))))$$

As before, the proof obligation is now replaced by a new proof obligation, this time arising from **ALUAdd** (because this event assigns $TRUE$ to $ALUOutSet$):

$$eqv(SP + 1, stack1 \cup \{SP + 1 \mapsto ALURegA + ALURegB\},$$
$$cons(hd(stack) + hd(tl(stack)), tl(tl(stack))))$$

This proof obligation differs from the previous one because $ALUOutReg$ is assigned to be the sum of $ALURegA$ and $ALURegB$:

    ALUAdd $\widehat{=}$
      WHEN
        $ALURegASet = TRUE \land$
        $ALURegBSet = TRUE \land$
        $ALUOutSet = FALSE$
      THEN
        $ALUOutReg := ALURegA + ALURegB \;\|$
        $ALUOutSet := TRUE$
      END

The second of the three subgoals is affected by the generalised substitution in this event:

- $ALURegA + ALURegB = hd(stack) + hd(tl(stack))$,

We are required to look further to discover the values assigned to *ALURegA* and *ALURegB*. First we add the proof obligation to the invariant:

$$ALURegASet = TRUE \ \wedge \ ALURegBSet = TRUE \ \wedge \ ALUOutSet = FALSE \ \Rightarrow$$
$$eqv(SP + 1, stack1 \cup \{SP + 1 \mapsto ALURegA + ALURegB\},$$
$$cons(hd(stack) + hd(tl(stack)), tl(tl(stack))))$$

This situation is a bit more interesting because there are two possible paths that could reach a state in which *ALURegASet* and *ALURegBSet* are true: (i) if *ALURegASet* is true and an occurrence of **writeALURegB** sets *ALURegBSet* to *TRUE*; (ii) if *ALURegBSet* is true and an occurrence of **writeALURegA** sets *ALURegASet* to *TRUE*.

<table>
<tr><td>

writeALURegA $\widehat{=}$
  WHEN
    stackDataInSet = TRUE $\wedge$
    ALURegASet = FALSE
  THEN
    ALURegA := stackDataIn $\|$
    stackDataInSet := FALSE $\|$
    ALURegASet := TRUE
  END

</td><td>

writeALURegB $\widehat{=}$
  WHEN
    stackDataInSet = TRUE $\wedge$
    ALURegBSet = FALSE
  THEN
    ALURegB := stackDataIn $\|$
    stackDataInSet := FALSE $\|$
    ALURegBSet := TRUE
  END

</td></tr>
</table>

Hence, two proof obligations are generated, which we add to the invariant with the appropriate assumptions:

(i) *stackDataInSet = TRUE* $\wedge$ *ALURegASet = TRUE* $\Rightarrow$
$$eqv(SP + 1, stack1 \cup \{SP + 1 \mapsto ALURegA + stackDataIn\},$$
$$cons(hd(stack) + hd(tl(stack)), tl(tl(stack))))$$

which has a further impact on the second of the three subgoals:

- $ALURegA + stackDataIn = hd(stack) + hd(tl(stack)),$

(ii) *stackDataInSet = TRUE* $\wedge$ *ALURegBSet = TRUE* $\Rightarrow$
$$eqv(SP + 1, stack1 \cup \{SP + 1 \mapsto stackDataIn + ALURegB\},$$
$$cons(hd(stack) + hd(tl(stack)), tl(tl(stack))))$$

which also has an impact on the second of the three subgoals:

- $stackDataIn + ALURegB = hd(stack) + hd(tl(stack)).$

Note, the assumption *stackDataInSet = TRUE* in (i) implies that *ALURegBSet* is *FALSE* (which is required to enable the guard of **writeALURegB**). Similarly, the assumption *stackDataInSet = TRUE* in (ii) implies that *ALURegASet* is *FALSE*.

    The addition of the above implications to the invariant forces us to consider the behaviour of **readStackAndDec** which precedes the occurrences of both **writeALURegA** and **writeALURegB**. Every time this event occurs, it assigns the top element of the (concrete) stack to *stackDataIn* and decrements the stack pointer *SP*:

> **readStackAndDec** $\hat{=}$
>   **WHEN**
>   *iadd_status1 = ACTIVE* $\wedge$
>   *SP* $\in$ dom (*stack1*) $\wedge$
>   *stackDataInSet = FALSE* $\wedge$
>   (*ALURegASet = FALSE* $\vee$
>   *ALURegBSet = FALSE*)
>   **THEN**
>   *SP* := *SP* $-$ *1* $\parallel$
>   *stackDataIn* := *stack1* (*SP*) $\parallel$
>   *stack1* := $\{$ *SP* $\}$ $\lhd$ *stack1* $\parallel$
>   *stackDataInSet* := *TRUE*
>   **END**

This event occurs twice during each execution of the (concrete) *iadd* operation, so multiple cases have to be considered:

1. *stackDataInSet = FALSE* $\wedge$ *ALURegASet = TRUE* $\wedge$ *ALURegBSet = FALSE*.
   In this state, **readStackAndDec** is enabled to assign a value to *ALURegB*. It is sufficient to discharge the proof obligation generated in this state if we can prove the following subgoals:

   * $(SP - 1) + 1$ is the 'next' unoccupied position in $\{SP\} \lhd stack1$;
   * $ALURegA + stack1(SP) = hd(stack) + hd(tl(stack))$;
   * $eqv(SP - 1,\ \{SP\} \lhd stack1,\ tl(tl(stack)))$.

   The first of these proof goals simplifies to:

   * SP is the 'next' unoccupied position in $\{SP\} \lhd stack1$.

2. *stackDataInSet = FALSE* $\wedge$ *ALURegASet = FALSE* $\wedge$ *ALURegBSet = TRUE*.
   In this state, **readStackAndDec** is enabled to assign a value to *ALURegA*. It is sufficient to discharge the proof obligation generated in this state if we can prove the following subgoals:

   * SP is the 'next' unoccupied position in $\{SP\} \lhd stack1$;
   * $stack1(SP) + ALURegB = hd(stack) + hd(tl(stack))$;
   * $eqv(SP - 1,\ \{SP\} \lhd stack1,\ tl(tl(stack)))$.

3. *stackDataInSet = FALSE* $\wedge$ *ALURegASet = FALSE* $\wedge$ *ALURegBSet = FALSE*.
   In this state, **readStackAndDec** is enabled to assign a value (non-deterministically) to either *ALURegA* or *ALURegB*. It is sufficient to discharge the proof obligation generated in this state if we can prove the following subgoals:

   * $SP - 1$ is the 'next' unoccupied position in $\{SP\} \lhd (\{SP - 1\} \lhd stack1)$;
   * $stack1(SP) + stack1(SP - 1) = hd(stack) + hd(tl(stack))$;
   * $eqv((SP - 1) - 1,\ \{SP\} \lhd (\{SP - 1\} \lhd stack1),\ tl(tl(stack)))$.

   The third subgoal can be simplified to:

   * $eqv(SP - 2,\ \{SP\} \lhd (\{SP - 1\} \lhd stack1),\ tl(tl(stack)))$.

Finally we are at a point where we can complete the proof. If we add the following clause to the invariant, then we can prove all of the subgoals:

$$stackDataInSet = FALSE \wedge ALURegASet = FALSE \wedge ALURegBSet = FALSE \Rightarrow$$
$$eqv(SP, stack1, stack)$$

For example, if $eqv(SP, stack1, stack)$ is true then removing two elements from both stacks results in equivalent stacks, i.e. $eqv(SP - 2 \ \{SP\} \lhd (\{SP - 1\} \lhd stack1), \ tl(tl(stack))))$, which confirms the third subgoal. Note that, by adding the final implication to the invariant, we have in fact weakened our original gluing invariant:

$$iadd\_status1 = INACTIVE \ \Rightarrow \ eqv(SP, stack1, stack)$$

In this analysis we have used the tool to generate the invariant for us. In discharging the proof obligations, all but two of the proof obligations were proven automatically. This kind of approach is not easy to follow when written down (even when it's simplified, as in the description above) so it is not very practical for hand-written proofs. However, the tool support that accompanies Event-B keeps track of all outstanding proof obligations, and provides an easy user interface and theorem proving support for interactive proofs.

3.3. Other Issues

One outstanding issue concerns parameters: the Score processor's microcoded instruction set includes instructions that take input parameters. Unlike operations in the B Method, events in Event-B do not allow input parameters. Instead, the **ANY** clause introduced in Section 1 can be used to model instruction parameters. In terms of proof, a non-deterministic substitution of the form:

$$\textbf{ANY } t \ \textbf{WHERE} \ \cdots$$

will typically generate proof obligations of the form:

$$\forall t. \ \cdots$$

That is, a proof must consider all possible instantiations of the parameters that are modelled by the local variable t.

4. Other Approaches

The most substantial body of work in this area to date has been done by Panagiotis Manolios. His technique for modelling and verifying hardware motivated the investigation undertaken in this paper. The general purpose theorem proving system ACL2 [2] provides the mechanical support for his approach. Lisp is used as the modelling language, in which models of a similar level of abstraction to our own are constructed. In particular, instruction set architecture (ISA) models and microarchitecture (MA) models are defined using Lisp primitives [7].

In order to prove a correspondence between an ISA model and an MA model, a *refinement map* from MA states to ISA states is constructed which, in essence, says how to view an MA state as an ISA state. Typically, the map will 'forget' some of the details of the MA state in order to recover a corresponding ISA state. If, using this mapping, it is possible to derive a well-founded equivalence bisimulation relation (see [7]) then the models can be seen to be equivalent. Note that this (equivalence) notion of refinement differs from that of Event-B because, in the latter case, the behaviours of the concrete model should be more constrained (or less non-deterministic) than the abstract model. However, there is a similarity between the two approaches because this notion of bisimulation only allows finite stuttering. This corresponds to Event-B's notion of divergence freedom: events introduced in a refinement (i.e. those events that are hidden at the abstract level) cannot take infinite control. Otherwise, this would correspond to infinite internal activity (i.e. infinite stuttering) at the abstract level.

To overcome the difficulties associated with using automated theorem provers (in particular, the level of interaction), Manolios has enlisted the help of the UCLID tool [12] which makes use of SAT solving technology and BDD's to prove refinements between models [8]. In a similar way, users of Event-B can call upon the model checking tool ProB [6] to provide more automated tool assistance in the development of Event-B models. It has the capability to animate specifications, analyse invariants, and check refinements of finite state models.

5. Conclusion

In this paper we have applied Event-B refinement to the verification of a Java processor. In particular, we have demonstrated a proof of an example bytecode with respect to its microcoded instruction implementation. We have chosen to use Event-B in this investigation because it has an off-the-shelf (and free) formal development tool with a dedicated refinement technique. Hence, our proposed approach has been tailored to make full use of the tool.

Of course, the process of verification must be repeated for each bytecode but, since the microcoded instructions will be used repeatedly, existing invariants (such as those derived in Section 3.2) can be reused in different contexts. Hence, we can expect subsequent proofs to be less time consuming.

The Event-B tool is being developed with extensibility in mind. The decision to use Eclipse as an environment for the tool is based on its plug-in capability. For example, in addition to ProB, other tools such as a UML to B translator [11] are being built to interact directly with the Event-B tool. This will provide alternative 'front ends' to the tool to enable formal development via other more familiar notations. Hence, it is likely that tools such as these will play a part in the development of future hardware projects rather than for post hoc verification.

AWE has been involved in using formal methods in hardware development for the last 15 years, and is keen to investigate formal techniques to make the production of rigorous hardware achievable. For instance, the development of computerised control systems requires verified hardware. Since no commercial processors have been available to meet this requirement, in-house hardware has been developed. Some early work, in collaboration with B-Core (UK), added hardware component libraries and a VHDL hardware description language code generator to the B Toolkit [5]. All hardware specifications written using this approach (called B-VHDL) mimic the structure of traditional VHDL programs and, hence, give a very low-level view of a development. The work presented in this paper investigates the applicability of the latest B technologies at a higher level of abstraction. Currently, a collaboration between AWE and the University of Surrey is investigating routes from high-level specifications (such as those presented in this paper) down to clocked physical hardware. In addition to B, other formal notations such as CSP are being used to specify and refine combined software/hardware models.

Acknowledgements

The authors thank the anonymous referees for their insightful comments.

References

[1] Abrial J. R.: *The B Book: Assigning Programs to Meanings*, Cambridge University Press (1996).

[2] ACL2, `http://www.cs.utexas.edu/users/moore/acl2/`.

[3] Atelier B, `http://www.atelierb.societe.com`.

[4] B Core (UK) Ltd, `http://www.b-core.com`.

[5] Ifill W., Sorensen I., Schneider S.: *The use of B to Specify, Design and Verify Hardware*. In *High Integrity Software*, Kluwer Academic Publishers, 2001.

[6] Leuschel M., Butler M.: *ProB: A Model Checker for B*, FME 2003: Formal Methods, LNCS 2805, Springer, 2003.

[7] Manolios P.: *Refinement and Theorem Proving*, International School on Formal Methods for the Design of Computer, Communication, and Software Systems: Hardware Verification, Springer, 2006.

[8] Manolios P., Srinivasan S.: *A Complete Compositional Framework for the Efficient Verification of Pipelined Machines*, ACM-IEEE International Conference on Computer Aided Design, 2005.

[9] Métayer C., Abrial J. R., Voisin L.: *Event-B Language*, RODIN deliverable 3.2, `http://rodin.cs.ncl.ac.uk` (2005).

[10] Schneider S.: *The B Method: An Introduction*, Palgrave (2001).

[11] Snook C., Butler M.: *UML-B: Formal Modeling and Design Aided by UML*, ACM Transactions on Software Engineering and Methodology (TOSEM), Volume 15, Issue 1, 2006.

[12] UCLID, `http://www.cs.cmu.edu/~uclid/`.

Communicating Process Architectures 2007
Alistair McEwan, Steve Schneider, Wilson Ifill and Peter Welch (Eds.)
IOS Press, 2007

Advanced System Simulation, Emulation and Test (ASSET)

Gregory L. WICKSTROM
Sandia National Laboratories, Albuquerque NM, USA
`glwicks@sandia.gov`

Abstract. Maturing embeddable real-time concepts into deployable high consequence systems faces numerous challenges. Although overcoming these challenges can be aided by commercially available processes, toolsets, and components, they often fall short of meeting the needs at hand. This paper will review the development of a framework being assembled to address many of the shortcomings while attempting to leverage commercial capabilities as appropriate.

Keywords. System simulation, system test, real-time.

Introduction

The needs for component development at Sandia National Laboratories vary widely between applications that have effectively no resource constraints to those that are highly constrained. The type of constraints also varies widely between applications that include but are not limited to: power consumption, volume, weight, timing, emissions, implementation technologies, safety, security, reliability, as well as physical environments such as temperature, and radiation. For those applications whose constraints are less restrictive, such as those for an office or many industrial settings, numerous development frameworks exist commercially to aid in streamlining the development process. However, for applications with relatively harsh constraints, commercial solutions often have gaps in the development processes and capabilities that prevent their use or limit their value.

This paper will describe a framework under development to streamline the development process of those more constrained applications. In general, the primary focus will be for electronically-based systems with some amount of decision making logic in the system. However, some of the framework has a wider applicability.

Section 1 of this paper will first summarize some of the problem areas for constrained systems. Section 2 supplies an executive overview of the ASSET development framework with a summary of each of its elements and their corresponding function. The remaining sections give more detail of the ASSET elements, ending with a summary and discussion of future work.

Note that the work presented in this paper was required to support the development of a product to be deployed under rigorous schedules and limited personnel resources, then subsequently reused on future projects. A relatively significant amount of time and effort was spent early on investigating development technologies and tools, but no single tool or combined toolsets offered the needed productivity improvements. As a result, we set out to develop what was needed and this paper is intended to describe the resulting development infrastructure. As of yet, we have not had the resources to compare this framework with others that may have been simultaneously developed.

1. Problem Descriptions

A large number of issues must be addressed during development of systems of interacting components. This is equally true for the development of the components themselves. This section touches on some of the development problems and why commercial solutions may not be sufficient as an aid in their solution.

1.1 Real-Time Systems

Correct real-time software requires not only that system functions are correct, the timing of those functions must be correct as well. Hardware-based processor interrupts help to address this additional dimension in requirements by allowing a non-linear flow of program execution. They enable the ability to handle asynchronous input events or periodically timed functions with hardware-enforced timing accuracy. However, correctly managing interrupts introduces additional complexities and likelihood of errors. Real-Time Operating Systems (RTOS) abstract the use of interrupts to handle system timing issues, relieving much of the burden of time management from the programmer. These technologies are highly invaluable in addressing the bulk of real-time system needs, but may be prohibited by some safety and/or security related applications. For our applications the use of an RTOS is prohibited because we are required to perform extensive analysis of all deployed code, which would in turn require access to proprietary commercial source code. Should the source code for an RTOS become available, the detailed analysis of such a system was estimated to be too costly. Our applications were also prohibited from using interrupts by an external certification agency due to system analysis complexity.

1.2 HW/SW Co-simulation

Embedded systems have an additional problem of defining, implementing, and testing the interface between control software and the hardware to which it is connected, i.e. control and status registers associated with digital control hardware. Commercial emulators exist to help debug problem areas, but the drawback is that one must face the potentially expensive prospect of first building the hardware that will be connected to the emulator. Once testing begins, the hardware is likely to be changed because all of its requirements were not yet known. This chicken-and-egg problem can be alleviated by simulating both the hardware and the software with numerous forms of Hardware Description Languages (HDL). Using such capabilities one can use an instruction set simulator that simulates the execution of the controlling software interacting with a model of the hardware. However, the cost of instruction set simulators can be prohibitively high[1], and are generally too slow to be used for anything other than software driver development for the hardware. Testing higher level application software remains a difficult problem since it expects to exchange data with an arbitrary number of other system components. This problem may be addressed with behavioral stubs to those components, but this creates its own problems as the development of those stubs is generally asymmetric with the rest of the development process.

1.3 Hardware Debugging

At some point the hardware and software are mature enough to be integrated and tested in the physical world. Again, this may aided by the use of a hardware emulator. For real-time systems this presents a problem in that a user may halt the emulator at any time to query

---

[1] One seat of an instruction set simulator investigated for use with our VisualHDL simulator cost $40,000US in 1998.

internal state values, but the remaining part of the system keeps on running. This is likely to prevent the user from continuing from the breakpoint and have the system behave nominally. If a hardware bug is detected in a part of the system that is not emulated, the visibility into that part of the system is extremely limited, and the problem has to be inferred from whatever evidence can be collected from its interface pins. Again the instruction set simulator may be used with the HDL to determine the problem, but that solution remains highly problem dependent and is dependent on the fidelity of the other component simulation stubs.

1.4 Demonstration Concepts May Not Be Deployable For Constrained Systems

An ideal development system would allow a user to do full hardware/software co-simulation and do it fast enough to allow for software development, and then deploy the software to the target hardware environment. This would require not only accurate models of the hardware but accurate models of the rest of the system as well. Commercial solutions with such capabilities do exist and are highly effective, but they generally require the use of real-time operating systems and/or proprietary digital logic and/or require the target system to be deployed in some commercial standard form[2]. However, the necessary support infrastructure may not meet the necessary deployment requirements with respect to power, volume, operating environments, etc.

1.5 Moving From Models to Hardware

Commercial modeling tools allow engineers to learn about their system and discover problems early. They may also allow component simulation to assure interface requirements are well understood and defined. Once the system is well understood, often a separate set of engineers with different skill sets are challenged with building the specified system using a largely separate set of tools tuned for the target technology. Unfortunately, the potential to detect errors in the translation from models to hardware is limited since the testing infrastructure is fundamentally different between the two implementation paradigms.

1.6 Model / Implementation Equivalence Checking

The equivalence between the model of a component and its hardware realization is of key importance to model-based engineering. Although commercial tools may assist in the role of equivalence verification and validation they generally limited to performing their analysis within the same or similar toolsets. For example, property checking of models represented at different levels of fidelity may contribute to confidence of functional equivalence between the two. However, these comparisons are made more difficult when toolsets differ between the high level model of a system and its low level representation. For example, if a system is modeled with SysML or Simulink and then implemented in VHDL it's often difficult to identify properties with identical semantics between the two representations.

---

[2] National Instruments offers powerful modelling and deployment capabilities but to deploy the resulting design requires their real-time OS and/or FPGA (Field Programmable Gate Array) within the target system, and generally the systems target CompactPCI-based systems. Simulink offers similar capabilities but requires the target platform to host a real-time OS.

1.7 System Testing Issues

Product development is only one part deploying a product. For each of the system components, a tester is developed to verify it meets it requirements before being placed into the next higher level assembly. Once a full system is assembled that too must be tested before deployment. Finally, inevitably product will come back from the field and need to be tested in order to be repaired. Each of the described testers has its own development cycle and associated development issues. Also it is not uncommon for each tester developer to have different testing architectures and infrastructures even though there may be more similarities than dissimilarities in the data they are collecting and analyzing. These inefficiencies heavily contribute to the overall testing costs which can easily surpass those associated with developing the system under test. Improvements in streamlining the development of testers may significantly reduce the overall system development costs.

1.8 Summary

The problems of developing and deploying systems are numerous. The commercial market recognizes this fact and offers solutions that make a profit while simultaneously profiting the customer. However, the profit motive drives commercial suppliers to target high volume users, and much of that work has relatively benign operating constraints. So while the tools offered may benefit the masses, they often do not aid developers of more constrained systems. Although academic research often works on unique problems not addressed by industry it is usually focused on solving difficult abstract problems, and not how to merge the resulting solutions with the remainder of a more concrete development cycle.

2. ASSET Development Framework

The ASSET development framework has a number of focus areas geared for streamlining the development of highly constrained components and systems. A high level view of this framework is illustrated below in figure 1.

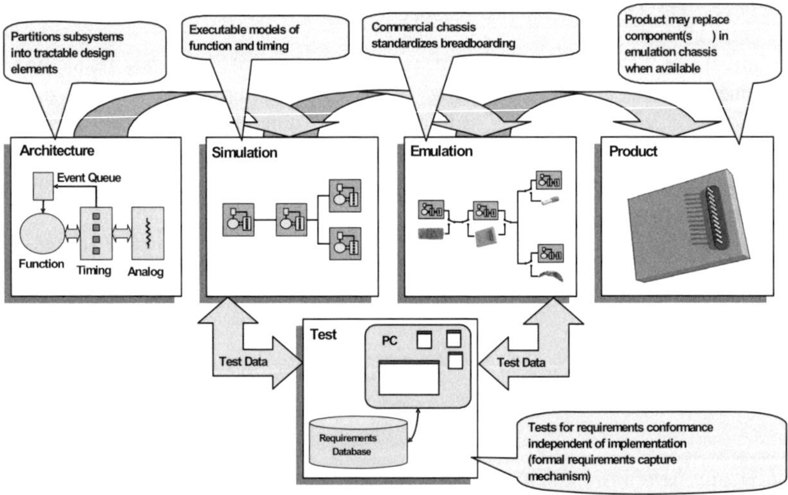

Figure 1: the ASSET Development Framework

A summary of this illustration begins with the notion of developing reusable architectures for component specification and implementation. A simulator may then be able to model many instances of those architectures all interacting simultaneously. The simulation capability is such that the modeled fidelity of each component can be increased such that it is a relatively seamless transition to hardware emulation prototypes of the components and system. The emulation environment is architected such that when final product components become available, they can be swapped with their corresponding emulated component in the system. Finally, since it is expected that the simulated system behaves identically to the hardware system, it is theoretically possible to develop a test infrastructure that is abstracted from the implementation of the system that is it testing, and effectively become an automated equivalence checker between the two implementations.

It is the *Architecture* defined in the leftmost block that facilitates much of the seamless integration between the remaining blocks in the system. The event driven nature of the *Architecture* enabled the development of a very simple *Simulation* tool, and the easy translation to hardware-based *Emulation* capabilities. Of course the ultimate implementation of a component may not implement the *Architecture* as shown, e.g. electromechanical components, but it remains valuable in terms of requirements development, *Simulation*, and *Emulation*.

3. Reusable Architectures

The architecture portion of ASSET attempts to identify common needs between component development efforts and create reusable architectures to address those common needs. To date, two have been identified and are discussed below.

3.1 Small Embedded Event Driven (SEED) Systems

The left-most block of Figure 1 referred to as the SEED architecture illustrates that any embedded system must perform some functions, perform those functions with some set of timing requirements, and somehow interface with other system elements.

The bulk of component requirements are satisfied by the *Function* and *Timing* elements. For our applications, the *Timing* blocks are implemented in digital electronics, since synchronous digital designs have deterministic timing characteristics and can react quickly to input conditions. This block effectively queues and handles Input/Output (I/O) requests which serve to insolate real-time requirements from the *Function* block. Since the behavior of a system tends to have late changing requirements that often encompasses complex data associations and decision making logic, the *Function* is typically implemented in software. As such, that block represents all elements required to support a general computing function. However, should the behavioral requirements be relatively simple and stable, that block may be implemented in custom digital logic.

The Event Queue is a hardware First-In-First-Out (FIFO) collection of events generated by the I/O hardware that must be read and handled by the *Function* block. This eliminates the need for interrupts and dictates the *Functional* block be represented as a state machine.

The *Analog* element of the architecture may be thought of as a translator between the digital subsystem and the analog world with which it connects.

The theory of operation of a SEED-based system wishing to output data to other system elements is:

1. The *Functional* block computes data to be sent.
2. The *Functional* block sends data to the *Timing* block and configures the logic to send the data with some timing characteristics at some specified time in the future.
3. The *Timing* block hardware stores the data and invokes the sending operation at the specified time with the specified timing characteristics.
4. The *Analog* block converts the digital signals to whatever medium required at the physical interface.

For asynchronous input data from other system elements:

1. The *Analog* block converts the physical communication medium to digital signals for the *Timing* block.
2. The *Timing* block demodulates the timing characteristics and stores the data as necessary.
3. The *Timing* block notifies the *Function* block to retrieve and process the stored data by posting an event into the *Event Queue*.
4. The *Function* block reads the event from the *Event Queue*, which instructs it to process data from a specific I/O module.
5. During the processing of that event the *Function* block may interact with other I/O elements as necessary.

Note that this architecture closely parallels the I/O subsystem that exists in most computer systems today, only is scaled to very small systems (often within a single integrated circuit). The primary difference is that asynchronous communication from external systems to the controlling software is generally done through the use of pre-emptive interrupts from the hardware I/O block. The SEED architecture instead uses the *Event Queue* to eliminate the need for interrupts. Unfortunately, this implementation precludes the powerful ability for pre-emption of the *Functional* software, which could introduce unacceptable latency in the processing of events. For the systems we have built, this limitation was mitigated by incorporating additional functionality within the I/O hardware.

In short, the organization of the SEED architecture is such that the digital I/O timing logic offloads the real-time responsibilities from the software, eliminating the need for an RTOS or the use of interrupts. A more detailed description of the SEED architecture may be found in [1].

3.2 The Score Processor

The SEED concept has been used to develop and deploy numerous components and systems with various processors serving as the *Function* element. It has proven useful as a tool to eliminate the need for a RTOS and interrupts. However, the general problem of writing correct software for the *Function* block remained. To reduce the likelihood of errors by developing with more traditional languages like assembly, C, or C++, we opted to use Java for its security and safety properties. Since Java is based on a Virtual Machine (VM) that consumes more memory that is typically available in our systems ($\approx$ 32K ROM, $\approx$ 8K RAM), a hardware Java processor is required. No commercial solution existed that met our harsh physical requirements with respect to temperature and radiation, so a decision was made to develop a custom Java processor called the Scaleable core (Score). The majority of commercial Java processors are a hybrid of existing processor cores with extensions to an existing instruction set to aid in the implementation of a Java Virtual Machine (JVM). With these implementations, supporting tools invoke the new instructions when applicable, and trap to libraries in the native instruction set to support bytecodes not directly supported. The Score has no notion of a native instruction set and was targeted at supporting the semantics of the Java bytecodes directly in the hardware microcode. The result was a Micro-Coded Architecture (MCA) whose core is a simple state machine where functionality is defined in micro-code. This reusable architecture has been scaled up to implement nearly all of the JVM specification. It has also been scaled down and used as the core controller for simple serial protocol communication engines.

In summary, the SEED architecture is used to appropriately partition component implementations into *Function*, *Timing*, and *Analog* elements. The MCA is an architecture that may be leveraged into the *Function* portion of the SEED architecture.

4. Orchestra Simulator

Orchestra is a custom developed discrete event simulator developed by Sandia National Laboratories. It was originally designed to run simulations of systems whose components leveraged the SEED architecture and was later expanded to support non-SEED based simulations as well. Java was used as the simulator's implementation language as well as the modeling language used to describe component behavior.

4.1 Motivation

HDL simulators offer the ability to do relatively abstract simulation through the use of instruction set simulators, and then refine the level of detail needed to automatically generate gate level representations of those processors that can execute software. However, since HDL toolsets are focused on hardware design they fail to provide a useable software development environment. Higher level commercial system simulation tools offer a capability to develop system level concepts and even form the basis for front-end software development, but they don't effectively enable development of a hardware/software interface, nor do they tend to handle the notion of time. What is desired is a hardware/software co-simulation ability that's abstract enough to do application level development and use existing software development tools, but also addresses low-level hardware interface issues.

4.2 Theory of Operation

Figure 2 illustrates the basic elements and relationships required to discuss the theory of operation for Orchestra. At its root, the Orchestra *Simulator* simply manages the interface to a queue of simulation *Events* that are sorted by time. *Simulation Modules* represent those elements in the system being modeled and have an arbitrary number of *Ports* used to communicate with other *Simulation Modules* through *Connections*. Each *Connection* provides interconnectivity to an arbitrary number of *Ports*. *Simulation Modules* may communicate with one another by asserting values on or getting values from their associated *Ports*. The values being passed on the *Connections* may take many forms, including discrete values or arbitrary objects. A duration of communication time may be specified when a value is asserted on a *Port*, in which case the receiving *Simulation Module* will be notified at the end of the specified time.

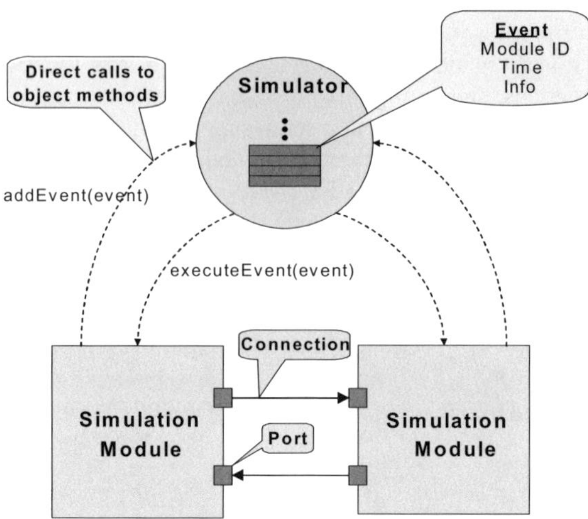

Figure 2: an Orchestra Simulation

At its root, the Orchestra *Simulator* simply manages the interface to a queue of simulation *Events* that are sorted by time. *Simulation Modules* represent those elements in the system being modeled and have an arbitrary number of *Ports* used to communicate with other *Simulation Modules* through *Connections*. Each *Connection* provides inter-connectivity to an arbitrary number of *Ports*. *Simulation Modules* may communicate with one another by asserting values on or getting values from their associated *Ports*. The values being passed on the *Connections* may take many forms, including discrete values or arbitrary objects. A duration of communication time may be specified when a value is asserted on a *Port*, in which case the receiving *Simulation Module* will be notified at the end of the specified time.

Simulation starts when the *Simulator* informs each *Simulation Module* to begin execution. During this process, one or more *Events* will be placed into the time sorted queue. Once all modules have been started the first *Event* will be pulled from the queue, a global time variable will be updated and the module that posted the *Event* will be called back with that *Event* as an argument. The *Simulation Module* will then process that *Event* to completion and return back to the *Simulator*, which will pull the next *Event* from the queue and start the process over again. Note that during as the *Simulation Module* processes of each of the *Events,* it is likely that more *Events* will be generated and placed into the queue. When the queue is empty the simulation is complete.

To provide a capability to model the passage of time within the execution of any *Simulation Module*, the *Simulation Module* may request any number of callbacks from the S*imulator* at any time in the future. The *Event* object used in these transactions may also store an object that may be used for any purposes defined by the *Simulation Module*. Often these objects specify what the *Simulation Module* is to do when the *Event* is received.

Communication between *Simulation Modules* is initiated by the assertion of a value on a *Port*, and the *Port* in turn asserts the value on its associated *Connection*. The *Connection* then asserts the value on each of the *Ports* stored within its connection list. In the case that a communication time was specified by the *Simulation Module* while asserting the value, the *Connection* will interact with the *Simulator* to delay the calling of *Ports* stored within its list.

4.3 A User's Perspective of Writing Models

Since Orchestra provides the operating infrastructure, the author of a model must simply implement the behaviors of each of the *Simulation Modules*. In its simplest form the modeler must implement following four functions as described below:

1. *initialize()* will be called by the *Simulator* when the system simulation is to begin.
2. *portChanged(Port p)* is called whenever a *Port* value has been changed by its associated *Connection*.
3. *executeEvent(Event e)* is called by the *Simulator* whenever the time for a previously posted *Event* has been reached. The argument is simply that *Event* which was passed to the *Simulator* during its posting.

Figure 3 illustrates an example of a simple system comprising of two interacting *Simulation Modules*.

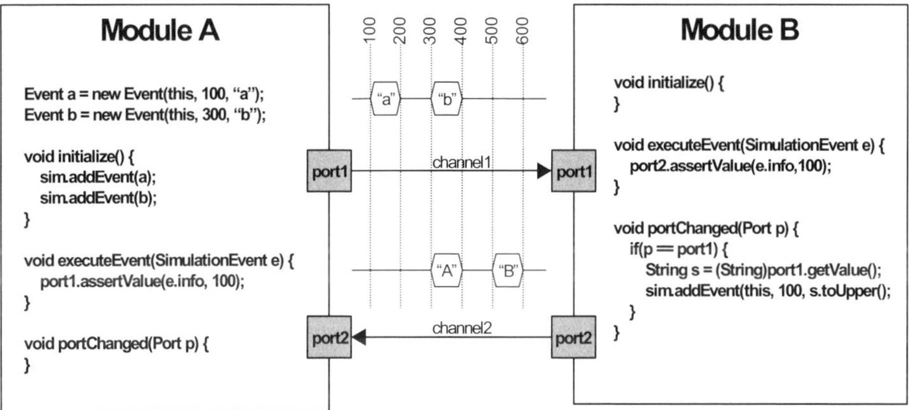

Figure 3: Simulation Model Examples

In this system model *ModuleA* sends two String objects of lower case letters to *ModuleB* which converts them to upper case and sends them back. Each message consumes 100 time units for the communication transaction and each message on a connection is separated by 100 time units as well.

4.4 Refining Models to Deployable Systems

As illustrated in Figure 3, component models may first be started at a high level of abstraction, where the model is comprised of one or more state machines to handle events coming in through the *executeEvent()* or *portChanged()* methods. This is illustrated abstractly in Figure 4.

The *Simulation Module* implementation as described above may take on any abstract implementation that meets the required interface protocol to other components. That implementation may be refined to one that separates the functional requirements from the timing requirements as encouraged by the SEED architecture. An example of that refinement process is illustrated in Figures 5 and 6.

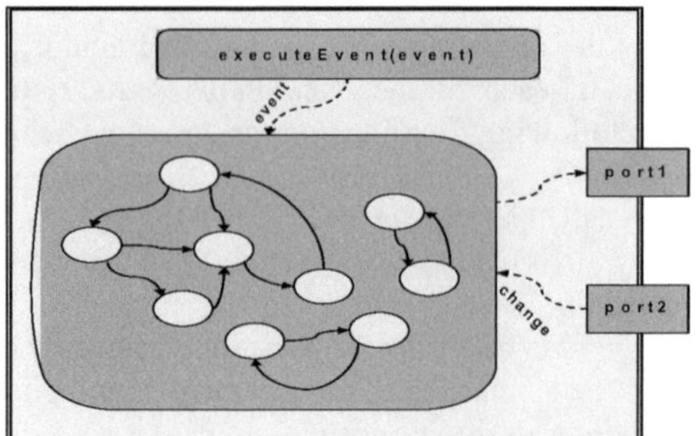

Figure 4: a High-Level Component Model

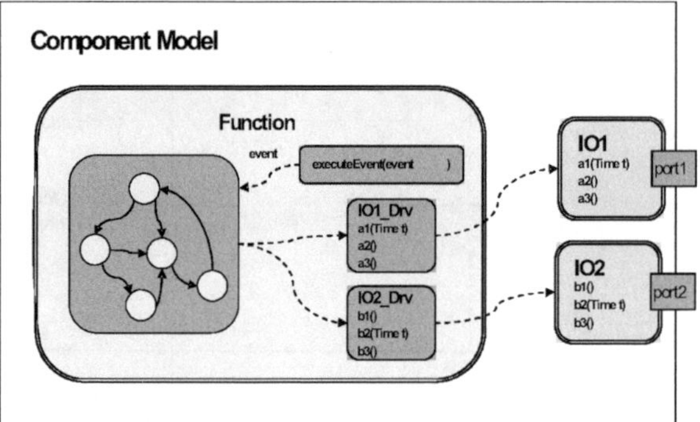

Figure 5: a First Refinement toward Hardware

In Figures 5 and 6, the doubly outlined elements represent S*imulation Modules* that interact with the simulator, and the singly outlined elements represent simple objects. **Figure 5** shows where all handling of *Ports* has been removed from the *Function* model and is handled by separate *Simulation I/O Modules* whose methods are directly called (shown with dotted lines) by the *Functional Simulation Module*. Furthermore, the interfaces to those external I/O models have been abstracted by a driver (*Drv*) objects within the *Function* model. During the evolution from the High-Level Model, the number and complexity of the state machines within the *Function* model is likely to be reduced since it is no longer handling port changes. In most cases, the port handling state changes are simply moved to the I/O models. Just as in the case of SEED-based hardware, the I/O modules will need to post events into a queue that will be handled by the *Function* model. At this level of refinement, the simulator's queue is used in lieu of an actual model of a SEED hardware *Event Queue*.

At this level of refinement a designer defines those functions that each *I/O module* must perform. The *I/O module* class must then interact with the *Simulator* and the *Ports* to implement the interface timing requirements. The level of functionality in each *I/O module* can vary in complexity, but should be limited to that which can be realized in digital hardware. Once this partitioning has been verified to work as it interacts with the system level model, a next level of refinement can begin as shown below in **Figure 6**.

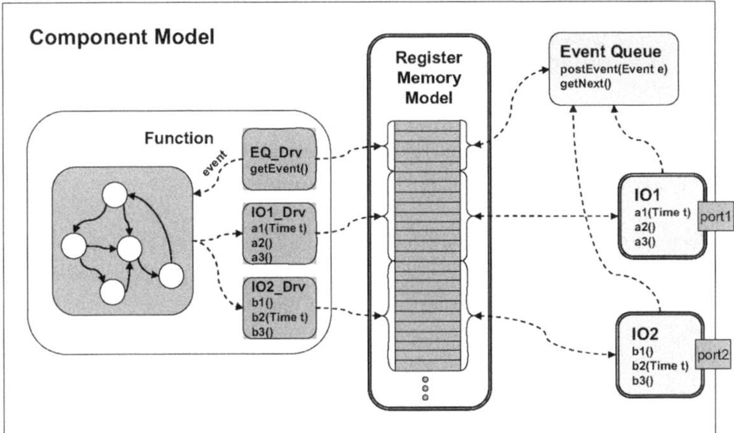

Figure 6: Register Level Refinement

Note that everything from the *Register Memory Model* to the right will ultimately be implemented in hardware. The *Function* element includes the software and the necessary general purpose computing hardware that includes the processor, RAM, ROM, etc.

The most noticeable change is the addition of the Register Memory Model (RMM), and the implementation of the I/O driver objects must also change to interact with it as appropriate. The addresses of the memory elements are selected to be identical to their hardware counterparts. This allows the I/O driver objects to remain completely isolated from the implementation of the I/O modules with which it is communicating, i.e. software models or physical hardware. The software models of the *I/O modules* also need to be slightly changed since now it is register changes that must cause them to perform the necessary I/O behavior rather that direct method calls. The final refinement is to model the SEED *Event Queue* along with its register map. Note that as the I/O system is refined to high levels of detail the controlling state machine, has remained unchanged. It is at this point that the *Functional* model operates entirely through the register maps. It is completely unaware of the simulator and fully represents the final application code, so full software debugging may commence.

The last step to deploying emulation hardware is to take the register map definitions and modeled *I/O module* functionality and implement it in a hardware description language that can be synthesized into physical hardware. Once that has been achieved, the debugged software can simply be moved to physical hardware and re-executed. Note that the logical step from Orchestra models of *I/O modules* to VHDL designs may be large; however, the implementation of the models may be refined to a point where the conversion is largely mechanical. Also note that demonstrating the equivalence between the Orchestra models, and ultimately the VHDL design is of utmost importance to assure that the models behave the same as the hardware. A number of techniques have been developed to gain this assurance, and is roughly described in section 5.4.

4.5 Capabilities

The previous sections have described the basics of how abstract models are defined and refined to a deployable implementation, but has only touched on Orchestra's capabilities. A partial list below summarizes some of its advanced capabilities:

- *VHDL standard logic vector connections* – allows connections to model contention states and pull-up and pull-down resistors.
- *Message-based or state-based communication* – enables various types of object communication.
- *Built-in bus modeling* – allows for abstract system buses to be modeled with automatic collision detection and notification of contention states.
- *Port Listeners* – to streamline port interactions.
- *Simulation module properties* – enables automatic generation of GUI interfaces to view and control state information both before and during simulation.
- *Hierarchical simulation modules* – for design sharing and abstraction.
- *GUI interconnection infrastructure* – that allows for independently developed GUI control and state views to transparently connect to simulation modules.
- *Analog modeling* – offers an automatic resistive network solver for voltage, current, and power estimates.
- In addition enhanced core simulation capabilities Orchestra also offers a number of pre-built reusable classes for things like seven segment displays, memory and system state viewers, etc.

5. Score Processor

A major element in the architecture portion of ASSET is the development of the Score microprocessor. Although any processor may implement the *Function* portion of the SEED architecture, the Score is directly supported by the ASSET development process since Orchestra models for the processor already exist. In fact, the processor itself was developed within the Orchestra.

5.1 Motivation

Numerous commercial processors and embedded microcontrollers exist for consumer and industrial use, but few are available that meet the harshest military standards that include radiation environments. Furthermore, full system analysis may be required for safety and/or security systems where proprietary information and processor complexity can make such analysis unattainable. These and numerous other reasons factored into the decision to create a custom processor as the functional element of SEED-based system components.

For many embedded system components, attaining correct software proves to be more expensive than the hardware development effort. Once again the SEED architecture helps to reduce the software complexity, but there was still a desire to select a modern language to reduce chances of deploying software errors. After evaluating a number of languages that most suited the needs at hand (including the notion of developing a custom language), Java was selected for its safety, security, and popularity in the commercial market. Unfortunately, the fact that Java relied on a virtual machine with a large memory footprint and large computational needs to execute that virtual machine invalidated its use for memory and computationally constrained devices. To address these concerns, it was decided to implement the Java Virtual Machine (JVM) in hardware to eliminate the overhead of a software VM implementation. A custom design also enabled the ability to target an implementation technology that would meet the most stringent military environmental requirements.

5.2 Implementation

A decision was made to directly follow the semantics of the language and its underlying bytecodes[3] without adding any features that violated its core operational concepts [2]. This decision was made difficult by the fact that Java abstracts away the notion of memory locations, so there are no bytecodes that can read or write to an arbitrary memory address. Although this characteristic is a strong safety/security argument for the language, it posed a problem with register-based I/O as described in section 4.4 of this paper. The notion of object-oriented I/O was developed to overcome this hurdle. An object is no more than a set of state variables stored in heap memory, so the basic idea is to overlay the storage of I/O objects directly on top of the control and status registers of hardware-based I/O functions. During the execution of the **new**[4] bytecode the system must recognize the object as being I/O related and select the appropriate memory segment to be allocated for that object. With such a scheme the source code writer simply creates a class with member variables that have the same size and order as those defined for the register map of the hardware, and the system automatically aligns it with the hardware. In this way I/O objects are handled by the Java machine as any other object.

The primary goal of the processor implementation was to simplify the hardware as much as possible for ease of analysis, and reduce gate count size. The core processor was reduced to just that hardware necessary to support a generic Micro-Coded Architecture (MCA) with any number of state registers. State information is rotated through an Arithmetic Logic Unit (ALU) and back into the state memory to effect state changes. **Figure 7** below provides a high level view of this architecture.

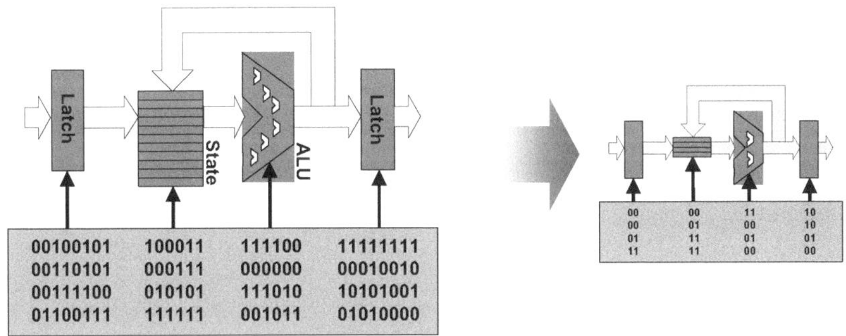

Figure 7: the Score/MCA Architecture

First note that the left image is simply a scaled version of the MCA than that shown on the right. The latches to hold the address and data values for the program, stack, and heap memories, which are not shown in this figure. The table of ones and zeroes below the logic elements represents the microcode control table and associated micro-code control logic.

The theory of operation is that on each clock cycle a row of the micro-code table is driven into the logic that commands and routes data through the hardware. It is important to note that the hardware architecture itself is in no way specialized for Java execution and

---

[3] A bytecode is similar to an opcode for a traditional microprocessor. However, unlike traditional opcodes bytecodes may have an arbitrary level of complexity. For example the **pop** bytecode simply pops a value off the stack, and the **instanceof** bytecode will search the entire application inheritance/interface hierarchies.

[4] **new** is a bytecode that allocates and initializes memory on the heap for Java objects.

that supported JVM elements are implemented in the micro-code. Today, the micro-code table is roughly 1300 lines deep by 92 bits wide.

The ALU is comprised of a number of internal ALUs that perform a single function such as adding, multiplication, shifting logic, etc. The number of internal ALU required is defined by the application being targeted.

Depending on required throughput and logic size, functionality may be completely eliminated from the system or traded between specialized hardware support in an internal ALU and micro-code manipulation of more generalized ALU operations. In fact, it is a long term goal that the ASSET tool suite be able to automatically configure the smallest required set of hardware for any given Java program. For example if a Java application does not contain any division bytecodes, the divide portion of the ALU would be omitted from the hardware VHDL files used to generate the hardware. The corresponding micro-code that exists solely for the purpose of division would automatically be removed as well.

5.3 Supporting Tools

It may be noted that although the hardware is extremely simple, the development of the micro-code table could be extremely complex. To address this problem an automated micro-code table generator was developed. With it, the necessary control signals and their bit widths can be specified for each of the hardware elements. Meaningful and typed constant names can also be defined, and shown in the leftmost text box of Figure 8. Low level discrete instructions like "write value to location x in the state memory" can then be defined and possibly combined into the same line of micro-code[5]. Many of these instructions can be combined into macros[6] as shown in the centre text box. The rightmost text box shows the highest level micro-code. Note that this level of specification very closely matches the commercial specification of the **iadd** bytecode specified by Sun [2].

```
Control Field Type Definitions          Macros                              Microcode (bytecodes)
    type ALUStart(1)                    pop(                                    iadd:
    type ALUControl(3)                      loadStackAddr(SP)                   pop(T1)
                                            readStack()                         pop(T2)
Type Values                                 writeReg(STACK, %reg)               add(T1, T2, T3)
    ALUStart INACTIVE    = "0"               decrementReg(SP)                   push(T3)
    ALUStart ACTIVE      = "1"           )

    ALUControl DEFAULT   = 0            push(
    ALUControl ADD       = 0                incrementReg(SP)
    ALUControl SUB       = 0                loadStackAddr(SP)
                                            loadStackData(%reg)
                                            writeStack()
                                        )

                                        add(
                                            loadALU(%regA, %regB)
                                            aluOp(ADD)
                                            writeALUResult(%reg)
                                        )
```

Figure 8: Micro-code Table Generator (example)

---

[5] Combining more operations into a single line of micro-code is more space and time efficient. It reduces the number of lines of micro-code and each line is executed in a single clock cycle.

[6] All macros can be nested and will be fully unrolled by the micro-code assembler as necessary.

For every JVM implementation, there must be a corresponding class loader. Typically a class loader translates the output of a Java compiler, class files, to a data format that can be more efficiently executed than the standard class file format, which was optimized for size rather than execution speed. The class loader for the Score processor reads the class files and generates an image that can be used to program a Read Only Memory (ROM). It is from this ROM that the Score processor executes. In addition to translating the class files, the class loader can automatically and safely removes unused objects and methods from the ROM. In addition to being able to generate ROMable images it can generate a human readable translation of the ROM image along with heap image maps and provide statistical information to identify potential space saving program changes.

Finally, in order to fully debug Score processor software within the Orchestra system simulator, two models of the processor have been develop that execute from the ROM image in the same way as the hardware processor. The first of these Virtual Scores (VScore) uses the same algorithm to process each of the bytecodes, but does so as abstractly as possible to optimize for simulation speed. This model only approximates the bytecode timing within the simulator, but has been measured to be roughly 97% timing accurate for a typical Java program. The second version of the VScore models the processor hardware itself and is driven from the same micro-code table programmed into the processor hardware. It is completely clock cycle accurate but runs between one and two orders of magnitude slower than the abstract version. Typically, application software is developed and debugged with the abstract VScore, and only when the system is believed to be finished is it retested using the clock cycle accurate version. However, the clock cycle accurate version must be used to validate changes to the micro-code.

5.4 Validation

Verifying equivalence between a model and the physical system it represents always poses a challenge. There must also be high assurance that the processor models and associated micro-code, I/O modules, and the class loader are all performing their associated tasks correctly.

There are two major concerns with respect to verifying correct operation of the Score processor: conformance to the Java specification, and equivalence between the VScore and the hardware implementation.

To address Java conformance, a set of automated tests are performed first on the VScore then on a hardware version of Score. It is simply a Java application that causes all supported Java features to be exercised and compared to a set of hard-coded answers within that same application. This test sequence is largely being developed independently by the University of Omaha at Nebraska [5] and is being designed to test as many strange corner cases associated with Java as can be identified.

Another set of Java compliance tests are performed that may only be executed on VScore implementations using a bytecode Validator. The *Validator* is a block of code that connects to the VScore and has complete visibility into all the VScore internals. At the beginning of each bytecode, the Validator will build fences in the program memory, stack, and heap. Each fence is a data structure that includes a lower bound, an upper bound, and an operation (read and/or write). The program memory fences also include the type of data that can be accessed (constant pool element, method table element, etc.). As a bytecode executes, the memory models will check each memory access to ensure that it is a valid operation within the bounds of the relevant fences. If not, an error is reported.

Functional equivalence between VScore and hardware implementations is accomplished by instrumenting both the Java VScore processor and the VHDL Score processor to generate a state change document during execution of the aforementioned Java compliance tests. The documents contain all state register change values and the time of occurrence and are compared for equivalence. Once this has been achieved, the VHDL model can then be re-executed to record chip level I/O which can then be applied as test vectors for hardware chip verification [3].

I/O module verification is accomplished in much the same way. Application software is written to exercise the various registers of the module under test to exercise its functional and timing characteristics. The input and output lines of that module are looped back to another module that has the ability to time stamp its outputs and stimulate its inputs. All expected function and timing behavior is then verified by the application software driving the modules. That same application software is executed in both in simulation and the physical hardware, so both are guaranteed to operate identically with respect to those functions tested.

Finally, the class loader must be verified to produce a ROM image whose semantics are equivalent to the class files generated by the compiler [4]. This task is made easier by the fact that the ROM image was purposefully designed to be reverse engineered. That is, the application tree can be easily extracted from the ROM. This allows for two stages of verification. The first is verification of the ROM is by a tool called the ROM Integrity Checker (RIC), is a valid one. All data structures must be well formed, and testable data values must be within limits. At this stage roughly 60 properties are checked for correctness. Upon successful completion of that stage, application trees are extracted from the class files and its associated ROM image and compared for semantic equivalence by another tool, called the Class Loader Integrity Checker (CLIC). To test the validity of this test methodology using RIC and CLIC, a program was written to exhaustively corrupt each individual bit of a ROM image containing a non-trivial application program and run it through the staged verification sequence. All corruptions were detected.

6. System Emulation

As stated in section 2 of this paper, a goal of the ASSET development framework is to enable a relatively seamless transition from modeling to hardware development. Much of the work necessary to achieve this goal is accomplished through the refinement process described in the section 4.4. What remains is to provide a hardware infrastructure to minimize bread boarding efforts and allow a seamless transition of behavioral code from the simulation world to a physical one.

6.1 Motivation

During system hardware development it is not uncommon to distributed the responsibility of building hardware prototypes of components to different groups. It is also not uncommon for each group to adopt differing development approaches. Although this does not hinder the ability to satisfy the stated requirements, it is inefficient as each group will spend valuable resources to find differing solutions to the same problems. For example, physical form factors, processor selection, power sources, and testing infrastructures are all common needs that must be defined and implemented for each component. The goal of the ASSET emulation support is to predefine a development infrastructure for common needs, and let the development teams focus their efforts on the differences.

6.2 Implementation

Given that the SEED architecture removes the most stringent timing requirements from the software, the processor selection becomes less important for hardware demonstration purposes. Any commercial processor may be used to host the functional behaviors as long as it meets the minimum throughput requirements. Furthermore, the logic representing the I/O modules may too take on any form, as long as it meets the requirements established during the modeling phase. It is advantageous to use reprogrammable logic for this component since the physical portion of the system may be reused by simply reprogramming new I/O module designs into the logic as necessary. The analog portion of system components tends to be unique from component to component, so it is beneficial to keep the analog interface electronics physically separated from the rest of the electronics.

The generic nature of the SEED architecture allows for commercially defined infrastructures to be leveraged for bread boarding. Figures 9 and 10 shows a system with three components and how it may be partitioned logically to fit within a commercial chassis that utilizes a commercial CPU. The CPU may use any multithreaded operating system that is capable of running the component behavior developed within the simulator. Each of the functional elements is executed as a separate thread on the host CPU card and communication to its associated digital I/O timing card is performed over the backplane. The timing cards each have a dedicated discrete digital bus to a physically separate analog card which makes the necessary translations to interface with other system elements.

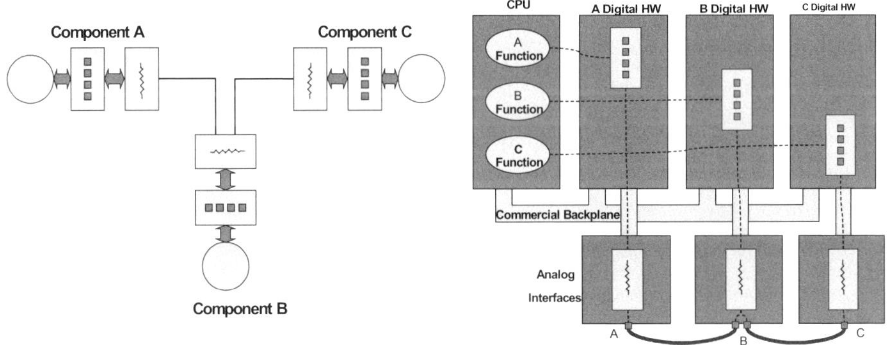

Figure 9: An Example System

Figure 10: A Commercial Chassis Implementation

The figures below illustrate a physical representation of the same system. Figure 11 shows all three components represented as a set of bread boards that fit within a commercial chassis form factor. Figure 12 illustrates how to connect a final product into the emulated system when it becomes available.

7. System Testing

The final major element of the ASSET development framework is the standardization of component and system level testing. Although all the of the topics in this section have been demonstrated or deployed in some fashion, this part of the ASSET development framework is the least mature.

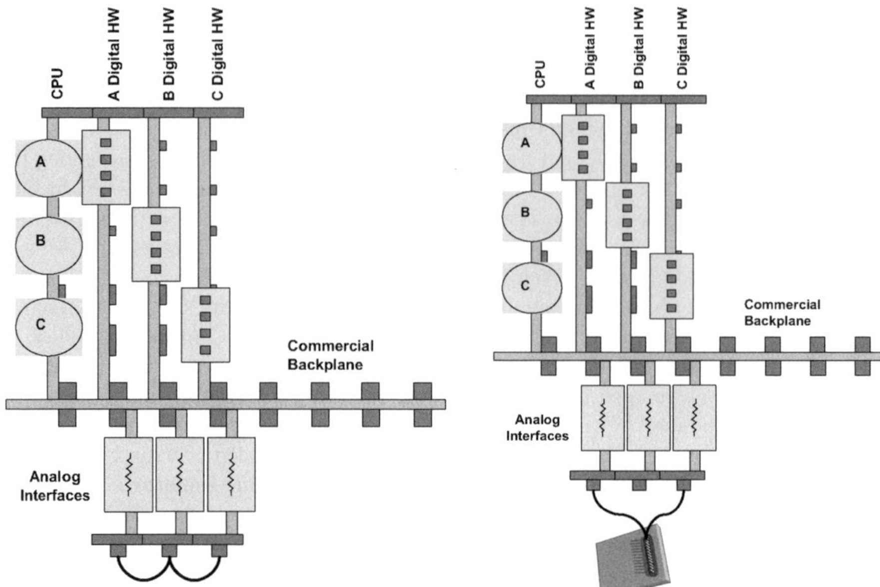

Figure 11: A Chassis-Based Bread Board

Figure 12: Incorporating Production
Components

7.1 Motivation

Just as independent development of system components by separate groups leads to inefficiencies, so does independent development of component and system level testers. Figure 13 below highlights some of the inefficiencies associated with independent tester development.

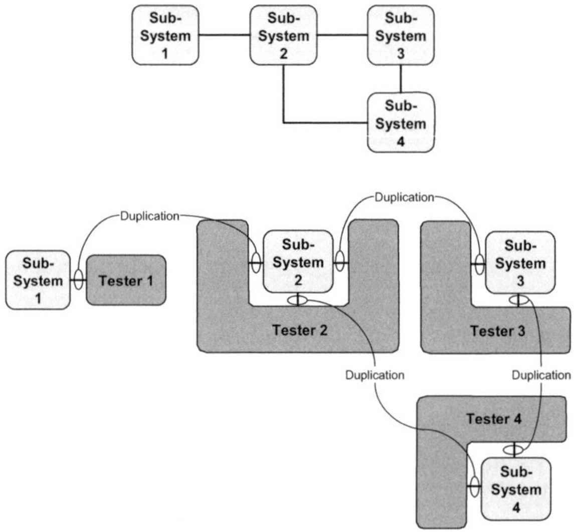

Figure 13: Traditional Development Testers

In addition to recreating the interfaces, each tester is likely to be measuring and validating voltage, current, timing, and communication content. The implementation of this validation work is likely to vary widely unless some effort is made to develop standards for such testing.

To exacerbate the problem, most tester development does not begin in earnest until late in the development cycle as they're have not typically needed until hardware is available to test. These resource and schedule inefficiencies may be addressed if the notion of testers and testing is viewed from a different perspective.

7.2 Testing Approach

Traditional testing approaches tap into the interconnecting wires of system components, collect waveform and timing data, and extract content from that data to perform validation work. However, from the high level view offered in **Figure 1** an observation may be made about the relationship between the model of a system and its associated system hardware. The observation is that they are simply different implementations of the same system. The information content, timing, and the communication waveforms between interacting components of both the model and the physical system are identical. Since the relevant information is being generated by the system model long before hardware becomes available, a testing infrastructure for systems and their components can be assembled and verified much earlier in the development process. In fact, the evaluator of test data need not know of the source of that data. The interface presented to the test platform should abstract the underlying source of the data and include a way for the tester to stimulate the system under test and collect the sampled data for analysis. This view allows a large portion of the testing infrastructure to be developed and debugged in the modeling world.

The ASSET development process offers another benefit. The development of emulation hardware has the side effect of building hardware stimulus for each of the system components when building emulations of its surrounding components. The idea is to leverage the existing emulation capabilities by allowing the test control to manipulate the behaviors of each system component to more fully exercise the interfaces of the other components in the system. Figure 14 below illustrates the extension of the emulator to support component and system level testing.

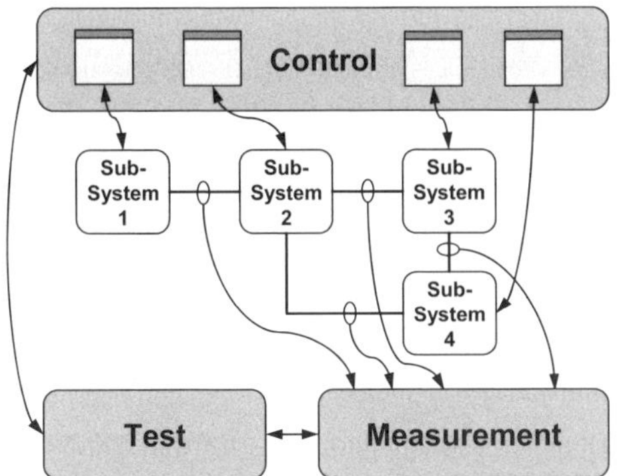

Figure 14: Leveraging of Emulation into Testing

This illustration shows the potential for the emulated components to be controlled graphically by a user or automatically by the *Test* block. The *Measurement* block is a passive sensing system only and has no control capabilities.

Although the extension of component emulators to support component and system level testing seems a natural evolution of previously developed hardware, it may be fundamentally different than the testers illustrated in Figure 13. This approach to testing completely separates the control infrastructure from the measurement capabilities, whereas many traditional testers tightly interweave control and measurement into the tester.

7.3 Integrating Commercial Test Systems

In **Figure 1**, only conceptual *Test* block connectivity was shown. In order to fully abstract the implementation being validated, an intermediate translator must perform the abstraction functions and are shown in Figure 15.

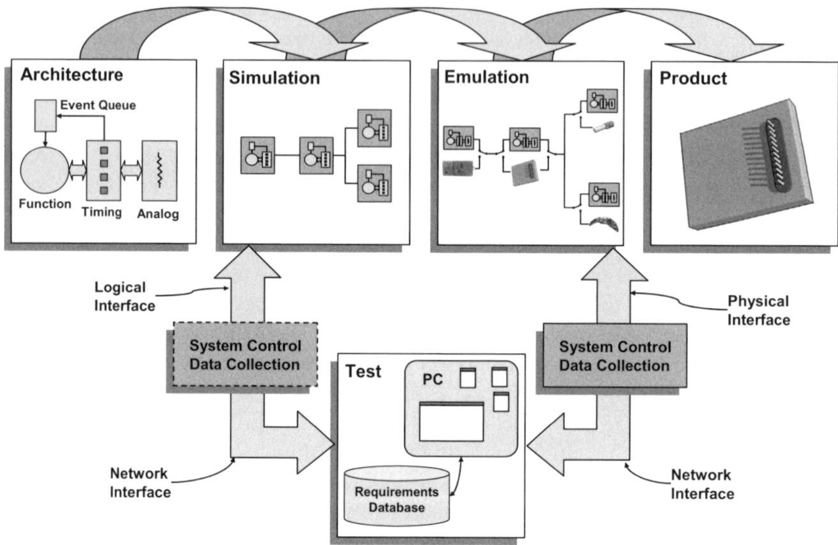

Figure 15: an Expanded View of Test Connectivity

In this illustration, the *System Control & Data Collection* blocks take on two implementations. One that interacts with the simulator to drive the system and collect the necessary data, and the other drives the physical hardware and utilizes commercial measurement capabilities to collect data for analysis. Each must then present an identical interface to the *Test* block. For the systems we've built to date, that standardized interface appeared as separate but functionally identical nodes on a network.

8. Summary

This paper began with a number of issues that must be addressed during development of systems of interacting components. It also recognized that commercial toolsets were valuable for overcoming such issues, but were more limited in value with respect to the development of constrained systems. Over time ASSET development framework evolved to solutions to many of the shortcomings offered by commercial tools. Those solutions include:

- Architectures and technologies that address problems in developing deployable real-time embedded components for physically and environmentally restrictive scenarios.
- An ability to model any number of interacting components at an arbitrary level of detail that includes a hardware/software co-development environment with full debugging capabilities in either paradigm.
- A hardware bread-boarding infrastructure that allows for relatively seamless transition from modeling to prototyping and integration of production components when they become available.
- A testing approach that reduces cost by heavily leveraging previous work and serves as a functional equivalence validator between modeled systems and the resulting physical realization.

9. Future Work

The ASSET capabilities have been applied to a number of components and systems, but they continue to evolve as necessary. Some of the current and planned future work includes:

- An ability to enable component and/or system specification in a neutral language such as XML with automatic generation of hardware and software for specific implementations.
- The Score processor development efforts include research with the University of Nebraska at Omaha to implement a provably correct micro-code optimizer as well as a provably correct class loader [4].
- Work is also under way with the Atomic Weapons Establishment (AWE) in Aldermaston, UK, to prove correctness of the Java micro-code as well as the Score hardware on which it relies.
- A Graphical User Interface (GUI) is being added to the Orchestra simulator to allow users to create and interact with system models graphically.
- To date the emulation system is based on the VMEbus standard. A similar capability is being developed to utilize the CompactPCI chassis.
- The ASSET testing concepts are being evaluated to assure they can meet component, production, and deployment testing needs.

Acknowledgements

Sandia is a multiprogram laboratory operated by Sandia Corporation, a Lockheed Martin Company for the United States Department of Energy's National Nuclear Security Administration under contract DE-AC04-94AL85000.

References

[1] G. L. Wickstrom, A Flexible Real-Time Architecture. *In Proceedings of the 5th IEEE International Symposium on High Assurance Systems Engineering (HASE),* pages 99-106, 2000.

[2] Lindholm, T. and F. Yellin, *The Java Virtual Machine Secification Second Edition,* Addison-Wesley, 1999

[3] G. L. Wickstrom, J. Davis, S. E. Morrison, S. Roach, and V. L. Winter. The SSP: An example of high-assurance system engineering. *In Proceedings of the 8th IEEE International Symposium on High Assurance Systems Engineering (HASE),* pages 167-177, 2004.

[4] Venners, B., *Inside The Java Virtual Machine Second Edition*, McGraw-Hill, 1999.

[5] V. Winter Model-driven Transformation-based Generation of Java Stress Tests. *Electronic Notes in Theoretical Computer Science (ENTCS), 174(1), pages 99-114, 2007.*

[6] V. Winter, J. Beranek, F. Fraij, S. Roach, and G. Wickstrom. A Transformational Perspective into the Core of an Abstract Class Loader for the SSP. *ACM Trans. on Embedded Computing Sys.*, 5(4), 2006.

Communicating Process Architectures 2007
Alistair A. McEwan, Steve Schneider, Wilson Ifill, and Peter Welch
IOS Press, 2007

Development of a Family of Multi-Core Devices Using Hierarchical Abstraction

Andrew DULLER [1], Alan GRAY, Daniel TOWNER,
Jamie ILES, Gajinder PANESAR, and Will ROBBINS

picoChip Designs Ltd., Bath, UK

Abstract. picoChip has produced a range of commercially deployed multi-core devices, all of which have the same on-chip deterministic communications structure (the picoBus) but vary widely in the number and type of cores which make up the devices. Systems are developed from processes connected using unidirectional signals. Individual processes are described using standard C or assembly language and are grouped together in a hierarchical description of the overall system. This paper discusses how families of chips may be developed by "hardening" structures in the hierarchy of an existing software system. Hardening is the process of replacing sets of communicating processes with an equivalent hardware accelerator, without changing the interface to that sub-system. Initial development is performed using a completely software implementation, which has advantages in terms of "time to market". When cost/power reductions are required, the proposed hardening process can be used to convert certain parts of a design into fixed hardware. These can then be included in the next generation of the device. The same tool chain is used for all devices and this means that verification of the hardware accelerator against the original system is simplified. The methodology discussed has been used to produce a family of devices which have been deployed in a wide range of wireless applications around the world.

Keywords. picoArray, Wireless communications, Multi-core

Introduction

The area of wireless communications is one that is in constant flux and the applications place enormous demands on the underlying hardware technology. This is both in terms of processing requirements and the need to produce cost reduced solutions for mass markets. picoChip[2] initially addressed the need for processing power and flexibility by producing a completely programmable multi-core device, the picoArray PC101. Flexibility is important when the communications standards are evolving rapidly and programmability is crucial to reduce the initial time to market. However, there are costs associated with flexibility and programmability and therefore picoChip has addressed this through the PC102 device and PC20x family of devices. In each case, blocks of processing were identified which required considerable amounts of programmable hardware and were common across a range of wireless communications standards. These blocks were "hardened" into accelerator blocks in subsequent devices, producing great cost savings to the end user.

The picoArray is a tiled processor architecture in which hundreds of processors are connected together using a deterministic interconnect [1,2,3,4]. The level of parallelism is rela-

---

[1]Corresponding Author: *A.W.G. Duller, picoChip Designs Limited, 2nd Floor Suite, Riverside Buildings, 108 Walcot Street, Bath BA1 5BG, United Kingdom.* Tel: +44 1225 469744; E-mail: `andy.duller@picochip.com`.

[2]picoChip, picoArray and picoBus are trademarks of picoChip Designs Ltd. (registered trademarks in the United Kingdom).

tively fine grained, with each processor having a small amount of local memory. Each processor runs a single process in its own memory space, and they use "signals" to synchronise and communicate. Multiple picoArray devices may be connected together to form systems containing thousands of processors by using peripherals which effectively extend the on-chip bus structure to adjacent chips.

The picoArray tools support an input language, picoVhdl, which is a combination of VHDL [5], ANSI/ISO C and assembly language. Individual processes are written in C and assembler, while structural VHDL is used to describe how processes are connected together using signals. Signals are strongly typed, have specified bandwidths and are allocated fixed communication slots, leading to deterministic communications behaviour. They may be synchronous or asynchronous, point-to-point or point-to-multi-point. Processes are statically created — no runtime creation of processes is possible. Thus, after a system has been compiled, the complete set of processes and their connections is known, and the system will behave deterministically.

The remainder of this paper is structured as follows. Section 1 contains an overview of the picoArray devices. Section 2 outlines the whole tool chain, while section 3 covers simulation in more detail. The "Behavioural Simulation Instance" (BSI) is introduced in section 4 and the methodology of using BSIs for architectural "hardening" is then described in section 5. Section 6 gives two examples of the process.

1. The picoArray Concept

1.1. The picoArray Architecture

picoChip's thrid generation devices - PC202/PC203/PC205 - are based around the picoArray tiled processor architecture in which 248 processors (3-way VLIW[3], Harvard architecture with local memory) and a range of accelerator units are interconnected by a 32-bit picoBus and programmable switches. In addition, an ARM926EJ-S[4] is included on the device and can share data memory with the picoArray .

The term Array Element (AE) is used to describe either processors or accelerator units (i.e., there are 263 AEs in the PC20x array). There are three processor variants which share the same basic structure: Standard AE (STAN), Control AE (CTRL) and Memory AE (MEM). Memory configuration and the number of communications ports varies between AE types.

1.2. Inter-Processor Communications

Within the picoArray core, AEs are organised in a two dimensional grid, and communicate via a network of 32-bit buses (the picoBus) and programmable bus switches. AEs are connected to the picoBus by hardware ports which provide buffering as well as an interface to the bus. Programmable AEs interact with the ports using *put* and *get* instructions in the instruction set.

The inter-processor communication protocol is based on a time division multiplexing (TDM) scheme, where data transfers between processor hardware ports occur during time slots, scheduled automatically by the tools, and controlled using the bus switches. The bus switch programming and the scheduling of data transfers is fixed at compile time, and requires no run-time arbitration. Figure 1 shows an example in which the switches have been set to form two different signals between processors. Signals may be point-to-point, or point-

---

[3]Very Long Instruction Word instruction level parallelism
[4]ARM is a registered trademark of ARM Ltd.

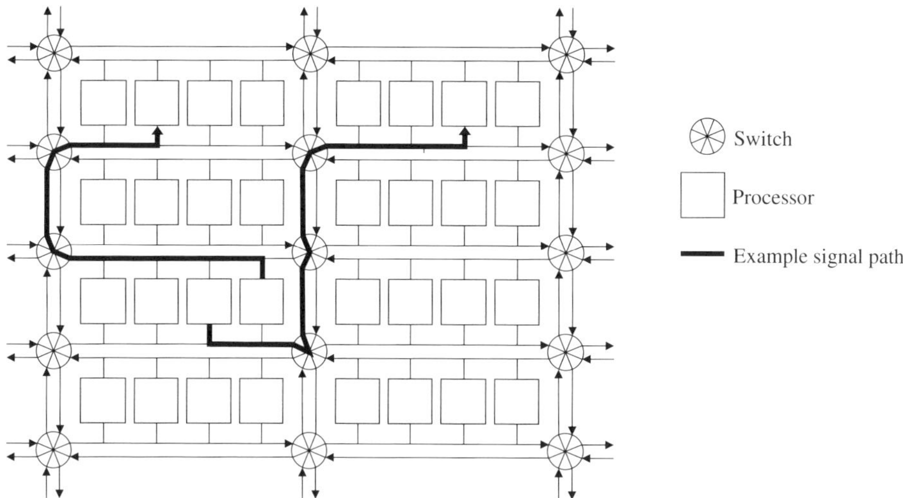

Figure 1. picoArray Interconnect

to-multi-point. The data transfer does not take place until all the processor ports involved in the transfer are ready. The theoretical absolute maximium internal data bandwidth for the signals is 2.73 Terabits per second (267 processors x 2 buses x 32-bits x 160MHz clock).

The default signal transfer mode is synchronous; data is not transfered until both the sender and receiver ports are ready for the transfer. If either is not ready the transfer will be retried at the next available time slot. Using this protocol ensures that no data can be lost. There is also an asynchronous signal mode where transfer of data is not handshaken and in consequence data can be lost by being overwritten in the buffers without being read. If communications is not able to occur then the processors involved will sleep, reducing power consumption.

1.3. Connectivity

In most systems using the picoArray architecture, there is a need to control the system at a higher level, often using a standard microprocessor, operating system and peripherals. The PC102 and PC203 devices both have an external microprocessor interface that allows a host processor to control the picoArray system. In the PC202 and PC205 devices, there is an on-chip ARM926 that acts as the host processor and provides the same control.

The picoArray provides a number of data connectivity interfaces which can be configured in one of two ways. They may either be used as inter-picoArray interfaces (IPI) to allow multiple picoArray devices to be connected, forming a larger system, or they can be configured as asynchronous data interfaces (ADI) to allow high bandwidth data exchange between the picoArray and external data streams e.g a DAC.

2. Basic Tool Flow

The picoArray is programmed using picoVhdl, which is a mixture of VHDL [5], ANSI/ISO C and assembly language. The VHDL is used to describe the structure of the overall system, including the relationship between processes and the signals which connect them together. Each individual process is programmed in conventional C or in assembly language. A simple example is given below.

```
   entity Producer is                        -- Declare a producer
     port (channel:out integer32@8);         -- 32-bit output signal
   end entity Producer;                       --   with @8 rate

 5 architecture ASM of Producer is           -- Define the 'Producer' in ASM
   begin MEM                                  -- use a 'MEM' processor type
     CODE                                     -- Start code block
       COPY.0 0,R0 \ COPY.1 1,R1             -- Note use of  VLIW
     loopStart:
10     PUT R[0,1],channel \ ADD.0 R0,1,R0    -- Note communication
       BRA loopStart
     ENDCODE;
   end;                                       -- End Producer definition.

15 entity Consumer is                         -- Declare a consumer
     port (channel:in integer32@8);          -- 32-bit input signal
   end;

   architecture C of Consumer is             -- Define the 'Consumer' in C
20 begin STAN                                 -- Use a 'STAN' processor
     CODE
     long array[10];                          -- Normal C code

     int main() {                             -- 'main' function - provides
25     int i = 0;                             --     entry point

       while (1) {
         array[i] = getchannel();             -- Note use of communication.
         i = (i + 1) % 10;
30     }

       return 0;
     }
     ENDCODE;
35 end Consumer;                              -- End Consumer definition

   use work.all;                              -- Use previous declarations
   entity Example is                          -- Declare overall system
   end;
40
   architecture STRUCTURAL of Example is     -- Structural definition
     signal valueChannel: integer32@8;       -- One 32-bit signal...
   begin
     producerObject: entity Producer         -- ...connects Producer
45     port map (channel=>valueChannel);
     consumerObject: entity Consumer         -- ...to Consumer
       port map (channel=>valueChannel);
   end;
```

The tool chain converts the input picoVhdl into a form suitable for execution on one or more picoArray devices. It comprises a compiler, an assembler, a VHDL parser, a design partitioning tool, a place-and-switch tool, a cycle-accurate simulator and a debugger. The relationship between these is shown in figure 2. The following sections briefly examine each of these tools in turn.

2.1. picoVhdl Parser (Analyzer, Elaborator, Assembler)

The VHDL parser is the main entry point for the user's source code. A complete VHDL design is given to the parser, which coordinates the compilation and assembly of the code

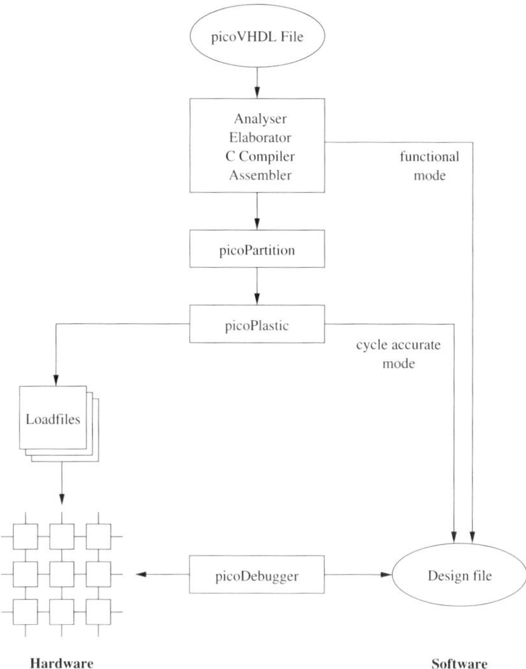

Figure 2. Tool Flow

for each of the individual processes. An internal representation of the machine code for each processor and its signals is created.

2.2. C Compiler

The C compiler is a port of the GNU Compiler Collection (GCC) [6]. Intrinsic functions have been provided to support communication, but the compiler otherwise supports conventional ANSI/ISO C. GCC is designed primarily for 32-bit general purpose processors capable of using large amounts of memory, making it a challenge to support 16-bit embedded processors with just a few kilobytes of memory. The compiler uses a Deterministic Finite Automata scheduling algorithm [7] to generate efficient VLIW schedules.

2.3. Design Simulation

The simulator can operate directly from the output of the picoVhdl parser, since there is no need to determine how a design must be partitioned between chips, or how processes are allocated to processors. Thus, the simulator can be used in two modes, either before or after the partitioning phase. More details of simulation modes are given in section 3.

2.4. Chip Partitioning

If a design requires more processors than are available in a single picoArray, the design must be partitioned across multiple chips. This process is currently manual, with the user specifying which processes map to which chip, although the splitting of signals between the chips is automated.

2.5. Place and Switch

Once a design has been partitioned between chips an automatic process akin to place and route in an ASIC design has to be performed for each device. This assigns a specific processor to each instance in the design and routes all of the signals which link instances together. The routing must use the given bandwidth requirements of signals while routing. The routing algorithm should also address the power requirements of a design, by reducing the number of bus segments that signals have to traverse, enabling unused bus segments to be switched off. This process is performed using the `picoPlastic` (PLace And Switch To IC) tool.

When a successful place and switch has been achieved a "load file" can be produced which can be loaded directly on to the hardware.

2.6. Debugging

The debugging tools allow an entire design to be easily debugged, either as a simulation or using real hardware. The tools support common debugging operations such as setting breakpoints, single and multi-step execution, halt-on-error, status display, and memory/register tracing. For flexibility, both graphical and command-line interfaces are provided. For more details on the methodology adopted for debugging and verification, please refer to [8].

3. Simulation of picoArray Systems

The simulator core is cycle-based, with simulated time advancing in units of the 160MHz picoArray clock. A simulation consists of a collection of models connected via signals. The models can represent a number of things:

- programmable AEs.
- peripheral and accelerator AEs.
- user defined behavioural models.
- test bench behavioural models.

For the programmable AEs in the system, the simulation accurately models the processing of the instructions and the connections to the picoBus via hardware ports. The remaining three categories are all modelled using Behavioural Simulation Instances (described in section 4) which provide an interface to the picoBus while allowing an arbitrary function to be performed.

Simulation can be used in two ways:

Functional In this mode the user's design is seen as a collection of AEs connected via unidirectional signals. The communication across the signals is assumed to be achievable in a single clock cycle and there is no limit to the number of AEs that can comprise a system. In addition, each AE is capable of using the maximum amount of instruction memory (64k since they are 16-bit processors). Furthermore, in this mode all three types of behavioural model can be included. These attributes mean that such simulations need not be executable on picoArray hardware. The importance of this mode is twofold. Firstly, to allow exploration of algorithms prior to decomposing the design to make it amenable for use on a picoArray . Secondly, to allow the "hardening" process to be explored (see section 5).

Back annotated This mode allows the modelling of a design once it has been mapped to a real hardware system. This can consist of a number of picoArray devices connected via IPI connections. In this case, the simulation of the design will have knowledge of the actual propagation delays across the picoBus and will also model the delays inherent in the IPI connections between devices. For this mode of simulation the only types of behavioural model permitted are those for the peripheral and accelerator AEs, since these can also be mapped directly to hardware.

4. Behavioural Simulation Instance

A "Behavioural Simulation Instance" (BSI) is an instance of a C++ class which provides a model of an abstract function in a form which can be used as part of a cycle-based simulation. In its most basic form a BSI comprises a core C++ function called from an interface layer which models its communication with the picoBus via hardware ports, as shown in figure 3. It is created from a picoVhdl entity containing C++ code sections which describes the construction and initialization of the instance, and its behaviour when the simulation is clocked. The C++ has access to the data from the hardware port models via communication functions similar to those provided by the C compiler. A program generator combines these individual code sections with "boilerplate" code to form the complete C++ class.

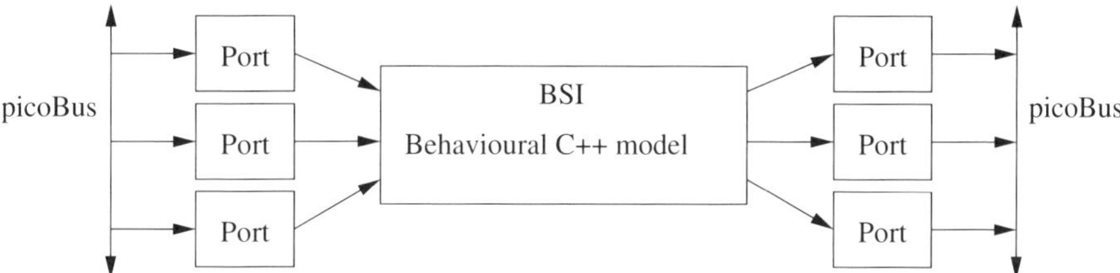

Figure 3. Behavioural Simulation Instance

4.1. A Simple Example BSI

The following example is about the most trivial useful BSI it is possible to produce. Its function is to accept null-terminated character strings on an input port and send them to the standard output of the simulation, each string being stamped with the simulation time at which its first bytes were received and with a name which identifies the instance receiving it.

```
   entity Console is
     generic (name:string:="CONSOLE"; -- Identifier for the messages
              slotRate:integer:=2);   -- rate of the input signal
     port (data:in integer32@slotRate);
 5   end entity Console;

     architecture BEHAVIOURAL of Console is
     begin NONE
       SIM_DATA CODE
10     char buf[1024];       // Buffer for the string
       int index;            // Insertion point in the buffer
       uint64_t latchCycles; // Remembers start time of message
       ENDCODE;

15     SIM_START CODE
       index = 0;
       latchCycles = 0;
       ENDCODE;
```

```
20   SIM_CASE data CODE
     if (index == 0)
       latchCycles = getSimTime();
     integer32 data = getdata();
     for (int i=0; i<4; i++)
25   {
       buf[index++] = data & 0xff;
       data >>= 8;
     }
     if (buf[index-1] == 0)
30   {
       printf("(%llu): %s: %s", latchCycles, name, buf);
       index = 0;
     }
     ENDCODE;
35   end Console;
```

The C++ code at lines 10-12 of the example defines the member data which each instance will have, and the code at lines 16 and 17 initialises this data at the start of simulation. The code at lines 21-33 is called every time data is available in the buffers of the input hardware port. The call to the communication function 'getdata' at line 23 reads an item from the port.

5. Decomposition and Hardening

Using a BSI, an arbitrary function in C++ can be "connected" to the picoBus and allow design abstraction to be performed. BSIs can be used in a number of ways. Since any function can be run inside a single BSI it would be possible for it to constitute an entire system. This model could then be used as a "golden reference" against which to compare subsequent designs. The single BSI solution can then be broken down into simpler BSIs and/or a collection of programmable AEs. At each stage it is possible to compare the performance of the original "golden reference" against that of the decomposed system. Eventually, the system must consist entirely of programmable AEs and accelerator blocks, if it is to be executed on real hardware. However, during development of new picoArray devices it is also possible to envisage new accelerator blocks which are presently collections of programmable AEs. The simulation behaviour of these new accelerators can be provided by BSIs, and thus tested in a "system" environment, and this behaviour can also be used to verify the behaviour of the new hardware block as it is being developed. Importantly, the verification can be performed using the same test benches as were created for the software implementation.

The basic process of hardening is undertaken using the following method. It starts from one or more software reference designs which could implement a variety of wireless standards or a number of different scenarios for a single wireless standard. The latter is important as a given standard will have different hardware requirements for each scenario (e.g. a 4 user femtocell base station, which may be in a private house, will have far lower processing requirements than a 64 user base station in an office building).

- The design is initially partitioned into a number of blocks based on minimizing the picoBus communications between the various blocks.
- Blocks smaller than a minimum size are combined to form large blocks.
- The partitioning is then revised depending on reuse possibilities of identified blocks.

A minimum block size (typically, about 10 programmable AEs) is used, as hardening a small block is unlikely to be efficient due to the overhead of hardware port buffering. In addition, layout of a very large number of small hardened blocks would make silicon design more difficult. This process of combining blocks to delimit new hardware accelerators could be thought of as running the top-down design decomposition of BSIs described above in reverse, but with the difference that the choice of partitioning may well be different.

Once a specific block has been identified for hardening, silicon design may proceed in one of two ways. Either a BSI may be produced for the block, assuming one did not exist already, and RTL design done primarily using the BSI as a reference. Or RTL design may proceed directly from a specification of the block, and use the software implementation as a reference. In either case, the internal structure of the RTL is not related to the other implementations, giving the silicon designer maximum freedom, and verification is done using simulation, both in the common unit testbenches and in full system contexts.

The above process would then be repeated for all of the designs that were being considered to produce the best set of hardened blocks for these given designs.

6. Examples of Accelerator Development

In the original PC101 picoArray, all of the AEs were programmable and the only "accelerator" support in the device was a set of special purpose instructions which helped with wireless applications. This flexibility had enormous advantages when systems were being developed for wireless standards which were in flux, and the main goal was to provide the required functionality in the shortest time.

In subsequent implementations and products, however, considerations of cost and power consumption increased in importance relative to flexibility. Therefore, the decision was taken in subsequent device families to provide optimised hardware for some important functions whose definition was sufficiently stable and where the performance gain was substantial.

For PC102, this policy led to the provision of multiple instances of a single accelerator type, called a FAU, which was designed to support a range of correlation and error correction algorithms. The use of this accelerator for Viterbi decoding is illustrated in section 6.1.

For PC20x, a wider range of functions were hardened but fewer instances of each accelerator were provided, as this device family is focused on a narrower range of applications and hence the requirements are more precisely known. Section 6.2 illustrates the hardening of an FFT function.

6.1. Viterbi Decoder

One of the common error correcting methods used in wireless communications is the Viterbi decoder. This example follows the stages in the hardening of this function. The example Viterbi decoder operates at 10Mbps and is instanced together with a testbench to exercise it.

The testbench comprises a random data generator, noise generator and output checking, all themselves implemented in software using other AEs. Control parameters for the test, and result status indication, are communicated to the user via the host processor. This testbench uses 11 AEs (4 MEM, 7 STAN) in addition to the host processor.

On PC101, the Viterbi decoder itself was also implemented entirely in software, and requires 48 AEs (1 MEM, 47 STAN). Figure 4 shows a schematic of this design, produced by the "Design Browser", a graphical design viewing tool based on the directed graph drawing package *dot* [9] (it should be noted that for all of the schematics shown, it is only the complexity of the design that is of interest, the labelling of the AEs is arbitrary and of no interest). Signal flow is predominantly from left to right here, and also in figures 5 and 6. The complexity and picoBus bandwidth requirements of this design are considerable.

On PC102, the hardware accelerator was used to implement the core trellis decode function. The modified version of the Viterbi decoder is shown in figure 5. The decoder itself now requires 4 instances of the hardware accelerator and only 8 other AEs (1 MEM, 7 STAN), a saving of almost 40 AEs.

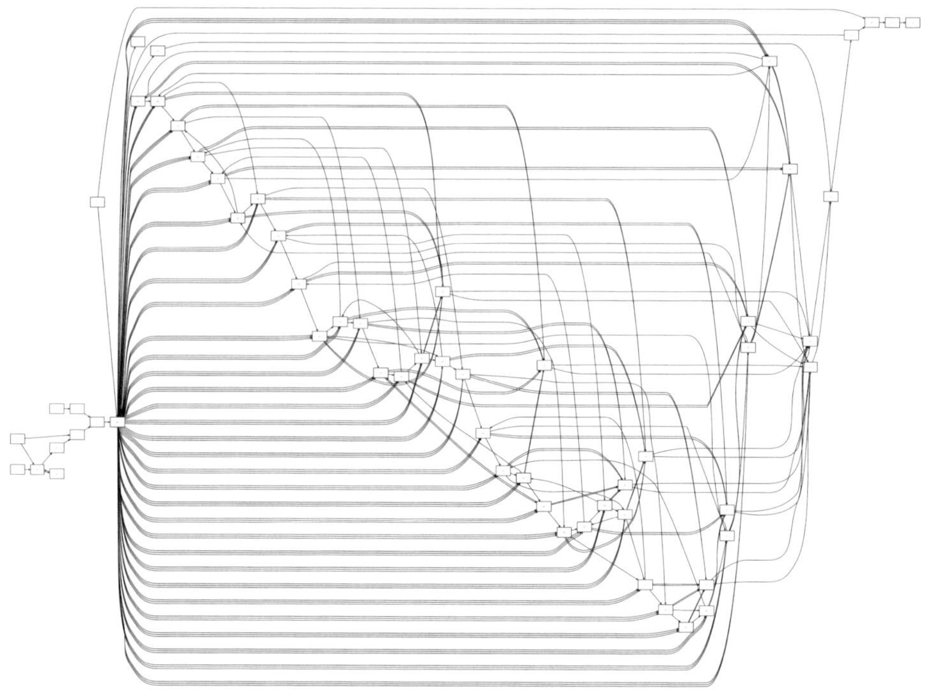

Figure 4. Software implementation of Viterbi decoder and testbench

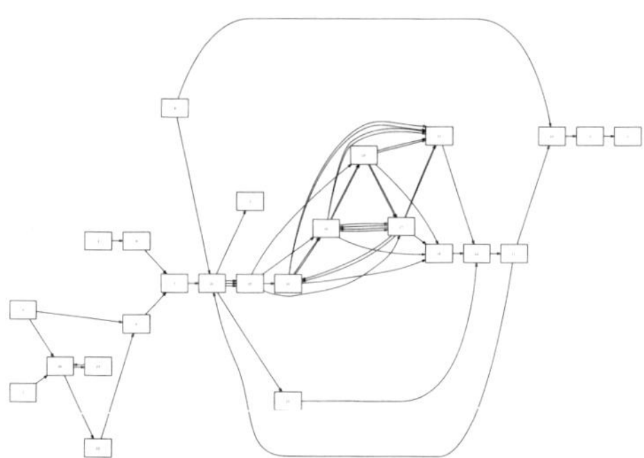

Figure 5. Partially hardened implementation of Viterbi decoder and testbench

Finally, on PC20x a hardware accelerator is provided which implements the complete Viterbi decoder function. This is shown in figure 6. Here the Viterbi decoder is reduced to a single instance. Moreover, the accelerator is actually capable of operating at over 40Mbps, and is able to support multiple standards including IEEE 802.16-2004 [10] multi-user mode Viterbi decoding largely autonomously, which means that its use represents an even greater saving of resources in a more demanding application than this example.

Table 1 provides more quantitative detail on this hardening process, giving estimates of transistor counts for each of the two example Viterbi decoders discussed. Area and power estimates are not included as different fabrication processes were used for different picoArray devices, rendering comparisons meaningless. Meaningful, however, is the 40Mbps case: similar functionality is compared and the transistor reduction count is a factor of 23.

In addition, it is obvious that the results of the hardening process are not unique and that the change from PC101 to PC102 produced a factor of 2 reduction in transistor count but resulted in a far more flexible solution. The FAU hardening allows a range of wireless standards to be performed. The full hardening of the Viterbi into a single block in the PC20x produced far more transistor count reduction but it can only perform the specific Viterbi functions for which it was designed.

Table 1. Viterbi decoder transistor estimates (all numbers are millions of transistors)

	MEMs @1.0M trans.	STANs @250k trans.	FAUs @1.0M trans.	Viterbi AEs @4.0M trans.	Total
		10Mbps Viterbi			
PC101	1	47			11.75
PC102	1	8	4		6.75
PC20x				1	4
		40Mbps Viterbi			
PC101					N/A
PC102	39	147	18		93.75
PC20x				1	4

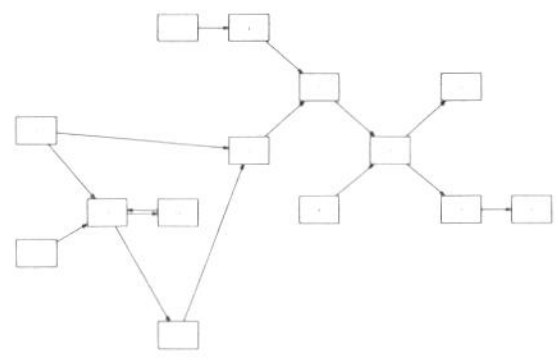

Figure 6. Fully hardened implementation of Viterbi decoder and testbench

6.2. FFT

Figure 7 shows the software implementation of two independent 256 point FFTs, capable of a data rate of 80Msps, on PC102. This requires a total of 96 AEs (44 MEM, 52 STAN), including a trivial testbench of 4 MEMs which are used to interface with a file reading and writing mechanism in `picoDebugger`. Each FFT operates on 16 bit complex data. In figures 7 and 8 signal flow is predominantly from top to bottom.

On PC20x a hardware accelerator is provided which is capable of the equivalent function (and in fact is more flexible). The same design, including the same minimal testbench, is shown in figure 8.

For the hardening of the FFT functionality the transistor counts are shown in table 2.

Table 2. Dual FFT transistor estimates (all numbers are millions of transistors)

	MEMs @1.0M trans.	STANs @250k trans.	FFT AEs @2.0M trans.	Total
PC102	40	52		53
PC20x			1	2

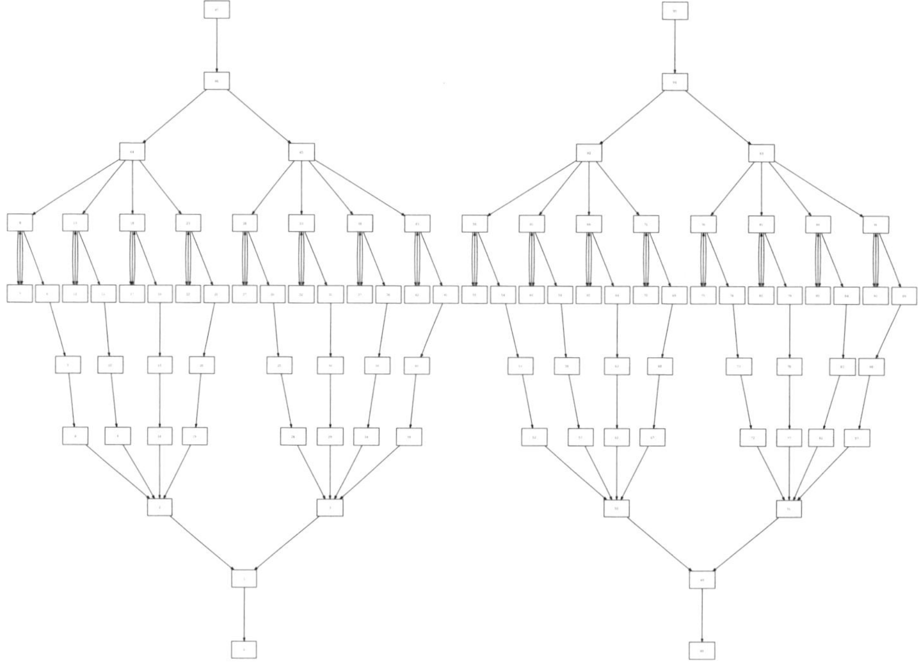

Figure 7. Software implementation of dual FFT and testbench

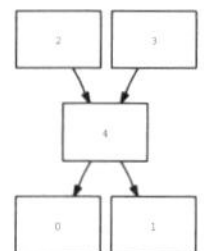

Figure 8. Fully hardened implementation of dual FFT and testbench

7. Conclusion

In order to address the target markets in wireless communications picoChip has created a family of picoArray devices which provide the computational power required by these applications and allow designers to trade off flexibility and cost. This family of devices is now in production and has been deployed in a wide range of wireless applications by a number of companies. This paper has explained the basic process of behavioural modelling that has been developed to aid in the decomposition of designs and to allow the exploration of future architectures. Importantly, all of the blocks are interfaced using the same picoBus interface and consequently the programming paradigm remains the same, which means that employing the hardened blocks is simply a matter of removing the programmable version, inserting the hardened block and re-routing the design using the picoPlastic tool.

As was shown for the Viterbi hardening there are many ways that the hardening can be done, which allows a variation in the trade off between transistor count and flexibility. The behavioural model based hardening process allows a range of these options to be explored before devices are fabricated.

The architectural "hardening" process has been used to produce a progression of commercially deployed devices and we have briefly shown how this has been used in the development of two specific accelerators, the Viterbi decoder and the FFT. The advantage of this architectural "hardening" is to allow large reductions in system cost to be realised whilst still allowing the unified picoArray programming and development environment to be used.

References

[1] Andrew Duller, Gajinder Panesar, and Daniel Towner. Parallel Processing — the picoChip way! In J.F. Broenink and G.H. Hilderink, editors, *Communicating Processing Architectures 2003*, pages 125–138, 2003.

[2] Peter Claydon. A Massively Parallel Array Processor. In *Embedded Processor Forum*, 2003.

[3] G. Panesar, D. Towner, A. Duller, A. Gray, and W. Robbins. Deterministic parallel processing. *Int Journal of Parallel Processing*, 34(4):pages 323–341, 2006.

[4] G. Panesar. Multicore products - not an oxymoron. GSPx Multicore Conference, Santa Clara, 2006.

[5] Peter Ashenden. The Designer's Guide to VHDL. Morgan Kaufmann, ISBN 1-55860-270-4, 1996.

[6] Richard Stallman. Using and porting the GNU compiler collection. ISBN 059510035X, `http://gcc.gnu.org/onlinedocs/gcc/`, 2000.

[7] Vladimir Makarov. The finite state automaton based pipeline hazard recognizer and instruction scheduler in GCC. The 2003 GCC Developers' Summit Conference Proceedings, `http://www.linux.org.uk/~ajh/gcc/gccsummit-2003-proceedings.pdf`, May 2003.

[8] Daniel Towner, Gajinder Panesar, Andrew Duller, Alan Gray, and Will Robbins. Debugging and Verification of Parallel Systems — the picoChip way! In Ian East, Jeremy Martin, Peter Welch, David Duce, and Mark Green, editors, *Communicating Processing Architectures 2004*, pages 71–83, 2004.

[9] Emden R. Gansner, Eleftherios Koutsofios, Stephen C. North, and Kiem-Phong Vo. A technique for drawing directed graphs. *Software Engineering*, 19(3):pages 214–230, 1993.

[10] IEEE. *802.16-2004 IEEE Standard for Local and metropolitan area networks.*

Communicating Process Architectures 2007
Alistair A. McEwan, Steve Schneider, Wilson Ifill, and Peter Welch
IOS Press, 2007

479

Domain Specific Transformations for Hardware Ray Tracing

Tim TODMAN and Wayne LUK [1],

Imperial College, London, U.K.

Abstract. We present domain-specific transformations of the ray-tracing algorithm targeting reconfigurable hardware devices. Ray tracing is a computationally intensive algorithm used to produce photorealistic images of three-diemnsional scenes. We show how the proposed transformations can adapt the basic ray-tracing algorithm to a breadth-first style, and give estimates for the hardware needed for realtime raytracing.

Keywords. Ray tracing, Reconfigurable hardware, transformations

Introduction

Ray tracing [1] is a method used in computer graphics for rendering images of three-dimensional scenes. It has also seen use in fields as diverse as seismology and acoustics. For computer graphics, it has several advantages over the hidden-surface polygon renderers used in most graphics hardware. It can integrate many optical effects into one simple method, and is particularly good for shiny or transparent objects. Ray tracing is much slower than hidden-surface methods, though it has a lower time complexity in the number of scene objects (sub-linear vs. linear).

We present a study of developing real-time, interactive ray tracing using advanced reconfigurable hardware such as Field Programmable Gate Arrays (FPGAs). Interactive means at least 25 frames per second with response within two frames to user inputs. Previous work on hardware for ray tracing has been limited. ART makes a rendering appliance which uses ray tracing for the television and movie industries [2] for non-interactive work. Interactive raytracing has been achieved on large multiprocessors and workstation networks [3], whereas we target a single machine aided by reconfigurable hardware. Woop et al [4] have demonstrated realtime raytracing on programmable graphics hardware. In contrast, we are concerned here with using a framework to help automate the the process of transforming raytracing into a form suited to programmable graphics hardware, or other implementation technologies.

Our work is intended to make three contributions:

- We identify and map the time-consuming part of ray tracing to hardware.
- We transform the ray tracing algorithm to improve efficiency of hardware execution.
- We estimate the hardware required for complex scenes.

This paper proposes a way to refactor the ray-tracing algorithm to make it easier to exploit hardware parallelism. The basic ray-tracing algorithm is massively parallel – each pixel is independent – but there is conditional processing within each pixel, so some feedback is necessary. Our approach may apply to other data-parallel applications requiring limited conditional processing at each node, such as image processing.

---

[1]Corresponding Author: *Tim Todman, Department of Computing, Imperial College, London.* E-mail: `t.todman@imperial.ac.uk`

1. The Ray Tracing Algorithm

Ray tracing is an algorithm used for rendering high quality, photorealistic images of three-dimensional scenes. It was one of the first approaches to integrate support for modelling many optical effects within a single algorithm. The basic scheme affords such effects as specular (mirror) reflection and refraction, multiple light sources and shadows. Various researchers have developed extensions that allow the rendering of such effects as motion blur, soft shadows and caustics (light focussed by transparent objects). Ray tracing can also be used in partnership with the radiosity method, which can model complementary effects. The algorithm has also found use in modelling radar and sound systems and in visualisation of scientific data. An excellent introduction to ray tracing is given in the first chapter of Glassner's book [1].

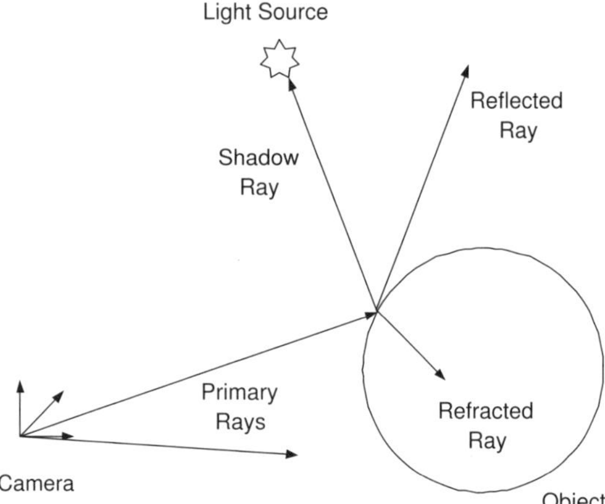

Figure 1. The ray tracing algorithm

The basic ray tracing algorithm is simple (figure 1): for each pixel of the image to be produced, rays are traced backward from a virtual camera into the scene of interest – these rays are known as the primary or eye rays. If a ray strikes no objects, the colour of the corresponding pixel will be that of the scene's background. If the ray does strike an object, the point of intersection is calculated. The illumination at the point of intersection is found by firing shadow rays from that point to all light sources in the scene, each light source making a contribution only if no objects block the light source. Rays are also fired recursively from the intersection point to account for specular reflection and refraction. The colour of the corresponding pixel is calculated according to a shading model, using surface properties of the object intersected and the calculated normal at the intersection point. A number of shading models are available for different kinds of surfaces. Ray tracing can also be used to model Constructive Solid Geometry (CSG) objects – these are composites formed by boolean operations (intersection, union and difference) on simpler objects.

The biggest problem with the ray tracing algorithm is its computational complexity – for each pixel, in the worst case, rays must be intersected with each object in the scene. Any of these rays that hit objects will then give rise to further rays. Even in a simple scene consisting of few objects at a standard resolution, several million ray-object intersection tests may occur per frame. Various software schemes have been proposed to deal with this problem. Following the classification of Haines [5], these schemes may speed the algorithm by speeding ray-object intersections, by reducing the number of ray-object intersections needed to produce the image, by reducing the number of rays or by using generalised forms of ray such as cones. These schemes can achieve impressive speedups – two orders of magnitude have been reported. However, they can greatly complicate the basic algorithm and their effectiveness can be strongly dependent on the scene to be rendered and how they are prepared

Ray-object intersection calculations dominate the time needed for ray tracing [1], and are the natural choice for hardware assistance. We have used fixed and floating-point number formats. The format size can be customised to suit performance and quality requirements. Floating-point numbers can cope with calculations involving a larger dynamic range for large scenes. Fixed-point numbers are faster and smaller in current FPGAs. For either format, image fidelity (improved by using more fractional bits) can be traded for speed. In our tests, a 32-bit fixed-point format (8 integer bits, 24 fractional bits) worked well.

We have implemented a ray-sphere intersector using Celoxica's Handel-C compiler v2.1, on an RC1000-PP board containing a single Xilinx Virtex XCV1000-5 device. The intersector includes 7 multipliers, 9 add/subtracters and one square root. Using our own operator implementations, it runs at 16 MHz, producing a result every three cycles. We assume that the ray direction is a unit vector which allows for narrow paths and operators.

C++ code for a basic ray-sphere intersector is shown below:

```
int sphereIntersect(Sphere * sphere, Ray * ray, Hit * hit) {
   int numHits = 0;
   Vector v = vecSub(sphere->centre, ray->start);
   float b = vecDot(v, ray->dir);
   float disc = (b * b) + sphere->rad2 - vecDot(v, v);
   if (disc > eps) {
      float sqrtDisc = sqrt(disc);
      float dist1 = b + sqrtDisc;
      float dist2 = b - sqrtDisc;
      if (dist2 > 0) {
         numHits = 2;
         hit[0].obj = sphere;
         hit[0].dist = dist2;
         hit[1].obj = sphere;
         hit[1].dist = dist1;
      }
      else if (dist1 > 0) {
         numHits = 1;
         hit[0].obj = sphere;
         hit[0].dist = dist1;
      }
   }
   return numHits;
}
```

Our Handel-C intersector uses the Handel-C par statement to run independent computations (such as the terms of disc above) in parallel. Figure 2 shows the dataflow within the ray-sphere intersector and the corresponding dataflow in our implementation on the RC1000-PP FPGA board. Using the dataflow as a guide, the intersector is pipelined to read a ray and emit a result every three cycles.

With the breath-first transformation in this paper, the application is free to exploit concurrency between software and hardware. A basic design runs as follows:

- Software: generate ray batch n, write to shared memory banks, signal to hardware to start processing.
- Hardware: receive signal from software, intersect ray batch n, write to shared memory banks, signal to software.
- Software: receive signal from hardware, process results of batch n

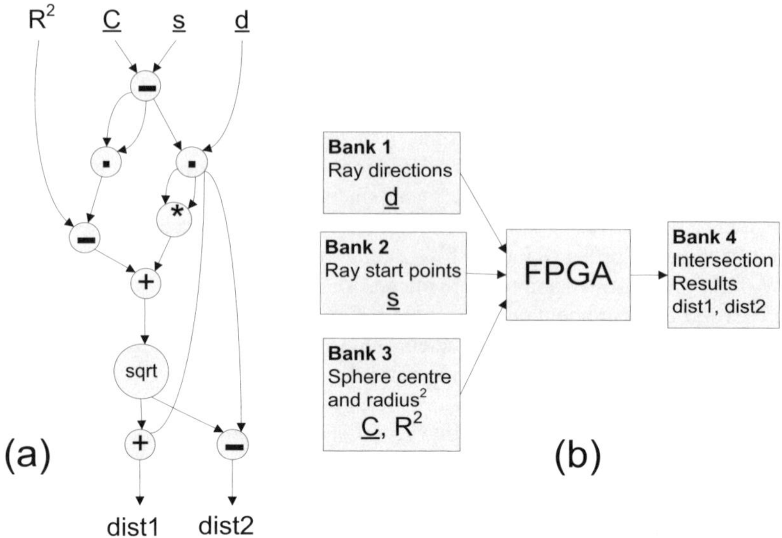

Figure 2. Dataflow in ray tracer: (a) within ray-sphere intersector - note "." means vector dot product, "*" means scalar multiplication; (b) between FPGA and external memory banks on RC1000-PP board.

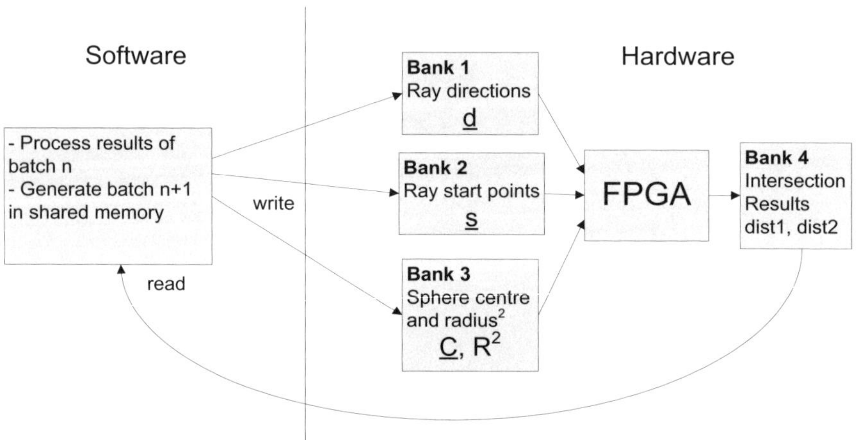

Figure 3. Dataflow between software and hardware.

Figure 3 shows a process diagram. We use the RC1000-PP's programming interface to synchronise between software and hardware, and to communicate data between the two using the shared memory banks. Because each screen pixel is independent in the basic ray-tracing algorithm, generation and processing can run concurrently with intersection. In this case the hardware will intersect batch $n-1$ while the software is processing the results of batch $n-2$ and generating batch n.

2. Transformation Strategies

Using our hardware in a standard ray tracing algorithm results in poor performance. The standard algorithm adopts a depth-first approach: a ray is generated for each pixel, the result of this ray may lead to more rays which may themselves lead to other rays. Processing these rays results in many bus transactions per scene, which is particularly inefficient with a slow bus such as PCI.

Our improvement involves a breadth-first approach, sending the first rays for each pixel in a large batch, then the second rays, etc. The bus sees a few large transactions instead of many small ones; hardware pipeline overheads are less significant with the longer data sets. This results in better performance – for our test animations, the breadth-first approach takes 5.6 seconds per frame, with the depth-first approach taking 16 seconds.

The breadth-first approach has two costs in software. First, it needs data structures to store partial results that were stored on the stack. The cost is 128 bytes per pixel, or about 40 MB for a 640 by 480 resolution screen. Second, because it needs to cycle through all this storage for every frame, it has rather poor caching behaviour. In our tests, a software-only ray tracer performed up to 50% slower using the breadth-first approach compared to the depth-first one. A hybrid between depth and breath-first is also possible, allowing the trading of software for hardware performance.

The ray tracing algorithm is simple enough to be expressed in about a page of code, in its most simple form. However, even this would exceed the capacity of current reconfigurable hardware as it contains several vector multiplications and additions that need to be of relatively high precision (at least 32 bits), or the image will have visible errors. We have investigated ways of partitioning the ray tracing algorithm between hardware and software and what restructuring of the algorithm is necessary to achieve the best performance. Profiling the algorithm on some simple test scenes has confirmed the view of the ray tracing literature: the ray-object intersection tests contribute most to the overall run time. In complex scenes, with a ray tracer that makes few optimisations to reduce the number of tests, this testing can account for up to 95% of the total time [1]. Other time-consuming parts of the algorithm are the pixel shading calculations and the CSG-tree intersection routines. These parts require more data and use more complex, control-dominant algorithms than the intersection calculations, making them poorer choices for hardware implementation. This is because most available reconfigurable hardware is best suited to simple, regular algorithms – irregular algorithms are better mapped to conventional microprocessors, which are optimised for them.

The closeness of coupling between the reconfigurable hardware and the host processor running the software is very important to the performance of our ray tracing implementation. If the bus connecting the two is slow, like the PCI bus common in desktop PCs, this bus will tend to become the bottleneck. To use the bus efficiently, communications over it must be marshalled into large groups, as busses operate less efficiently when transferring many small pieces of data. Unfortunately, the ray-object intersection tests are tightly interwoven with the rest of the algorithm. To make the best use of a slow bus, the algorithm has to be restructured so those ray-object intersection tests are scheduled in large blocks.

Ray tracing can be thought of as a kind of search algorithm, where the objects in the scene are continually searched to find the closest along the direction of each ray. The point of intersection with the closest object is then used to generate the next ray and so becomes the start of the next search. The basic ray tracing algorithm uses a depth-first search strategy (figure 4). This allows the main part of the algorithm to be compactly expressed as a pair of mutually recursive routines, but means that ray-object intersections have to be performed one at a time because of the dependences from each search to the next. Our restructuring changes this to a breadth-first strategy, which buffers independent rays into large groups that can be efficiently sent across slow busses (figure 5). For each scene to be traced, the primary rays are

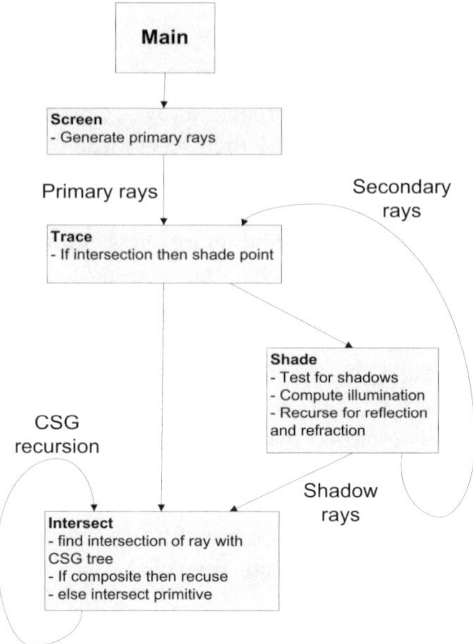

Figure 4. Call Graph of Depth-First Ray Tracing (after Heckbert [6])

traced, then the shadow rays from the objects intersecting the primary rays, then reflection and refraction rays and so on. These rays are independent and can be intersected in large groups.

Breadth-first strategies have previously been used by Hanrahan [7] for exploiting the coherence between similar rays, by Muller et al [8] and Nakamaru and Ohno [9] to reduce disk thrashing when accessing very large scenes and for ray tracing on vector and parallel processors (Parker et al. [3], Plunkett and Bailey [10]). Our work is the first that we know of that has applied it in the context of hardware acceleration. The disadvantage of breadth-first strategies is that they take far more memory than depth-first ones. The entire ray tree for each pixel has to be stored, with each tree node containing the location of the intersection point corresponding to that node, a pointer to the object intersected the surface normal, the incoming ray direction, pointers to its children and the calculated colour. For example, for a 640 by 480 pixel screen, with an average ray depth of 1.5 (a typical number), some 27 Mbytes are necessary. There are also buffers for the rays to be sent to the hardware, to record which ray trees sent those rays and to receive the intersection results from the hardware. By contrast, the depth-first strategy maintains all its working data on the stack, in about 2Kbytes in our implementation. The depth-first strategy thus has much better data locality than the breadth-first, and runs somewhat faster in software as a result (some 30% faster for simple scenes in our implementation).

The depth and breadth-first strategies are simply two logical extremes, of course. Where the reconfigurable resource is more closely coupled to the host, the bandwidth between them becomes less of a bottleneck and so smaller groups of intersections can be used. This will improve the data locality and thus the caching performance of the application, offsetting any inefficient use of the bus. Where the reconfigurable resource is most closely coupled to the host, for example where reconfigurable execution units are present in the host, the depth-first strategy will be preferred. Very closely coupled reconfigurable resources tend to be smaller than more loosely coupled ones, as they have to share silicon with the host. In this case, it will

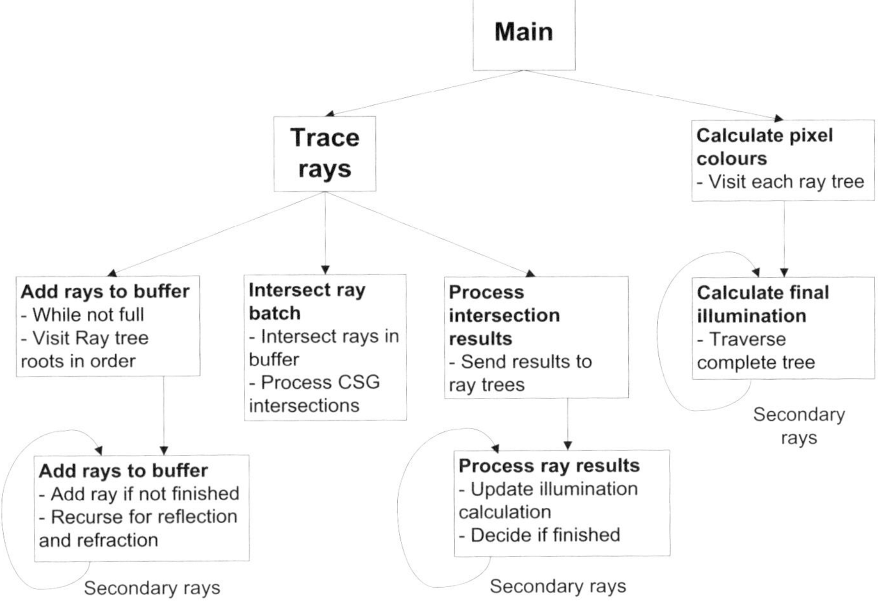

Figure 5. Call Graph of Breadth-First Ray Tracing

probably be only possible to implement part of the intersection calculation on the hardware. The most sensible parts to implement are the actual vector arithmetic operations, leaving the conventional part of the host to control these operations.

The architecture of the reconfigurable resource has less effect on the choice of partitioning. Most kinds of reconfigurable architecture are poor for control-dominant applications, as the frequent branching means that large parts of the hardware will be left unused for much of the application run-time. Even those architectures where each cell is a miniature, near complete processor, such as Raw machines [11] are less suited to control-dominant applications than conventional microprocessors as they usually have simplified control units and no centralised control – they are best suited to data-parallel applications rather than control-parallel ones. All the time-intensive parts of the ray tracing algorithm require arithmetic or comparisons of large numbers. None of the parts is so much more suited to fine-grain architectures, for example, that it would be more profitable to implement it than the intersection calculations.

3. Algorithm Description

In this section we break down the transformation of ray tracing from its usual depth-first style to a breadth-first approach into component parts. We show which transformations could be done using existing SUIF [12] passes and which would need custom passes.

SUIF (Stanford University Intermediate Format) is a software framework for developing parallelising compilers. Suif consists of a front end to translate C++ source code into an intermediate format, libraries of standard analysing and transforming compiler passes, and means for developers to build their own passes. Developers are free to choose which passes to use and their running order. SUIF can be adapted to work with C-like languages like Handel-C by using annotations on the intermediate format to represent Handel-C extensions such as variable bit widths and par statements. SUIF provides dataflow analyses to extract information about opportunities for parallelisation, as well as a basic paralleliser. SUIF also includes many utility passes – for example, the "skweel" pass does loop transformations such as skewing, reversal and interchange.

We use a basic version of the ray-tracing algorithm for clarity, consisting of a pair of mutually recursive procedures called for each screen pixel. In the code below we have already inlined the shade function into the main trace function – a standard compiler transformation:

```
const int maxLevel = 5; float minWeight = 0.01;

for (int i = 0; i < screenHeight; i++) {
   for (int j = 0; i < screenWidth; j++) {   /* for each pixel */
      Ray ray = makePrimaryRay(i,j);
      screen[i][j] = trace(1, 1.0, ray);
   }
}


Colour trace(int level, float weight, Ray ray) {
   Isect hit, closestHit; Object closestObj;
   bool anyHits = false;
   closestHit.dist = FLT_MAX; // large initial value
   for each object in scene {
      Isect hit;
      numHits = intersect(ray, object, hit);
      if (numHits > 0 && hit.dist < closestHit.dist) {
         closestObj =  object; closestHit = hit;
         anyHits = true;
      }
   }
   if (anyHits) {
      /* inlined shade function */
      Point p = ray.start + ray.dir * closestHit.dist;
      Vector n = calculateNormal(p, closestObj);
      Colour col = {0, 0, 0};
      for each light {
         Vector l = unitVector(light.pos - p);
         float nDotL = vecDot(n, l);
         if (nDotL > 0 && shadow(l, vecDist(light.pos, p)) > 0)
            colour += nDotL * light.colour;
      }
      if (level + 1 < maxLevel) {
         Ray tray.start = p;
         Surface * surf = closestHit.obj.surf;
         float newWeight = surf->kspec * weight;
         if (newWeight > minWeight) {
            tray.dir = reflectionDirection(ray.dir, n);
            return colour + surf->kspec * trace(level + 1, newWeight, tray);
         }
      }
      else
         return colour;
   }
   else
      return shadeBackground(ray);
}
```

As can be seen, the time-consuming intersection calculations (intersect and shadow) are tightly coupled to the rest of the algorithm. The goal of the transformation is to isolate these so they can be performed in large groups.

4. Algorithm Transformation

The top level of the raytracing algorithm consists of two mutually recursive functions: *trace* and *shade*. Trace, the entry function, intersects the ray with the objects, then calls the shade function to obtain the colour.

The algorithm transformation consists of 15 steps. The first step is to convert trace's recursion to an iteration and inline it into the loop. Although trace is tail-recursive, explicit stacks (arrays in the code) are made for all the parameters (except level, which is the iteration variable) to aid optimisation later. The return values for each recursion are accumulated into a scalar, which is allowed by the tail recursion optimisation. We have inferred a for-loop from the iteration. To give the for-loop the same behaviour as the recursion it eliminates, a new variable, finished, has been introduced to guard the loop body. When finished is true, the recursion would have finished. Clearly this pass is specialised and would need to be written for this application. SUIF can infer for-loops from the do-while loops that the tail-recursion elimination produces, but the rest of the transformations would need to be specially written. Of course, conventional scalar optimisers can perform tail-recursion elimination.

```
for (int i = 0; i < screenHeight; i++) {
   for (int j = 0; i < screenWidth; j++) {
      Ray ray = makePrimaryRay(i,j);
      bool finished = false;
      float weight[maxLevel]; Ray ray[maxLevel];
      Colour result= {0.0, 0.0, 0.0};
      for (int level = 1; level < maxLevel; level++) {
         /* inlined trace() */
         if (!finished) {
            Isect hit, closestHit;
            closestHit.dist = FLT_MAX;
            Object closestObj;
            weight[1] = 1.0;
            for each object in scene {
               numHits = intersect(ray[level], object, hit);
               if (numHits > 0 && hit.dist < closestHit.dist) {
                  closestHit = hit; closestObj = object;
               }
            }
            if any hits {
               Point p = ray.start + ray.dir * closestHit.dist;
               Vector n = calculateNormal(p, closestObj);
               for each light {
                  Vector l = unitVector(light.pos - p);
                  float nDotL = vecDot(n, l);
                  if (nDotL > 0.0 && shadow(l, vecDist(light.pos, p)) > 0)
                     colour += nDotL * light.colour;
               }
               if (level + 1 < maxLevel) {
                  Ray tray; tray.start = p;
                  Surface * surf = hit.obj->surf;
                  if (surf->kspec * weight[level] > minWeight) {
                     tray.dir = reflectionDirection(ray[level].dir, n);
                     weight[level + 1] = surf->kspec * weight[level];
                     ray[level + 1] = tray;
                  }
               }
            }
            else
               finished = true;
```

```
          result += weight[level] * surf->kspec * colour;
        }
        else
          result += weight[level] * shadeBackground(ray);
      }
    }
    screen[i][j] = result;
  }
}
```

Various other optimisations are not shown, such as hoisting initialisation out of the loop, propagating the values in the transmitted ray `tray` and hence eliminating `tray` itself. All these can be achieved with the standard SUIF *porky* phase [13]. This phase combines several utility transformations, such as constant propagation and the hoisting of initialisation code out of loops. Now that the loop has been made explicit, it can be interchanged with its outer loops. Loop interchange is a bread-and-butter transformation for SUIF and it would guard the `makePrimary` call and initialisation of finished, to make the transform correct. Each variable that persists between iterations of the inner loop needs to become an array, to keep access to that variable private to that pixel. SUIF would also normalise the outer loop indices. The result looks like:

```
for (int level = 0; level < maxLevel - 1; level++) {
  for (int i = 0; i < screenHeight; i++) {
    for (int j = 0; i < screenWidth; j++) {
      if (level == 0) {
        ray[level][i][j] = makePrimaryRay(i,j);
        finished[i][j] = false;
      }
      if (!finished[i][j]) {
        Isect hit, closestHit; ... etc.
      }
    }
  }
}
```

The following is a possible specification for the transform in the style of the CML-pre language [14], with the addition of a SUBSTITUTE block for describing various variable substitutions:

```
PATTERN {
  VAR x, y, z;
  for (x=EXPR(1); BOUND(1, x); STEP_EXPR(2, x)) {
    for (y=EXPR(1); BOUND(2, y); STEP_EXPR(3, y)) {
      for (z=EXPR(1); BOUND(3, z); STEP_EXPR(4, z)) {
        STMTLIST(1);
      }
    }
  }
}
CONDITIONS {
  stmtlist_has_no_unsafe_jumps(1);
}

RESULT {
  VAR x, y;
  for (y=EXPR(1); BOUND(2, y); STEP_EXPR(3, y)) {
    for (x=EXPR(1); BOUND (1, x); STEP_EXPR(2, x)) {
      STMTLIST(1);
```

```
        }
    }
}
SUBSTITUTE {
    for each VAR v in STMTLIST(1)
        causing dep ("* direction = (<,>) between STMTLIST(1) and STMTLIST(1)")
    VAR v_array[BOUND(1, x).max][BOUND(2, y).max] for VAR v
    v_array[x][y] for v
}
```

The substitution removes the dependencies by replacing accesses to the relevant variables by array accesses that are private to each pixel. This step has unlocked the parallelism in
the inner loop. Note that the inner loop body only has dependencies to itself in the outermost
loop. These dependencies correspond to the dependencies along each ray tree (in this case,
the ray "tree" only has one branch at each level). Ray intersections can now be mined from
the inner loop and scheduled in batches, allowing efficient communication across the bus and
scope for deeply pipelined hardware intersection units. One problem is that the parts we want
to parallelise are guarded by the finished variable for each loop. We cannot simply strip mine
the inner loop as in conventional compiler restructuring, because after the primary rays there
will be less than one ray per pixel on average.

The maximum number of intersections that can be carried out at once is determined by
the size of the available memory on the hardware target divided by the size of each intersection result. In an architecture with multiple memory banks, we will typically want to use one
bank for intersections results and one for each kind of input (rays, objects). The pass writer
should use these parameters, available from the architecture description, to determine the size
of buffers used in communicating with the hardware.

Here is the start of one possible solution:

```
for (int level = 0; level < maxLevel - 1; level++) {
    numRaysToIntersect = 0;
    int oldi = 0, oldj = 0;
    bool exit = false;
    for (int i = oldi; i < screenHeight && !exit; i++) {
        for (int j = oldj; j < screenWidth && !exit; j++) {
            if (level == 0) {
                ray[level][i][j] = makePrimaryRay(i,j);/*guarded by if*/
                finished[i][j] = false;
            }
            if (!finished[i][j]) {
                Point p; Vector n; Isect hit, closestHit;
                Object closestObj; Colour result= {0.0, 0.0, 0.0};
                float weight[maxLevel]; Ray ray[maxLevel];
                weight[1] = 1.0;
                for each object in scene { /* mine rays */
                    raysToIntersect[currRay] = ray;
                    objectsToIntersect[currRay] = object;
                    numRaysToIntersect++;
                    rayPixelX[currRay] = x;
                    rayPixelY[currRay] = y;
                }
                if (numRaysToIntersect >
                    maxRaysToIntersect - number of objects in scene) {
                    exit = true;
                }
            }
        }
    }
```

```
for (int rayNum = 0; rayNum < numRaysToIntersect; rayNum++) {
   numHitsBuffer[rayNum] =
   intersect(raysToIntersect[rayNum], objectsToIntersect[rayNum],
      hits[rayNum]);
}
for (int rayNum = 0; rayNum < numRaysToIntersect; rayNum++) {
   for each pixel in ray buffer {
      find closest ray intersect
      if any hits {
         p[i][j] = ray[i][j].start + hit.dist * ray[i][j].dir;
      }
   }
}
}
```

This incomplete restructuring shows the general ideas. Rays are intersected in batches, mined from each unfinished pixel. The original `for`-loops are split up to allow the rays to be intersected in batches.

5. Hardware Estimate for Complex Scenes

So far our hardware runs at about 16MHz and produces a result every three cycles – about five million intersections per second (ips). In contrast our software implementation achieves about two million ips on an 800MHz Pentium III.

In this section we examine the performance we achieve with the hardware developed so far and what hardware would be needed to generate real-time ray-traced images for interactive applications. We calculate, for the current hardware:

- the number of objects which could be animated at 25 frames per second, 400x400 resolution
- the frame rate for a single object at the same resolution

We then assess what hardware would be needed for a realistic scene, such as a scene with five objects comprising 10 primitives apiece. With suitably complex primitives such as cones, cylinders and ellipsoids, this is enough to model even a relatively complex object such as an aircraft. Methods using triangles as primitives (such as hardware Z-buffering) would require many more of their primitives to model objects with the same fidelity – perhaps several thousand.

The calculations assume that all hardware runs at its peak rate (no allowance for pipeline filling, or other set-up operations) and that corresponding resources are available to perform the rest of the ray tracing algorithm at the same rate. They also assume that exhaustive ray tracing (as above) is still necessary, with no software optimisations to reduce the number of rays cast. This means that each ray must be intersected with each object, and a ray must be traced for each pixel:

$$ipp = n + ((n * m) + n) * a = n * (1 + a * (1 + m))$$

where ipp denotes the intersections per pixel, n denotes the number of objects, m denotes the number of light sources, and a denotes average recursion depth.

Using multiplier cores we expect to be able to run at a speed up to 25MHz, the maximum speed at which the RC1000-PP's memory can be clocked. With five memory banks, a result could be produced every cycle. For real-time rendering of complex scenes, say 2,000 objects at 640 by 480 resolution, 25 frames per second, we would need $2000 \times 640 \times 480 \times 25 = 15.4 \times 10^9$ ips. This omits time for bus transfers, but use of existing software optimisations for ray tracing would greatly reduce the number of intersections needed. If we assume these optimisations can yield a 15-fold reduction in the number of intersections, this number is in range of a system containing ten Virtex devices, with intersectors clocked at 100MHz, with 100MHz RAM. Two of the RAM banks, used for output of results, would need to be as large as possible (at least 128MB) to minimise bus transfers.

6. Exploiting Run-Time Reconfiguration

Many FPGAs can be reconfigured at run time. In the following, we outline two ways to support hardware reconfigurability.

Firstly, reconfiguration can be used to alter the tradeoff between rendering speed and image quality. In fast-moving sequences, or those with many visible objects, the frame rate can be maintained by selectively narrowing the number format and lowering the image quality. A narrower format could also be used for secondary rays (reflections and shadows) as observers are less sensitive to the fidelity of these compared to the directly-visible objects.

Secondly, reconfiguration allows the balancing of hardware resources with the changing proportions of different objects within a frame. We suggest two ways to use reconfiguration for this purpose.

The first way is based on the pipeline morphing technique [15] and thus needs partially reconfigurable devices. Each device would contain a single pipeline, with each stage morphing as appropriate for its current object.

The second way is suited to mainstream devices like the Xilinx Virtex. Current devices cannot be reconfigured fast enough for each frame – the Virtex takes 23.6ms to reconfigure completely [16], which leaves little time left for computation and communication (at 25 frames per second, only 40ms is available). Rendering the frames in sequences at a speed slightly above the display frame rate can save enough time over the sequence to spend on reconfiguration. The sequences should be short or the configuration will poorly match the object types at the end of the sequence.

7. Summary

This paper presents a case study implementing ray tracing on reconfigurable hardware. We map ray-sphere intersection, the most time-consuming part of the algorithm, to the hardware. Although the hardware is used several million times per frame, this results in poor performance because each use sends a few bytes to and from the hardware across the bus, resulting in very poor transmission rates.

We suggest two approaches to improve performance: first, more of the algorithm could be placed onto hardware. Second, the algorithm could be transformed into a breadth-first approach, in contrast with its original depth-first style. We choose the second approach as current hardware is less well-suited to control-dominant applications, like the rest of the ray tracing algorithm. Transforming to a breadth-first approach is more than just adding simple buffering, due to the way that ray tracing uses the results from previous intersections in further intersection calculations. Although the breadth-first approach has much poorer performance in software, due to poorer caching behaviour, the combined software and hardware performs much better than the depth-first version.

Finally, we show how to transform the algorithm to the breadth-first approach using a mixture of well-known and custom compiler transformations, both of which can be potentially automated using the framework proposed in this paper.

We propose several extensions to the raytracing work: incorporating our work into other raytracing packages and supporting other primitives.

Our work could be incorporated into existing raytracing packages such as PovRay [17]. Although very large, PovRay's implementation broadly follows the pattern we consider in our example transformation. The automated transform would need to be updated for PovRay's extra features, but could still be used as a base, without needing to be totally rewritten.

Styles [18] has already implemented the ray-triangle intersection algorithm on reconfigurable hardware. We can implement other intersection algorithms for quadrics and bicubic patches. For bicubic patches, which are often implemented using iterative algorithms, reconfigurable hardware has the disadvantage of needing a fixed-length pipeline, for the worst-case number of iterations. However, we can vary the number precision within the pipeline so the first, initial estimates use less precision and hence less hardware.

Future work could also put more of the algorithm into hardware. First candidates would be primary ray generation and sorting of the closest intersection. Also, some contemporary hardware such as the Xilinx Virtex II Pro [19] includes a small instruction processor within the FPGA. This overcomes our objections to putting the more control-dominated parts of the algorithm into hardware.

References

[1] A. Glassner (ed), *An Introduction to Ray Tracing*, Academic Press 1989.

[2] Advanced Rendering Technology web site, `http://www.artvps.com/`.

[3] S. Parker, W. Martin, P. Sloan, P. Shirley, B. Smits, C. Hansen, "Interactive Ray Tracing", in *Proceedings 1999 Symposium on Interactive 3D Graphics*, ACM Press, April 1999.

[4] Sven Woop, Jrg Schmittler, and Philipp Slusallek, "RPU: A Programmable Ray Processing Unit for Real-time Ray Tracing" in *Proceedings SIGGRAPH 2005*, IEEE, 2005

[5] E. Haines, "Essential Ray Tracing Algorithms", in [1].

[6] P. Heckbert, "Writing a Ray Tracer", in [1].

[7] P. Hanrahan, "Using Caching and Breadth-first Search to Speed Up Ray-Tracing", in *Proceedings of Graphics Interface '86*, May 1986, pp56–61.

[8] H. Muller, J. Winckler, "Distributed Image Synthesis With Breadth-First Ray Tracing and the Ray-Z-buffer", in B. Monien, T. Ottmann (eds), *Data Structures and Efficient Algorithms - Final Report on the DFG Special Initiative*, Springer-Verlag LNCS 594, 1992, pp125–147.

[9] K. Nakamaru, Y. Ohno, "Breadth-First Ray Tracing Using Uniform Spatial Subdivision", in *IEEE Transactions on Visualization and Computer Graphics* Vol. 3, No. 4, IEEE, 1997.

[10] D. Plunkett, M. Bailey, "The Vectorization of a Ray-Tracing Algorithm for Increased Speed", *IEEE Computer Graphics and Applications* Vol. 5, No. 8, 1985.

[11] E. Waingold, M. Taylor, D. Srikrishna, V. Sarkar, W. Lee, V. Lee, J. Kim, M. Frank, P. Finch, R. Barua, J. Babb, S. Amarasinghe, A. Agarwal, "Baring It All to Software: Raw Machines". *Computer* Vol. 30, No. 9, IEEE, 1997, pp86–93.

[12] SUIF 2 web site, `http://suif.stanford.edu/suif/suif1/index.html`.

[13] SUIF Manual page for porky, Available at `http://suif.stanford.edu/suif/suif1/docs/man_porky.1.html`.

[14] M. Boekhold, I. Karkowski, H. Corporaal, A. Cilio, "A Programmable ANSI C Transformation Engine", in S. Jähnichen (ed), *Compiler construction: 7th International Conference*, Springer-Verlag LNCS 1575, 1999, pp292–295.

[15] W. Luk et al., "Pipeline Morphing and Virtual Pipelines", in *Field Programmable Logic and Applications: 7th International Workshop*, Springer-Verlag LNCS 1304, 1997, pp111–120.

[16] H. Styles, W. Luk, "Customising Graphics Applications: Techniques and Programming Interface", *Proceedings IEEE Symposium on Field-Programmable Custom Computing Machines*, IEEE Computer Society Press, 2000, pp77–87.

[17] PovRay web site, `www.povray.org`.

[18] H. Styles, W. Luk, "Accelerating Radiosity Calculations using Reconfigurable Platforms", *Proceedings IEEE Symposium on Field-Programmable Custom Computing Machines*, 2002, pp279–281.

[19] Xilinx, "Introduction to the Virtex-II FPGA Family", `http://www.xilinx.com/products/virtex/handbook/ug002_intro.pdf`.

Communicating Process Architectures 2007
Alistair McEwan, Steve Schneider, Wilson Ifill and Peter Welch (Eds.)
IOS Press, 2007

493

A Reconfigurable System-on-Chip Architecture for Pico-Satellite Missions

Tanya VLADIMIROVA and Xiaofeng WU

Surrey Space Centre
Department of Electronic Engineering
University of Surrey, Guildford, GU2 7XH, UK
{ T.Vladimirova, X.Wu}@surrey.ac.uk

Abstract. Spacecraft operate in the unique space environment and are exposed to various types of radiation. Radiation effects can damage the on-board electronic circuits, particularly silicon devices. There is a pressing need for a remote upgrading capability which will allow electronic circuits on-board satellites to self-repair and evolve their functionality. One approach to addressing this need is to utilize the hardware reconfigurability of Field Programmable Gate Arrays. FPGAs nowadays are suitable for implementation of complex on-board system-on-chip designs. Leading-edge technology enables innovative solutions, permitting lighter pico-satellite systems to be designed. This paper presents a reconfigurable system-on-chip architecture for pico-satellite on-board data processing and control. The SoC adopts a modular bus-centric architecture using the AMBA bus and consists of soft intellectual property cores. In addition the SoC is capable of remote partial reconfiguration at run time.

Keywords. System-on-a-chip architecture, pico-satellite, partial run-time reconfiguration.

Introduction

The miniaturisation of the satellite platform is an active field of research and commercial activities. Small satellites are characterised by low cost and rapid time-to-market development (often ranging from six to thirty-six months) when compared with the conventional space industry. An approach to classifying satellites in terms of deployed mass has been generally adopted, as detailed in Table 1.

Leading-edge technology enables innovative solutions, permitting lighter satellite systems to be designed inside smaller volumes. Very small satellites, having a mass less than one kilogram, have the potential to enable a new class of distributed space missions by merging the concepts of distributed satellite systems and terrestrial wireless sensor networks. Many new distributed space mission concepts require hundreds to thousands of satellites for real-time, distributed, multi-point sensing to accomplish advanced remote sensing and science objectives.

Spacecraft operate in the unique space environment and are exposed to various types of radiation. Radiation effects can damage the on-board electronic circuits, particularly silicon devices. As satellites are not available for physical repairs or modifications after launch there is a pressing need for a remote upgrading capability which will allow electronic circuits on-board satellites to self-repair and evolve their functionality.

The Field Programmable Gate Array (FPGA) technology enables reconfiguration and evolution of hardware designs composed of soft intellectual property (IP) cores on board satellites. FPGAs nowadays are suitable for implementation of complex on-board system-on-chip (SoC) designs, for example a complete on-board controller could be implemented using Xilinx Virtex FPGAs [1,2]. A disadvantage of SRAM-based devices is that they are vulnerable to the high levels of radiation in the space environment [3]. Heavy ions from cosmic rays can easily deposit enough charge in or near an SRAM cell to cause a single-bit error, or single event upset (SEU). Because SRAM FPGAs store their logic configuration in SRAM switches, they are susceptible to configuration upsets, meaning that the routing and functionality of the circuit can be corrupted.

Table 1. Classifying satellites by mass.

Large satellites	> 1000 kg
Medium satellites	500 – 1000 kg
Mini-satellites	100 – 500 kg
Micro-satellites	10 – 100 kg
Nano-satellites	1 – 10 kg

The ESPACENET (Evolvable Networks of Intelligent and Secure Integrated and Distributed Reconfigurable System-On-Chip Sensor Nodes for Aerospace Based Monitoring and Diagnostics) project, targets the development of a robust space sensor network based on flexible pico-satellite nodes [4]. Important feature of the network is its reconfigurability which will be manifested at two levels – node level and system level. The pico-satellite nodes will include reconfigurable SoC devices to process data from various sensing elements. A generic SoC controller design encompassing a number of soft IP cores and driver capability, will be utilized. This will enable SoC customization at run time to suit best the processing requirements of the network.

In this paper we present a modular SoC design based on the LEON microprocessor core and the Advanced Microcontroller Bus Architecture (AMBA) for payload computing, targeting the CubeSat pico-satellite platform [5]. The SoC is capable of partial run-time reconfiguration, which can be used to mitigate radiation effects by repairing the damaged area while the test of the modules continue their operation. At the same time this SoC architecture is also used to evolve on-chip circuits in order to adapt to changes in the satellite mission. The paper is organized as follows. Section 1 describes the proposed system-on-chip architecture. Section 2 details the design tools and methodologies for remote on-board partial run-time reconfiguration. Section 3 presents a case study verifying the feasibility of the design and discusses self-repair and hardware evolution on-chip support for future satellite missions. Section 4 concludes the paper.

1. A Reconfigurable SoC Architecture

The proposed SoC design is targeted at the Xilinx Virtex series of FPGAs. The central processing unit of the SoC is the LEON3 microprocessor, which is a SPARC V8 soft intellectual property core written in VHDL [6]. The SoC is an AMBA centric design and subsystems of the OBC of the spacecraft can be added to the LEON3 processor providing that they are AMBA interfaced.

1.1 The AMBA Bus

The AMBA bus [7] defined by ARM is a widely used open standard for an on-chip bus system. This standard aims to ease the component design, by allowing the combination of interchangeable modules in the SoC design. It promotes the reuse of intellectual property (IP) cores, so that at least a part of the SoC design can become a composition, rather than a complete rewrite every time.

The AMBA standard defines different groups of buses, which are typically used in a hierarchical fashion. A typical microprocessor design (Figure 1) usually consists of a system bus; either the older version, the Advanced System Bus (ASB), or the Advanced High-performance Bus (AHB). All high performance components are connected to the system bus. Low speed components are connected to the peripheral bus, the Advanced Peripheral Bus (APB).

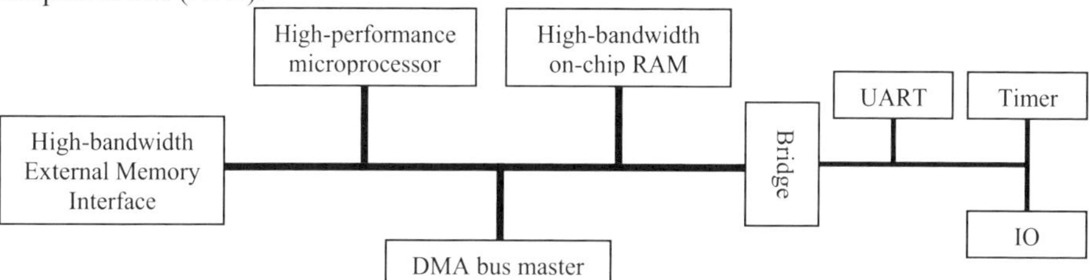

Figure 1. AMBA-based SoC architecture

The system buses ASB and AHB are designed for high performance connection of processors, dedicated hardware and on chip memory. They allow: multiple bus masters; pipelined operations; and burst transfers. The peripheral bus APB on the other hand is designed for low power peripherals with a low complexity bus interface. The APB can be connected via a bridge to both system buses AHB and ASB. The APB bridge acts as a master on the APB bus and all peripheral devices are slaves. The bridge appears as a single slave device on the system bus; it handles the APB control signals, and performs retiming and buffering.

Between the two system buses the AHB delivers a higher performance than its older counterpart ASB. The AHB features: retry and split transactions; single clock edge operation; non-tristate implementation; and wider data bus configuration (e.g. 64 bits and 128 bits). Retry and split transactions are introduced to reduce the bus utilization. Both can be used in case the slave does not have the requested data immediately available. In case of a retry transaction, the master retries the transaction after and own arbitrary delay. On the other hand in a split transaction the master waits for a signal from the slave that the split transaction can be completed.

A key task for today's SoC designers is to ensure that each component in the system obeys the interconnecting bus protocol. It is therefore of paramount importance to exhaustively prove that an IP peripheral core in the SoC architecture obeys the AHB protocol rules. SolidPC is a software package based on the static functional verification tool Solidify™ that can check the AMBA bus compliance for register transfer level (RTL) designs [8].

1.2 LEON3 Based System-on-a-Chip

The SPARC V8 is a RISC architecture with typical features like large number of registers and few and simple instruction formats. However, the LEON3 IP core is more than a SPARC compatible CPU. It is also equipped with various modules that interconnect through two types of the AMBA bus (AHB and APB), e.g. Ethernet, SpaceWire, PCI, UART etc. Figure 2 shows the diagram of the SoC architecture. Different subsystems will be considered for specific satellite missions, for example a high-level data link controller (HDLC) interface for downlink and uplink data transmission, a compression core, an encryption hardware accelerator, etc..

So far we have introduced the soft IP cores of the SoC architecture, however, the Xilinx FPGAs also provide on-chip hard-wired cores, e.g. Block SelectRAM (BRAM), multipliers. Starting from the Virtex II series, Xilinx Virtex FPGAs have integrated an internal configuration access port (ICAP) into the programmable fabric, which enables the user to write software programs that modify the circuit structure and functionality at run-time for an embedded processor. The ICAP is actually a subset of the SelectMAP interface [9], which is used to configure Xilinx FPGAs. Hard processor IP cores (PowerPC) are also available in some Virtex II Pro and Virtex IV FPGAs.

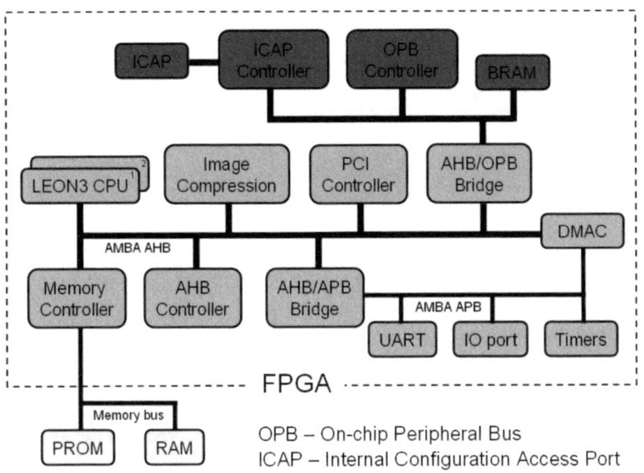

Figure 2. The SoC architecture of the OBC

The on-chip peripheral bus (OPB) is used to connect all the ICAP modules. The ICAP is connected to the LEON3 processor via the OPB-to-AHB bridge. Once the FPGA is initially configured, the ICAP is used as an interface to reconfigure the FPGA. The ICAP device driver is included in the embedded Linux operating system [10]. The control logic for reading and writing data to the ICAP is implemented in the LEON3 processor as a software driver. The BRAM is used as a configuration cache. Because Virtex II FPGAs support reconfiguration only by frames, the BRAM must be big enough to hold one frame each time. The bitstream of each SoC component can be stored on board in a Flash memory. The bitstream of a new or upgraded SoC component can be uploaded through the satellite uplink from the ground station.

2. Partial Run-time Reconfiguration with Xilinx FPGAs

Partial run-time reconfiguration is the ability to update only a part of the logic in an FPGA without affecting the functionality of the unchanged section of the FPGA. This allows designers to plug in components for the purpose of adding new functional blocks, upgrading and improving existing ones as well as correcting malfunctions. This capability can also be used to mitigate radiation effects by repairing the areas affected by soft failures.

2.1 Mitigation of Radiation Effects

Radiation effects in SRAM-based FPGAs have been a topic of active investigation over the last couple of years. M. Ohlsson [11] studied the sensitivities of SRAM-based FPGAs to atmospheric high-energy neutrons. FPGAs were irradiated by 0-11, 14 and 100 MeV neutrons and showed a very low SEU susceptibility. P. Graham [12] classified the radiation effects in SRAM FPGAs and showed that SEUs can result in five main categories of design changes: mux select lines, programmable interconnect point states, buffer enables, LUT values, and control bit values.

A number of SEU mitigation strategies for SRAM-based FPGAs have been developed [13]. Scrubbing is the periodic readback of the FPGA's configuration memory followed by comparing of the memory content to a known good copy and writing back any corrections required. By periodically scrubbing an FPGA, configuration errors present in the FPGA can be corrected. Triple module redundancy (TMR) is an effective technique creating fault tolerant logic circuits. In TMR, the design logic is tripled and a majority voter is added at the output. Recently, Xilinx [14] have provided a design tool, XTMR that automatically implements TMR in Xilinx FPGA designs, protecting from SEUs the voting circuits. However, designs with TMR are at least three times as large as non-TMR designs, and suffer from speed degradation as well. Power consumption is also tripled along with the logic.

Xilinx produced two design flows for partial run-time reconfiguration: module-based and difference-based [15]. Difference based partial reconfiguration is accomplished by making a small change to a design, and then generating a bitstream based only on the difference between the two designs. For the difference-based design flow, the JBits development environment [16] is widely used to create partial bitstreams, which can be committed to FPGAs via the Xilinx hardware interface (XHWIF). Module-based partial reconfiguration is used with the proposed SoC design as described in the section 2.2 below.

SEU effects due to radiation can be mitigated if we can detect the faulty area in the SoC and partially reconfigure that area with the correct bitstream using either module-based or difference-based partial run-time reconfiguration techniques. We can produce bitstreams for each peripheral module, and store them in the on-board memory. Hence, when a peripheral is affected by SEUs, the bitstream of this peripheral will be reloaded from the memory and written into the FPGA's configuration memory.

2.2 Module-Based Partial Run-Time Reconfiguration

For the module-based design flow, partial bitstreams can be created using the Xilinx PlanAhead tool [17], which can then be committed to FPGAs using the SelectMAP interface or the on-chip ICAP module. The OPB interface to the ICAP module permits connection of this peripheral to the MicroBlaze soft core processor or the PowerPC hard core processor inside the FPGA. J. Williams [10] developed an ICAP device driver for the uCLinux kernel, running on the MicroBlaze processor.

With the modular design flow, a design is divided into modules, which can be developed in parallel and merged into one FPGA design later. Modular design also allows modifying a module while leaving the other modules stable, intact and unchangeable. The communication between the reconfigurable and fixed modules happens on the ABMA bus through the special bus macro developed by Xilinx as shown in Figure 3.

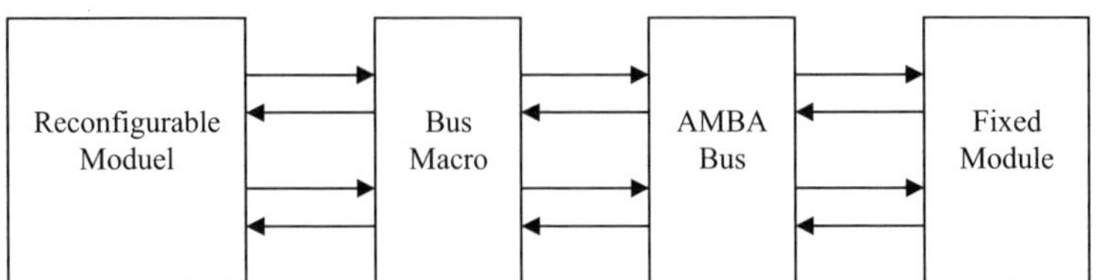

Figure 3. Module-based partial run-time reconfiguration

An example of the physical implementation of the bus macro is shown in Figure 4. It is a pre-defined, pre-routed component, which consists of eight tristates buffers and can provide a 4-bit bus. The direction of each signal on the bus is decided by the three-state input that is active low. For example, if LT[0] and RT[0] are set to " 0" and " 1" respectively, then the signal direction is left-to-right and the bus provides the connection between RO[0] and LI[0].

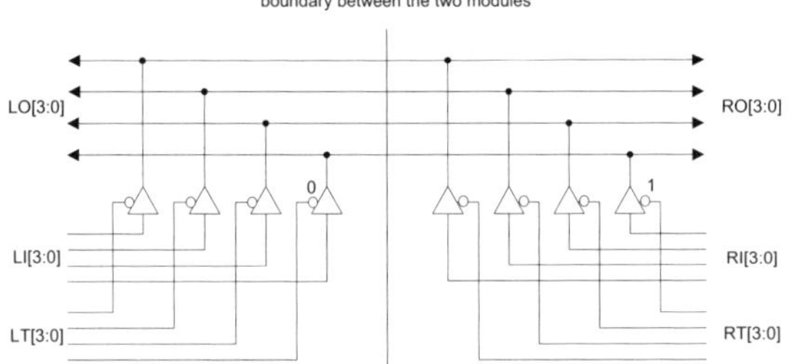

Figure 4. Implementation of the bus macro

2.3 Remote Partial Run-Time Reconfiguration

The proposed SoC is also capable of remote partial run-time reconfiguration. This can be realised by sending the partial bitstream from a remote location (either the ground station or other satellites) to the satellite. The satellite stores this bitstream in its on-board memory and uses it to reconfigure the corresponding area on the FPGA.

LEO satellites are only visible from the ground station for 10 minutes two to three times a day, which can be a problem for the ground-to-satellite reconfiguration. Furthermore, the uplink rate of small satellites is often of low baud rate – from 9.6 Kbit/s to 19.2 Kbit/s, although with some small satellite platforms, the uplink can reach 128 Kbit/s. The SoC configuration file for the Virtex II FPGAs is over 10 Mbit (10 Mbit for a XC2V3000, 20 Mbit for a XC2V6000). It may be difficult to upload such a big file via the low uplink rate, although the partial configuration file is much smaller than the complete one.

One solution to the low uplink baud rate problem is to compress the configuration file before uploading and to decompress it after uploading. A configuration file can be compressed up to around 25% of the original file even using a routine text compression algorithm. For example, the LEON3 processor [6], which is the main IP core in the SoC design, results in a bitstream of 1,311,870 bytes when targeting the XC2V3000 FPGA. After compression using WinRar, the size is reduced to 376,063 bytes, which is 28.67% of the original size. The worst situation occurs when the complete configuration file needs to be uploaded. In this case it would take about 23 seconds to upload the compressed file, excluding the control signal transmission overhead.

The partial run-time reconfiguration capability is aimed at a pico-satellite constellation where satellites are inter-networked using inter-satellite links (ISL). If a configuration file is too big to be transferred in time, we may adopt a distributed reconfiguration scheme. When a large configuration file needs to be transmitted from the ground stations, it is split into a number of smaller portions, which are transmitted to satellites being in range of the ground station. After these files are received on-board, they are sent to the destination pico-satellite over inter-satellite links. Then the data are fused to reconfigure the SoC. For this scheme, middleware is required for the communication between the servers and clients. The IEEE 802.11 wireless communication standard is currently under investigation for the inter-satellite links.

3. On-Board Computer SoC Reconfiguration

In this section we present an example to demonstrate partial run-time reconfiguration by adding a direct memory access controller (DMAC) to the SoC OBC. In addition SEU self-repair and hardware evolution support for the SoC is discussed.

3.1 DMA Controller Case Study

The proposed SoC has several high data rate interface modules. The SpaceWire interface with a data rate up to 400 Mbit/s is used to connect to other on-board devices. The HDLC interface with up to 10 Mbit/s is employed for uplink and downlink data transmission to the ground station. The DMAC handles the data transfer between the main memory and the peripherals bypassing the CPU. At the Surrey Space Centre a soft DMAC IP core was developed for the AMBA interface [18]. Figure 5 shows the block diagram of the DMAC and its interconnection with the peripherals.

The CPU allocates a memory block and assigns it to the DMAC and then writes the transfer mode and the peripheral device address to the DMAC registers. After configuring the DMAC there are two possibilities to trigger the data transfer process: 1) the CPU sends a start command to the DMAC; 2) the transfer will be triggered via hardware handshake between the DMAC and the peripheral device. In the latter case the device must be DMA-capable by providing appropriate hardware handshake signals. The minimal hardware handshake between the DMAC and the peripheral device consists of a request signal. In addition, an acknowledge signal is normally used additionally. If a peripheral device receives data from "outside" it asserts the request signal DREQ. The DMAC transfers the received data from the peripheral device controller to the memory and asserts the acknowledge signal DACK. When the transfer is completed a state bit will be set in the DMAC or the DMAC causes an interrupt.

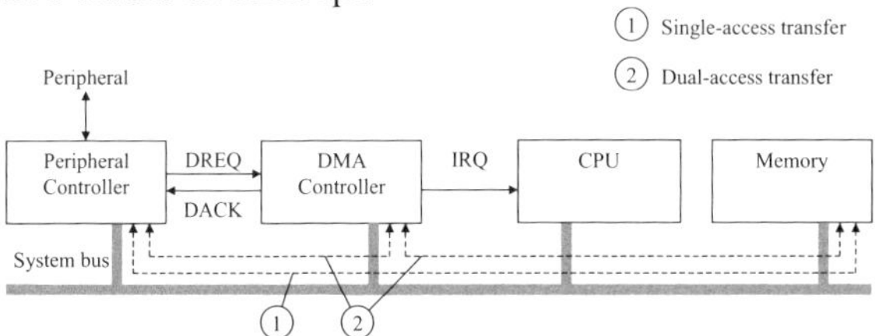

Figure 5. Interconnection between the DMAC and the peripherals [18]

There are two types of data transfer – single access and dual access. In the single-access transfer the DMAC activates the control and address bus signals, the peripheral device puts its data on the data bus and the memory reads the data, or the memory puts its data on the data bus and the peripheral device reads it. In the dual-access transfer the DMAC reads the data from a peripheral device or memory and buffers it internally and then writes the data to memory or to a peripheral device.

We first implement a partial SoC, which consists of the LEON3 processor, the ICAP, and the BRAM, into the Virtex II FPGA. Then we add the DMA controller into the partial SoC while it is running. Synplify Pro is used to produce the netlists of the partial SoC, the DMA controller, and the complete SoC that consists of both the partial SoC and the DMA controller. The resultant netlists are floorplanned using the PlanAhead tool. The reason to floorplan the complete SoC is that it provides a reference for the placing of the individual components. Hence it ensures that the dynamic circuit (i.e. the DMAC) is correctly interfaced to the static circuit (i.e. the partial SoC). Figure 6 illustrates the design partitioning between the resultant static and dynamic circuits. Bus macros are inserted to interface signals between the static and dynamic circuit partitions on the AMBA bus.

We download the partial SoC bitstream to the FPGA and store the DMAC bitstream dma.bit in the memory. At the same time the image of the SnapGear Linux is downloaded to the bootloader. After system boot the ICAP device is automatically registered as /dev/icap. We can manually reconfigure the SoC through the debugging window on the PC terminal. The reconfiguration can be achieved simply by executing the following command:

$ cat dma.bit > /dev/icap

Now the DMAC is added to the SoC and it is ready to transfer data between the peripherals and the memory. In order to check whether or not the DMAC works we connect the SoC to a PC via the RS232 interface. We create a data block with arbitrary values and send the data block size and the data block to the RS232 interface.

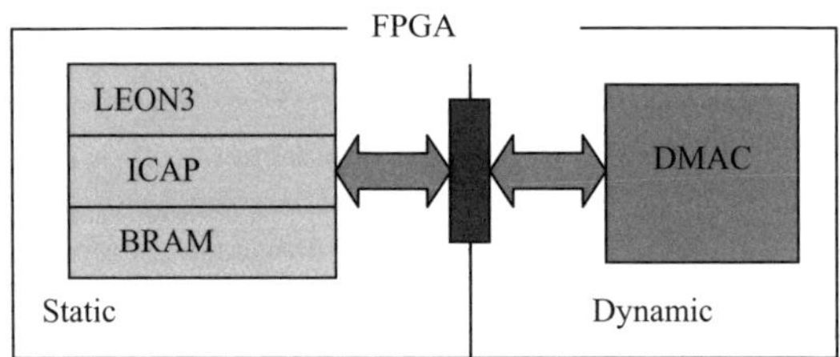

Figure 6 .Block-diagram of the design partitioning for partial reconfiguration

The LEON3 processor receives the block size from the serial interface and configures the DMAC according to this size. After initiation of the DMA transfer the LEON UART sends a DMA request with each received byte. So the DMAC controller reads each received byte from the UART and transfers it to the main memory. Furthermore the processor calculates a check sum for all received values. The results are printed to the debugging window through the serial interface as shown in Figure 7.

3.2 SEU Self-Repair and Hardware Evolution Support

Work is in progress on providing on-chip support for hardware evolution, which will enable the SoC to autonomously self-repair and update its modules. The hardware evolution will be achieved by an evolutionary algorithm (EA), which will create a new bitstream to update the payload SoC. For that the SoC design will be expanded with hard PowerPC processor cores to run the evolutionary algorithm or an evolvable soft IP core which will allow the architecture to evolve into optimized configurations.

```
$ ./dmatest
I am waiting for data!
Wait for end of transfer!
    The DMA controller transferred 1000 bytes from
    the UART to the memory.
I am calculating check sum.
The check sum is 248.
I am waiting for data.
```

Figure 7. Testing of the DMAC by transferring 1000 bytes from the UART to the memory

Evolutionary algorithms, based on different models of the biological evolution theory are one of the most popular general-purpose optimizers. They can successfully work with any cost function. The main drawback of EAs is the huge computational burden as in each iteration they evaluate a "population" of several test solutions. The evaluation is a time consuming task, which needs many iterations to converge. During evaluation the whole population of many test solutions must be calculated and all values of cost functions are used in the "survive-contest" to choose candidates, which will be further improved. Each test solution is independent from the others, so this stage is a natural place where parallelization can be applied. In a distributed environment the computational resource for evaluation of an "individual" could be located on a separate node, which could communicate the results to the node where the main EA runs [19]. On-board hardware evolution will be greatly assisted by a distributed implementation of EA accross a network of pico-satellites connected by intersatellite links.

4. Conclusions

Future space missions are envisioned as highly autonomous, intelligent and distributed multi-spacecraft missions consisting of miniaturized satellite nodes. Constellations of very small satellites can be used to implement virtual satellite missions, which are a cost-effective and flexible alternative approach to building large spacecraft. There is a pressing need for remote upgrading capability in satellites after they are launched. FPGAs provide flexibility of design; shorter time-to-market; lower cost; reconfigurability etc., which makes them suitable for use on board very small satellites. The implementation of an on-board computer together with its peripherals on a single reconfigurable FPGA provides the possibility for conditional maintenance, self-repair and upgrade.

In this paper we present a SoC architecture for on-board partial run-time reconfiguration to enable system-level functional changes on board satellites, ensuring correct operation, longer life and higher quality of service while satellites are running in space. The SoC design is an attempt to build a generic on-board computer, which takes advantage of high-density SRAM-based FPGAs from Xilinx. The SoC is designed based on the AMBA bus, from which it is able to dynamically add/remove modules. Distributed computing over inter-satellite links will enable on-board hardware evolution in future pico-satellite networks.

References

[1] H.Tiggeler, T.Vladimirova, D.Zheng, J.Gaisler. A System-on-a-Chip for Small Satellite Data Processing and Control, *Proceedings of 3rd Military and Aerospace Applications of Programmable Devices and Technologies International Conference (MAPLD'2000)*, P-20, September 2000, Laurel, Maryland US, NASA.

[2] T.Vladimirova and M.N.Sweeting. System-on-a-Chip Development for Small Satellite On-Board Data Handling – AIAA Journal of Aerospace Computing, Information, and Communication, Vol. 1, No. 1, pp. 36-43, January 2004, AIAA

[3] M.Caffrey, P.Graham, E.Johnson, M.Wirthlin, N.Rollins, and C.Carmichael. Single-Event Upsets in SRAM FPGAs, *Proceedings of 5th Military and Aerospace Applications of Programmable Devices and Technologies International Conference (MAPLD'2002)*, P8, 2002, Laurel, Maryland, USA.

[4] N.Haridas, E.Yang, A.T.Erdogan, T.Arslan, N.Barton, A.J.Walton, J.S.Thompson, A.Stoica, T.Vladimirova, X.Wu, K.D. McDonald-Maier, W.G.J. Howells. ESPACENET: A Joint Project for Evolvable and Reconfigurable Sensor Networks with Application to Aerospace–Based Monitoring and Diagnostics – Proceedings of 6th International Conference on Recent Advances in Soft Computing (RASC2006), Ed. K.Sirlantzis, pp. 410-415, 10-12 July 2006, Canterbury.

[5] T.Vladimirova, X.Wu, A.-H.Jallad and C.P.Bridges. Distributed Computing in Reconfigurable Picosatellite Networks, to appear in Proceedings of 2007 NASA/ESA Conference on Adaptive Hardware and Systems, August 5-8, 2007, Edinburgh.

[6] J.Gaisler. GRLIB IP Library User's Manual (Version 1.0.4). Gaisler Research, 2005.

[7] AMBA Specification (Rev 2.0), ARM Ltd., 1999

[8] SolidPC Datasheet, www.saros.co.uk/amba

[9] B.Blodget, P.James-Roxby, E.Keller, S.McMillan, and P.Sundararajan. A Self-reconfiguration Platform, *Proceeding of 13th International Conference on Field-Programmable Logic and Applications, FPL'2003*, pp. 565-574. 2003, Lisbon, Portugal.

[10] J.A.Williams, and N.W.Bergmann. Embedded Linux as a Platform for Dynamically Self-Reconfiguring Systems-On-Chip, *Proceedings of the International Conference on Engineering of Reconfigurable Systems and Algorithms (ERSA 2004)*, 2004, Las Vegas, Nevada, USA.

[11] M.Ohlsson, P.Dyreklev, K.Johansson, and P.Alfke. Neutron Single Event Upsets in SRAM-based FPGAs, *Proceedings of IEEE Nuclear and Space Radiation Effects Conference (NSREC'1998)*, 1998, Newport Beach, California, USA.

[12] P.Graham, M.Caffrey, J.Zimmerman, P.Sundararajan, E.Johnson, and C.Patterson. Consequences and Categories of SRAM FPGA Configuration SEUs, *Proceedings of the 6th Military and Aerospace Applications of Programmable Devices and Technologies International Conference (MAPLD'2003)*, C6, 2003, Washington DC, USA.

[13] M.Stettler, M.Caffrey, P.Graham, and J.Krone. Radiation effects and mitigation strategies for modern FPGAs, *Proceedings of 10th Workshop on Electronics for LHC Experiments and Future Experiments*, 2004, Boston, USA.

[14] The First Triple Module Redundancy Development Tool for reconfigurable FPGAs, Datasheet, Xilinx, http://www.xilinx.com/esp/mil_aero/collateral/tmrtool_sellsheet_wr.pdf.

[15] Two Flows for Partial Reconfiguration: Module Based or Difference Based, Application Note, Xilinx, http://www.xilinx.com/bvdocs/appnotes/xapp290.pdf.

[16] S.Guccione, D.Levi and P.Sundararajan. JBits: Java Based Interface for Reconfigurable Computing, *Proceedings of the 2nd Military and Aerospace Applications of Programmable Devices and Technologies International Conference (MAPLD'1999)*, P-27, 1999, Laurel, Maryland, USA.

[17] PlanAhead 8.1 Design and Analysis Tool: Maximum Performance in Less Time, Datasheet, Xilinx, http://www.xilinx.com/publications/prod_mktg/pn0010825.pdf

[18] M.Meier, T.Vladimirova, T.Plant, A.da Silva Curiel. DMA Controller for a Credit-Card Size Satellite Onboard Computer, *Proceedings of the 7th Military and Aerospace Applications of Programmable Devices and Technologies International Conference (MAPLD'2004)*, P-208, 2004, Washington, US, NASA.

[19] G.Jones. Genetic and Evolutionary Algorithms, http://www.wiley.co.uk/ecc/samples/sample10.pdf

Transactional CSP Processes

Gail CASSAR[a] and Patrick ABELA[b]
Department of Computer Science and AI, University of Malta[1]
Ixaris Systems (Malta) Ltd[2]
gailcassar@gmail.com[a], patrick.abela@ixaris.com[b]

Abstract. *Long-lived transactions (LLTs)* are transactions intended to be executed over an extended period of time ranging from seconds to days. Traditional transactions maintain data integrity through ACID properties which ensure that: a transaction will achieve an 'all-or-nothing' effect (atomicity); system will be in a legal state before a transaction begins and after it ends (consistency); a transaction is treated independently from any other transactions (isolation); once a transaction commits, its effects are not lost (durability). However, it is impractical and undesirable to maintain full ACID properties throughout the whole duration of a long lived transaction. Transaction models for LLTs, relax the ACID properties by organizing a long-lived transaction as a series of *activities*. Each activity is a discrete transactional unit of work which releases transactional locks upon its execution. Activities are executed in sequence and can either commit, rollback or suspend execution of the transaction. The long-lived transaction commits if all its activities complete successfully. If any of the activities fail, the long lived transaction should roll back by undoing any work done by already completed activities.

Unless an activity requires the result of a previously committed activity, there is no constraint which specifies that the various activities belonging to a long lived transaction execute sequentially. Our proposed research focuses on combining long-lived transactions and CSP such that independent activities execute in parallel thus achieving flexibility and better performance for long lived transactions.

Very much as the occam CSP-based constructs, SEQ and PAR, allow processes to be executed sequentially or concurrently, the proposed SEQ_LLT and PAR_LLT constructs can be used to specify the sequential or concurrent execution of transactions. Two activities that are coordinated with the SEQ_LLT construct are evaluated in such a way that the second activity is executed only after the first activity commits. This corresponds to the SEQ construct which, from a concurrency perspective, executes in such a way that the second process starts its execution after the first process is complete. Similarly, PAR_LLT specifies that activities can start their execution, independently from whether any other activities have committed their transaction or not. We also use the same synchronization mechanisms provided by CSP to have concurrent activities communicate with one another. An activity which 'waits' on a channel for communication with another concurrent activity is automatically suspended (and its transactional locks released) until it receives a message from another activity. A prototype implementation of the described constructs and some example applications have been implemented on SmartPay LLT (a platform loosely based on JSR95 developed by Ixaris Systems). This work has been part of an undergraduate dissertation at the University of Malta.

Keywords. transaction processing, parallel transactions, long-lived transactions, compensating actions, CSP.

---

[1] Casa Roma, Sir Augustus Bartolo Street, Ta' Xbiex, Malta. Tel: +356 21314514; Fax: + 356 21314514
[2] Department of Computer Science and AI, Room 202, Computer Science Building, University of Malta, Msida MSD06 Tel: +356 21315046; Fax: +356 21320539

Communicating Process Architectures 2007
Alistair McEwan, Steve Schneider, Wilson Ifill and Peter Welch (Eds.)
IOS Press, 2007

Algebras of Actions in Concurrent Processes

Mark BURGIN [a] and Marc L. SMITH [b]

[a] *Department of Computer Science, Univ. of California, Los Angeles*
Los Angeles, California 90095, USA
`mburgin@math.ucla.edu`
[b] *Department of Computer Science, Vassar College*
Poughkeepsie, New York 12604, USA
`mlsmith@cs.vassar.edu`

Abstract. We introduce a high-level metamodel, EAP (event-action-process), for reasoning about concurrent processes. EAP shares with CSP notions of observable events and processes, but as its name suggests, EAP is also concerned with actions. Actions represent an intermediate level of event composition that provide the basis for a hierarchical structure that builds up from individual, observable events, to processes that may themselves be units of composition. EAP's composition hierarchy corresponds to the reality that intermediate units of composition exist, and that these intermediate units don't always fall neatly within process boundaries.

One prominent example of an intermediate unit of composition, or action, is threads. Threads of execution are capable of crossing process boundaries, and one popular programming paradigm, object-oriented programming, encourages this approach to concurrent program design. While we may advocate for more disciplined, process-oriented design, the demand for better models for reasoning about threads remains.

On a more theoretical level, traces of a computation are also actions. Traces are event structures, composed by the CSP observer, according to a set of rules for recording the history of a computation. In one of the author's model for view-centric reasoning (VCR), the CSP observer is permitted to record simultaneous events without interleaving; and in previous joint work by the authors, the extended VCR (EVCR) model permits the CSP observer to record events with duration, so that events may overlap entirely, partially, or not at all. Sequential composition may be viewed as a special case of parallel composition—one of many forms of composition we wish to be better able to reason about.

Since such diverse types of composition exist, at the event, action, and process levels; and because such problematic actions as threads exist in real systems, we must find more appropriate models to reason about such systems. To this end, we are developing algebras at different levels of compositionality to address these goals. In particular, we are interested in a corresponding hierarchy of algebras, at the event, action, and process levels.

The present focus of our efforts is at the action level, since these are the least well understood. This talk presents fundamental notions of actions and examples of actions in the context of real systems. A diversity of possible compositions at the action level will be revealed and discussed, as well as our progress on the action algebra itself.

Keywords: event, action, process, composition, interleaving, true concurrency

Communicating Process Architectures 2007
Alistair A. McEwan, Steve Schneider, Wilson Ifill, and Peter Welch
IOS Press, 2007

Using `occam-`π Primitives with the Cell Broadband Engine

Damian J. DIMMICH

Computing Laboratory, University of Kent, U.K.
E-mail: `damian@transterpreter.org`,

Abstract. The Cell Broadband Engine has a unique non-heterogeneous architecture, consisting of an on-chip network of one general purpose PowerPC processor (the PPU), and eight dedicated vector processing units (the SPUs). These processors are interconnected by a high speed ring bus, enabling the use of different logical network topologies. When programming the Cell Broadband Engine using languages such as C, a developer is faced with a number of challenges. For instance, parallel execution and synchronisation between processors, as well as concurrency on individual processors, must be explicitly, and carefully, managed. It is our belief that languages explicitly supporting concurrency are able to offer much better abstractions for programming architectures such as the Cell Broadband Engine.

Support for running `occam-`π programs on the Cell Broadband Engine has existed in the Transterpreter for some time. This support has however not featured efficient inter-processor communication and barrier synchronisation, or automatic deadlock detection. We discuss some of the changes required to the `occam-`π scheduler to support these features on the Cell Broadband Engine. The underlying on-chip communication and synchronisation mechanisms are explored in the development of these new scheduling algorithms. Benchmarks of the communications performance are provided, as well as a discussion of how to use the `occam-`π language to distribute a program onto a Cell Broadband Engine's processors. The Transterpreter runtime, which already has support for the Cell Broadband Engine, is used as the platform for these experiments.

The Transterpreter can be found at `www.transterpreter.org`.

Communicating Process Architectures 2007
Alistair A. McEwan, Steve Schneider, Wilson Ifill, and Peter Welch
IOS Press, 2007

509

Shared-Memory Multi-Processor Scheduling Algorithms for CCSP

Carl G. RITSON

Computing Laboratory, University of Kent,
Canterbury, Kent, CT2 7NF, England.

cgr@kent.ac.uk

Abstract. CCSP is a monolithic C library which acts as the run-time kernel for occam-π programs compiled with the Kent Retargetable occam Compiler (KRoC). Over the past decade, it has grown to encompass many new and powerful features to support the occam-π language as that has evolved – and continues to evolve – from classical occam. However, despite this wealth of development, the general methodology by which processes are scheduled and executed has changed little from its transputer inspired origins.

This talk looks at applying previous research and new ideas to the CCSP scheduler in an effort to exploit fully the potential of new mass-market multicore processor systems. The key objective is to introduce support for shared-memory multicore systems, whilst maintaining the low scheduling overheads that occam-π users have come to expect. Fundamental to this objective are wait-free data-structures, per-processor run-queues, and a strong will to consolidate and simplify the existing code base.

Keywords. occam-pi, concurrency, CSP, wait-free, multi-processor

Communicating Process Architectures 2007
Alistair McEwan, Steve Schneider, Wilson Ifill and Peter Welch (Eds.)
IOS Press, 2007

511

Compiling occam to C with Tock

Adam T. SAMPSON
Computing Laboratory, University of Kent
Canterbury, Kent, CT2 7NZ, UK
`ats@offog.org`

Abstract. Tock is a new occam compiler from the University of Kent, the latest result of many years' research into compiling concurrent languages. The existing occam compiler generates bytecode which is then translated into native instructions; this reduces opportunities for native code optimisation and limits portability. Tock translates occam into C using the CIF concurrent runtime interface, which can be compiled to efficient native code by any compiler supporting the C99 language standard. The resulting programs combine the safety and featherweight concurrency of occam with the performance and portability of C. Unlike previous attempts at translating occam to C, Tock's output resembles hand-written CIF code; this eases debugging and takes better advantage of the C compiler's optimisation facilities. Written in the purely functional language Haskell, Tock uses monadic combinator parsing and generic data structure traversal to provide a flexible environment for experimenting with new compiler and language features.

Keywords: occam, C99, Haskell, CIF, compilers, concurrency, optimisation

Communicating Process Architectures 2007
Alistair A. McEwan, Steve Schneider, Wilson Ifill, and Peter Welch
IOS Press, 2007

Author Index

Abela, P.	503	Moores, J.	349
Allen, A.R.	299	Moschoyiannis, S.K.	267
Anshus, O.	229	Orlic, B.	119, 207
Barnes, F.R.M.	323	Paige, R.F.	33
Bjørndalen, J.M.	229	Panesar, G.	465
Broenink, J.F.	119, 207	Passama, R.	89
Brooke, P.J.	33	Poppleton, M.R.	67
Brown, N.	183, 349	Razavi, A.R.	267
Burgin, M.	505	Ritson, C.G.	249, 323, 509
Cassar, G.	503	Robbins, W.	465
Chalmers, K.	163, 349	Romdhani, I.	163
Dimmich, D.J.	507	Royer, J.-C.	89
Duller, A.	465	Sampson, A.T.	511
East, I.	109	Schneider, S.	v, 379, 399
Evans, N.	425	Schou Jørgensen, U.	287
Faust, O.	299	Simpson, J.	339
Fernandes, F.	89	Singh, S.	371
Grant, N.	425	Smith, M.L.	505
Gray, A.	465	Sputh, B.H.C.	299, 349
Hoare, C.A.R.	1	Suenson, E.	287
Huntbach, M.	51	Teig, Ø.	313
Ifill, W.	v, 399	Todman, T.	479
Iles, J.	465	Towner, D.	465
Jacobsen, C.L.	339	Vinter, B.	229
Jadud, M.C.	339	Vladimirova, T.	493
Kerridge, J.	149, 163	Welch, P.H.	v, 249, 349
Krause, P.J.	267	Wickstrom, G.L.	443
Luk, W.	479	Wu, X.	493
May, D.	21	Yang, L.	67
McEwan, A.A.	v, 379